Subcellular Biochemistry

Volume 24

Proteins: Structure, Function, and Engineering

SUBCELLULAR BIOCHEMISTRY

Recent Volumes in This Series:

Volume 15 **Virally Infected Cells**
Edited by J. R. Harris

Volume 16 **Intracellular Transfer of Lipid Molecules**
Edited by H. J. Hilderson

Volume 17 **Plant Genetic Engineering**
Edited by B. B. Biswas and J. R. Harris

Volume 18 **Intracellular Parasites**
Edited by J. L. Avila and J. R. Harris

Volume 19 **Endocytic Components: Identification and Characterization**
Edited by J. J. M. Bergeron and J. R. Harris

Volume 20 **Mycoplasma Cell Membranes**
Edited by S. Rottem and I. Kahane

Volume 21 **Endoplasmic Reticulum**
Edited by N. Borgese and J. R. Harris

Volume 22 **Membrane Biogenesis**
Edited by A. H. Maddy and J. R. Harris

Volume 23 **Physicochemical Methods in the Study of Biomembranes**
Edited by Herwig J. Hilderson and Gregory B. Ralston

Volume 24 **Proteins: Structure, Function, and Engineering**
Edited by B. B. Biswas and Siddhartha Roy

Subcellular Biochemistry

Volume 24

Proteins: Structure, Function, and Engineering

Edited by

B. B. Biswas and
Siddhartha Roy
Bose Institute
Calcutta, India

Springer Science+Business Media, LLC

The Library of Congress cataloged the first volume of this title as follows:

Sub-cellular biochemistry.
London, New York, Plenum Press.
v. illus. 23 cm. quarterly.
Began with Sept. 1971 issue. Cf. New serial titles.
1. Cytochemistry—Periodicals. 2. Cell organelles—Periodicals.
QH611.S84 574.8'76 73-643479

This limited facsimile edition has been issued for the purpose of keeping this title available to the scientific community.

ISBN 978-1-4899-1729-4 ISBN 978-1-4899-1727-0 (eBook)
DOI 10.1007/978-1-4899-1727-0

This series is a continuation of the journal *Sub-Cellular Biochemistry,* Volumes 1 to 4 of which were published quarterly from 1972 to 1975

Originally published by Plenum Press, New York in 1995
Softcover reprint of the hardcover 1st edition 1995

10 9 8 7 6 5 4 3 2

Contributors

Sankar Adhya Laboratory of Molecular Biology, National Cancer Institute, National Institutes of Health, Bethesda, Maryland 20892

Azat Ya. Badretdinov Institute of Protein Research, Russian Academy of Sciences, 142292, Pushchino, Moscow Region, Russian Federation

B. Bhattacharyya Department of Biochemistry, Bose Institute, Calcutta 700 054, India

Sylvain Blanquet Laboratoire de Biochimie, Unité de Recherche Associeé au CNRS n° 240, Ecole Polytechnique, F-91128, Palaiseau Cedex, France

Roberta F. Colman Department of Chemistry and Biochemistry, University of Delaware, Newark, Delaware 19716

Anthony L. Fink Department of Chemistry and Biochemistry, University of California, Santa Cruz, California 95064

Alexei V. Finkelstein Institute of Protein Research, Russian Academy of Sciences, 142292, Pushchino, Moscow Region, Russian Federation

Susan Garges Laboratory of Molecular Biology, National Cancer Institute, National Institutes of Health, Bethesda, Maryland 20892

Alexander M. Gutin Institute of Protein Research, Russian Academy of Sciences, 142292, Pushchino, Moscow Region, Russian Federation

Morio Ikehara Protein Engineering Research Institute, Osaka 565, Japan

Shigenori Kanaya Protein Engineering Research Institute, Osaka 565, Japan

Sherwin S. Lehrer Boston Biomedical Research Institute, and Department of Neurology, Harvard Medical School, Boston, Massachusetts 02114

Yves Mechulam Laboratoire de Biochimie, Unité de Recherche Associeé au CNRS n° 240, Ecole Polytechnique, F-91128, Palaiseau Cedex, France

Thierry Meinnel Laboratoire de Biochimie, Unité de Recherche Associeé au CNRS n° 240, Ecole Polytechnique, F-91128, Palaiseau Cedex, France

Edith Wilson Miles Enzyme Structure and Function Section, Laboratory of Biochemical Pharmacology, NIDDK, National Institutes of Health, Bethesda, Maryland 20892

Takashi Miura Division of Cell Biology and Biophysics, School of Biological Sciences, University of Missouri, Kansas City, Missouri 64110-2499

Siddhartha Roy Department of Biophysics, Bose Institute, Calcutta 700 054, India

Sangryeol Ryu Laboratory of Molecular Biology, National Cancer Institute, National Institutes of Health, Bethesda, Maryland 20892

Dan L. Sackett Laboratory of Biochemical Pharmacology, NIDDK, National Institutes of Health, Bethesda, Maryland 20892

Jose M. Sanchez-Ruiz Departamento de Quimica Fisica, Facultad de Ciencias, Universidad de Granada, 18071-Granada, Spain

George J. Thomas, Jr. Division of Cell Biology and Biophysics, School of Biological Sciences, University of Missouri, Kansas City, Missouri 64110-2499

Preface

A hallmark for complete understanding of protein structure–function relationships is the ability to design proteins with novel functionalities. A first and very important step toward this goal is the determination of protein structure by X-ray crystallography or nuclear magnetic resonance. But a complete understanding is only possible when information obtained from many other techniques, such as site-directed mutagenesis, calorimetry, and fluorescence spectroscopy, is combined.

Although it is almost certain that general principles of structural organization of proteins exist, our understanding of these principles is currently incomplete. It is likely that formulation of such principles may be assisted to a great extent from study of many different proteins. In many ways, each protein is unique and a treasure trove of information. The almost seemingly endless variety of protein structure and function is ultimately responsible for generating the rich mosaic of life on this planet. A complete understanding of each protein or class of proteins in its individuality may be required for understanding life itself.

We have organized the volume on three themes: emerging principles, emerging techniques, and a survey of important classes of proteins. The area of protein structure, function, and engineering is one of the most active and exciting ones and more demanding than ever before. Many review articles appear regularly and many areas are updated frequently. The personal impressions and angles, however, often may not be evident. Thus, we have requested the authors to highlight their own research or new concepts in their chapters. This we hope will bring a distinctly different flavor to the volume.

The first section of the volume deals with some emerging concepts of protein structure and contains two chapters, one by Finkelstein and co-workers and another by Fink. The chapter by Finkelstein and co-workers describes the Boltzmann-like distribution of substructures encountered in proteins. The chapter by Fink deals with compact intermediate states of protein folding, which have now been universally recognized as intermediates of protein folding pathways of many, if not all, proteins. Understanding its nature may be crucial for understanding the protein folding problem.

The second section deals with some emerging techniques that may have great impact in this field in the near future. Raman spectroscopy has been widely used in chemistry, but only with recent developments in instrumentation is it coming into its own in the area of structural biology. The chapter by Miura and Thomas describes the applications of Raman spectroscopy in proteins. Fluorescence spectroscopy, on the other hand, has been widely used in biology from very early days. The continual improvements in instrumentation, introduction of time-resolved techniques, and advances in the knowledge of protein chemistry has maintained a steady advance. The two chapters by Roy and Bhattacharyya and Lehrer survey the applications of fluorescence spectroscopy and excimer formation to protein structure, respectively.

With the rapid appearance of protein structures, the importance of thermodynamic measurements is increasingly being appreciated. Differential scanning calorimetry has made an impact in the understanding of forces that are involved within and between macromolecules. Sanchez-Ruiz describes calorimetric techniques and their applications in the analysis of protein structure and function. Affinity labeling is a technique that still holds great promise, not only for protein structure function analysis, but for the development of chemotherapeutic agents as well. The chapter by Colman deals with affinity labeling and related areas and its applications to proteins.

The third section focuses on individual proteins that have already been studied in significant details. The chapter by Miles describes the attempt to understand a very complex, multisubstrate and allosteric enzyme, tryptophan synthetase, in molecular terms. Tubulin is a protein of central importance in the eukaryotic cells. The carboxy terminii of α- and β-subunits of tubulin have been shown to regulate many aspects of tubulin function. Sackett's chapter deals with this aspect of tubulin structure and function.

The last few years have seen an explosion of information on protein–nucleic acid interaction. Crystal structures of many protein–DNA and protein–RNA complexes have been determined and new proteins have been discovered that are involved in nucleic acid recognition, metabolism, and modification. Adhya and co-workers describe conformational changes and their role in the functioning of a global regulatory protein: the cyclic AMP receptor protein. Mechulam and co-workers survey the recent rapid advances in the field of aminoacyl–tRNA synthetases. Kanaya and Ikehara describe the recent advances in elucidation of structure and function of ribonuclease H, which hydrolyzes RNA–DNA hybrids.

Difficulties of structure determinations have hampered the growth of structural biology in the past. That may not be the limitation in the very near future, however, as almost one new structure is being reported every week and the pace is quickening. The challenge in the future is twofold: first, to organize this information and formulate general principles, and second, to correlate the structural information to the function of the protein, perhaps with the help of emerging techniques such as spectroscopy, fast kinetics, and protein engineering. We hope that this volume will help the structural biologists to meet this future challenge.

B. B. Biswas
Siddhartha Roy

Calcutta, India

Contents

Chapter 1
Boltzmann-like Statistics of Protein Architectures: Origins and Consequences
Alexei V. Finkelstein, Alexander M. Gutin, and Azat Ya. Badretdinov

1. Introduction 1
   1.1. Popular and Rare Folding Patterns 1
   1.2. Boltzmann-like Statistics of Protein Elements 3
   1.3. Paradox of Boltzmann-like Statistics of Protein Structures 6
2. Theory 6
   2.1. Energy of Folds 7
   2.2. Energy Spectra of Heteropolymer Globules 7
   2.3. Energy Threshold 8
   2.4. Critical Temperature 10
   2.5. Stable Unique Folds 10
3. Random Amino Acid Sequences That Stabilize 11
   3.1. Unique Folds 11
   3.2. A Given Interaction 12
   3.3. A Given Folding Pattern 17
   3.4. Origins of Boltzmann-like Statistics of Globular Proteins 20
   3.5. An Estimate of Critical Temperature 21
4. Consequences for Protein Structure Prediction and Folding 21
5. Conclusion 24
6. References 24

Chapter 2
Compact Intermediate States in Protein Folding
Anthony L. Fink

1. Introduction 27
2. Compact Intermediates under Equilibrium Conditions 30
   2.1. α-Lactalbumin 32
   2.2. DnaK 33
   2.3. Apomyoglobin 35
   2.4. β-Lactamase 37
   2.5. Carbonic Anhydrase 39
   2.6. Other Proteins 39
3. Compact Intermediates Formed under Transient Folding Conditions 40
4. Forces Involved in Compact Intermediate States 41
5. Structural Models for Compact Intermediates 44
   5.1. Theory-Based Models 44
   5.2. Experiment-Based Models 45
   5.3. Structural Model for Compact Intermediates 46
6. Possible Physiological Roles of Compact Intermediates 47
   6.1. Membrane Fusion and Translocation 47
   6.2. *In Vivo* Folding of Proteins 47
7. References 48

Chapter 3
Raman Spectroscopy of Proteins and Their Assemblies
Takashi Miura and George J. Thomas, Jr.

1. Introduction 55
   1.1. Scope of This Chapter 56
   1.2. Advantages and Disadvantages of the Method 56
   1.3. Nature of the Data of Raman Spectra 58
2. Experimental Considerations 64
   2.1. Instrumentation 64
   2.2. Sample Handling 65
3. Analysis of Secondary Structure and Environments of Side Chains 66
   3.1. Main Chain Conformations 66
   3.2. Side-Chain Conformations and Local Environments 70
4. Applications to Protein Structures and Assemblies 77
   4.1. Thioredoxin 77

4.2. Viruses 83
4.3. Raman Microspectroscopy 88
4.4. Dynamic Processes 88
5. Perspectives 95
6. References 96

Chapter 4

Fluorescence Spectroscopic Studies of Proteins

Siddhartha Roy and B. Bhattacharyya

1. Introduction 101
2. Study of Ligand Binding 102
2.1. General Discussions on Method 102
2.2. Tubulin–Colchicine Interaction 103
2.3. Magnesium Binding to Tubulin 106
3. Protein Association–Dissociation 107
3.1. Methods 107
3.2. Tetramer–Dimer Equilibrium of λ-Repressor 107
3.3. Dimer–Monomer Equilibrium of Tubulin and Tubulin-S 108
4. Conformational Changes 108
4.1. General 108
4.2. Conformational Changes in λ-Repressor 109
4.3. Substrate-Induced Conformational Changes in Glutaminyl–tRNA Synthetase 109
5. Folding and Unfolding of Proteins 110
5.1. General 110
5.2. Urea-Induced Equilibrium Denaturation of λ-Repressor 111
6. Distance Information 111
6.1. General 111
6.2. Donor Quenching 111
7. References 112

Chapter 5

Pyrene Excimer Fluorescence as a Probe of Protein Conformational Change

Sherwin S. Lehrer

1. Introduction 115
2. Pyrene Excimer Formation in "Good" Solvents 116

3. Pyrene Excimer Formation in Aqueous Solution 119
3.1. Proteins 119
3.2. Pyrene–Dithiothreitol 122
4. Labeling of Proteins 125
5. Examples of Protein Studies 126
5.1. Tropomyosin 126
5.2. Other Muscle Proteins 127
5.3. Other Proteins 128
6. Prospects for Future Studies 129
7. References 130

Chapter 6
Differential Scanning Calorimetry of Proteins
Jose M. Sanchez-Ruiz

1. Introduction 133
2. The Equilibrium Thermodynamics Analysis of DSC Data 137
2.1. The Two-State Equilibrium Model 137
2.2. Multistate Equilibrium 140
3. The Thermodynamic Stability of Proteins 143
3.1. The Protein Stability Curve 144
3.2. The Molecular Interpretation of the Denaturational Heat Capacity Increment 148
3.3. The Convergence Behavior of the Denaturational Enthalpy and Entropy Changes 151
3.4. The Convergence Temperature Puzzle 156
4. Kinetic Analysis of DSC Thermograms 158
4.1. The Effect of the Calorimeter Time Response 158
4.2. The Effect of Slow Equilibrium 159
4.3. The Effect of Irreversibility 161
5. References 172

Chapter 7
Affinity Labeling and Related Approaches to Targeting Specific Enzyme Sites
Roberta F. Colman

1. General Considerations 177
1.1. Approach of Affinity Labeling 177
1.2. Representative Types of Reactive Compounds 179
2. Specific Examples 183
2.1. Rabbit Muscle Pyruvate Kinase: Affinity Labeling of the Active Site 183

2.2. Rat Liver Glutathione S-Transferases: Affinity Labeling of the Hydrophobic Substrate Site 187
2.3. Pig Heart NADP-Dependent Isocitrate Dehydrogenase: Affinity Cleavage at the Metal–Isocitrate Site 191
2.4. Bovine Liver Glutamate Dehydrogenase: Affinity Labeling of Regulatory Sites with Mono- and Bifunctional Reactive Nucleotides 195
2.5. *Salmonella typhimurium* NADP-Dependent Glutamate Dehydrogenase: Affinity Labeling and Site-Directed Mutagenesis at the Catalytic Coenzyme Site 198
3. Concluding Remarks 202
4. References 203

Chapter 8

Tryptophan Synthase: Structure, Function, and Protein Engineering

Edith Wilson Miles

1. Introduction 207
2. Structure and Function of the α Subunit 209
2.1. Three-Dimensional Structure 209
2.2. Folding and Conformational Stability 212
2.3. Active Site and Mutagenesis Studies 213
2.4. Reaction Mechanism 214
3. Structure and Function of the β Subunit 215
3.1. Three-Dimensional Structure 215
3.2. Proteolysis, Folding, and Assembly 216
3.3. Effects of Pyridoxal Phosphate on Conformation 217
3.4. Active Site and Reaction Mechanism 219
3.5. Protein Engineering 224
4. Structure and Function of the $\alpha_2\beta_2$ Complex 228
4.1. Subunit Association 228
4.2. Effects of Mutation on Subunit Association 229
4.3. Effects of Ligands and Subunit Association on Thermal Stability 232
4.4. Metabolite Channeling 235
4.5. Information Transfer from the α Subunit to the β Subunit 237
4.6. Information Transfer from the β Subunit to the α Subunit 240
4.7. Coordination of the Activities of the α and β Subunits 242
5. References 245

Chapter 9

Structure and Function in the Tubulin Dimer and the Role of the Acidic Carboxyl Terminus

Dan L. Sackett

1. Introduction ... 255
   1.1. Tubulin Is the Subunit Protein of Microtubules ... 256
   1.2. Tubulin Occurs in Cells in Multiple Structural Forms ... 256
   1.3. Tubulin Forms Fall into Three Families: Alpha, Beta, and Gamma ... 257
   1.4. Multiple Genes Are Known for Alpha and Beta Tubulin: Isotype Classes ... 261
   1.5. Posttranslational Modifications Increase Tubulin Heterogeneity ... 262
   1.6. Tubulin Contains Bound Guanine Nucleotides That Regulate Polymerization ... 263
   1.7. Tubulin Binds Several Classes of Pharmacologic Agents ... 264
   1.8. Many Proteins Bind to Tubulin ... 265
2. Structure of the Tubulin Dimer ... 266
   2.1. Size of the Dimer ... 267
   2.2. Shape of the Dimer ... 268
   2.3. Proteolysis of the Dimer ... 269
   2.4. Conformation of the C-Termini ... 278
   2.5. Covalent Modification of Tubulin C-Termini and Functional Changes ... 281
3. Models of the Tubulin Dimer ... 286
   3.1. General Models of Dimer ... 286
   3.2. The Structure of the C-Termini ... 288
4. References ... 291

Chapter 10

Role of Allosteric Changes in Cyclic AMP Receptor Protein Function

Sankar Adhya, Sangryeol Ryu, and Susan Garges

1. Introduction ... 303
2. Three-Dimensional Structure and Function of CRP ... 305
3. CRP–DNA Interactions ... 306
4. Transcription Activation ... 308
   4.1. Contact with RNA Polymerase ... 308

4.2. Change the DNA Structure 309
5. Conformation Changes in CRP 310
5.1. DNA-Induced Allostery 310
5.2. cAMP-Induced Allostery 311
6. Sites of Allosteric Changes in CRP Induced by cAMP 313
6.1. F α-Helix Protrusion 314
6.2. Hinge Reorientation 314
6.3. Subunit–Subunit Alignment 317
6.4. Domain–Domain Adjustment 317
7. Conclusion 318
8. References 318

Chapter 11

A Family of RNA-Binding Enzymes: The Aminoacyl–tRNA Synthetases

Yves Mechulam, Thierry Meinnel, and Sylvain Blanquet

1. Functions of Aminoacyl–tRNA Synthetases in the Living Cells 323
1.1. Early Characterization of Aminoacyl–tRNA Synthetases 323
1.2. Aminoacyl–tRNA Synthetases and the Synthesis of Proteins 324
1.3. The Synthesis of Adenylated *bis* (5′-Nucleosidyl)Polyphosphates 327
1.4. Other Functions of Aminoacyl–tRNA Synthetases 327
2. From the Three-Dimensional Structures to the Catalytic Mechanisms 328
2.1. Two Classes of aaRS as Revealed by Three-Dimensional Structures 328
2.2. Functional Domains in aaRS 333
2.3. The Catalytic Centers of aaRS 335
3. Motifs in RNA Structures Recognized by aaRS 340
3.1. Strategies for the Identification of Identity Elements in tRNAs 341
3.2. Identity Elements among the Various RNAs 346
3.3. tRNA-like Structures Recognized by aaRS 349
4. tRNA Binding Sites on the aaRS 352
4.1. Structure of tRNA: aaRS Crystalline Complexes 353
4.2. Mapping of the tRNA-Binding Area on the aaRS Using Genetic and Biochemical Approaches 358
5. Conclusions 360
6. References 361

Chapter 12

Functions and Structures of Ribonuclease H Enzymes

Shigenori Kanaya and Morio Ikehara

1. Introduction 377
2. *E. coli* RNase HI 378
 2.1. Physiological Functions 378
 2.2. Enzymatic Activity 380
 2.3. Overproduction 381
 2.4. Physicochemical Properties 382
 2.5. Three-Dimensional Structure 383
 2.6. Alignment of Secondary Structure Elements 385
 2.7. NMR Studies 390
 2.8. Role of Amino Acid Residues 393
 2.9. Mechanisms for Substrate Binding and Hydrolysis 399
3. Other RNase H Enzymes 404
 3.1. Retroviral Reverse Transcriptase 404
 3.2. Bacterial RNases H 407
 3.3. Yeast RNases H 408
 3.4. RNases H from Higher Eukaryotes 409
4. Protein Engineering of RNase H 410
 4.1. DNA:Protein Hybrid RNase H 410
 4.2. Protein Stabilization 412
5. References 412

Index 423

Chapter 1

Boltzmann-like Statistics of Protein Architectures

Origins and Consequences

Alexei V. Finkelstein, Alexander M. Gutin, and
Azat Ya. Badretdinov

1. INTRODUCTION

Inspection of globular proteins shows that the most common protein structures are those that have some advantage in stability. This would cause no wonder if the advantages were not so small and the difference in occurrence was not so great.

1.1. Popular and Rare Folding Patterns

A small number of folding patterns describe in outline most of the known protein domains, the same patterns being found in proteins quite different from the biochemical and philogenetical points of view (Rao and Rossmann, 1973; Levitt and Chothia, 1976; Richardson, 1977, 1981; Ptitsyn and Finkelstein,

---

Alexei V. Finkelstein, Alexander M. Gutin, and Azat Ya. Badretdinov Institute of Protein Research, Russian Academy of Sciences, 142292, Pushchino, Moscow Region, Russian Federation. *Present address for AMG:* Department of Chemistry, Harvard University, Cambridge, Massachusetts 02138.

Subcellular Biochemistry, Volume 24. Proteins: Structure, Function, and Engineering, edited by B. B. Biswas and Siddhartha Roy. Plenum Press, New York, 1995.

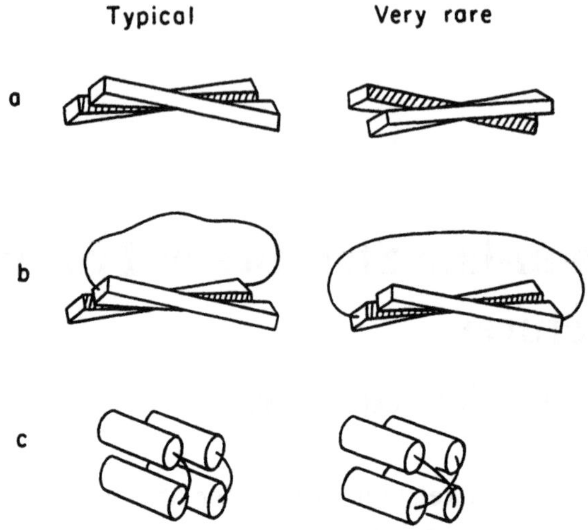

FIGURE 1. Typical protein folding patterns in comparison with similar but rare ones. β-strands are shown as thick ribbons, α-helices as cylinders, loops as solid lines. (a) The right-handed twist of β-sheets is predominant; the left-handed one contradicts a spontaneous twist of extended chains of L amino acids. (b) The right-handed connection between the parallel β-strands is predominant; it demands less loop bending than the left-handed one. (c) Crossing of loops is very rare; it either dehydrates a peptide group of a loop or demands additional loop bending. All these "defects" cost a few kcal/mole.

1980; Chothia and Finkelstein, 1990; Branden and Tooze, 1991). Rossmann folds, triose phosphate isomerase (TIM) barrels, helical bundles, Greek keys, and jelly rolls are examples of the most popular folding patterns. It has been shown that the widespread architectures are those that have some advantages in thermodynamic stability (Finkelstein and Ptitsyn, 1987).

For example, practically all β-sheets in proteins have a right-handed twist (Fig. 1a), which is slightly more energetically favorable (for nature L amino acids) than a left-handed one (Chothia, 1973). Further, a right-handed connection of parallel β-strands is a standard detail of the most popular folds, while a left-handed one is extremely rare (Fig. 1b). A right-handed connection in a sheet with a right-handed twist demands less loop bending (Sternberg and Thornton, 1976); this loop "costs" ~ 2 kcal/mole less than a left-handed one due to polypeptide chain rigidity (Finkelstein and Ptitsyn, 1987). The same predominance of one pathway of a loop over another has been also observed for other standard connections of secondary structure elements (Chothia, 1984; Efimov, 1984, 1991; Kajava, 1992). In each case the predominant way of a loop is that which demands its smaller bending. Similarly, loop crossing (Fig. 1c) is rare in proteins, and it seems that this "defect" is prohibited because the crossing either buries and dehydrates a loop peptide group (which costs about 5 kcal/mole) or demands additional loop bending to avoid this dehydration, which also costs a few kcal/mole (Finkelstein and Ptitsyn, 1987).

Thus, the most popular protein folds have obvious, though small, advantages in stability. It is not clear, however, why these small advantages provide for the observed rigid limitations in patterns of protein folding.

First, a defect costs only a few kcal/mole, while total protein energy is hundred times more, and different sequences can readily add or subtract ~ 50 kcal/mole to the energy of a fold (Novotny *et al.*, 1984). The additional 2 or 5 kcal caused by a mutation cannot even denature a protein (see, e.g., Branden and Tooze, 1991; Gregoret and Sauer, 1993), because its reserve of stability is usually about 10 kcal/mole (Privalov, 1979). This poses a natural question: Why does a small energy of a defect play any role in the selection of folds and why it cannot be compensated by an "appropriate" amino acid sequence?

Second, the effects based on loop rigidity must concern mainly the entropy, because the main reason for polymer elasticity is that an additional bending decreases the number of possible chain conformations (Birstein and Ptitsyn, 1966; Flory, 1969). However, the entropy of a unique protein fold is zero in any case. Do the "entropic" arguments against some protein architectures have any sense at all?

1.2. Boltzmann-like Statistics of Protein Elements

Before we start to answer these questions, it should be remembered that a similar phenomenon is observed for small elements of protein structures: the elements of low energy occur much more often than others in three-dimensional structures of globular proteins (Pohl, 1971, 1980). Here again, a small energy difference plays a significant selective role because the observed occurrence–energy relationship resembles Boltzmann statistics,

$$\text{Occurrence} \sim \exp(-\text{Energy}/RT_*) \qquad (1)$$

where T_* is a "conformational temperature" (Pohl, 1971) of protein statistics, which seems to be more or less close to room temperature, and R is the gas constant. This statistic was observed for different elements of protein structure (Fig. 2): for distributions of the ϕ, ψ, χ angles (Pohl, 1971); for distribution of residues between the surface and interior of globules (Miller *et al.*, 1987), as well as between different secondary structures and different positions in secondary structure (Finkelstein *et al.*, 1977; Serrano *et al.*, 1992; Horovitz *et al.*, 1992); for occurrence of *cis* and *trans* prolines (MacArthur and Thornton, 1991), of ion pairs (Bryant and Lawrence, 1991), of empty cavities (Rashin *et al.*, 1986) in protein globules, and so on.

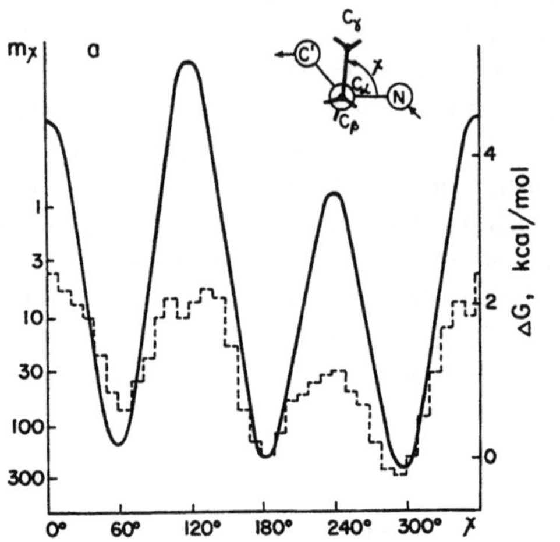

FIGURE 2. (a) Correlation between torsional energies. Broken line: $\Delta G_1 = -RT\ln(m_\chi)$: the "potential" from protein statistics; m_χ is the number of those side chains with one C_γ atom that have $\chi_1 = \chi \pm 5°$; statistical data are taken from McGregor *et al.* (1987). Solid line: ΔG_2: energy of the $C_\gamma H_3$ side chain computed with potentials of De Santis and Liquori (1971). ΔG_1 (180°) and ΔG_2 (180°) are taken as zero. (b) Correlation between helix–coil transition free energies. $-RT\ln f_{AA}$: "free energy" from protein statistics (Richardson and Richardson, 1988). $RT = 0.6$ kcal/mole at 300 K. The partition coefficient $f = (N^\alpha_{AA}/\Sigma N^\alpha_{AA})/(N_{AA}/\Sigma N_{AA})$, N^α_{AA} and N_{AA} being the numbers of helical and all chain residues of a given type. $\Delta\Delta G_{AA-Gly}$: experimental difference between helix–coil transition free energies of amino acid (AA) and Gly according to O'Neil and DeGrado (1990). Letter "P" marks proline, whose $\Delta\Delta G$ value was determined rather approximately; this point is not included in the following regression analysis. (c) Correlation between transfer free energies (adapted from Miller *et al.*, 1987). $-RT\ln f$: surface/interior transfer "free energy" from protein statistics. The partition coefficient $f = (N_s/\Sigma N_s)/(N_b/\Sigma N_b)$, N_s and N_b being the numbers of surface and buried residues of a given type. ΔG_2: experimental organic solvent → water transfer free energy. All plots show a definite (though, far from perfect) proportion between energy and the logarithm of occurrence. Dashed lines in the plots (b) and (c) correspond to the best (least-square) approximation of ΔG by $-(R\ln f)\cdot T_* + b$; solid lines in the same plots correspond to the best approximation of $R\ln f$ by $-(\Delta G)\cdot T_*^{-1} + b$ (the adjustable parameters T_* and b in these two cases are different when correlation is not perfect (see, e.g., Hudson, 1964). The difference between these lines reflects a deviation from strictly Boltzmann-like statistics. An estimate of conformational temperature T_*, which one can extract from these plots ($T_* \approx 200$–600 K) is rather rough, as (1) the statistics in (a) is biased to refining programs of X-ray analysis; (2) the statistics in (b) is biased to definition of two states of a residue: "buried" and "surface"; and (3) the experimental data in (c) refer to a particular amino acid environment.

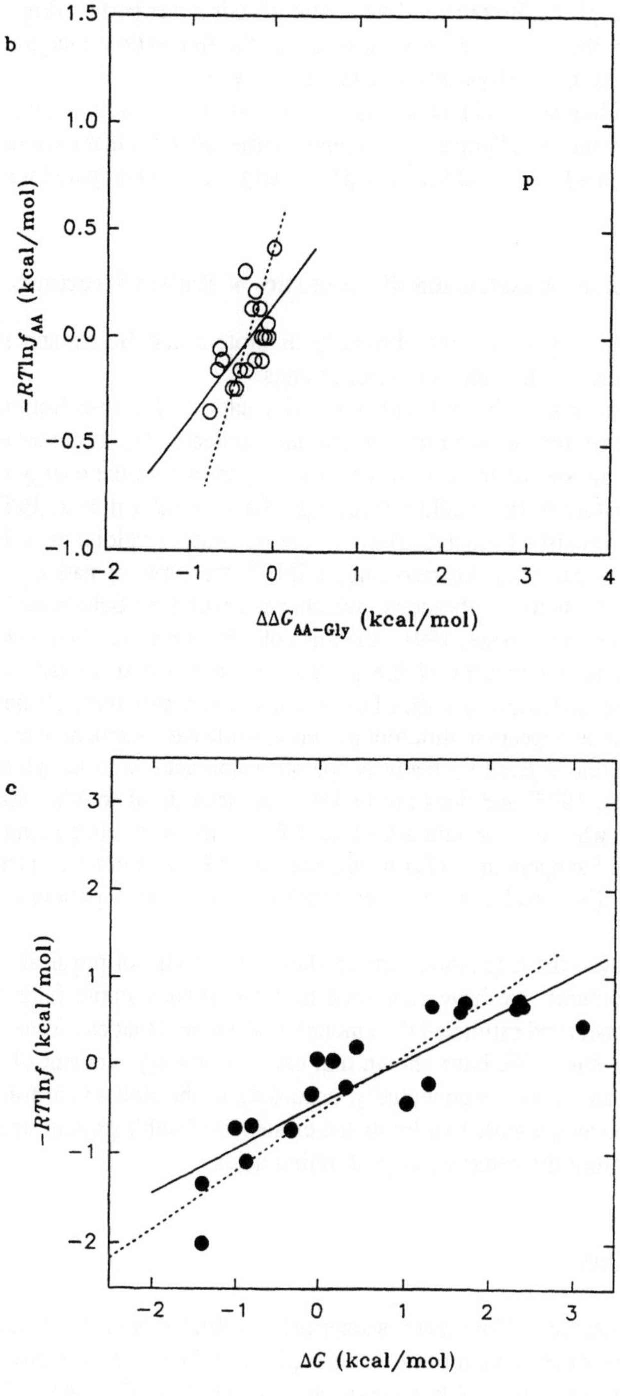
b
1.5
1.0
0.5
0.0
−0.5
−1.0
−2
−1
0
1
2
3
4
p
$-RT\ln f_{AA}$ (kcal/mol)
$\Delta\Delta G_{AA-Gly}$ (kcal/mol)
c
3
2
1
0
−1
−2
−2
−1
0
1
2
3
$-RT\ln f$ (kcal/mol)
ΔG (kcal/mol)

This empirical Boltzmann-like relationship is never perfect (Fig. 2), but it is so common that it is even used to estimate the interaction energies in protein globules (see, e.g., Miyazava and Jernigan, 1985).

If a relationship (1) concerns also the overall protein folds, this could explain why the most "popular" folding patterns are those that have no structural defects: even a few kilocalories is a great energy loss as compared with $RT_* \sim 1$ kcal/mole.

1.3. Paradox of Boltzmann-like Statistics of Protein Structures

Generally speaking, the similarity of protein and Boltzmann statistics is very puzzling, as they mean different things.

Boltzmann statistics is a statistics of fluctuations. It results from the equilibrium of *transitions* between the ground and excited states. It describes the time each molecule spends in each of the states depending on the energy of the state and temperature of the medium (see, e.g., Landau and Lifshitz, 1959).

Protein statistics concerns only the *ground* states of globules. It is based on X-ray or nuclear magnetic resonance (NMR) structures of native proteins and neglects their thermal vibrations, which are small (see Schulz and Schirmer, 1979; Branden and Tooze, 1991): the links of a protein chain do not wander from the surface to the interior of the protein or from one secondary structure to another. The statistics describes how often a given structural element (e.g., a given χ_1 angle) appears in different proteins, while each element (each χ_1-rotamer) of a protein is *fixed* by many cooperative interactions in the globule (Gelin and Karplus, 1975) and does not undergo any transitions to other states.

Then, why does protein statistics look so similar to Boltzmann statistics? What is the "temperature" (T_*) in this statistics? Does it concern just the structural details (for which it has been established) or the overall protein architectures as well?

To answer these questions and to clarify the origins of physical selection of protein structures, we have examined the spontaneous stable folds of random heteropolymers and estimated the amount of sequences that can form stable folds of different kinds. We have shown that each low-energy element of a structure increases this amount exponentially, according to the Boltzmann formula, with T_* being the temperature that limits the existence of stable protein structures and originates from the heterogeneity of amino acids.

2. THEORY

The question, "How many sequences can form this or that?" is stimulated by the recent development of the general physical theory of structures formed by random heteropolymers (Bryngelson and Wolynes, 1987, 1990; Shakhnovich and Gutin, 1989, 1990b). Investigations of this kind are related to globular proteins, since their amino acid sequences (unlike, e.g., the periodic chains of fibrous proteins) look like random heteropolymers that are "edited" only slightly (Ptitsyn, 1985). In this connection it is noteworthy that, according to the theory (Shakhnovich and Gutin, 1990b), a significant part of randomly synthesized heteropolymers can have a thermodynamically dominant fold.

In this section we propose a basic analytical theory of Boltzmann-like statistics of protein structures that is illustrated by the results of computer experiments on the "physical selection" of protein structures.

The reader who is not interested in theoretical details can skip Sections 2.1–2.5.

2.1. Energy of Folds

Let us consider a heteropolymer of $N \gg 1$ links which can form M different globular folds. The energy of a chain fold* can be presented in the form

$$E_f = \sum_i \delta_i^f \epsilon_i^a \tag{2}$$

This sum is taken over every possible interaction i (contacts, bends, etc.); for those inherent in the considered fold f the value of δ_i^f is 1, for others $\delta_i^f = 0$. The potential energy of each interaction, ϵ_i^a, depends on amino acid sequence a. A fold that has the lowest energy for a given sequence is called its "native" fold.

Our aim is to estimate a relationship between the energy of some interaction and the chance to observe it in native chain folds. This chance is determined by statistical properties of the low-energy edge of energy spectra (Gutin *et al.*, 1992).

2.2. Energy Spectra of Heteropolymer Globules

The following consideration is based on the random energy model, or REM (Derrida, 1981), which is successfully used in theories of disordered systems. According to Bryngelson and Wolynes (1987, 1990) and Shakhnovich and Gutin (1989, 1990a,b), REM is particularly applicable to heteropolymers such as protein chains.

REM describes heterogeneous systems, each part of which can be changed (mutated) independently of the mutations of other parts or, better to say, it

*In this context, "fold" corresponds to a local minimum of protein energy. It is assumed that each interaction either contributes all its potential energy to the energy of a fold or contributes nothing at all. The term *energy* is used in the text only for simplicity: actually, one has to speak about "mean force potentials," as far the solvent-mediated (hydrophobic and electrostatic) interactions play a significant role in protein stability. To be particular, the energies of interactions are counted off their values in the unfolded chain.

describes a set of systems which result from these independent mutations (like a set of random heteropolymers).[†] REM assumes that a system has a large, but finite, number of states (folds, in the case of polymers) which can be compared with local energy minima.

REM neglects any correlation between the energies of different structures, i.e., it assumes that different folds can acquire their energies irrespective of one another because they have no common elements. This approximation is reasonable for dense globular folds, because the number of the same interactions in two different folds of this kind is, as a rule, low (Shakhnovich and Gutin, 1989).

REM assumes a Gaussian distribution of fold energies (Fig. 3), because such an energy is the sum of many independent terms [see Eq. (2)], and a Gaussian distribution is typical for such sums (see, e.g., Hudson, 1964). This means that each of the folds (as concerns heteropolymers) acquires an energy E for a fraction of sequences, which is

$$p(E) = (2\pi\sigma^2)^{-1/2} \cdot \exp(-(E - \bar{E})^2/2\sigma^2) \tag{3}$$

Here $\bar{E}$ is the mean fold energy, and σ^2 a dispersion of fold energies. The values of $\bar{E}$ and σ are assumed to be the same for all folds (in Section 3.3., we consider what happens if this assumption is not valid). The density of energy levels, in average over the sequences, is

$$\bar{m}_E = M \cdot p(E), \tag{4}$$

where M is the total number of globular folds. M is a great number because it grows exponentially with chain length N.

2.3. Energy Threshold

In average (over all the random sequences), the lowest-energy line of a spectrum appears at such an energy E_c that the average number of spectrum lines below E_c is unity:

$$\bar{M}_{E_c} = \int_{-\infty}^{E_c} \bar{m}_E dE = 1 \tag{5}$$

[†]Generally, the energies of different interactions are not absolutely independent. For example: the energy of a contact between residues "a" and "b" has some correlation with the energy of a contact between residues "a" and "c"; α- and β-stabilizing properties of amino acid residues correlate with their hydrophobicity (i.e., with the energies of their long-range interactions). However, the approximation that treats these energies as independent values that change with mutations irrespective of one another is very convenient for theory and rather accurate when the number of link types is great (Shakhnovich and Gutin, 1989). Below we present a computer experiment that estimates the accuracy of this approximation.

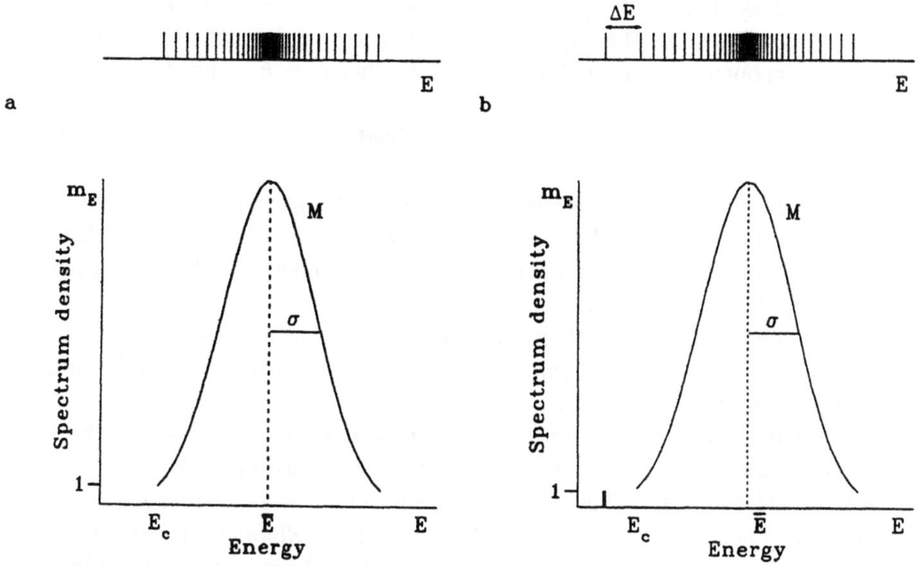

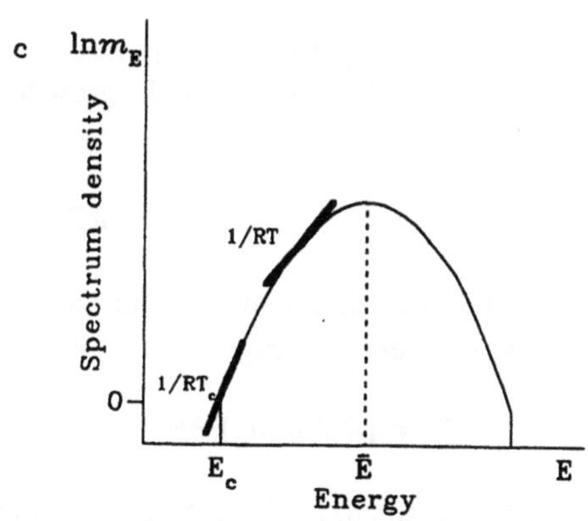

FIGURE 3. Distribution of energy levels for globular folds for typical random heteropolymers (a) and for those of them that can form stable unique folds (b). Each of M spectrum lines corresponds to a fold. Below: histograms of spectrum density. $\bar{E}$ is the mean energy of the spectrum; σ is its characteristic width: most levels occur in the range of $\bar{E} \pm \sigma$. (c) Logarithmic representation of density of continuous spectrum. The tangent to this plot determines the temperature corresponding to a given spectrum region (to draw the tangent, the abatement of spectrum density must be averaged over the energy region of $\sim\sigma$ width, according to conventional thermodynamics). The tangent corresponding to the beginning of the spectrum gives the critical temperature of chain freezing (see the text).

This integral easy to estimate using the approximation

$$\bar{m}_E = Mp(E) \approx Mp\,(E_c)\cdot\exp[(E - E_c)\cdot\partial\ln p/\partial E|_{E=E_c}] \tag{6}$$

which is valid when $E_c < \bar{E} - \sigma$ and $|E - E_c| < \sigma$, and taking the integral over $E - E_c$. Then

$$\bar{M}_{E_c} \approx Mp(E_c)\cdot(\partial\ln p/\partial E|_{E=E_c})^{-1} = 1 \tag{7}$$

Thus, energy threshold E_c is a point where $M \cdot \exp(-(E_c - \bar{E})^2/2\sigma^2) = \sqrt{2\pi} \cdot (\bar{E} - E_c)/\sigma$. As $M \gg \bar{M}_{E_c} = 1$, the value of $\bar{E} - E_c$ is much greater than σ, and the exponent changes with E_c much faster than $\bar{E} - E_c$. Therefore

$$E_c \approx \bar{E} - \sigma\sqrt{2\ln M} \tag{8}$$

This value of the energy threshold is characteristic for the vast majority of sequences, because $\bar{m}_E$ and $\bar{M}_E$ grow rapidly with E when $\bar{M}_E \ll M$. Therefore, only a narrow ($\sim \sigma^2/(\bar{E} - E_c) = \sigma/\sqrt{2\ln M}$) energy region divides the districts where $\bar{M}_E \ll 1$ and $\bar{M}_E \gg 1$.

The physical sense of $\bar{M}_E$ is quite different in these two districts.

When $\bar{M}_E \ll 1$, $\bar{M}_E$ determines the small fraction of sequences which still have some folds of such a low energy.

When $\bar{M}_E \gg 1$, a random heteropolymer chain has about $\bar{M}_E$ spectrum lines below the energy E (actually, the majority of the chains have from $\bar{M}_E - (\bar{M}_E)^{1/2}$ to $\bar{M}_E + (\bar{M}_E)^{1/2}$ lines). Thus, $\bar{m}_E = \partial \bar{M}_E/\partial E$ now reflects a spectrum density for an ordinary heteropolymer chain, and $S_E = R\ln(m_E) \approx R\ln(\bar{m}_E)$ is the entropy of a chain which has an energy $E > E_c$. In this region the spectrum is virtually a continuous one.

2.4. Critical Temperature

The temperature of any macroscopic system is directly determined by the growth of its entropy with the growth of energy (see, e.g., Landau and Lifshitz, 1959):

$$RT = (\partial S_E/\partial E)^{-1} \tag{9}$$

In this way one can estimate the temperature of any region of a continuous energy spectrum (Fig. 3c). A critical temperature

$$T_c = [R\ \partial \ln(\bar{m}_E)/\partial E|_{E \to E_c}]^{-1} \approx \sigma^2/[R(\bar{E} - E_c)] = \sigma/(R\sqrt{2\ln M}) \tag{10}$$

corresponds to the beginning of the spectrum. Below T_c, the lowest-energy folds of the chains are "frozen out"; above T_c, these folds are not stable thermodynamically. T_c depends on the heterogeneity of residues and folds rather than on the chain length because σ^2 is proportional to this heterogeneity, and σ^2, $\ln M$, and $\bar{E} - E_c$ are all proportional to the protein size (Bryngelson and Wolynes 1987, 1990; Shakhnovich and Gutin 1989, 1990a,b).

2.5. Stable Unique Folds

Energy of the most low-energy folds formed by majority of random sequences is very close to E_c, the deviation being not greater than RT_c. These chains "melt" at $T \leq T_c$, and this melting is a second-order phase transition (Shakhnovich and Gutin, 1989).

However, the best folds of some sequences are more stable: their energy is below E_c by a value $\Delta E > RT_c$. Denaturation of these heteropolymers is a first-order phase transition (Go, 1975; Go and Taketomi, 1979). Now the melting temperature is somewhat above T_c,

$$T_m = T_c\,[1 + \{(\Delta E/RT_c)/\ln M\}^{1/2}], \tag{11}$$

and the heat of melting is

$$\Delta E_m = \sigma\cdot(\Delta E/RT_c)^{1/2} \tag{12}$$

These effects are considerable only when $\Delta E \gg RT_c$; when $\Delta E \rightarrow 0$, the "all-or-none" transition disappears and one observes a second-order phase transition, characteristic for random heteropolymers.

3. RANDOM AMINO ACID SEQUENCES THAT STABILIZE

3.1. Unique Folds

Protein chains resemble that part of random heteropolymers which can form stable folds the energy of which is below E_c by $\Delta E \gg RT_c$ (see above).

What fraction of random sequences can form such stable folds?

According to Eqs. (6) and (10), one can represent $p(E)$ as

$$p(E) \approx p(E_c)\cdot\exp[(E - E_c)/RT_c] \tag{13}$$

Thus, the average expected number of folds with energy below $E_c - \Delta E$ is

$$\bar{M}_{E_c - \Delta E} = M \cdot RT_c p(E_c - \Delta E) = \exp(-\,\Delta E/RT_c) \tag{14}$$

[because $Mp(E_c)RT_c = 1$, see Eqs. (5), (6), and (10)]. When $\Delta E > RT_c$,

$$\exp(-\Delta E/RT_c)$$

is a good estimate of the fraction of random sequences that form the folds with energy below $E_c - \Delta E$. This is because fold energies are independent values (see above), and when one of them is unusually low, this does not affect the expected energies of others.

Thus, the folds with energy $E_c - \Delta E$ are separated from others by an energy gap of $\approx \Delta E$ (Fig. 3b).

The sequences that can form stable unique folds are not too rare (Shakhnovich and Gutin, 1990b). ΔE can be estimated as a "reserve" of protein stability, which is $\approx$ 10 kcal/mole (cf. Privalov, 1979); then, a fraction of sequences with stable unique folds is about 0.001–0.0001% of *all* the random sequences.

This means that a 100-residue protein can be formed "only" by 20^{95} of the 20^{100} random sequences.

3.2. A Given Interaction

Suppose that we know that the potential energy of some interaction i is ϵ_i, and ask, "What is a chance to observe this interaction in a native chain fold?"

To answer this question, let us consider all those sequences a where the interaction i has a fixed energy value $\epsilon_i^a = \epsilon_i$. One can imagine, for example, that we fix two amino acids forming some contact, while all the other $N-2$ positions of the chain can be occupied by any amino acids (thus, one considers 20^{N-2} sequences).

Let us divide all M folds into two groups: "+" folds include the interaction in question, and "−" folds do not. The interaction is observed when the energy of some "+" fold is below the energies of all other folds.

Each "−" fold acquires an energy E with a probability $p(E)$ given by Eq. (3). This probability is independent of ϵ_i, because the interaction i does not contribute to the energy of "−" folds*.

Each "+" fold acquires an energy E with a probability

$$p_{\epsilon_i}(E) = [2\pi(\sigma^2 - \sigma_i^2)]^{-1/2} \cdot \exp\{-[(E - \epsilon_i) - (\bar{E} - \bar{\epsilon}_i)]^2/2(\sigma^2 - \sigma_i^2)\} \tag{15}$$

which takes into account that the energy of interaction i has a fixed value ϵ_i, and that only $E - \epsilon_i$, the energy of all others interactions in a "+" fold, changes with the sequence. This results in the modifications $E \rightarrow E - \epsilon_i$, $\bar{E} \rightarrow \bar{E} - \bar{\epsilon}_i$, and $\sigma^2 \rightarrow \sigma^2 - \sigma_i^2$ in $p_{\epsilon_i}(E)$ as compared with $p(E)$. $\bar{\epsilon}_i$ is the mean (over all the sequences) energy of interaction i, and $\sigma_i^2 = \overline{\epsilon_i^2} - \bar{\epsilon}_i^2$ is the mean dispersion of energy of this interaction.

Let M_+ be the number of "+" folds, and $M_- = M - M_+$ the number of "−" folds. Then the fraction of sequences which stabilize the "+" folds is

$$\phi_+(\epsilon_i) = \int_{-\infty}^{+\infty} dE \cdot M_+ p_{\epsilon_i}(E) \cdot \left[\int_E^{+\infty} p_{\epsilon_i}(E')dE'\right]^{M_+ - 1} \cdot \left[\int_E^{+\infty} p(E')dE'\right]^{M_-} \tag{16}$$

*See footnote p. 8.

Here the term $M_+ p_{\epsilon_i}(E)$ is a probability that some of the "+" folds has energy E; the second term is a probability that all the other "+" folds have energy above E, and the last term is a probability that the energy of any "−" fold is above E. According to the random energy model, all these probabilities are independent ones.

The fraction of sequences that stabilize the "−" folds is

$$\phi_-(\epsilon_i) = \int_{-\infty}^{+\infty} dE \cdot M_- p(E) \cdot \left[\int_E^{+\infty} p_{\epsilon_i}(E')dE' \right]^{M_+} \cdot \left[\int_E^{+\infty} p(E')dE' \right]^{M_- - 1} \quad (17)$$

The values of these integrals are mainly determined by the small region around $E = E_c$ [see Eqs. (5)–(8)]. Indeed, the first terms of the integrals are much less than unity when E is below E_c, while the second and third terms are small when E is above E_c. The approximation

$$p_{\epsilon_i}(E) \approx p(E) \cdot \exp\left[(-\epsilon_i + \bar{\epsilon}_i) \cdot \partial \ln p/\partial E - \sigma_i^2 \cdot \partial \ln p/\partial(\sigma^2) \right] \approx \approx p(E) \cdot \exp\left\{ - \left[\epsilon_i - (\bar{\epsilon}_i - \sigma_i^2/2RT_c) \right] /RT_c \right\}, \quad (18)$$

helps to solve these equations; it is very accurate in the region $|E - E_c| \leq \sigma$ when $RT_c \ll \sigma$ [this follows from Eq. (10)], $|\epsilon_i - \bar{\epsilon}_i| \leq \sigma$, and $\sigma_i^2 \ll \sigma^2$. The terms $|\epsilon_i - \bar{\epsilon}_i|$ and σ_i^2 are relatively small because they are due to only one of the many interactions. We can also neglect the differences between M_+ and $M_+ - 1$, and between M_- and $M_- - 1$ in Eqs. (17) and (18), because M_+ and M_- are great.

Now, it is obvious that the probability to observe the interaction i in the native structures related to the probability of the opposite as

$$\phi_+ (\epsilon_i) / \phi_- (\epsilon_i) = (M_+/M_-) \cdot \exp\{-[\epsilon_i - (\bar{\epsilon}_i - \sigma_i^2/2RT_c)]/RT_c\} \sim \sim \exp(-\epsilon_i/RT_c) \quad (19)$$

and we come back to an Eq. (1) with $T_* = T_c$, because neither M_+/M_-, nor $\bar{\epsilon}_i - \sigma_i^2/2RT_c$ depend on the ϵ_i value.*

*When the energies of different interactions are correlated (i.e., when they do not change independently with mutations), the term $\Delta\epsilon_i = \epsilon_i - (\bar{\epsilon}_i - \sigma_i^2/2RT_c)$ obtains a general, though complicated form $\Delta\epsilon_i = \epsilon_i - \{[\bar{E} - \langle E - \epsilon_i \rangle] - [(\overline{E^2} - \bar{E}^2) - (\langle E^2 \rangle - \langle E \rangle^2)]/2RT_c)\}$, where $\langle .. \rangle$ means averaging over only those sequences a where $\epsilon_i^a = \epsilon_i$, and the overbar means averaging over all the sequences. Then all the Boltzmann-like relationships, still strict for $\Delta\epsilon_i$, are approximate for ϵ_i itself, as $\Delta\epsilon_i$ includes now, besides ϵ_i, traces of all the interactions coupled with the interaction i and dependent on its value ϵ_i.

In other words, the *conformational temperature* T_* of statistics of heteropolymers coincides with T_c, the *critical temperature* of their folding. T_* can be called also a "selective temperature" as it determines the severity of physical selection of any interaction in the lowest-energy folds of heteropolymer chains.

It should be stressed that this temperature T_* reflects the heterogeneity of interactions of polymer links and must not coincide with room temperature.

Note that the Eq. (19) gives a value of conformational temperature for "ordinary" random sequences. For those that form very stable folds with energy $E_c - \Delta E$ (see Section 3.1), the value of T_* is slightly below T_c [as well as T_m, see Eq. (11)] value because the expansion in Eq. (18) relates now to the energy $E_c - \Delta E$, and

$$T_* = (R\,\partial \ln p/\partial E|_{E=E_c-\Delta E})^{-1} \approx \sigma^2/[R(\bar{E} - E_c + \Delta E)] =$$

$$= T_c \cdot [1 + (\Delta E/RT_c)/(2\ln M)]^{-1} \tag{20}$$

[cf. Eq. (10)].

3.2.1. Computer Modeling of Protein Statistics

It is worthwhile to check the obtained results by a computer experiment. The reason is that (1) the analytical treatment is based on the random energy model (REM) which assumes that different folds acquire their energies irrespective of one another, while the folds of a real chain can have some common elements, and (2) the energies of different interactions are not absolutely independent as assumed by the REM.

We used a computer model of the protein chain shown in Fig. 4. In this model all the compact folds have the same number ($K = 28$) of contacts between nonneighboring links, while the number of bends varies from 14 to 23. For each

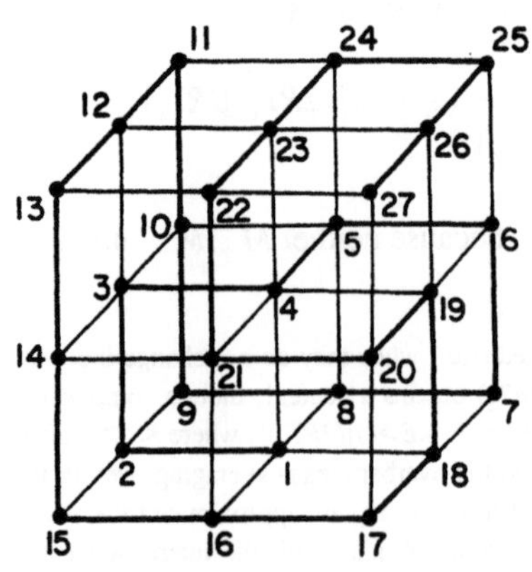

FIGURE 4. A simplified model of a protein globule: one of the 103,346 folds of a self-avoiding chain of 27 links at the 3×3×3 fragment of the cubic lattice (Shakhnovich and Gutin, 1990a). The fold energy E includes energies of contacts (such as those formed by links 7 and 18) and bends (like link 11).

primary structure (i.e., for each set of all contact and bend energies) we found the "native" conformation (one with the minimal energy) by an exhaustive enumeration of all conformations.

To investigate the statistics of interactions, we fixed the energy of one arbitrarily chosen contact or bend and generated a set of n_0 "random sequences" where random energy values were assigned to all other interactions. Then we found n, the number of those sequences that have the contact or bend in question in their native folds, and $n_0 - n$, the number of those that do not have any. According to Eq. (19), one should obtain a linear dependence between $-\ln[n/(n-n_0)]$ and the energy of the examined interaction with a proportionality coefficient equal to $1/RT_*$. This proportionality is clearly seen in Fig. 5. The conformational temperature T_*, which follows from these experiments, has virtually the same value for different interactions, independently of how the random energies are generated and is indeed very close to the critical temperature of folding that follows from the shape of energy spectrum for corresponding sequences.

To check the approximation that treats the energies of different interactions of a same link as independent values, we used the same self-avoiding chain shown in Fig. 4 and carried out a computer experiment with sequences of "black" and "white" beads. In this model each black link attracts any black one, each white link repulses any white one, while white and black links do not interact. Thus, black links play a role of hydrophobic residues. Then we calculated the number black–black, white–black, and white–white contacts in the lowest-energy folds of these chains (Fig. 6); when a chain had more than one lowest-energy fold, we choose one of them by random.

In this case the energies of different contacts are extremely correlated, because the variety of beads is very small. This leads to a deviation from a simple Boltzmann's formula [Eq. (1)]. Nevertheless, the general tendency is the same, and the T_* value calculated from the occurrence-to-energy dependence is close to T_c estimated from the abatement of spectrum density at the beginning of corresponding spectra.

The last experiment can shed some light on the deviations from the strict Boltzmann-like statistics observed in proteins (Fig. 2). Different interactions of an amino acid residue are somewhat correlated; for example, α- and β-stabilizing properties of amino acids correlate with their hydrophobicity (Ptitsyn and Finkelstein, 1980). Therefore, the internal energy of each interaction which enters the Eq. (1) is actually somewhat "renormalized" by the interactions that are coupled with the considered one.*

It is noteworthy that the deviations from Boltzmann's formula are well known in physics of condensed matter (actually, this formula is correct only for

*See footnote p. 13.

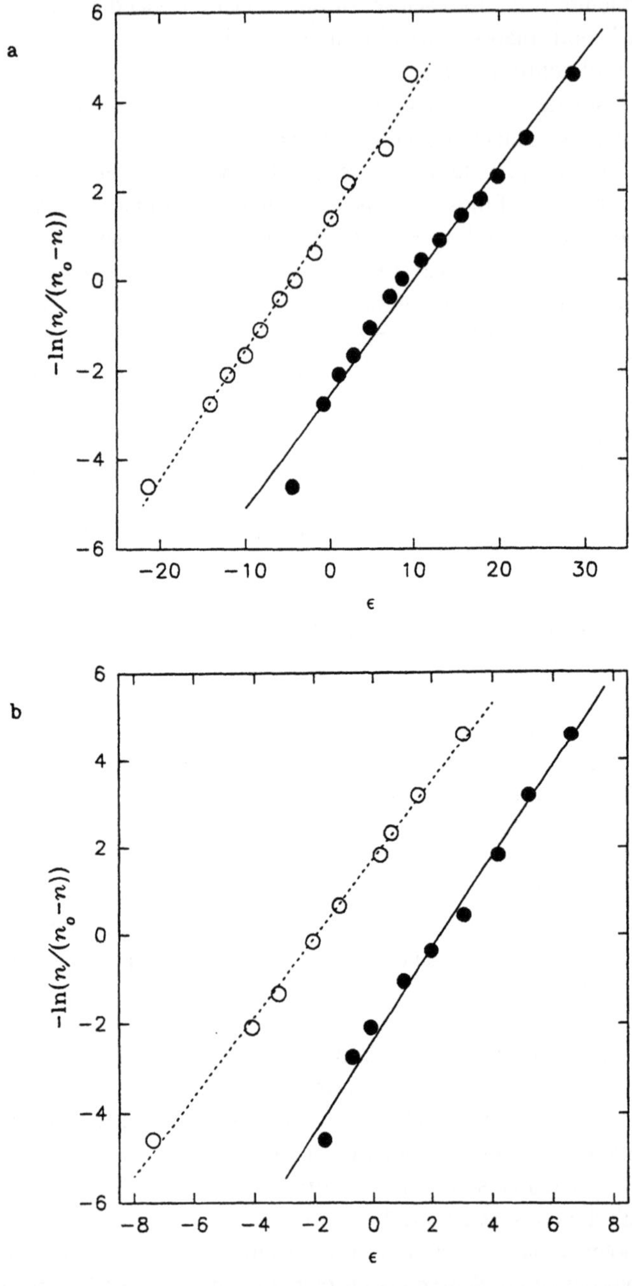

FIGURE 5. Occurence of a contact 9:18 (○) and of a bend 11 (●) in the "native" folds of random sequences; ϵ is the energy of the examined element, n_o = 100 is the number of explored random sequences, n the number of those which have a native fold with the examined element. Mean

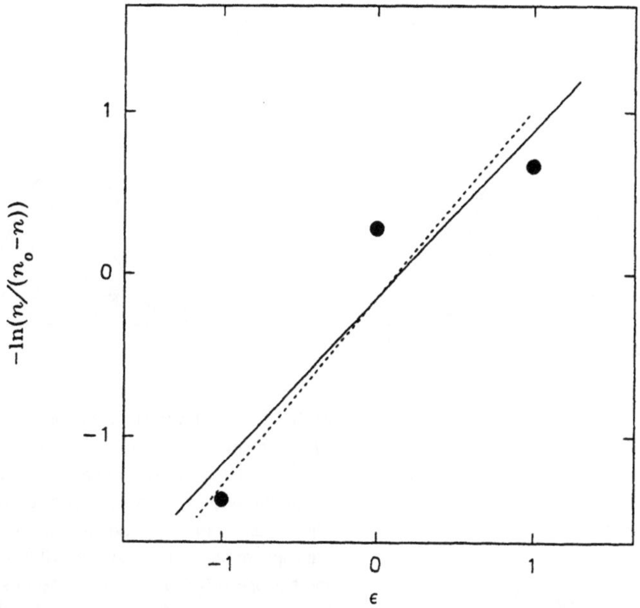

FIGURE 6. Occurrence of different contacts in the lowest-energy folds of random chains of "white" and "black" beads. The points refer to black–black, black–white, and white–white contacts. The deviation of these points from a straight line reflects the coupling of different interactions of the beads: n_o = the total number of possible contacts of given type in 100 explored sequences, n = the total number of contacts of given type in native folds of these sequences. The best approximation of ϵ by $-R\ln[n/(n_o - n)] + b$ is shown by -----; the best approximation of $R\ln[n/(n_o - n)]$ by $-\epsilon \cdot T_*^{-1} + b$ is shown by ——— (cf. Fig. 2b,c).

gases), and the reason for these deviations is the same: a correlation of different interactions (Landau and Lifshitz, 1959).

3.3. A Given Folding Pattern

Above we considered a simplest case when all the folds have the same overall properties. To understand the physical selection of protein folding patterns we have to consider a case when the folds are divided into groups with different properties (Finkelstein *et al.*, 1993). Let us consider the basic cases:

---

deviation of the contact energy from zero is taken as unity; the mean deviation of randomly generated bend energy from zero is 10 times greater (a) and 10 times less (b). In both cases, the linear (Boltzmann-like) dependence of $-\ln[n/(n_o - n)]$ on ϵ is evident. The slope of the best-fitted line gives the value of $1/RT_*$. The conformational temperature T_* obtained from these lines is virtually the same for the contact and bend, and close to T_c measured from the abatement of the logarithm of spectrum density at the beginning of the energy spectra of corresponding chains (cf. Fig. 3c).

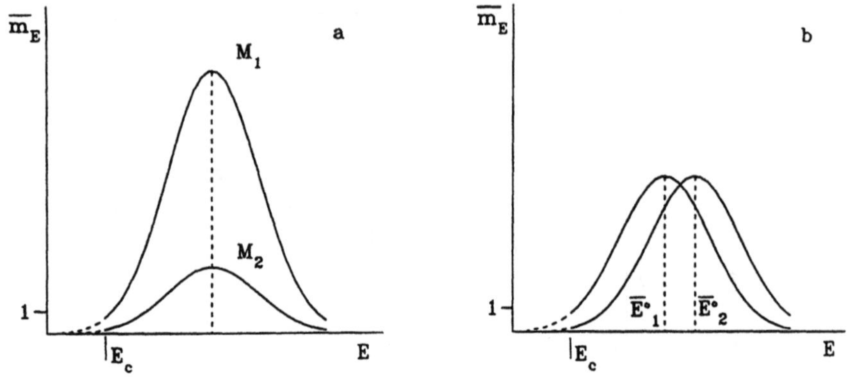

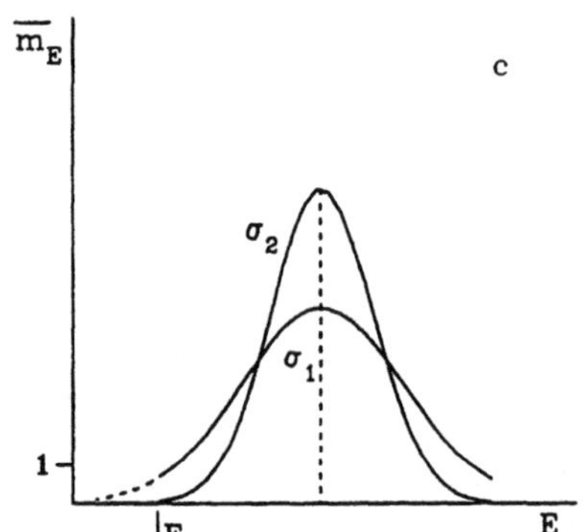

FIGURE 7. "Physical selection" of folding patterns: the smaller amount of low-energy folds gives the sequence a smaller chance to find its energy minimum within a handicapped group of folds. Three basic cases are shown. (a) Group 2 has an "entropic defect": it contains much fewer folds than another one ($M_2 \ll M_1$); other properties of the folds (mean energies and dispersions) are the same for both groups: $\bar{E}_1^\circ = \bar{E}_2^\circ$, and $\sigma_1 = \sigma_2$. (b) Group 2 consists of folds with an"energetic defect": $\bar{E}_2^\circ > \bar{E}_1^\circ$; other properties of the folds of both groups are the same: $M_1 = M_2$ and $\sigma_1 = \sigma_2$. (c) Group 2 includes folds with a smaller variety of interactions: $\sigma_2 < \sigma_1$; other properties of the folds of both groups are the same: $\bar{E}_1^\circ = \bar{E}_2^\circ$ and $M_1 = M_2$.

3.3.1. Entropic Defects

Suppose, all M folds are divided into two groups: the first consisting of M_1 folds and the second of and $M_2 \ll M_1$ ones (Fig. 7a). This corresponds, for example, to the division of folding patterns into those with right- and left-handed connections of β-strands (Fig. 1b): a left-handed connection includes a smaller number of chain conformations (see above). Let all energetic properties of the folds be the same. Then each fold has an equal probability to serve as the lowest-energy one for some random amino acid sequence. However, the number of right-handed folds is much greater than that of their competitors. Proportionally, the lowest-energy fold of a random chain has a much greater chance to happen to be a right-handed one.

This can also be explained from another point of view. The expected energy of the lowest-energy fold of group 1 is $\bar{E} - \sigma\sqrt{2\ln M_1}$ [see Eq. (6)], while the lowest-energy fold of group 2 usually has the energy of $\bar{E} - \sigma\sqrt{2\ln M_2}$. As $M_1 \gg$

M_2, the best fold of group 1 usually has a lower energy than the best fold of group 2. This shows that a smaller variety of folds gives a random sequence a smaller possibility to obtain a low-energy fold. Thus, an "entropic" defect (a lack of folds) results in an "energetic" one: this "defect" can discriminate protein structures, even though the entropy of a native protein globule is zero.

3.3.2. Energetic Defects

Suppose now that the folds of group 2 have some "energetic" defect which enlarges their mean energy $\bar{E}^0_2$ relatively to $\bar{E}^0_1$, the mean energy of the folds of group 1 (Fig. 7b). This corresponds, in particular, to division of folding patterns into those with different twists of a β-sheet (Fig. 1a) or with noncrossed and crossed loops (Fig. 1c). Let both groups have the same number of folds ($M_1 = M_2$) and the same variety of interactions (i.e., $\sigma_1 = \sigma_2$).

Then $\bar{m}'_E/\bar{m}''_E = \exp[-(E - \bar{E}^0_1)^2/2\sigma^2]\ /\ \exp[-(E - \bar{E}^0_2)^2/2\sigma^2]$ is the ratio of average numbers of folds of these groups with the energy E. Evidently, this ratio can be represented in a simple "Boltzmann-like" form:

$$\bar{m}'_E/\bar{m}''_E = \exp[-(\bar{E}^0_1 - \bar{E}^0_2)/RT_E] \tag{21}$$

where

$$T_E = \sigma^2/[R(\bar{E} - E)] \tag{22}$$

and $\bar{E}$ is equal to $(\bar{E}^0_2 + \bar{E}^0_1)/2$. At the beginning of the energy spectrum $T_E = T_c$ [see Eq. (10)], the same which governs a Boltzmann-like statistic of small elements of protein globules, and this ratio [Eq. (21)] show the proportion of random sequences that can stabilize the folds of competing groups.

Thus, even a small difference $\bar{E}^0_1 - \bar{E}^0_2$ can discriminate folding patterns. One must not compare it with the total energy of protein structure; to discriminate a pattern "inconvenient" for random sequences, $\bar{E}^0_1 - \bar{E}^0_2$ must exceed only RT_c, i.e., only ~1 kcal/mole.

Of course, *any* defect can be compensated by an appropriate amino acid sequence, but according to the above estimates the amount of these "lucky" sequences is low. This explains why the "popular" folds of globular proteins are those which have virtually no defects such as crossed loops, cavities, and so forth.

3.3.3. Defect of Variety

Let both groups have the same number of folds ($M_1 = M_2$), the same mean energy ($\bar{E}^0_1 = \bar{E}^0_2$), but different energy dispersions ($\sigma_1 > \sigma_2$): the energy spectrum of one group is broader than that of another (Fig. 7c).

Then $\bar{m}'_E/\bar{m}''_E = \exp[-(E - \bar{E}^0)^2/2\sigma_1^2] / \exp[-(E - \bar{E}^0)^2/2\sigma_2^2]$ is the ratio of average numbers of folds of these groups with the energy E. When $|\sigma_1 - \sigma_2| \ll \sigma$ [σ^2 is $(\sigma_1^2 + \sigma_2^2)/2$], and $E = E_c$, this ratio has a simple form

$$\bar{m}'_E/\bar{m}''_E = \exp[(\sigma_1^2 - \sigma_2^2)/2(RT_c)^2] \tag{23}$$

which shows a predominance of low-energy folds of that group which has a greater dispersion of interactions. In general, this effect discriminate noncompact folds.

3.3.4. Statistics of Folding Patterns

Summarizing the above equations, we see that the abundance of a folding pattern p is proportional to

$$\exp(-\tilde{F}_p/RT_*)$$

where

$$\tilde{F}_p = \bar{E}_p^0 - \sigma_p^2/2RT_* - RT_*\ln M_p \tag{24}$$

is the "selective free energy" of the pattern. It depends on M_p, the number of different folds within the pattern; on $\bar{E}_p^0$, the mean energy of these folds; on σ_p, the mean dispersion of their energies; and on the universal conformational termperature $T_* \approx T_c$.

Thus, Boltzmann-like statistics covers not only small details, but also the overall chain folds.

3.4. Origins of Boltzmann-like Statistics of Globular Proteins

One can see that each interaction of low energy increases exponentially the number of sequences that ensure the stability of folds with a given feature (residue-to-residue contact, bend, folding motif, etc.), and each high-energy interaction leads to an exponential decrease in this number. This is a general physical origin of Boltzmann-like statistics of the *ground* states of heterogeneous and mutable systems: the more sequences provide the stability of folds with a given feature, the more often this feature can be observed. This resembles the origin of Boltzmann and Gibbs statistics of thermodynamic systems, where each interaction of low energy increases the number of possible microstates of a thermostat (see, e.g., Landau and Lifshitz, 1959). The difference is that our "thermostat" is now a multitude of sequences that stabilize folds with a given feature.

Thus, Boltzmann-like statistics of globular proteins originates from spontaneous statistics of low-energy folds of random heteropolymers, as if the proteins obey the "multitude principle": the more random sequences provide the stability of folds with a given feature, the more often this feature can be observed in proteins.

3.5. An Estimate of Critical Temperature

T_c has the meaning of denaturation temperature for such a kind of protein melting that does not change the compactness of the globule (Shakhnovich and Gutin, 1989). This melting has been never observed up to now. However, a close analog of this melting is known: heat melting leads to the molten globule state, which is only slightly expanded as compared with the native one (Ptitsyn, 1992). When protein melting leads to a looser state (this takes place, in particular, when the solvent quality is improved by a denaturant), the temperature of denaturation drops (Tanford, 1968; Finkelstein and Shakhnovich, 1989). Thus, the minimal estimate of T_c ($\approx$ 350–400 K) is given by the highest temperature of protein denaturation. On the other hand, the temperature of protein melting in a vacuum (which is a limiting case of a poor solvent) is $\sim$ 500 K according to a theoretical estimate (Shakhnovich and Finkelstein, 1989). This can give a maximal estimate of T_c.

The value of T_c fluctuates with the sequence. The range of these T_c deviations from its mean value (like a range of thermodynamic fluctuations of the usual temperature of a small system) is $\sim \pm T_c \cdot(\text{number of residues})^{-1/2}$, i.e., $\sim \pm$ 30° for a domain of a normal size. It is noteworthy that melting temperatures of meso- and thermostable proteins differ by approximately the same value, while their primary and tertiary structures have no considerable differences (Branden and Tooze, 1991).

4. CONSEQUENCES FOR PROTEIN STRUCTURE PREDICTION AND FOLDING

The folding process and three-dimensional structure prediction have a common feature: they both have to operate on the basis of *only some part* of the interactions that fix the native structure of a protein.

In a case of protein folding, it is known that some fluctuating secondary structures and then some globular states appear at early stages of *in vitro* self-organization (for a review, see Kuwajima *et al.*, 1987; Udgaonkar and Baldwin, 1990; Bycroft *et al.*, 1990; Ptitsyn, 1991, 1992; Ptitsyn and Semisotnov, 1991; Matouschek *et al.*, 1992). These early intermediates demonstrate some nativelike features (in particular: secondary structure content and location). The intermediates are formed by hydrogen bonds and hydrophobic interactions much earlier than other forces start to work (such as specific van der Waals interactions responsible for tight packing). Being responsible for the stability of native protein structures (Shakhnovich and Finkelstein, 1989), these "late" interactions are rather powerful. The questions arise: Why do the "late" forces preserve some features of the intermediates formed only by the "early" interactions? (In other words: why do the observed semifolded states serve as intermediates rather than as the dead ends of folding?) What kind of a concordance of "early" and "late" interactions in proteins does this need?

Similar questions arise in protein structure prediction. Any prediction is based on some energetic parameters, and any set of parameters can give only some approximation of actual interactions. Thus, a prediction is always based on some part of the actual interactions, while another part remains unknown and can be only roughly estimated. As a result, the fold that has the lowest "calculated" energy can be very different from that which has the lowest actual energy. Native folds are among those whose calculated energy is low (Covell and Jernigan, 1990; Finkelstein and Reva, 1991), but how large is this set of "promising" folds that is worthy of consideration? And, in the main, how can the predictions be more or less successful, at least in some cases (Fasman, 1989; Thornton *et al.*, 1991), if they ignore a considerable part of actual interactions?

The answers to these questions follow from the above results. Let us consider the structures formed by some part of all the forces operating in the protein chain (this can be, for example, the forces operating at the early stages of folding). Let these forces be independent of others and let the temperature be equal to T_*. Under this temperature, a part of the forces cannot fix a unique fold and the molecule fluctuates. However, those of its fluctuating structures that dominate at the termperature T_* have (in proportion with that predominance) a predominant chance to enter the native fold, according to Eq. (1). Therefore, when a fluctuating protein chain at the early stages of folding demonstrates some definite structural features (e.g., some secondary structure type), these features are the most probable for the native fold (Fig. 8).

Earlier we assumed that the "early" forces are independent of the "late" ones, and concluded that the most nativelike intermediates are expected at the temperature T_*, which is close to that of protein melting. However, when some positive correlation of the "early" and "late" forces exists (such as the correlation of α- and β-stabilities of amino acids with their hydrophobicity), the temperature characteristic for the most nativelike intermediates drops. Now this temperature must be below T_* in proportion to the correlation of the forces, because the increment of the "early" interactions to protein statistics [Eq. (1)] is virtually increased by the "late" ones.

All the above concerns not only protein folding, but protein structure prediction as well, because any prediction is also based only on some part of the actual interactions: another part is hidden by the "noise" of inevitable errors in energetic parameters.

This makes energy minimization not the best tool for protein structure prediction from its amino acid sequence. Rather, one can rely on statistical mechanics: the Eq. (1) actually describes the basic case of protein structure prediction.

These thermodynamic calculations carried out at a temperature equal to or somewhat below that of protein melting give the best possible, though approximate, predictions (Fig. 8). This conclusion is a general one, independent of what interactions are taken into account and how precisely they are estimated, though the quality of prediction is definitely sensitive to the value of errors in energetic parameters.

Only in the case when errors are negligible, the lowest-energy fold can be singled out unambiguously—by statistical mechanics as well as by energy minimization. When errors are moderate, statistical mechanics singles out the common features of all the low-energy folds (Fig. 8); the most pronounced of them

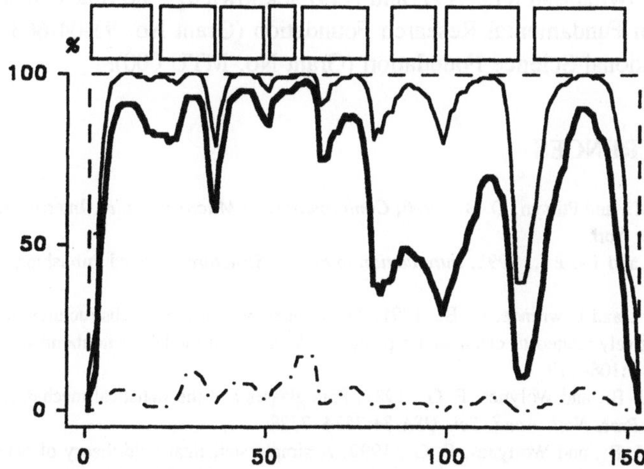

FIGURE 8. Probability of the α-helical state of residues of unfolded myoglobin chain calculated at different temperatures; the case $T = 0$ K corresponds to the energy minimum. The ordinate shows the average helicity of each residue in the chain. The algorithm of calculations is based on statistical mechanics: it corresponds to Zimm and Bragg's (1959) model of helix–coil transition. The energetic parameters of helix–coil transition are the same as in (Finkelstein *et al.*, 1991); it has been demonstrated that they give a possibility to calculate the helicity of polypeptides rather accurately. Rectangles show the positions of α-helices in the native protein (Watson, 1969). It is clearly seen that the secondary structure of the unfolded protein chain at a temperature $T \approx 300$ K reflects the secondary structure of native protein and that a reasonable prediction can be based on thermodynamic calculations done at $T \approx 300$ K and not on the minimization of the secondary structure energy which corresponds to $T = 0$ K and gives a single helix (- -, $T = 0$ K; ———, $T = 250$ K; ━━, $T = 300$ K; •-•-, $T = 350$ K).

must usually [according to Eq. (1)] coincide with features of the native one. At the same time the fold with the minimal "calculated" energy (e.g., a single helix in Fig. 8) can have little to do with the actual lowest-energy fold. Finally, when errors are great, these common features are slurred over, and any prediction is impossible.

5. CONCLUSION

We see that Boltzmann-like statistics is a general feature of stable folds of "random" heteropolymers, the "temperature" in this statistics emerging from the diversity of residues. The origin of this statistics is simply that the more sequences provide the stability of folds with a given feature, the more often this feature can be observed. As far as the chains of globular proteins (unlike, e.g., fibrous ones) that resemble random heteropolymers (Ptitsyn, 1985), this "multitude principle" is valid for them as well.

ACKNOWLEDGMENTS. AVF and AYB acknowledge the financial support of the Russian Fundamental Research Foundation (Grant No. 93-04-6636) and of the International Science Foundation (Grant No. MTU 000).

6. REFERENCES

Birstein, T. M., and Ptitsyn, O. B., 1966, *Conformations of Macromolecules*, Interscience Publishing, New York.

Branden, C., and Tooze J., 1991, *Introduction to Protein Structure*, Garland Publishing, New York, London.

Bryant, S. H., and Lawrence, C. E., 1991, The frequency of ion-pair substructures in proteins is quantitatively related to electrostatic potential: A statistical model for nonbonded interactions, *Proteins* **9**:108–119.

Bryngelson, J. B., and Wolynes, P. G., 1987, Spin glasses and the statistical mechanics of protein folding, *Proc. Natl. Acad. Sci. USA* **84**:7524–7528.

Bryngelson, J. B., and Wollynes, P. G., 1990, A simple statistical field-theory of heteropolymer collapse with application to protein folding, *Biopolymers* **30**:177–188.

Bycroft, M., Matouschek, A., Kellis, J. T., Jr., Serrano, L., and Fersht, A. R., 1990, Determination and characterization of a folding intermediate by NMR, *Nature* **346**:488–490.

Chothia, C., 1973, Conformation of twisted β pleated sheets in proteins, *J. Mol. Biol.* **75**:295–302.

Chothia, C., 1984, Principles that determine the structure of proteins, *Annu. Rev. Biochem.* **53**:537–572.

Chothia, C., and Finkelstein, A. V., 1990, The classification and origins of protein folding patterns, *Annu. Rev. Biochem.* **59**:1007–1039.

Covell, D. G., and Jernigan, R. L., 1990, Conformations of folded proteins in restricted spaces, *Biochemistry* **29**:3287–3294.

Derrida, B., 1981, Random-energy model: An exactly solvable model of disordered systems, *Physiol. Rev. B.* **24**:2613–2626.

De Santis, P., and Liquori, A. M., 1971, Conformation of Gramicidin S, *Biopolymers* **10**:699–710.

Efimov, A. V., 1984, A novel super-secondary structure of proteins and the relation between the structure and amino acid sequence, *FEBS Lett.* **166**:33–38.

Efimov, A. V., 1991, Structure of coiled β-β-hairpins and β-β-corners, *FEBS Lett.* **284**:288–292.

Fasman, G. D., 1989, *Prediction of Protein Structure and the Principles of Protein Conformation,* Plenum, New York.

Finkelstein, A. V., and Ptitsyn, O. B., 1987, Why do globular proteins fit the limited set of folding patterns? *Prog. Biophys. Mol. Biol.* **50**:171–190.

Finkelstein, A. V., and Reva, B. A., 1991, Search for the most stable folds of protein chains, *Nature* **351**:497–499.

Finkelstein, A. V., and Shakhnovich, E. I., 1989, Theory of cooperative transitions in protein molecules. II. Phase diagram for protein molecule in solution, *Biopolymers* **28**:1681–1694.

Finkelstein, A. V., Ptitsyn, O. B., and Kozitsyn, S. A., 1977, Theory of protein molecule self-organization. II. A comparison of calculated thermodynamic parameters of local secondary structures with experiment, *Biopolymers* **16**:497–524.

Finkelstein, A. V., Badretdinov, A. Ya., and Ptitsyn, O. B., 1991, Physical reasons for secondary structure stability, *Proteins* **10**:287–299.

Finkelstein, A. V., Gutin, A. M., and Badretdinov, A. Ya., 1993, Why are the same protein folds used to perform different functions? *FEBS Lett.* **325**:23–28.

Flory, P. J., 1969, *Statistical Mechanics of Chain Molecules,* Interscience, New York.

Gelin, B. R., and Karplus, M., 1975, Sidechain torsional potentials and motion of amino acids in proteins: Bovine pancreatic trypsin inhibitor, *Proc. Natl. Acad. Sci. USA* **72**:2002–2006.

Go, N., 1975, Theory of reversible denaturation of globular proteins, *Int. J. Pept. Protein Res.* **7**:313–323.

Go, N., and Taketomi, H., 1979, Studies on protein folding, unfolding and fluctuations by computer simulation. III. Effect of short-range interactions, *Int. J. Pept. Protein Res.* **13**:235–252.

Gregoret, L. M., and Sauer, R. T., 1993, Additivity of mutant effects assessed by binomial mutagenesis, *Proc. Natl. Acad. Sci. USA* **90**:4246–4250.

Gutin, A. M., Badretdinov, A. Ya., and Finkelstein, A. V., 1992, Why is the statistics of protein structures Boltzmann-like? *Molek. Biol* (in Russian) **26**:118–127.

Horovitz, A., Matthews, J. M., and Fersht, A. R., 1992, α-Helix stability in proteins. II. Factors that influence stability at an internal positions, *J. Mol. Biol.* **227**:560–568.

Hudson, D. J., 1964, *Statistics,* CERN, Geneva.

Kajava, A. V., 1992, Left-handed topology of super-secondary structure formed by aligned α-helix and β-hairpin, *FEBS Lett.* **302**:8–10.

Kuwajima, K, Yamaya, H., Miwa, S., Sugai, S., and Nagamura, T., 1987, Rapid formation of secondary structure framework in protein folding studied by stopped flow circular dichroism, *FEBS Lett.* **221**:115–118.

Landau, L. D., and Lifshitz, E. M., 1959, *Statistical Physics,* Pergamon, London.

Levitt, M., and Chothia, C., 1976, Structural patterns in globular proteins, *Nature* **261**:552–557.

MacArthur, M. W., and Thornton, J. M., 1991, Influence of proline residues on protein conformation, *J. Mol. Biol.* **218**:397–412.

Matouschek, A., Serrano, L., and Fersht, A. R., 1992, The folding of enzyme. IV. Structure of an intermediate in refolding of barnase studied by a protein engineering procedure, *J. Mol. Biol.* **224**:819–835.

McGregor, M. J., Islam, S. A., and Sternberg, M. J. E., 1987, Analysis of the relationship between side chain conformation and secondary structure in globular proteins, *J. Mol. Biol.* **198**:295–310.

Miller, S., Janin, J., Lesk, A. M., and Chothia, C., 1987, Interior and surface of monomeric proteins, *J. Mol. Biol.* **196**:641–656.

Miyazava, S., and Jernigan, R. L., 1985, Estimation of effective interresidue contact energies from crystal structures: Quasi-chemical approximation, *Macromolecules* **18**:534–552.

Novotny, J., Bruccoleri, R., and Karplus, M., 1984, An analysis of incorrectly folded protein models. Implication for structure prediction, *J. Mol. Biol.* **177**:787–818.

O'Neil, K. T., and DeGrado, W. F., 1990, A thermodynamic scale for the helix-forming tendencies of the commonly occurring amino acids, *Science* **250**:646–651.

Pohl, F. M., 1971, Empirical protein energy maps, *Nature New Biol.* **234**:277–279.

Pohl, F. M., 1980, Statistical analysis of protein structures, in *Protein Folding* (R. Jaenicke, ed.), pp. 183–196, Elsevier/North-Holland Biomedical Press, Amsterdam, New York.

Privalov, P. L., 1979, Stability of proteins. Small globular proteins. *Adv. Protein Chem.* **33**:167–241.

Ptitsyn, O. B., 1985, Random sequences and protein folding, *J. Molec. Struct. (Theochem.)* **123**:45–65.

Ptitsyn, O. B., 1991, How does protein synthesis give rise to the 3D-structure? *FEBS Lett.* **285**:176–181.

Ptitsyn, O. B., 1992, The molten globule state, in *Protein Folding* (T. E. Creighton, ed.), pp. 243–300, W. H. Freeman, New York.

Ptitsyn, O. B., and Finkelstein, A. V., 1980, Similarities of protein topologies: Evolutionary divergence, functional convergence or principles of folding? *Quart. Rev. Biophys.* **13**:339–386.

Ptitsyn, O. B., and Semisotnov, G. V., 1991, The mechanism of protein folding, in *Conformations and Forces in Protein Folding* (B. T. Nall and T. E. Creighton, eds.), pp. 155–168, American Association for Advancement of Science, Washington.

Rao, S. T., and Rossmann, M. G., 1973, Comparison of super-secondary structures in proteins, *J. Mol. Biol.* **109**:235–250.

Rashin, A. A., Ionif, M., and Honig, B., 1986, Internal cavities and buried waters in globular proteins, *Biochemistry* **25**:3619–3625.

Richardson, J. S., 1977, β Sheet topology and the relatedness of proteins, *Nature* **268**:495–500.

Richardson, J. S., 1981, Anatomy and taxonomy of protein structures, *Adv. Protein Chem.* **34**:167–339.

Richardson, J. S., and Richardson, D. C., 1988, Amino acid preferences for specific locations at the ends of α-helices, *Science* **240**:1648–1652.

Schulz, G. E., and Schirmer, R. H., 1979, *Principles of Protein Structure*, Springer-Verlag, New York.

Serrano, L., Sancho, J., Hirshberg, M., and Fersht, A. R., 1992, α-Helix stability in proteins. I. Empirical correlations concerning substitutions of side-chains at the N and C-caps and the replacement of alanine by glycine or serine at solvent-exposed surfaces, *J. Mol. Biol.* **227**:544–559.

Shakhnovich, E. I., and Finkelstein, A. V., 1989, Theory of cooperative transitions in protein molecules. I. Why denaturation of globular protein is the first order phase transition, *Biopolymers* **28**:1667–1680.

Shakhnovich, E. I., and Gutin, A. M., 1989, Formation of unique structure in polypeptide chains. Theoretical investigation with the aid of a replica approach, *Biophys. Chem.* **34**:187–199.

Shakhnovich, E. I., and Gutin, A. M., 1990a, Enumeration of all compact conformations of copolymers with random sequence of links, *J. Chem. Phys.* **93**:5967–5971.

Shakhnovich, E. I., and Gutin, A. M., 1990b, Implication of thermodynamics of protein folding for evolution of primary sequences, *Nature* **346**:773–775.

Sternberg, H. J. E., and Thornton, J. M., 1976, On the conformation of proteins: The handedness of β-strand–α-helix–β-strand unit, *J. Mol. Biol.* **105**:367–382.

Tanford, C., 1968, Protein denaturation, *Adv. Protein Chem.* **23**:121–218.

Thornton, J. M., Flores, T. P., Jones, D. T., and Swindells, M. B., 1991, Prediction of progress at last, *Nature* **354**:105–106.

Udgaonkar, J. B., and Baldwin, R. L., 1990, Early folding intermediate of ribonuclease A, *Proc. Natl. Acad. Sci. USA* **87**:8197–8201.

Watson, H. C., 1969, The stereochemistry of the protein myoglobin, *Prog. Stereochem.* **4**:299–333.

Zimm, B. H., and Bragg, J. R., 1959, Theory of the phase transition between helix and random coil in polypeptide chains, *J. Chem. Phys.* **31**:526–535.

Chapter 2

Compact Intermediates States in Protein Folding

Anthony L. Fink

1. INTRODUCTION

Recently there has been growing recognition of the existence and importance of compact intermediate states of proteins. Such species have been observed under both transient (refolding kinetics) and equilibrium conditions. It is clear that for many proteins most denaturing conditions do not lead to a fully unfolded protein (random coil), but rather to species with substantial secondary structure and substantial compactness, relative to the fully unfolded state. In addition, there is now good experimental data to demonstrate the existence of two classes of compact denatured states of proteins: compact intermediates, in the thermodynamic sense (i.e., a minimum in the free energy profile for the reaction), and compact substates of the unfolded state (Palleros *et al.*, 1993). It is important to note that it is often experimentally difficult to distinguish between these two types of compact denatured states, especially by spectral methods. Recent reviews of compact denatured states, and particularly the molten globule, include those of Dill and Shortle (1991), Ptitsyn (1987, 1992), Kuwajima (1989), Christensen and Pain (1991), and Baldwin and Roder (1991). Theoretical models

Anthony L. Fink Department of Chemistry and Biochemistry, University of California, Santa Cruz, California 95064.

Subcellular Biochemistry, Volume 24. Proteins: Structure, Function, and Engineering, edited by B. B. Biswas and Siddhartha Roy. Plenum Press, New York, 1995.

for the existence of two classes of denatured states have been presented by Dill and co-workers (Alonso *et al.*, 1991), Ptitsyn (1987, 1992), and Finkelstein and Shakhnovich (1989).

Rapid-mixing kinetics experiments using circular dichroism or nuclear magnetic resonance (NMR) reveal the formation of intermediates with substantial secondary structure and presumed compact conformations on a time scale of milliseconds or less (summarized in Lecomte and Matthews, 1993; Englander and Mayne, 1992). The lack of suitable methods for measuring the hydrodynamic size of proteins on such short time scales has made it difficult to measure the compactness of early-formed intermediates in refolding experiments. Considerable evidence suggests that the equilibrium compact intermediates may be good models for the transient intermediates formed during folding (Ikeguchi *et al.*, 1986a; Barrick and Baldwin, 1993; Baldwin, 1993).

Although there have been reports in the literature for a long time that denatured proteins, especially when denatured by heat or pH, may have residual structure (e.g., Aune *et al.*, 1967), it is only recently that there has been general recognition that denatured proteins may exist in a wide range of conformations in terms of compactness and residual secondary structure. Since some of these species are almost as compact as the native state and contain essentially as much secondary structure as the native state, the question arises as to the distinguishing features between such compact structured states and the native state on one hand and the more fully unfolded state on the other.

Many terms connected with denatured states have been used rather loosely, and to mean different things by different authors, leading to some confusion because of semantic problems. For example, *intermediate* has been used both in the sense of a thermodynamic intermediate state (i.e., a minimum in the potential energy surface) and also to indicate a structure or species with properties between two other conformations. The term *state* is often not used in the thermodynamic sense. In many cases in which compact intermediate or denatured states have been reported, insufficient experimental evidence was available to determine whether the species was a true intermediate in the thermodynamic sense or just a compact substate of the unfolded state. In fact, in many cases the hydrodynamic radius has not been measured [there is some evidence to indicate that the ellipticity at 222 nm correlates strongly with the compactness of protein conformations (Palleros *et al.*, 1993)].

The stability of a native folded protein is determined by the difference in free energy of the unfolded and folded conformations. Thus the free energy of the unfolded state contributes equally to that of the folded state in terms of the overall stability. Since Shortle's (1987) seminal observations regarding microsubstates of denatured staphylococcal nuclease the importance of structure in the denatured state to the stability of the protein has been the focus of considerable attention. An interesting approach to the question has been that designed on the principle of increasing the stability of the *denatured* state by site-directed mutagenesis and thus decreasing the apparent stability of the native state. Recently Bowler *et al.* (1993) have reported that replacing the surface lysine-73 of yeast iso-1-cytochrome *c* with hydrophobic residues resulted in a 1–1.5 kcal/mole decrease in stability of the folded protein, consistent with the idea that residue 73 in the mutants is more buried in the denatured state than in the native state and consistent with a structured denatured state. Confirmation that unfolded states may be compact is also found in time-resolved fluorescence energy transfer studies of staphylococcal nuclease (James *et al.*, 1992) that demonstrated that the denatured state of a fluorophore-labeled derivative is compact.

It is important to note that there are a *minimum* of four criteria necessary to confirm that a denatured state is a compact intermediate state, namely (1) significant secondary structure, (2) little or no nativelike tertiary structure, (3) compactness closer to the native state than the unfolded state, and (4) a significant energy (kinetic) barrier between the putative compact intermediate and the native and unfolded states (in order to distinguish the compact intermediate from a compact form of the unfolded state). Since both compact intermediates and compact unfolded states may share several features in common, such as substantial secondary structure, affinity for 8-anilinonaphthalene-1-sulfonic acid (ANS), and significant compactness, their distinction is not as straightforward as might be thought.

The most widely known compact denatured state of proteins is the molten globule. This term was first used in the literature to describe the acid-denatured state of cytochrome *c* by Ohgushi and Wada (1983), which resembled a compact state with nativelike secondary structure but fluctuating tertiary structure, which had recently been reported by Ptitsyn and co-workers for α-lactalbumin (Dolgikh *et al.*, 1981). It has now been used to describe the nonnative state of a number of proteins denatured under a variety of conditions in which residual structure, especially secondary structure, was observed. Although the compactness of the species has been measured in some cases, it has frequently not been determined. Thus the term *molten globule* has been used in a rather general way, at least by many authors, to describe any denatured state with significant secondary structure and a propensity to bind hydrophobic dyes such as ANS. In cases where the hydrodynamic radius was measured, some of these species were found to be quite compact, whereas others were considerably expanded relative to the native state. Some are undoubtedly true intermediate states, others are just compact forms (substates) of the unfolded state. The distinction between a true (thermodynamic) intermediate and a compact substate of the unfolded state can only be determined by experiments in which the existence of an energy barrier between the putative intermediate and the native and unfolded states is shown. Spectroscopic experiments that show residual structure and spectral transitions that appear cooperative, in the absence of thermodynamic or kinetic data, cannot distinguish the two classes of denatured states. Based on our current understanding of compact denatured states it seems reasonable to equate molten globules with compact intermediate states, as elaborated in this chapter.

Recent investigations have now made it clear that for both intermediates and unfolded states the degree of folding and compactness can vary over a wide range depending on the experimental conditions. A particularly good example of this is the *Escherichia coli* molecular chaperone DnaK. In the presence of low concentrations of guanidine HCl the native state is converted into a compact intermediate whose hydrodynamic radius increases monotonically with increasing denaturant; when first detected at low guanidine HCl (0.25 M) the Stokes radius (R_S) is 43 Å, when last observed at higher denaturant (1.4 M) the R_S is 55 Å (Palleros *et al.*, 1993). Similarly, the unfolded state, when first detected at relatively low denaturant concentration, has R_S = 53 Å, whereas at concentrations of guanidine HCl above 4 M its R_S is 69 Å. The compact intermediate is sufficiently stable and sufficiently well separated in terms of energy barriers from the native and unfolded states that it is readily observed under equilibrium conditions by size-exclusion high-pressure liquid chromatography (HPLC). In the vicinity of 0.6 to 0.9 M guanidine HCl it is the *only* species present and thus can be readily characterized by a variety of techniques (Palleros *et al.*, 1993).

2. COMPACT INTERMEDIATES UNDER EQUILIBRIUM CONDITIONS

The following properties, which are elaborated on subsequently, are necessary to characterize a denatured state as a compact intermediate: (1) a compactness closer to that of the native state than the unfolded state, measured by techniques such as intrinsic viscosity, dynamic light scattering, size exclusion chromatography, or small-angle X-ray scattering, (2) a substantial amount of secondary structure as revealed by techniques such as far-UV circular dichroism, NMR, or Fourier-transform infared (FTIR), (3) little or no nativelike tertiary structure as evidenced by the lack of a near-UV circular dichroism signal or a one-dimensional NMR spectrum indicative of substantial side-chain equivalence, and (4) significantly exposed hydrophobic surface area compared to the native state, leading to a propensity for aggregation and an affinity for binding hydrophobic dyes such as ANS. In addition, it appears that compact intermediate states have only very small differences in heat capacity from the unfolded state.

Compact intermediates have been most frequently observed under acid denaturation conditions. Many proteins, when brought to conditions of pH < 3 and moderate ionic strength, will exhibit the properties of a compact denatured state. The compact species observed at low pH are usually known as A states, and have many molten globulelike properties. A systematic investigation of the acid denaturation of several proteins revealed that their behavior could be classified into three types (Fink *et al.*, 1990), although these are all manifestations of the same underlying conformational phase diagram (Goto and Fink, 1990), reflecting different phase boundary positions. On HCl titration of a salt-free solution, some proteins, such as ubiquitin and T4 lysozyme, remain nativelike to pH 1 or lower; others, such as α-lactalbumin, transform into the A state (the molten globule state) at pH 3–4; whereas a substantial number, such as cytochrome *c* and β-lactamase, unfold in the vicinity of pH 3–4 to the acid-unfolded state and then, below pH 2, refold into the A state (Goto *et al.*, 1990a).

The acid-unfolded state (observed at low pH at very low ionic strength) may be converted into the A state by the addition of anions (Goto *et al.*, 1990a,b). Different anions vary in both their effectiveness (i.e., concentration required) as well as in the degree of folding they induce in the A state. Based on a limited number of measurements, the number of protons involved in the transition from native to acid-unfolded state is typically in the 2–3 range (Calciano *et al.*, 1993); similarly, the number of anions involved in triggering the transition from the acid-unfolded state to the A state is also in the 2–3 range (Goto *et al.*, 1990b; Calciano *et al.*, 1993). These small numbers of relatively tightly bound ions indicate that the conformational transitions are triggered by a very few key ionizations: protonation of carboxyls in the unfolding transition and anion binding to ammonium groups in the "refolding" transitions.

Although there have been relatively few detailed studies on thermally denatured proteins it is clear that, at least in some cases, such species are quite compact. For example, the hydrodynamic radius of thermally denatured ribonuclease A has been reported to be as little as 7% larger than that for the native state (Nicoli and Benedek, 1976) and to have a radius of gyration only 30% larger than that of the native state (Sosnick and Trewhella, 1992). Similarly reduced, denatured bovine pancreatic trypsin inhibitor (BPTI) has been shown to be relatively compact and, in fact, to become more compact as the temperature is raised from 1 to 35°C, consistent with hydrophobic stabilization (Amir *et al.*, 1992; Gottfried and Haas, 1992). DnaK and other members of the hsp70 family undergo a thermally induced transition to a compact intermediate with a midpoint around 42°C; at higher temperatures the proteins undergo a second unfolding transition (Palleros *et al.*, 1992).

Frequently the binding of fluorescent hydrophobic dyes, such as ANS, can be used to suggest the presence of compact intermediate states, due to the preferential binding of the dye to the intermediate compared to native and unfolded states. This method has been used at both extremes of pH and in the presence of denaturants (Semisotnov *et al.*, 1991; Tandon and Horowitz, 1989; Goto and Fink, 1989), as well as in transient folding experiments (e.g., Mann and Matthews, 1993; Ptitsyn *et al.*, 1990). It is important to note that the presence of ANS tends to increase the propensity of compact denatured states to aggregate, and that aggregation increases the ANS fluorescence emission.

For technical reasons it has been easier to study equilibrium compact intermediates compared to transient compact intermediates during the folding process. The best characterized equilibrium compact intermediate is the molten globule of α-lactalbumin (summarized in Dolgikh *et al.*, 1985; Kuwajima, 1989; and Ptitsyn, 1992). Its properties will be used to illustrate the typical characteristics of a compact intermediate. A number of other proteins which form compact intermediates and have played important roles in the elucidation of the characteristic properties of such intermediates will also be discussed in some detail.

2.1. α-Lactalbumin

α-Lactalbumin has been shown to adopt a molten globule conformation under a variety of conditions, including low pH, low denaturant concentration, and removal of Ca^{2+} at pH 7. The concentration of Ca^{2+} is critical in determining the conformational state of α-lactalbumin. In the presence of Ca^{2+} the native state is stabilized relative to the molten globule and the latter is not observed (Ikeguchi *et al.*, 1986b).

The α-lactalbumin molten globule exhibits nativelike secondary structure, as indicated by far-UV circular dichroism and infrared and optical rotary dispersion spectra, whereas the side chains appear in a solvent-exposed environment as determined by NMR, near-UV circular dichroism, and aromatic absorbance and fluorescence, although two of the four tryptophans are solvent-inaccessible (Lala and Kaul, 1992). The molten globule is only slightly less compact than the native state (approx. 10–20% expansion as measured by small-angle X-ray scattering, size-exclusion chromatography, and electrophoresis). The α-lactalbumin molten globule binds hydrophobic fluorescent dyes such as ANS and has a strong propensity to aggregate. No cooperative unfolding is observed for thermal denaturation of the A state (acidic pH) form of the α-lactalbumin molten globule, indicating that ΔC_p is similar for the intermediate and unfolded states; however, some cooperativity is seen with the guanidine-induced molten globule (Xie *et al.*, 1991), although the significance of this observation has been questioned (Yutani *et al.*, 1992). From fluorescence polarization studies the mobility of the tryptophan side chain in the molten globule is similar to that of the native state. Hydrogen/deuterium exchange kinetics indicate that exchange is much faster in the molten globule compared to the native state. Low-angle X-ray scattering studies indicate that the molten globule has a relatively dense core. Ewbank and Creighton (1993a,b) have shown that in the refolding of reduced α-lactalbumin the molten globule is stabilized by nonspecific interactions that result in the formation of a large number of nonnative disulfide bonds.

NMR analysis of the α-lactalbumin molten globule reveals that some regions of secondary structure (helixes) are nativelike and that some tertiary interactions are present (Baum *et al.*, 1989; Chyan *et al.*, 1993; Alexandrescu *et al.*, 1993). The NMR spectrum of the molten globule is clearly different from that of the native and unfolded states. Due to the widespread overlap and substantial line widths of many of the resonances in the molten globule, techniques such as magnetization transfer (with the native state) and amide protection were necessary to obtain structural information. Two-dimensional NMR studies of the A state of α-lactalbumin (Chyan *et al.*, 1993; Alexandrescu *et al.*, 1993) indicate that the chemical shifts of side chains are generally shifted upfield relative to those in the unfolded state. This observation, in conjunction with the observation of some nuclear overhauser effects (NOEs), indicates some side-chain interactions, consistent with a hydrophobic cluster. Nonnative interactions between aromatic residues in the 101–110 region were detected in the A state by NOEs. Amide protection experiments revealed the presence of the B and C helixes of the native state in the A state.

Similar NMR studies have now been reported for the molten globules of apomyoglobin (Hughson *et al.*, 1990), cytochrome *c* (Jeng *et al.*, 1990), barnase (Bycroft *et al.*, 1990), and ubiquitin (Harding *et al.*, 1991). In each case some nativelike secondary structure was observed in the intermediate, especially helixes.

2.2. DnaK

The significance of the following studies on DnaK (Palleros *et al.*, 1993) is the clear-cut distinction between a *compact intermediate* state and a *compact unfolded* state, as well as the dramatic effects of denaturant concentration on the compactness of the species. Guanidine HCl unfolds DnaK in a reversible manner. When the unfolding is monitored by tryptophan fluorescence there is a single unfolding transition centered about 0.5 M guanidine HCl; if followed by tyrosine fluorescence, two transitions are observed at 0.5 and 1.6 M denaturant. The unfolding transition is also biphasic when measured with far-UV circular dichroism: with transitions at 0.5 M (where only about 15% of the signal at 222 nm is lost) and 1.6 M guanidine HCl. The ellipticity remains essentially constant between 0.7 and 1.0 M guanidine HCl. The multistep unfolding is consistent with the presence of an intermediate in the vicinity of 0.8 M guanidine HCl. That this is indeed the case was made evident by the detection of three distinct species by size exclusion HPLC (SEC-HPLC).

In these experiments in which the column was equilibrated and eluted with varying concentrations of denaturant, DnaK (native) appeared as a single peak between 0 and 0.3 M guanidine HCl. Between 0.3 and 0.5 M guanidine HCl a second peak, corresponding to the intermediate, was observed, becoming more prominent as the guanidine HCl concentration increased. Between 0.6 and 1.0 M it was the only peak detected. Above 1.1. M the presence of a new peak, due to unfolded protein, was evident. Above 1.7 M only one peak, corresponding to the unfolded state, was seen. The Stokes radius, R_S, values for the intermediate and the unfolded states increased sharply as the denaturant concentration increased, indicating a significant increase in hydrodynamic size with increasing denaturant concentration. The C_m's obtained from SEC peak areas were in agreement with those obtained by spectroscopic methods. The Stokes radii determined from SEC-HPLC were confirmed by dynamic light scattering measurements and were in excellent agreement, namely, 41.5 Å for native DnaK, 46.5 Å for the compact intermediate (in 0.7 M guanidine HCl), and 69.5 Å for the unfolded state (in 4.0 M guanidine HCl).

Thus the intermediate is compact, possesses most of the secondary structure present in the native state, as determined by far-uv circular dichroism, but has lost the tertiary structure around its only tryptophan residue. No further change in tryptophan fluorescence is observed by increasing the guanidine HCl concentration to 6.0 M, indicating that in the intermediate the tryptophan is already fully exposed to the solvent. The two-step transition observed by tyrosine fluorescence intensity at 305 nm indicates that some of the tyrosines are solvent-exposed in the compact intermediate and others are not. The hydrophobic dye, *bis*-ANS, was found to bind preferentially to the intermediate state of DnaK. Independent measurements confirmed the slow kinetics of transformation between the compact intermediate and the native and unfolded states. Studies on the refolding kinetics results for DnaK indicate that a corresponding compact intermediate can be detected (R_S = 43 Å), which binds *bis*-ANS (D. Palleros, L. Shi, and A. Fink, unpublished observations).

One of the most important observations in this investigation is the fact that two different denatured states are seen to coexist by SEC-HPLC, which indicates that their interconversion rate is slower than the time scale of the chromatographic separation; therefore, a significant energy barrier must separate them. Additionally, the results indicate that the compactness of the intermediate state is very dependent on the concentration of denaturant (Fig. 1). At low guanidine HCl concentration its Stokes radius is less than 10% larger than that of the native state; however, as the guanidine HCl concentration increases, the Stokes radius also increases. At 1.34 M guanidine HCl the intermediate is almost as large as the unfolded state (at that denaturant concentration), 54 Å and 58 Å, respectively. This represents more than a 35% increase in radius of the intermediate compared to the native protein (and a 2.3-fold increase in volume, assuming a globular shape). The unfolded state also expands; the Stokes radius in 6 M guanidine HCl is 69 Å, which represents a 20% increase compared to the unfolded state near the midpoint of the I → U transition.

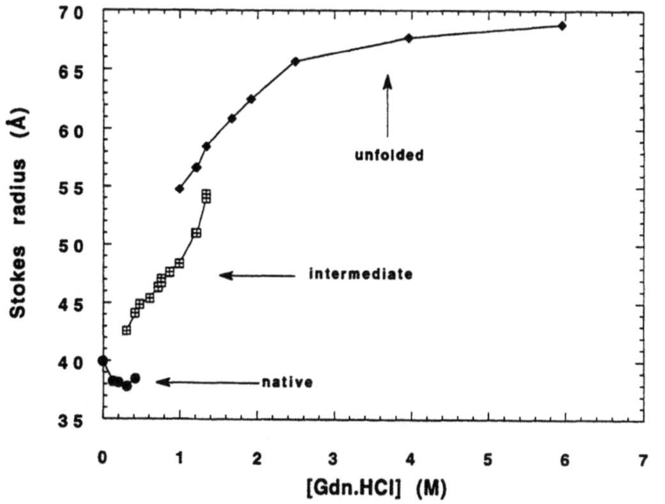

FIGURE 1. Variation in the compactness of the intermediate and unfolded states of DnaK. The effect of denaturant concentration on the Stokes radius of DnaK in the native (circles), compact intermediate (squares), and unfolded (diamonds) states. (From Palleros *et al.*, 1993).

The continuous increase in the molecular volume of the intermediate state, as well as the unfolded state, with increasing denaturant concentrations suggests that both the intermediate and unfolded states are *ensembles* of substates (Fig. 2). At low guanidine HCl concentrations, near the midpoint of the first transition, the most populated substates of the intermediate state are compact. However, as the denaturant concentration is increased, the populated substates shift to favor those expanded to almost the same size as the unfolded protein (Fig. 2). The existence of such swollen species was predicted by Finkelstein and Shakhnovich (1989). Based on the thermodynamics of protein–solvent interactions these authors showed that solvent penetration causes the molten globule to expand gradually without any phase transition. The swollen intermediate then undergoes a second-order phase transition to the coil (unfolded) conformation; their model does not account for variations in ordered structure. A structural model to account for these observations is given in Section 5.

2.3. Apomyoglobin

There have been several important investigations of compact intermediates of apomyoglobin and its mutants (Griko *et al.*, 1988; Goto and Fink, 1990; Goto *et al.*, 1990a,b; Hughson *et al.*, 1990, 1991; Barrick and Baldwin, 1993). The intermediate is stable at low pH under conditions of moderate ionic strength, and

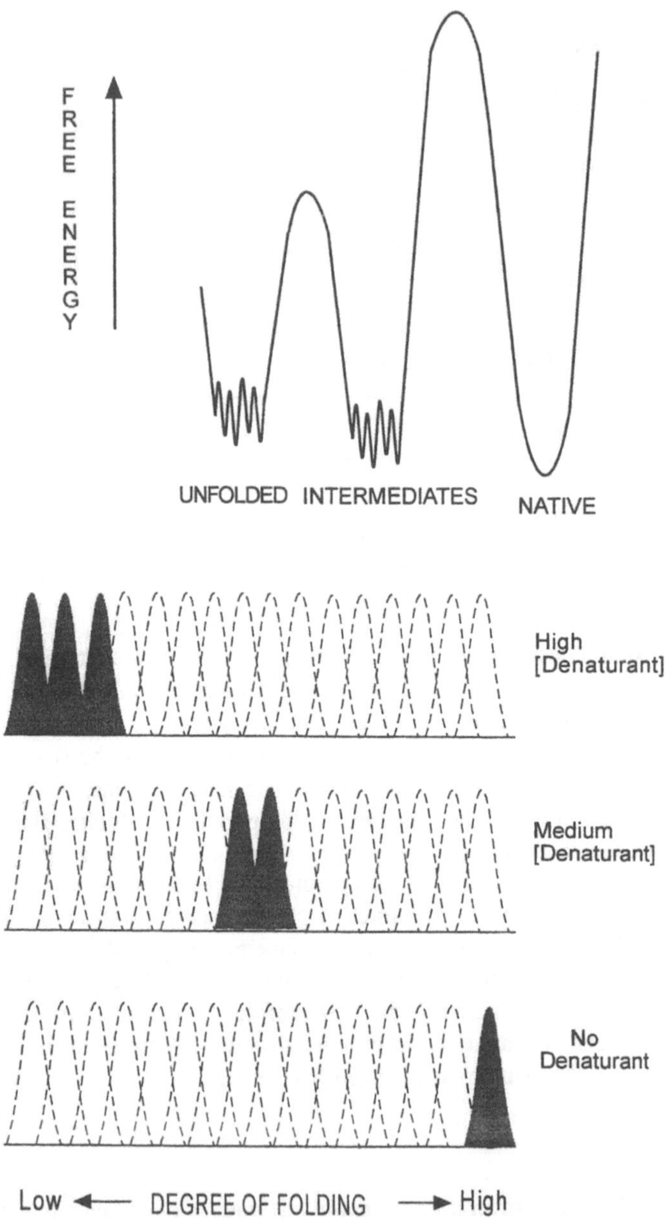

FIGURE 2. Substates of intermediate and unfolded states of proteins. (A) Hypothetical potential energy surfaces, showing the multiple small minima corresponding to substates, for both intermediate and unfolded states. (B) Changes in distribution of populated substates as a function of denaturant.

even in the pH 4–5 region at low salt. The detailed dependence of the conformational state on pH, chloride concentration, and urea has been reported in the form of phase diagrams (Goto and Fink, 1990; Barrick and Baldwin, 1993). At low ionic strength apomyoglobin is substantially unfolded in the vicinity of pH 2. The addition of anions induces a transition to the compact intermediate. The intermediate is characterized by substantial secondary structure, the absence of a near-UV circular dichroism signal, and a much reduced hydrodynamic radius compared to the unfolded state. Interestingly the tryptophan fluorescence signal has a similar λ_{max} to that of the native state (335 nm), indicating a substantially nonpolar environment, although the intensity is significantly quenched compared to that in the native state (Goto *et al.*, 1990a). The size (measured by dynamic light scattering) of the apomyoglobin compact intermediate induced by anions (from salts or acids) and the amount of secondary structure, as determined by the ellipticity at 222 nm or FTIR, vary depending on the nature of the anion (Goto *et al.*, 1990b; K. Oberg, S. Swedberg, S. Seshadri, and A. Fink, unpublished observations). The most structured intermediate among monovalent anions is found in the presence of trichloroacetate, the least structured is with chloride.

Interestingly, unlike the native state, the stability of the intermediate state is relatively insensitive to the effect of mutations (Hughson and Baldwin, 1989; Hughson *et al.*, 1991). Site-directed mutagenesis was used to construct mutants in which interactions between the A, G, and H helixes were disrupted: these three helices have been shown to be present in the apomyoglobin intermediate by NMR amide exchange experiments (Hughson *et al.*, 1990). The mutations destabilized the native state (relative to the intermediate) regardless of their nature, indicating that interactions between the A-H and G-H helixes did not contribute to the stability of the intermediate state. Mutations involving an increase in nonpolar area increased the stability of the intermediate, whereas those involving an increase in polar surface area or a decrease in nonpolar area led to a decrease in stability of the intermediate (as detected by the position of the midpoint of the native/intermediate transition). The authors conclude that nonspecific hydrophobic interactions are responsible for stabilizing the helices in the intermediate state.

2.4. β-Lactamase

The results of an investigation to measure side-chain mobility in the low-pH compact intermediate state of β-lactamase are surprising in that they suggest some regions in a molten globule are far from "molten." Compact intermediate states have been reported for β-lactamases from *Staphylococcal aureus* (Robson and Pain, 1976a,b; Ptitsyn *et al.*, 1990; Uversky *et al.*, 1992b), *Bacillus cereus* (Goto and Fink, 1989), and *B. licheniformis* (Calciano *et al.*, 1993) formed under conditions of either low or high pH or low denaturant concentration.

β-Lactamase from *B. licheniformis* forms a stable compact intermediate state at low pH and moderate salt concentration (the A state). This intermediate has essentially the same amount of secondary structure as the native state, as determined by far-UV circular dichroism, has no near-UV circular dichroism signal, binds ANS, and has a Stokes radius about 10% larger than that of the native state (Calciano *et al.*, 1993). The side-chain mobility of this compact intermediate was investigated by incorporating a spin-label probe (Calciano *et al.*, 1993). The spin label was attached to the thiol of the single cysteine residue introduced by site-directed mutagenesis at position 166 via a disulfide bond. The mutant protein and the spin-labeled derivative exhibited similar conformational properties to the wild type enzyme at acidic pH. The A state at pH 0.5 or at pH 2 in the presence of 8 mM trichloroacetate or 0.4 M chloride had comparable amounts of secondary structure to the native state, but lacked significant tertiary structure, as judged by the lack of near-UV circular dichroism.

The line widths in the electron spin resonance (ESR) spectrum for the acid-unfolded state were noticeably narrower than those for the native state, reflecting the greater mobility of the probe in the unfolded state; the ESR spectrum of the guanidinium-unfolded state was comparable to that of the acid-unfolded state. The ESR spectrum of the A state protein was essentially superimposable on that of the native protein and gave an identical rotational relaxation time. These data indicate that the environment of the spin label in the A state is very similar to that in the native state, and that it is much less mobile than in the unfolded state. Thus there must be substantial structure remaining in the vicinity of residue 166 in the A state, which is rather surprising, given the apparent lack of tertiary structure as observed by other methods.

The tryptophan fluorescence emission maximum in the A state is similar to that in the native state (Goto and Fink, 1989), although significantly quenched. This suggests that the tryptophan residues are in a nonpolar, buried environment in the A state; the quenching is likely to be due to increased mobility of the side chains leading to dynamic quenching. There are few reasonable structural models that are consistent with the limited expansion, the nativelike far-UV circular dichroism spectrum, the loss of the near-UV circular dichroism signal, and the limited mobility of the spin label. One model involves small movements of the secondary structural components so as to pull the side chains apart sufficiently to result in loss of their interdigitation while maintaining the secondary structure. This would account for the small expansion, the loss of the near-UV circular dichroism signal, the presence of the tryptophans in a nonpolar environment, and the retention of nativelike secondary structure as detected by far-UV circular dichroism. The ESR results could be interpreted in the light of this model as follows: since the spin label is larger than the wild type side chain, it will be in a more crowded environment than most side chains. The molecular expansion on forming the A state does not lead to sufficient increase in space in the local environment of the spin label to allow increased mobility. An alternate model, in which subdomains move apart slightly, leading to loss of the tertiary interactions between them, while maintaining secondary structure, some tertiary interactions (within subdomains), and leading to some expansion of the molecule, is described in Section 5.

2.5. Carbonic Anhydrase

There have been several studies of the equilibrium molten globule intermediate of carbonic anhydrase, formed either at low concentration of denaturant or at acidic pH (Dolgikh *et al.*, 1984; Jagannadham and Balasubramanian, 1985; Brazhnikov *et al.*, 1985; Mårtensson and Jonsson, 1993). The unique affinity of the intermediate for hydrophobic dyes such as ANS (Semisotnov *et al.*, 1991) and pyrene (Mitaku *et al.*, 1991) has been demonstrated. An intermediate, assumed to be similar to the molten globule, has also been detected and characterized in kinetics experiments during refolding (Ebert *et al.*, 1990; Semisotnov *et al.*, 1987; Dolgikh *et al.*, 1984). The aggregation of the equilibrium intermediate has also been investigated (Cleland and Wang, 1990); the aggregation may be prevented by polyethylene glycol, which forms a 1:1 complex with the molten globule (Cleland and Randolph, 1992). By incorporating cysteine residues at selected locations and measuring the reactivity of the thiol side chain in the intermediate, Mårtensson and Jonsson (1993) demonstrated that the intermediate retained a significant portion of the core β-sheet structure, as well as a very stable hydrophobic cluster, located between the central β-sheet core and secondary structure elements on the surface.

2.6. Other Proteins

Compact intermediates or molten globules have now been reported for many proteins under a variety of equilibrium conditions and by a variety of techniques: in many cases, however, direct measurements of compactness are lacking. Examples include Arc repressor at high pH (Peng *et al.*, 1993), retinol-binding protein at low pH (Bychkova *et al.*, 1992), stefin B at low pH, moderate guanidinium chloride, and thermal denaturation (Zerovnik *et al.*, 1992), β-globulin at low pH (Rajendran and Prakash, 1992), and human and nonhuman growth hormones (Brems and Havel, 1989; Lehrman *et al.*, 1991; Bastiras and Wallace, 1992; DeFelippis *et al.*, 1993), phosphoglycerate kinase (Ptitsyn *et al.*, 1990; Ballery *et al.*, 1993), aspartate aminotransferase (Leistler *et al.*, 1992), rhodanese (Horowitz and Xu, 1992), staphylococcal nuclease (Fink *et al.*, 1993), cytochrome *c* (Ohgushi and Wada, 1983; Goto *et al.*, 1990a,b; Jeng *et al.*, 1990;

Kuroda *et al.*, 1992; Kataoka *et al.*, 1993), interleukin-2 (Dryden and Weir, 1991), all at low pH. A number of others are listed in Fink *et al.* (1990) and in Ptitsyn (1992).

3. COMPACT INTERMEDIATES FORMED UNDER TRANSIENT FOLDING CONDITIONS

A number of investigations using stopped-flow circular dichroism or NMR have revealed the rapid formation of secondary structure in the early stages of protein refolding (millisecond or shorter time scale). Although it has not been possible to measure the compactness of these species due to technical difficulties, it is assumed that they are more compact than the unfolded state (the existence of substantial secondary inherently means that the conformation must be more compact). The initial demonstration of such a transient compact state was by Kuwajima and co-workers who demonstrated that an intermediate was formed very rapidly in the refolding of α-lactalbumin which had very similar far-UV circular dichroism properties to the A state determined under equilibrium conditions (Kuwajima *et al.*, 1985). Transient intermediates have now been observed using far-UV stopped-flow circular dichroism for several additional proteins, including cytochrome *c* and β-lactoglobulin (Kuwajima *et al.*, 1987), lysozyme and α-lactalbumin (Kuwajima *et al.*, 1985), tryptophan synthase subunits (Chaffotte *et al.*, 1992b; Goldberg *et al.*, 1990), staphylococcal nuclease (Sugawara *et al.*, 1991) and dihydrofolate reductase (Kuwajima *et al.*, 1991). In addition, the propensity of compact intermediates to bind ANS has been used to demonstrate the formation and breakdown of molten globulelike intermediates during refolding using stopped-flow fluorescence techniques (Ptitsyn *et al.*, 1990; Semisotnov *et al.*, 1991; Goldberg *et al.*, 1990; Varley *et al.*, 1993). On the basis of the apparent generality of this phenomenon, Ptitsyn, Pain, and coworkers (Ptitsyn *et al.*, 1990) have proposed that the molten globule is a common folding intermediate.

Combined rapid-mixing circular dichroism, fluorescence, and NMR experiments on the refolding of the all-β protein interleukin-1β (Varley *et al.*, 1993) indicate that although nativelike secondary structure appears within a few milliseconds, as judged by circular dichroism, the NMR amide H/D exchange experiments indicate that stable secondary structure is formed only on a time scale of > 25 sec. The initially formed intermediate (25 msec) bound ANS, suggesting it had at least some of the properties of a compact intermediate.

Several studies have been made on the kinetics of refolding of chicken lysozyme that reveal the transient presence of intermediates. Although a stable equilibrium compact intermediate has not been observed for this protein, the far-UV circular dichroism properties of the transient intermediate are similar to those of α-lactalbumin (which is homologous to lysozyme) (Kuwajima *et al.*, 1985; Ikeguchi *et al.*, 1986a). The initially observed intermediate in refolding studies, at 4 msec, has substantial secondary structure and negligible tertiary structure (Chaffotte *et al.*, 1992a).

Dobson and co-workers (Radford *et al.*, 1992) have also investigated the folding kinetics of chicken lysozyme, using a combination of NMR amide-pulsed H/D exchange and stopped-flow circular dichroism (far- and near-UV) experiments. Their data indicate that a near nativelike ellipticity at 225 nm is reached within the 2 msec dead time, followed by a further decrease in ellipticity and a slower return to the native signal. No amide exchange protection is observed within the first 2 msec. The results were interpreted to indicate parallel folding pathways with multiple intermediates. Kuwajima and co-workers have also observed early folding intermediates with β-lactoglobulin and chymotrypsinogen, with apparently greater secondary structure than that in the native state (Sugawara *et al.*, 1991).

The refolding of parvalbumin also involves a rapidly formed intermediate whose far-UV circular dichroism is very similar to that observed for the equilibrium intermediate under conditions of low pH; interestingly, the equilibrium intermediate retains some tertiary structure, as measured by near-UV circular dichroism (Kuwajima *et al.*, 1988).

4. FORCES INVOLVED IN COMPACT INTERMEDIATE STATES

Compact denatured states are most frequently observed under mildly denaturing conditions, such as low concentrations of denaturant, extremes of pH, high concentrations of certain salts, or high temperatures. The existence of a *compact* denatured state for a given protein and its degree of folding and compactness are determined by a combination of factors: intrinsic ones reflecting the underlying structure of the particular protein, i.e., its sequence (and such consequences as metal-binding sites, disulfide bonds, buried salt-bridges, and relative amount of nonpolar residues), and external ones reflecting the environment, such as pH, solutes, cosolvents, and temperature. The resulting conformation reflects a balance between folding and unfolding forces. Other factors that can lead to formation of compact intermediates, molten globules, or similar species include truncation of the end of the polypeptide chain [e.g., fragments of staphylococcal nuclease in which several residues have been removed from the C-terminus (Flanagan *et al.*, 1992) or removal of the N-terminal region as with tryptophan synthase (Chaffotte *et al.*, 1991)], reduction of disulfides (Hirose and Yamashita, 1991), and removal of metal ions (Ikeguchi *et al.*, 1986b).

Conformational phase diagrams (Fig. 3) have been constructed for the effects of ionic strength and pH and which account for the observed variety of

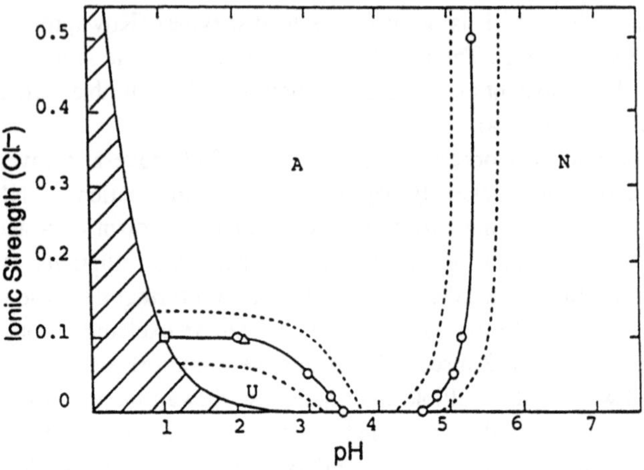

FIGURE 3. Conformational phase diagram for apomyoglobin. The phase boundaries represent the midpoints of the transitions between the native, A, and acid-unfolded states. The crosshatched area represents combinations of ionic strength and pH which are not attainable. (From Goto and Fink, 1990.)

behavior observed for proteins between neutral and acidic pH (Goto and Fink, 1990). Typically, three conformational states are observed: native, compact intermediate (A state), and unfolded. The particular protein and the experimental conditions determine which state (or states) will be present; in particular, the phase boundaries between the native and A state and the A state and the unfolded state are very dependent on the intrinsic properties of the protein, presumably determined ultimately by the amino acid sequence. For example, it is the position of these phase boundaries that determines that α-lactalbumin will transform directly to the molten globule (A state) as the pH is decreased, whereas lysozyme will remain native, and β-lactamase will initially unfold and then at lower pH refold to the compact A state.

For illustrative purposes it is useful to consider a parameter—the "degree of folding"—which is a measure of the compactness and amount of secondary structure in the conformation under consideration (Fig. 4). Based on existing results, compact intermediate states are found with compactness ranging from very close to the native state ($<10\%$ expansion in R_S) to as much as 50% expanded relative to the native state. Compact *unfolded* states may overlap compact *intermediate* states, in some cases being as little as 40% expanded relative to the native state (Palleros *et al.*, 1993). The position (conformation) a given protein under given conditions will adopt on this degree of folding axis is determined by a balance of intrinsic and external forces involving pressure to fold or unfold. Although certain conformations might be expected to be preferred

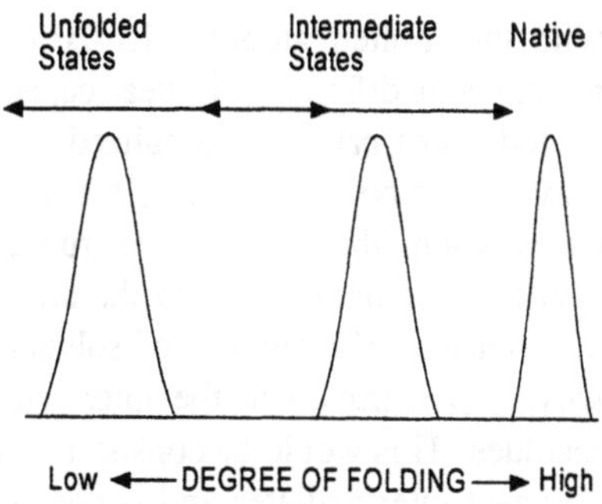

FIGURE 4. The relationship between native, compact intermediate, and unfolded states and the degree of folding. The degree of folding is a measure of the compactness and amount of secondary structure.

(i.e., minima on the potential energy surface for folding), the present picture suggests that such minima are quite broad, permitting significant variation in the properties of observed compact intermediates.

As noted, compact intermediates are frequently generated under conditions of low pH and moderate salt. The low pH behavior of proteins is controlled by a balance between electrostatic forces, which drive the unfolding due to coulombic charge repulsion between the positively charged ammonium groups of lysine, histidine, and arginine, and the intrinsic hydrophobic driving force for folding. In some cases the former is sufficient to unfold the protein in the vicinity of pH 2 and in others it is not. If a protein is unfolded at low pH due to electrostatic repulsion, the intramolecular charge repulsion may be overcome at higher ionic strength due to the screening of the positive charge by binding of anionic counterions. It is not apparent at present why the effectiveness of different anions in both bringing about the transition from unfolded to compact states and in the amount of structure induced by the anion in the compact intermediate vary in the way they do (Goto *et al.*, 1990b). It may reflect a combination of hydration potential, basicity, and the ability to form bifurcated interactions (e.g., carboxylate with arginine or perhaps with two lysines).

Similarly, whether a low-to-moderate amount of denaturant will induce a stable compact intermediate depends on the relative stabilities of the intermediate and the folded and unfolded states to the denaturant, which in turn reflects the nature of the interaction between the denaturant and the protein, and which is a consequence of the underlying amino acid sequence.

Various stable states of myoglobin (denaturant-unfolded, acid-unfolded, molten globule, native apomyoglobin, native myoglobin) have been used by Goto and co-workers (personal communication) as models for the progression from unfolded to native states. Key characteristics, such as compactness (from small-angle X-ray scattering) and heat capacity, were compared. The results show that the molecule becomes compact at the earliest stage (acid-unfolded apomyoglobin), whereas heat capacity changes occur incrementally, starting with the molten globule (15–20% of that of native myoglobin). The low-to-nonexistent difference in heat capacity between compact intermediates and unfolded states can be rationalized as follows: in some cases the compact species may in fact be a substate of the unfolded state (and not a true intermediate); in cases where the species is a true compact intermediate the absence of a heat capacity change relative to the unfolded state implies that there is insignificant difference in the amount of solvent-exposed surface. This in turn implies that solvent penetration in the intermediate must reach essentially all the nonpolar residues. This would be consistent with the lack of significant tertiary structure in compact intermediates and is consistent with the structural model proposed below.

5. STRUCTURAL MODELS FOR COMPACT INTERMEDIATES

One of the pressing questions regarding compact intermediate states is that of their structure. The difficulty lies in the variety of plausible models can be envisaged and the experimental challenges in obtaining the necessary detailed structural information. In this section both models derived from theory and from the interpretation of experimental results will be discussed.

5.1. Theory-Based Models

The difficulties associated with experimental characterization of the structures of compact intermediate states has led theoreticians to attempt to model such states. Finkelstein and Shakhnovich (1989), using polymer solution theory, predict that two forms of the molten globule, in addition to the unfolded state, will exist, namely a "wet," more compact and a "swollen," more expanded form of the molten globule. By applying mean-field theory to a denatured protein, which they considered as a stochastic polymer, they determined a theoretical phase diagram for stable protein states; particular attention was paid to the density and solvent penetration of the different states. The key disruption in going from the native state to the intermediate states was the breakdown of the tight packing in the core of the native state.

Dill and co-workers used statistical mechanics to model protein stability. By incorporating electrostatics contributions they were able to accurately predict the conformational phase diagram for apomyoglobin, showing the boundaries between native, compact intermediate, and unfolded states as a function of pH and ionic strength (Stigter and Dill, 1990). Their model, which predicts protein stability as a function of temperature, pH, and salt concentration, taking into account the physical properties of the constituent amino acids, predicts two stable forms of denatured states: a compact intermediate (or molten globule) and an expanded denatured state. Critical predictions of this theory include the major loss of hydrophobic contacts in the molten globule relative to the native state; that the main electrostatic contribution to stability is entropic; and that more hydrophobic polypeptides will have more prominent compact intermediate states.

Molecular dynamics simulations have been run on a few proteins (BPTI, lysozyme, and apomyoglobin) under conditions mimicking a molten globule (Daggett and Levitt, 1992; Mark and van Gunsteren, 1992; Tirado-Rives and Jorgensen, 1993). For example, molecular dynamics simulation of the thermal denaturation of chicken lysozyme showed a quasi-stable state in which significant rearrangement of the secondary structure had taken place: no evidence for distinct domains was observed (Mark and van Gunsteren, 1992).

5.2. Experiment-Based Models

Ptitsyn's models (Ptitsyn, 1992) assume a nonuniform expansion of the native state on denaturation, which allows retention of the "frame" of the native state in the molten globule. The frame is defined to include the central parts of helices and sheets that carry nonpolar side chains, contributing to the hydrophobic core of the native state. According to Ptitsyn's current model, water does not significantly penetrate this hydrophobic core in the molten globule, although the packing may become looser, permitting greater mobility of these side chains. This model—in which the hydrophobic core is retained, whereas much of the rest of the molecule is relatively unfolded—is claimed to reconcile experimental data that on the one hand indicate some nativelike properties and on the other reveal unfoldedlike properties. However, as discussed above, the lack of significant heat capacity difference between compact intermediate and unfolded state may indicate very substantial internal solvation of the intermediate.

It has been pointed out that whenever an unfolded protein is placed under nativelike conditions, it will rapidly collapse to a species with many of the properties associated with a molten globule (Dill and Shortle, 1991). There is some controversy regarding whether a hydrophobic collapse occurs before or after the initial formation of secondary structure during refolding or, as proposed by Dill (Chan and Dill, 1990), the formation of secondary structure is concurrent with and a consequence of the collapse. The strong correlation between the ellipticity at 222 nm and the volume of a protein in different conformational states (Palleros *et al.*, 1993) favors the latter, whereas the observation that apocytochrome *c* at neutral pH and low ionic strength is relatively compact but has no ordered secondary structure as judged by circular dichroism (Hamada *et al.*, 1993) favors the former model (this appears to be the only reported such case and may be a unique situation).

The main difference between the native and compact intermediate states appears to be disruption of the hydrophobic interactions binding the structural units together in the native state, with the consequent exposure of nonpolar surfaces to solvent. A model that would account for most of the experimental observations of molten globules follows.

5.3. Structural Model for Compact Intermediates

The native state is assumed to consist of a number of structural units, or building blocks, which are tightly packed together via tertiary interactions. These building blocks may consist of regions of secondary structure or other types of subdomains or autonomous folding units and will have significant intrinsic stability due to backbone and side-chain interactions, and would minimally consist of a helix or two-stranded beta-sheet and maximally of a domain. The transition from the native state to the compact intermediate state, which corresponds experimentally to the loss of tertiary structure but retention of secondary structure, would correspond to the separation of these structural units, with concomitant solvent penetration and expansion of the molecule as a whole. Thus the formation of the intermediate would involve disruption of the interactions *between* building blocks but would have little effect on side-chain interactions *within* the structural unit. This intermediate would also be expandable without losing its stability as more solvent molecules penetrate the space between building blocks. This model makes implicit the existence of two kinds of tertiary structure: that between building blocks, which would be the first to be disrupted by the solvent, and that within the building block, which would be more protected from interactions with the solvent and more resistant to denaturation.

A range of possible conformations for the compact intermediates or denatured states is anticipated, depending on the extent of the loss of side-chain contacts between building blocks. Once the contacts between building blocks are lost, the molecule will become a collection of structural units linked by flexible polypeptide chain links. It is likely that these conformations will be quite mobile; however, even with the loss of the building block contacts there may still be transient contacts between them. For small proteins with only a few structural units, only very compact intermediate conformations would be observed, since the molecule could not expand significantly without losing its stabilizing interactions within the building block.

Compact intermediates are thus pictured as an ensemble of substates ranging from those that are quite compact, with substantial contacts between all or most of the building block units, to those with relatively few contacts between the building blocks, which are thus quite expanded. Whereas the building block units themselves may retain a rather rigid, nativelike structure, there will be substantial mobility between them, as dictated by the freely rotating connecting links. Depending on the relative stabilizing or destabilizing forces of the external environment, as well as the intrinsic structural properties of the protein, the predominant substates may be quite compact and relatively nativelike or more expanded.

This model adequately accounts for the known properties of compact intermediates, in particular the retention of substantial secondary structure (within the building blocks), the loss of tertiary structure (contacts between structural units), the solvent penetration and expanded size relative to the native state, the possibility of nativelike side-chain environments (in intact structural units with strong neighboring side-chain interactions), and the variation in compactness and amount of secondary structure as a function of environment and protein.

6. POSSIBLE PHYSIOLOGICAL ROLES OF COMPACT INTERMEDIATES

Tantalizing evidence is accumulating to suggest that compact intermediates may play important roles in some physiological processes. It is quite likely that further examples will be discovered as we gain additional understanding about the properties of these states and as more penetrating investigations of the cell continue. For example, Hua *et al.* (1993) have recently shown that a monomeric analogue of insulin [des-pentapeptide (B26-B30)] has properties consistent with those expected for a compact intermediate (although the compactness was not measured). They raise the provocative possibility that the functional form of monomeric insulin in the blood is a molten globule.

6.1. Membrane Fusion and Translocation

Bychkova *et al.* (1988) and more recently van der Goot *et al.* (1992) have reviewed evidence that the molten globule state is involved in membrane translocation. The A state of α-lactalbumin has been shown to induce fusion of phospholipid vesicles (Kim and Kim, 1986). Low pH (as in lysosomes) has been shown to play an important role in membrane penetration by toxins such as the diphtheria toxin. It has been shown that diphtheria toxin forms a molten globule in the vicinity of pH 5 and that fusion with membranes correlates with formation of the molten globule conformation (Zhao and London, 1986; Blewitt *et al.*, 1985; Jiang *et al.*, 1991). Similarly, it has been shown that membrane insertion of colicin A correlates with its transition to the molten globule (van der Goot *et al.*, 1991; Lakey *et al.*, 1992).

6.2. *In Vivo* Folding of Proteins

Transient compact intermediates undoubtedly play a significant role in the folding of proteins *in vivo*. The current view of protein folding *in vivo* is that as the nascent polypeptide chain emerges from the ribosome, it will be bound to a member of the hsp70 family of molecular chaperones (Beckmann *et al.*, 1990). Subsequently, after release of the chain from the ribosome, and with the involvement of ATP, the nascent polypeptide is transferred to a member of the hsp60 chaperone family (Langer *et al.*, 1992) from which it is ultimately released as the folded protein, again with the requirement for ATP.

It seems reasonable that nascent polypeptide chains that are not complexed with hsp70 will collapse to a compact form as they grow on the ribosome. Presumably this structure will be highly mobile and continuously changing as additional residues emerge; in the later stages of the synthesis of large proteins it is likely that molten globulelike conformations would be found. It is not yet known whether hsp70 chaperones bind both unfolded and compact intermediate states of proteins, nor is it yet clear whether the bound protein is in an unfolded or compact intermediate conformation when bound to hsp70. It also remains to be shown whether the conformation of the nascent polypeptide, when being transferred from hsp70 to hsp60, is unfolded or a compact intermediate. There is some evidence that the conformation of the protein bound to hsp60 is a compact intermediate (Martin *et al.*, 1991).

7. REFERENCES

Alexandrescu, A. T., Evans, P. A., Pitkeathly, M., Baum, J., and Dobson, C. M., 1993, Structure and dynamics of the acid-denatured molten globule state of α-lactulbumin: A two-dimensional NMR study, *Biochemistry* **32:**1707–1718.

Alonso, D. O. V., Dill, K. A., and Stigter, D., 1991, The three states of globular proteins: Acid denaturation, *Biopolymers* **31:**1631–1749.

Amir, D., Krausz, S., and Haas, E., 1992, Detection of local structures in reduced unfolded bovine pancreatic trypsin inhibitor, *Proteins: Struct. Funct. Genet.* **13:**162–173.

Aune, K. C., Salahuddin, A., Zarlengo, M. H., and Tanford, C., 1967, Evidence for residual structure in acid-and heat-denatured proteins, *J. Biol. Chem.* **242:**4486–4489.

Baldwin, R. L., 1991, Molten globules: Specific or nonspecific folding intermediates? *Chemtracts, Biochem. Mol. Biol.* **2:**379–389.

Baldwin, R. L., 1993, Pulsed H/D exchange studies of folding intermediates, *Curr. Opin. Struct. Biol.* **3:**84–91.

Baldwin, R. L., and Roder, H., 1991, Characterizing protein folding intermediates, *Curr. Biol.* **1991:**218–220.

Ballery, N., Desmadril, M., Minard, P., and Yon, J. M., 1993, Characterization of an intermediate in the folding pathway of phosphoglycerate kinase: Chemical reactivity of genetically introduced cysteinyl residues during the folding process, *Biochemistry* **32:**708–714.

Barrick, D., and Baldwin, R. L., 1993, Three-state analysis of sperm whale apomyoglobin folding, *Biochemistry* **32:**3790–3796.

Bastiras, S., and Wallace, J. C., 1992, Equilibrium denaturation of recombinant porcine growth hormone, *Biochemistry* **31:**9304–9309.

Baum, J., Dobson, C. M., Evans, P. A., and Hanley, C., 1989, Characterization of a partly folded protein by NMR methods: studies on the molten globule state of guinea-pig α-lactalbumin, *Biochemisstry* **28:**7–13.

Beckmann, R. P., Mizzen, L. A., and Welch, W. J., 1990, Interactions of hsp70 with newly synthesized proteins: implications for protein folding and assembly, *Science* **248:**850–854.

Blewitt, M. G., Chung, L. A., and London, E., 1985, Effect of pH on the conformation of diphtheria toxin and its implications for membrane penetration, *Biochemistry* **24:**5458–5464.

Bowler, B. E., May, K., Zaragoza, T., York, P., Dong, A., and Caughey, W. S., 1993, Destabilizing effects of replacing a surface lysine of cytochrome *c* with aromatic amino acids: Implications for the denatured state, *Biochemistry* **32:**183–190.

Brazhnikov, E. V., Chirgadze, YU. N., Dolgikh, D. A., and Ptitsyn, O. B., 1985, Noncooperative temperature melting of a globular protein without specific tertiary structure: Acid form of bovine carbonic anhydrase B, *Biopolymers* **24:**1899–1907.

Brems, D. N., and Havel, H. A., 1989, Folding of bovine growth hormone is consistent with molten globule hypothesis, *Proteins: Struct. Funct. Genet.* **5:**93–95.

Bychkova, V. E., Pain, R. H., and Ptitsyn, O. B., 1988, The "molten globule" state is involved in the translocation of proteins across membranes? *FEBS Lett.* **238:**231–234.

Bychkova, V. E., Berni, R., Rossi, G. L., Kutyshenko, V. P., Ptitsyn, O. B., 1992, Retinol-binding protein is in the molten globule state at low pH, *Biochemistry* **31:**7566–7571.

Bycroft, M., Matouschek, A., Kellis, J. T., Jr., Serrano, L., and Fersht, A. R., 1990, Detection and characterization of a folding intermediate in barnase by NMR, *Nature* **346:**488–490.

Calciano, L. J., Escobar, W. A., Millhauser, G. L., Miick, S. M., Rubaloff, J., Todd, A. P., and Fink, A. L., 1993, Side-chain mobility of the β-lactamase A state probed by electron spin resonance spectroscopy, *Biochemistry* **32:**5644–5649.

Chaffotte, A., Guillou, Y., Delepierre, M., Hinz, H.-J., and Goldberg, M. E., 1991, The isolated C-terminal (F2) fragment of the *Echerichia coli* tryptophan synthase B2-subunit folds into a stable, organized nonnative conformation, *Biochemistry* **30:**8067–8074.

Chaffotte, A. F., Guillou, Y., and Goldberg, M. E., 1992a, Kinetic resolution of peptide bond and side chain far-UV circular dichroism during the folding of hen egg white lysoszyme, *Biochemistry* **31:**9694–9702.

Chaffotte, A. F., Cadieux, C., Guillou, Y., and Goldberg, M. E., 1992b, A Possible initial folding intermediate: The C-terminal proteolytic domain of tryptophan synthase β chains folds in less than 4 milliseconds into a condensed state with non-native-like secondary structure, *Biochemistry* **31:**4303–4308.

Chan, H. S., and Dill, K. A., 1990, Origins of structure in globular proteins, *Proc. Natl. Acad. Sci. USA* **87:**6388–6392.

Christensen, H., and Pain, R. H., 1991, Molten globule intermediates and protein folding, *Eur. Biophys. J.* **19:**221–230.

Chyan, C.-L., Wormald, C., Dobson, C. M., Evans, P. A., and Baum, J., 1993, Structure and stability of the molten globule state of guinea-pig α-lactalbumin: A hydrogen exchange study, *Biochemistry* **32:**5681–5691.

Cleland, J. L., and Randolph, T. W., 1992, Mechanism of polyethylene glycol interaction with the molten globule folding intermediate of bovine carbonic anhydrase B, *J Biol. Chem.* **267:**3147–3153.

Cleland, J. L., and Wang, D. I. C., 1990, Refolding and aggregation of bovine carbonic anhydrase B: Quasi-elastic light scattering analysis, *Biochemistry* **29:**11072–11078.

Daggett, V., and Levitt, M., 1992, A model of the molten globule state from molecular dynamics simulations, *Proc. Natl. Acad. Sci. USA* **89:**5142–5146.

DeFelippis, M. R., Alter, L. A., Pekar, A. H., Havel, H. A., and Brems, D. N., 1993, Evidence for a self-associating equilibrium intermediate during folding of human growth hormone, *Biochemistry* **32:**1555–1562.

Dill, K. A., and Shortle, D., 1991, Denatured states of proteins, *Annu. Rev. Biochem.* **60:**795–825.

Dolgikh, D. A., Gilmanshin, R. I., Brazhnikov, E. V., Bychkova, V. E., Semisotnov, G. V.,

Venyaminov, S. Y., and Ptitsyn, O. B., 1981, α-Lactalbumin: Compact state with fluctuating tertiary structure? *FEBS Lett.* **136**:311–315.

Dolgikh, D. A,. Kolomiets, A. P., Bolotina, I. A., and Ptitsyn, O. B., 1984, "Molten-globule" state accumulates in carbonic anhydrase folding, *FEBS Lett.* **165**:88–92.

Dolgikh, D., Abaturov, L., Bolotina, I., Brazhnikov, E., Bychkova, V., Gilmanshin, R., Lebedev, Y., Semisotnov, G., Tiktopulo, E., and Ptitsyn, O. B., 1985, Compact state of a protein molecule with pronounced small-scale mobility: Bovine α-lactalbumin, *Eur. Biophys. J.* **13**:109–121.

Dryden, D., and Weir, M. P., 1991, Evidence for an acid-induced molten globule state in interleukin-2; a fluorescence and circular dichroism study, *Biochim. Biophys. Acta* **1078**:94–100.

Ebert, B., Semisotnov, G. V., and Rodionova, N. A., 1990, Studies on globular protein refolding kinetics by ESR stopped flow spectroscopy, *Studia Biophysica* **37**:125–132.

Englander, S. W., and Mayne, L., 1992, Protein folding studied using hydrogen-exchange labeling and two-dimensional NMR, *Annu. Rev. Biophys. Biomol. Struct.* **21**:243–265.

Ewbank, J. J., and Creighton, T. E., 1993a, Pathway of disulfide-coupled unfolding and refolding of bovine α-lactalbumin, *Biochemistry* **32**:3677–3693.

Ewbank, J. J., and Creighton, T. E., 1993b, Structural characterization of the disulfide folding intermediates of bovine α-lactalbumin, *Biochemistry* **32**:3694–3707.

Fink, A. L., Calciano, L. J., Goto, Y., and Palleros, D. R., 1990, Acid-denatured states of proteins, in *Current Research in Protein Chemistry* (J. Villafranca, ed.), pp. 417–424, Academic Press, New York.

Fink, A. L., Calciano, L. J., Goto, Y., Nishimura, M., and Swedberg, S. A., 1993, Characterization of the stable, acid-induced, molten globule-like state of staphylococcal nuclease, *Protein Sci.* **2**:1155–1160.

Finkelstein, A. V., and Shakhnovich, E. I., 1989, Theory of cooperative transitions in protein molecules. II. Phase diagram for a protein molecule in solution, *Biopolymers* **28**:1681–1694.

Flanagan, J. M., Kataoka, M., Shortle, D., and Engelman, D. M., 1992, Truncated staphylococcal nuclease is compact but disordered, *Proc. Natl. Acad. Sci. USA* **89**:748–752.

Goldberg, M. E., Semisotnov, G. V., Friguet, B., Kuwajima, K., Ptitsyn, O. B., and Sugai, S., 1990, An early immunoreactive folding intermediate of the tryptophan synthase β_2 subunit is a "molten globule," *FEBS Lett.* **263**:51–56.

Goto, Y., and Fink, A. L., 1989, Conformational states of β-lactamase: Molten-globule states at acidic and alkaline pH with high salt, *Biochemistry* **28**:945–952.

Goto, Y., and Fink, A. L., 1990, Phase diagram for acidic conformational states of apomyoglobin, *J. Mol. Biol.* **214**:803–805.

Goto, Y., Calciano, L. J., and Fink, A. L., 1990a, Acid-induced folding of proteins, *Proc. Natl. Acad. Sci. USA* **87**:573–577.

Goto, Y., Takahashi, N., and Fink, A. L., 1990b, Mechanism of acid-induced folding of proteins, *Biochemistry* **29**:3480–3488.

Gottfried, D. S., and Haas, E., 1992, Nonlocal interactions stabilize compact folding intermediates in reduced unfolded bovine pancreatic trypsin inhibitor, *Biochemistry* **31**:12353–12362.

Griko, Y. V., Privalov, P. L., Venyaminov, S. Y., and Kutyshenko, V., 1988, Thermodynamic study of the apomyoglobin structure, *J. Mol. Biol.* **202**:127–138.

Hamada, D., Hoshino, M., Kataoka, M., Fink, A. L., and Goto, Y., 1993, Intermediate conformational states of apocytochrome *c*, *Biochemistry* **32**:10351–10358.

Harding, M. M., Williams, D. H., and Woolfson, D. N., 1991, Characterization of a partially denatured state of a protein by two-dimensional NMR: Reduction of the hydrophobic interactions in ubiquitin, *Biochemistry* **30**:3120–3128.

Hirose, M., and Yamashita, H., 1991, Partially folded state of the disulfide-reduced N terminal half-molecule of ovotransferrin as a renaturation intermediate, *J. Biol. Chem.* **266**:14631–14638.

Horowitz, P. M., and Xu, R., 1992, Acid pH-induced conformational changes in bovine liver rhodanese, *J. Biol. Chem.* **267:**19464–19469.

Hua, Q. X., Ladbury, J. E., and Weiss, M. A., 1993, Dynamics of monomeric insulin analogue: Testing the molten-globule hypothesis, *Biochemistry* **32:**1433–1442.

Hughson, F. M., and Baldwin, R. L., 1989, Use of site-directed mutagenesis to destabilize native apomyoglobin relative to folding intermediates, *Biochemistry* **28:**4415–4422.

Hughson, F. M., Wright, P. E., Baldwin, R. L., 1990, Structural characterization of a partly folded apomyoglobin intermediate, *Science* **249:**1544–1548.

Hughson, F. M., Barrick, D., and Baldwin, R. L., 1991, Probing the stability of a partly folded apomyoglobin intermediate by side-directed mutagenesis, *Biochemistry* **30:**4113–4118.

Ikeguchi, M., Kuwajima, K., Mitani, M,. and Sugai, S., 1986a, Evidence for identity between the equilibrium unfolding intermediate and a transient folding intermediate: A comparative study of the folding reactions of α-lactalbumin and lysozyme, *Biochemistry* **25:**6965–6972.

Ikeguchi, M., Kuwajima, K., and Sugai, S., 1986b, Ca^{2+}-induced alteration in the unfolding behavior of α-lactalbumin, *J. Biochem.* **99:**1191–1201.

Jagannadham, M. V., and Balasubramanian, D., 1985, The molten globular intermediate form in the folding pathway of human carbonic anhydrase B, *FEBS Lett.* **188:**326–330.

James, E., Wu, P. G., Stites, W., and Brand, L., 1992, Compact denatured state of a staphylococcal nuclease mutant by guanidinium as determined by resonance energy transfer, *Biochemistry* **31:**10217–10225.

Jeng, M.-F., Englander, S. W., Elöve, G. A., Wand, A. J., and Roder, H., 1990, Structural description of acid-denatured cytochrome *c* by hydrogen exchange and 2D NMR, *Biochemistry* **29:**10433–10437.

Jiang, J. X., Abrams, F. S., and London, E., 1991, Folding changes in membrane-inserted diphtheria toxin that may play important roles in its translocation, *Biochemistry* **30:**3857–3864.

Kataoka, M., Hagihara, Y., Mihara, K., and Goto, Y., 1993, Molten globule of cytochrome *c* studied by small angle x-ray scattering, *J. Mol. Biol.* **229:**591–596.

Kim, J., and Kim, H., 1986, Fusion of phospholipid vesicles induced by α-lactalbumin at acidic pH, *J. Am. Chem. Soc.* **25:**7867–7874.

Kuroda, Y., Kidokoro, S.-I., and Wada, A., 1992, Thermodynamic characterization of cytochrome *c* at low pH: Observation of the molten globule state and of the cold denaturation process, *J. Mol. Biol.* **223:**1139–1153.

Kuwajima, K., 1989, The molten globule state as a clue for understanding the folding and cooperativity of globular–protein structure, *Proteins: Struct. Funct. Genet.* **6:**87–103.

Kuwajima, K., Hiraoka, Y., Ikeguchi, M., and Sugai, S., 1985, Comparison of the transient folding intermediates in lysozyme and α-lactalbumin, *Biochemistry* **24:**874–881.

Kuwajima, K., Yamaya, H., Miwa, S., Sugai, S., and Nagamura, T., 1987, Rapid formation of secondary structure framework in protein folding studied by stopped-flow circular dichroism, *FEBS Lett.* **221:**115–118.

Kuwajima, K., Sakuraoka, A., Fueki, S., Yoneyama, M., and Sugai, S., 1988, Folding of carp parvalbumin studied by equilibrium and kinetic circular dichroism spectra, *Biochemistry* **27:**7419–7428.

Kuwajima, K., Garvey, E. P., Finn, B. E., Matthews, C. R., and Sugai, S., 1991, Transient intermediates in the folding of dihydrofolate reductase as detected by far-ultraviolet circular dichroism spectroscopy, *Biochemistry* **30:**7693–7703.

Lakey, J. H., González-Mañas, J. M., van der Goot, F. G., and Pattus, F., 1992, The membrane insertion of colicins, *FEBS Lett.* **307:**26–29.

Lala, A. K., and Kaul, P., 1992, Increased exposure of hydrophobic surface in molten globule state of α-lactalbumin, *J. Biol. Chem.* **267:**19914–19918.

Langer, T., Lu, C,. Echols, H., Flanagan, J., Hayer, M. K., and Hartl, F. U., 1992, Successive action of DnaK, DnaJ, and GroEL along the pathway of chaperone-mediated protein folding, *Nature* **356**:383–689.

Lecomte, J. T. J., and Matthews, C. R., 1993, Unraveling the mechanism of protein folding: New tricks for an old problem, *Protein Eng.* **6**:1–10.

Lehrman, S. R., Duls, J. L., Havel, H. A., Haskell, R. J., Putman, S. D., and Tomich, C.-S. C., 1991, Site-directed mutagenesis to probe protein folding: Evidence that the formation and aggregation of a bovine growth hormone folding intermediate are dissociable processes, *Biochemistry* **30**:5777–5784.

Leistler, B., Herold, M., and Kirschner, K., 1992, Collapsed intermediates in the reconstitution of dimeric aspartate aminotransferase from *Escherichia coli*, *Eur. J. Biochem.* **205**:603–611.

Mann, C. J., and Matthews, C. R., 1993, Structure and stability of an early folding intermediate of *Escherichia coli trp* aporepressor measured by far-UV stopped-flow circular dichroism and 8-anilino-1-naphthalene sulfonate binding, *Biochemistry* **32**:5282–5290.

Mark, A. E., and, van Gunsteren, W. F., 1992, Simulation of the thermal denaturation of hen egg white lysozyme: Trapping the molten globule state, *Biochemistry* **31**:7745–7748.

Mårtensson, L.-G., and Jonsson, B.-H., 1993, Characterization of folding intermediates of human carbonic anhydrase II: Probing substructure by chemical labeling of SH groups introduced by site-directed mutagenesis, *Biochemistry* **32**:224–231.

Martin, J., Langer, T., Boteva, R., Schramel, A., Hoewich, A. L., and Hartl, F. U., 1991, Chaperonin-mediated protein folding at the surface of GroEL through a "molten globule"-like intermediate, *Nature* **352**:36–42.

Mitaku, S., Ishido, S., Hirano, Y., Itoh, H., Kataoka, R., and Saitô, N., 1991, Hydrophobic core of molten-globule state of bovine carbonic anhydrase B, *Biophys. Chem.* **40**:217–222.

Nicoli, D. F., and Benedek, G. B., 1976, Study of thermal denaturation of lysozyme and other globular proteins, *Biopolymers* **151**:2421–2437.

Ohgushi, M., and Wada, A., 1983, "Molten-globule state": A compact form of globular proteins with mobile side-chains, *FEBS Lett.* **164**:21–24.

Palleros, D. R., Reid, K. L., McCarty, J. S., Walker, G. C., and Fink, A. L., 1992, DnaK, hsp73, and their molten globules, *J. Biol. Chem.* **267**:5279–5285.

Palleros, D. R., Shi, L., Reid, K. L., and Fink, A. L., 1993, Three-state denaturation of DnaK induced by guanidine hydrochloride: Evidence for an expandable intermediate, *Biochemistry* **32**:4314–4321.

Peng, X., Jonas, J., and Silva, J. L., 1993, Molten-globule conformation of arc repressor monomers determined by high-pressure ^{1}H NMR spectroscopy, *Proc. Natl. Acad. Sci. USA* **90**:1776–1780.

Ptitsyn, O. B., 1987, Protein folding: Hypotheses and experiments, *Protein Chem.* **6**:273–293.

Ptitsyn, O. B., 1992, The molten globule state, in *Protein Folding* (T. E. Creighton, ed.), pp. 243–300, W. H. Freeman, New York.

Ptitsyn, O. B., Pain, R. H., Semisotnov, G. V., Zerovnik, E., and Razgulyaev, O. I., 1990, Evidence for a molten globule state as a general intermediate in protein folding, *FEBS Lett.* **262**:20–24.

Radford, S. E., Dobson, C. M., and Evans, P. A., 1992, The folding of hen lysozyme involves partially structured intermediates and multiple pathways, *Nature* **358**:302–307.

Rajendran, S., and Prakash, V., 1992, Association of proteins in acidic solutions—a case study with β-globulin, *Int. J. Biol. Macromol.* **14**:298–304.

Robson, B., and Pain, R., 1976a, The mechanism of folding of globular proteins, *Biochem. J.* **155**:325–330.

Robson, B., and Pain, R., 1976b, The mechanism of folding of globular proteins, *Biochem. J.* **155**:331–344.

Semisotnov, G. V., Rodionova, N. A., Kutyshenko, V. P., Ebert, B., Blanck, J., and Ptitsyn, O. B., 1987, Sequential mechanism of refolding of carbonic anhydrase B, *FEBS Lett.* **224**:9–13.

Semisotnov, G. V., Rodionova, N. A., Razgulyaev, O. I., Uversky, V. N., Gripas, A. F., and

Gilmanshin, R. I., 1991, Study of the "molten globule" intermediate state in protein folding by a hydrophobic fluorescent probe, *Biopolymers* **31:**119–128.

Shortle, D., 1987, Equilibrium and kinetic intermediates in protein folding: An alternative explanation, in *Protein Structure, Folding, and Design 2* (Dale L. Oxender, ed.), pp. 353–361, Alan R. Liss, New York.

Sosnick, T. R., and Trewhella, J., 1992, Denatured states of ribonuclease a have compact dimensions and residual secondary structure, *Biochemistry* **31:**8329–8335.

Stigter, D., and Dill, K. A., 1990, Charge effects on folded and unfolded proteins, *Biochemistry* **29:**1262–1271.

Sugawara, T., Kuwajima, K., and Sugai, S., 1991, Folding of staphylococcal nuclease A studied by equilibrium and kinetic circular dichroism spectra, *Biochemistry* **30:**2698–2706.

Tandon, S,. and Horowitz, P. M., 1989, Reversible folding of rhodanese, *J. Biol. Chem.* **264:**9859–9866.

Tirado-Rives, J., and Jorgensen, W. L., 1993, Molecular dynamics simulations of the unfolding of apomyoglobin in water, *Biochemistry* **32:**4175–4184.

Uversky, V. N., Leontiev, V. V., and Gudkov, A. T., 1992a, Triple point mutatin Asp 10 → His, Asn 101 → Asp, Arg 148 → Ser in T4 phage lysozyme leads to the molten globule, *Protein Engineer.* **5:**781–783.

Uversky, V. N., Semisotnov, G. V., Pain, R. H., and Ptytsyn, O. B., 1992b, "All-or-none" mechanism of the molten globule unfolding, *FEBS Lett.* **314:**89–92.

van der Goot, F. G., González-Mañas, J. M., Lakey, J. H., and Pattus, F., 1991, A "molten globule" membrane-insertion intermediate of the poreforming domain of colicin A, *Nature* **354:**408–410.

van der Goot, F. G., Lakey, J. H., and Pattus, F., 1992, The molten globule intermediate for protein insertion or translocation through membranes, *Trends Cell Biol.* **2:**343–348.

Varley, P., Gronenborn, A. M., Christensen, H., Wingfield, P. T., Pain, R. H., and Clore, G. M., 1993, Kinetics of folding of the all-β sheet protein interleukin-1β, *Science* **260:**1110–1113.

Xie, D., Bhakuni, V., and Freire, E., 1991, Calorimetric determination of the energetics of the molten globule intermediate in protein folding: Apo α-lactalbumin, *Biochemistry* **30:**10673–10678.

Yutani, K., Ogasahara, K., and Kuwajima, K., 1992, Absence of the thermal transition in apo α-lactalbumin in the molten globule state: A study by differential scanning in microcalorimetry, *J. Mol. Biol.* **228:**347–350.

Zerovnik, E., Jerala, R., Kroon-Zitko, L., Turk, V., and Pain, R. H., 1992, Denaturation of Stefin B by GuHCl, pH and heat; evidence for molten globule intermediates, *Biol. Chem. Hoppe-Seyler* **373:**453–458.

Zhao, J.-M., and London, E., 1986, Similarity of the conformation of diphtheria toxin at high temperature to that in the membrane-penetrating low-pH state, *Proc. Natl. Acad. Sci. USA* **83:**2002–2006.

Chapter 3

Raman Spectroscopy of Proteins and Their Assemblies

Takashi Miura and George J. Thomas, Jr.

1. INTRODUCTION

Laser Raman spectroscopy, like infrared spectroscopy, is a method for determining molecular structure by measuring the energies (frequencies) of molecular vibrations. Although the two methods differ fundamentally in the mechanisms of interaction between radiation and matter, one obtains in both cases a *vibrational spectrum* consisting of a number of discrete bands, the frequencies and intensities of which are determined by the nuclear masses in motion, the equilibrium molecular geometry, and the molecular force field. An important advantage of Raman over infrared spectroscopy for biological applications is the virtual transparency of water (both H_2O and D_2O) in the Raman effect. This greatly simplifies the analysis of aqueous solutions and facilitates the investigation of hydrogen–isotope exchange phenomena. Changes in molecular geometry—particularly the conformational transitions characteristic of biological macromolecules—can produce large shifts in Raman band positions, often referred to as *frequency shifts*, empowering the technique in the diagnosis of protein secondary structure, determination of

---

Takashi Miura and George J. Thomas, Jr. Division of Cell Biology and Biophysics, School of Biological Sciences, University of Missouri, Kansas City, Missouri 64110-2499. *Present Address for T.M.:* Tohoku University, Pharmaceutical Institute, Aobayama, Sendai: 980, Japan.

Subcellular Biochemistry, Volume 24. Proteins: Structure, Function, and Engineering, edited by B. B. Biswas and Siddhartha Roy. Plenum Press, New York, 1995.

side-chain configurations, and detection of interacting side-chain groups. Since the molecular geometry and force field may be sensitive to interactions between molecules, the Raman method also has the potential for investigating intermolecular interactions, including the formation of biologically important protein complexes. Raman spectroscopy is gaining wide use as a method for probing protein structure, dynamics, assembly, and recognition.

1.1. Scope of This Chapter

In this chapter we present a brief discussion of the background of the Raman effect and consider the scope and limitations of Raman spectroscopy for the analysis of protein structures and detection of biologically important interactions of proteins. We also survey relationships between the data of Raman spectra (frequencies, intensities, and polarizations) and structural characteristics of protein main chains and side chains. Representative examples are selected from the large number of recent applications of Raman spectroscopy to proteins and related peptide and amino acid model compounds. Although applications of both off-resonance and resonance Raman spectroscopy are considered, emphasis is on the former. A more exhaustive review of resonance Raman applications to proteins has been published recently (Austin *et al.*, 1993).

1.2. Advantages and Disadvantages of the Method

Among the advantages of Raman spectroscopy for investigating structures of proteins and their complexes, the following are noteworthy:

1. The method is applicable to virtually any sample morphology. For proteins, this includes solutions (aqueous and nonaqueous), suspensions, precipitates, gels, films, fibers, single crystals, and polycrystalline and amorphous solids. Data obtained from a given sample in one morphological state are generally transferable to another morphological state of the same sample. This has important practical benefits; for example, in comparing the molecular structure of a protein in the crystal with that prevailing in solution.

2. Both normal water (H_2O) and heavy water (D_2O) generate very weak Raman spectra, thus producing relatively little interference with the Raman spectrum of the dissolved solute. This constitutes a significant advantage over Fourier-transform infrared (FTIR) absorption spectroscopy, where both H_2O and D_2O are highly problematic solvents. The innocuous Raman characteristics of H_2O and D_2O likewise facilitate the exploitation of Raman spectroscopy for monitoring hydrogen–isotope exchange processes in proteins and their assemblies, as well as for purposes of measuring isotope shifts to establish or confirm definitive vibrational spectroscopic assignments.

3. The sample volume required in a high-performance Raman spectrophotometer is relatively small ($\approx$1 μl). This represents an advantage over many other structural methods, including X-ray crystallography and magnetic resonance spectroscopy.

4. Laser Raman spectroscopy requires no molecular labels or probes and is generally nondestructive of protein samples. Accordingly, the native structure need not be modified for examination, and the protein may be recovered and assayed for biological activity following examination.

5. In the case of crystalline or pseudocrystalline lattices, polarized Raman methods can be employed to determine the directionality of the protein main chain within the lattice or to determine configurations of specific side chains.

6. Raman scattering occurs on a time scale that is very short ($\approx 10^{-15}$ sec) in comparison to the time scales of fluorescence ($>10^{-9}$ sec) and nuclear magnetic resonance phenomena ($\approx 10^{-6}$ sec). Thus, in principle, a more extensive range of biological processes can be temporally resolved by Raman spectroscopy. For example, photoreactions of rhodopsin and bacteriorhodopsin are conveniently time-resolved using Raman methods.

7. The fundamental selection rule for Raman spectroscopy, i.e., that relatively high Raman intensity accrues from molecular vibrations with which there is associated a large change in molecular polarizability, favors the electron-rich moieties of proteins. Thus, the Raman spectrum of a typical protein is dominated by spectral bands assignable to peptide groups (CONH), aromatic side chains (Trp, Phe, Tyr), sulfur-containing substituents (Met, Cys), side-chain carboxyls (Asp, Glu), and so forth. Raman intensities associated with hydrogenic groups (e.g., CH, CH_2, CH_3), although individually weak, are collectively significant in proteins, because of the large numbers of such groups present in the side chains.

8. Raman intensities are enhanced dramatically (several orders of magnitude) when the energy of the incident photon is selected in resonance with a molecular electronic transition of a protein chromophore. This constitutes the resonance Raman (RR) effect. Therefore, structural information about the chromophore can be obtained from relatively dilute protein solutions using the RR approach. This technique is particularly valuable for proteins containing chromophores which absorb in the visible (metallo-proteins, retinal-proteins, etc.) or near ultraviolet (nucleotide-binding proteins, nucleoprotein complexes, etc.).

9. Finally, there exists a large database of Raman spectra of polypeptides, amino acids, and related model compounds for which reliable band assignments, normal mode analyses, and spectra-structure correlations have been made. This facilitates interpretation of the often complex Raman spectra obtained from proteins and their assemblies.

Notable disadvantages of Raman spectroscopy for protein structure analysis are the following:

1. The Raman bands of structurally identical or near identical moieties often overlap one another. Unlike the case of nuclear magnetic resonance spectroscopy, the degree of spectral resolution cannot generally be increased through use of more sophisticated instrumentation or by variation of the electromagnetic field. To improve resolution of overlapping bands, the experimentalist must employ a strategy based upon chemical or molecular biological modification, such as isotope editing or site-directed mutagenesis.

2. The Raman process of inelastic light scattering is inherently weak compared to other light absorption and emission processes. Thus, considerable effort in sample purification and care in sample handling are necessary to avoid even minute traces of fluorescent impurities or other chromophores that may interfere with collection of the Raman scattering.

3. The inherent weakness of the Raman scattering mechanism in comparison to other mechanisms of interaction of radiation with matter imposes a requirement for sophisticated and relatively costly instrumentation.

4. Although small sample volumes are sufficient for Raman analysis, high-protein concentrations (1–10 weight percent) are required to obtain off-resonance Raman spectra of good quality. Accordingly, proteins that are sparingly soluble or that tend to aggregate at high concentrations may not be amenable to off-resonance Raman analysis.

5. Finally, presently available molecular force fields and theoretical methods are not adequate for calculating a priori the Raman spectrum of a protein.

Various tactics for optimizing the advantages and circumventing the disadvantages of the Raman method have been developed in recent years. Several examples are illustrated in the discussion of applications that follows.

1.3. Nature of the Data of Raman Spectra

The Raman effect consists essentially of inelastic scattering of light, whereby quanta of energy are transferred from a beam of monochromatic radiation (laser) to the target molecule (protein) upon which the beam impinges. Only a small fraction of the incident photons is inelastically scattered; the vast majority are scattered elastically (Rayleigh scattering) or quasi-elastically (Brillouin scattering). Furthermore, the quantity of energy transferred in each Raman event is only a small fraction of the total energy of the incident photon. The quanta are gained by the protein in the form of vibrational energies. The collection of discrete vibrational energies transferred in these processes (*Raman frequencies*), as well as their relative scattering probabilities (*Raman intensities*) and tensor characteristics (*Raman polarizations*), comprise the vibrational Raman spectrum of the protein. The spectrum is a unique and sensitive function of the intramolecular geometry and intra- and intermolecular force fields. The Raman spectrum of a protein is therefore a distinctive signature of the molecular structure and environment.

The Raman spectrum of a protein is obtained experimentally by measuring, with a highly sensitive spectrophotometer, the energies (frequencies) and relative numbers (intensities) of photons that have been scattered by the target protein. The scattered photons exhibiting frequencies less than the frequency of the impingent laser photons reflect the vibrational quanta transferred to the protein, and thus identify the (Stokes) Raman frequencies of the protein. Other factors that determine Raman frequencies, intensities, and polarizations are discussed below.

In practice, the spectrum is displayed as a series of bands, i.e., the Raman intensities are plotted as a function of the scattered frequencies. The frequencies are usually expressed in wave number (cm^{-1}) units and the intensities in arbitrary units. Since the vibrational energy transfers are small in comparison to the total energy of the incident photon, the frequencies of incident and scattered photons are clustered within a relatively narrow region of the electromagnetic spectrum (Fig. 1). For example, in a typical experiment, the green emission of an argon laser (wavelength = 514.5 nm; absolute wave number = 19,436 cm^{-1}) might be used to excite a protein Raman spectrum. The spectrum would be detected by monitoring the intensities of scattered photons with frequencies (wave numbers) less than 19,436 cm^{-1}. Typically, one might measure the intensities of scattered photons exhibiting frequencies between 19,236 and 17,686 cm^{-1}, corresponding to vibrational energies between 200 and 1750 cm^{-1}. This region (200–1750 cm^{-1}) of the spectrum would contain virtually all of the vibrational information

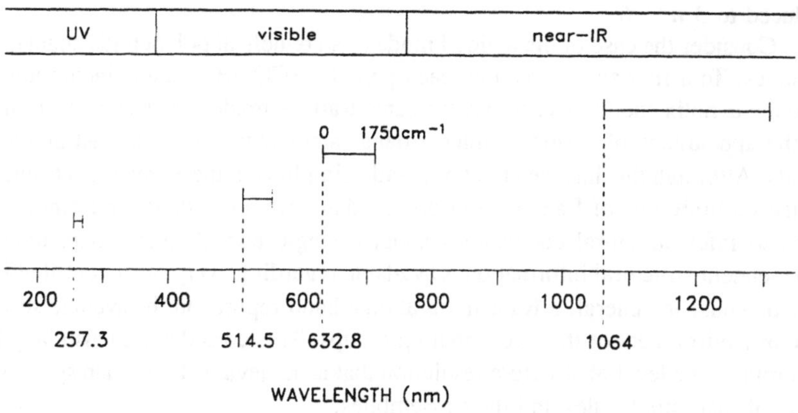

FIGURE 1. The region 0-1750 cm^{-1} of the Raman scattering spectrum portrayed (from left to right) for excitation at 257.3 nm (Ar^+ laser, second harmonic of the 514.5 nm line), 514.5 nm (Ar^+ laser, green), 632.8 nm (He/Ne laser, red), and 1064 nm (Nd:YAG laser, near infrared). The data are represented on a linear wavelength scale.

on the protein molecule, except for hydrogenic stretching modes, overtones, or combination modes, which occur at higher frequencies (1800–4000 cm^{-1}).

1.3.1. Raman Frequencies, Intensities, and Polarizations

For a protein of N atoms, $3N - 6$ normal modes of vibration are possible. In principle, each of the $3N - 6$ modes extends over the entire molecule and, because the whole protein molecule lacks symmetry, each is capable of leading to a Raman transition. In practice, however, most of the modes are highly localized; meaning that, to a good approximation, each represents a vibration of a small group of atoms largely isolated from the vibrations of other molecular subgroups. When the group of atoms is sufficiently small and identifiable, the vibration can be referred to as a *group vibration* and its Raman band designated as a *group frequency*. Many group frequencies of the protein main chain and of various side chains have been identified on the basis of extensive Raman and infrared studies of model peptides and amino acids. Many of the assignments are supported by normal coordinate calculations.

A second practial simplification of the Raman spectrum results from the degeneracy or near-degeneracy of many vibrational modes. Each molecular subgroup of a given type, which exists in an identical or nearly identical conformation and environment within the macromolecule, should yield the same set of group frequencies. A corresponding Raman band will then represent a superposition of contributions from each of the identical or nearly identical groups. Thus, if the N atoms of the protein are distributed among s sets of identical groups, each containing n atoms, then the collection of $3ns - 6$ hypothetical spectral bands is reduced to $3n - 6$.

Consider the case of the amide I mode of an α-helical polypeptide chain of s residues. To a first approximation, each peptide −CONH− that is incorporated into the α-helix should give nearly the same Raman amide I frequency, resulting in the appearance of a single amide I band rather than $s - 1$ distinct amide I bands. Although this has the effect of greatly simplifying the Raman spectrum, it reflects a limitation of Raman spectroscopy that such a resultant band cannot be used to infer structural conclusions about a single contributing group, unless independent structural information is available regarding every other contributing group. Thus, in general, a Raman vibrational band reports on the average structure or environment of the s contributing groups. This consideration also largely determines the level of structure resolution that is achievable by Raman spectroscopy of macromolecules and their assemblies.

As implied above, the spectral intensity of a Raman band is a measure of the efficiency of energy exchange between the incident laser photon and the vibrational mode in question. The following explanation, which is based on classical theory, is helpful for understanding the mechanism that determines the intensity of Raman scattering.

When monochromatic light of frequency ν_0 irradiates a molecule, the oscillating electric field E induces a small discrepancy between centers of negative (electronic) and positive (nuclear) charge, and a dipole moment is thereby induced in the molecule. (The induced dipole does not depend upon the existence of a permanent molecular dipole.) This phenomenon is called *polarization*. To a good approximation, the induced dipole moment P is proportional to the incident electric field strength E:

$$P = \alpha E \tag{1}$$

where the proportionality factor α is the molecular electric polarizability tensor. Each component of P (P_i, where $i = x, y, z$) is related to the three components of the electric field strength by equations of the form:

$$P_i = \Sigma \alpha_{ij} E_j \tag{2}$$

where the sum is over i and j ($= x, y, z$). The coefficients α_{ij} are the elements of the polarizability tensor, and the subscripts i and j represent, respectively, the components of P and E to which it relates. In consequence of the fact that both α and E are time-dependent (note that α depends upon the internuclear separations that vary with the frequency ν_v of molecular vibration, while the electric field strength oscillates at frequency ν_0), it can be shown (Carey, 1982) that P has terms which depend upon ν_0, $\nu_0 + \nu_v$ and $\nu_0 - \nu_v$ as follows:

$$P_i(t) = \Sigma[(\alpha_{ij}{}^0) \mid E \mid \cdot \cos 2\pi\nu_0 t + (\tfrac{1}{2}\alpha_{ij}{}') \mid qE \mid \cdot \cos 2\pi(\nu_0 \pm \nu_v)t] \tag{3}$$

where $\alpha_{ij}{}^0$ is the component ij of the molecular polarizability at the equilibrium (nonvibrating) configuration, $\alpha_{ij}{}'$ is the first derivative of the polarizability component taken with respect to the vibrational coordinate q, and $|E|$ is the amplitude of E. The first term on the right in Eq. (3) is a component of P_i oscillating with frequency ν_0 that, according to classical electromagnetic theory, represents energy radiated (scattered) with the same frequency as the incident light. This is called *Rayleigh* scattering. The second term on the right of Eq. (3) contains components of P_i oscillating at frequencies $\nu_0 \pm \nu_v$, i.e., representing radiation scattered with frequencies that are either a vibrational quantum larger (*anti-Stokes Raman* scattering) or smaller (*Stokes Raman* scattering) than the incident frequency. Although both anti-Stokes and Stokes Raman effects can be observed experimentally, the latter is favored by the Boltzmann distribution law and in practice is the easier to detect. Note that this simple classical description [Eq.

(3)] also indicates that the Raman effect requires a *change* of polarizability with vibration, i.e., $\alpha_{ij}' \neq 0$. If the polarizability does not change during a molecular vibration, that vibration generates no Raman scattering and is said to be Raman-inactive.

As a consequence of Eq. (3), the following additional generalizations can be made regarding Raman intensities anticipated from vibrations of protein subgroups. Vibrations localized in the planar trans-peptide group are expected to yield relatively high Raman intensities because of the large polarizability changes associated with in-plane stretching vibrations involving peptide C=O and C−N bonds. In-plane vibrations of the rings of aromatic side chains (Trp, Phe, Tyr) are also expected to be intense in the Raman spectrum. Conversely, out-of-plane vibrations, including ring deformations and out-of-plane bending motions of exocyclic substituents, are very weak in the Raman effect and usually are not observed in spectra of biological molecules. Stretching vibrations of C−C, C−N, and C−O bonds, especially concerted or symmetrical displacements of side-chain skeletons or carboxylates, lead also to intense Raman bands. Corresponding antisymmetrical and asymmetrical modes are not usually intense in the Raman (but may exhibit high infrared intensities). Bending and stretching modes of hydrogenic substituents (C−H, N−H, O−H) are generally weak in the Raman. However, the collective Raman intensities resulting from the large numbers of such groupings in a protein can be high. Intense Raman bands are associated with all group vibrations that involve displacements of heavy atoms, such as sulfur. Thus, relatively intense Raman bands are associated with C−S stretching modes of Met and Cys, S−S stretching of cystine, S−H stretching of cysteine, and Zn−S stretching in zinc metalloproteins.

In the resonance-Raman effect, additional factors govern the spectroscopic transition probabilities. Therefore, the Raman band intensity patterns observed in resonance conditions are quite different from those observed off-resonance (Nishimura *et al.*, 1978; Austin *et al.*, 1993). It should also be recognized that breakdown of the group frequency approximation can invalidate many of the above generalizations.

Polarization is a unique and useful characteristic of Raman scattered light. Although the incident laser beam is plane polarized, Eq. (2) indicates that the Raman scattered light is not generally polarized in the same plane. The change of the plane of polarization upon scattering is dealt with quantitatively by defining the depolarization ratio ρ ($\equiv I_{\perp}/_{\parallel}$), which is the ratio of intensities of light scattered along directions perpendicular ($I_{\perp}$) and parallel ($I_{\parallel}$) to the direction of polarization of the incident light. The value of ρ is determined experimentally by inserting a polarization dicriminator between the sample and photon detector. In isotropic ensembles of molecules (solutions), measurements of ρ are helpful in distinguishing normal modes of vibration that are totally or locally symmetric (i.e., which retain the symmetry of the molecule or subgroup and for which $\rho <$

0.75) from those modes that are nontotally symmetric ($\rho = 0.75$). In anisotropic ensembles of molecules (single crystals, oriented membrane multilayers, or oriented fibers), the Raman polarizations can yield information on the orientations of molecules or their subgroups. Experimental techniques required for measurement of the Raman polarizations of crystals and fibers are more complicated than those employed for isotropic ensembles (see, for example, Tsuboi *et al.*, 1991a,b; Benevides *et al.*, 1993). Applications of polarized Raman spectroscopy to biological molecules have been reviewed by Thomas and Tsuboi (1993).

1.3.2. Analysis of the Data

The successful application of vibrational spectroscopy in protein analysis presumes definitive band assignments and requires a combination of experimental and theoretical approaches. Tactics employed toward this objective include: (1) comparison of Raman frequencies and intensities with corresponding infrared data; (2) determination of vibrational frequency shifts accompanying stable-isotope substitutions, $^{1}H \rightarrow {}^{2}H(D)$, $^{14}N \rightarrow {}^{15}N$, $^{16}O \rightarrow {}^{18}O$ and $^{12}C \rightarrow {}^{13}C$; (3) effects of pH, temperature, and other environmental factors on the spectra; (4) detailed and well-analyzed Raman and infrared spectra of smaller, more symmetrical molecules that are structurally related to protein constituents; (5) *ab initio* molecular orbital calculation of intramolecular force fields and normal coordinate calculations based on them; (6) depolarization ratios of the Raman bands of isotropic solutions; (7) polarized Raman spectra of oriented samples; and (8) Raman excitation profiles, which correlate Raman intensities with laser wavelengths of excitation in the visible and ultraviolet.

Digitally computed Raman difference spectra can facilitate the visualization of changes in protein Raman bands accompanying changes in sample temperature, molecular environment, or the like. However, because the Raman spectrum is not an absorption spectrum, its intensities are not governed by Beer's Law and the comparison of two independently recorded spectra by subtraction can be subject to considerable uncertainty. This is dealt with by comparing intensities in the two spectra only after normalization of both the minuend and subtrahend to a reliable internal intensity standard. A normalization procedure should be deemed reliable only after careful consideration of the origin of the band in question and preferably after validation on a suitable model system. In the absence of a reliable internal normalization basis, a process of trial and error may be the only recourse for computation of a Raman difference spectrum.

Similar concerns apply to the procedure of *solvent correcting* the Raman spectrum of a protein solution. It should be kept in mind that the protein solute may influence the structure of the aqueous solvent (and therefore its Raman spectrum) as much as the water molecules may influence the structure of the dissolved protein. Accordingly, it is virtually impossible to completely compensate the Raman contributions of the aqueous solvent in the solution spectrum of a protein by subtracting therefrom the Raman spectrum of the pure solvent.

The computational power of microcomputers has ushered in an era of easy manipulation of all types of experimental data, including Raman spectra. Particularly popular procedures are the method of Fourier deconvolution, which is usually applied to enhance the resolution of overlapping bands, and the method of least-squares curve fitting, which can be used to fit a complex band shape to an arbitrary number of simpler band components. While the utility of such procedures cannot be questioned, considerable caution must be exercised in their application. The Raman spectroscopist is well advised to avoid these computational approaches whenever the collection of additional experimental data can serve the same objective.

2. EXPERIMENTAL CONSIDERATIONS

2.1. Instrumentation

The instrumentation required for measuring a Raman spectrum consists of a monochromatic light source (laser) for excitation of the Raman scattering, a diffraction grating spectrometer for analysis of the scattered frequencies and a detector for capturing the scattered photons. Laser wavelengths from the near ultraviolet (≈200 nm) extending to the near infrared (≈1000 nm) have been employed. The choice of excitation wavelength is determined by the absorption characteristics of the sample and the objectives of the experiment. Inert gas lasers providing excitation wavelengths in the visible are most often employed for off-resonance Raman spectroscopy of nonabsorbing proteins and for resonance Raman spectroscopy of proteins that absorb in the visible. In the latter case, the Raman spectrum is dominated by resonance-enhanced bands of the visible chromophore, usually a metallo-porphyrin or retinal cofactor. Typical excitation wavelengths include 488.0 and 514.5 nm of argon and 632.8 nm of helium–neon lasers. Intracavity frequency-doubled argon lasers, which provide continuous-wave (CW) laser emission in the ultraviolet (264, 257, 248, 244, 239, and 229 nm), have become available recently for ultraviolet resonance Raman (UVRR) spectroscopy of proteins that absorb in the ultraviolet. In the UVRR spectra of such proteins, the spectrum is dominated by the resonance-enhanced Raman bands of the aromatic amino acid side chains.

Selection of the excitation wavelength is also dictated by the emission characteristics of the sample. In the case of resonance or near-resonance (preresonance) conditions, the relative efficiencies of Raman and fluorescence phenomena depend strongly on the excitation wavelength (Asher, 1993). Near-infrared (IR) excitation is also effective in reducing sample fluorescence. On the other hand, near-IR excitation of Raman spectra is disadvantaged by the fact that the light scattering intensity is proportional to the fourth power of the frequency. This deterrent, while significant, has recently been circumvented by the development of Fourier transform-Raman (FT-Raman) methods.

Applications of FT-Raman to snake venom toxins (Chiou *et al.*, 1992) and bacteriorhodopsin (Rath *et al.*, 1993) have been described. Using near-IR excitation for FT-Raman spectroscopy of bacteriorhodopsin yields a spectrum dominated by Raman bands of the retinal chromophore (due to the preresonance effect), while the Raman scattering arising from the apoprotein is very weak. The advantage of near-IR FT-Raman over conventional resonance Raman (with visible excitation) is the avoidance of photoreactions which populate intermediate states in the photocycle (see Section 4.4.2).

A Raman spectrometer must be capable of efficiently rejecting the very intense Rayleigh scattering that accompanies the Raman scattering [Eq. (3)]. For this purpose, one employs either a highly dispersive (double) grating configuration in the analyzing spectrometer or a narrow cutoff filter (holographic edge filter) before the detector. The latter are available commercially for use with wavelengths in the visible and near-IR regions. However, holographic edge filters are not available in the UV region, and thus options for rejection of stray light and elastic scattering are severely limited for UVRR applications.

Detectors for Raman spectroscopy are gradually evolving from the single-channel photomultiplier and multichannel diode array technologies to the more efficient charge-coupled device (CCD) detector. In recent years, the liquid-nitrogen-cooled CCD camera, which virtually eliminates dark current noise and permits rapid multichannel detection, has come into wide use for detection of Raman scattered photons.

More detailed discussions of instrumentation requirements for off-resonance Raman (Carey, 1982) and UVRR (Austin *et al.*, 1993) applications have been given.

2.2. Sample Handling

Sample handling procedures for Raman spectroscopy of proteins have been surveyed by Carey (1982) and Thomas and Kyogoku (1977). Special procedures for viral proteins and their assemblies have also been discussed (Thomas, 1987). Among the most stringent requirements for collection of useful Raman data from proteinaceous materials are the purity and optical homogeneity of the sample.

Typically, a native protein is acquired as a relatively dilute solution which requires concentration many-fold in order to yield measurable Raman signals. This can be achieved by lyophilization of the solution, provided that the protein is not denatured or inactivated by the freezing process. In this procedure, it is important to eliminate or dilute beforehand (e.g, by dialysis) any buffer and/or salts that are present in order to avoid concentrating them along with the protein. It is helpful to keep in mind that at the protein concentrations employed for off-resonance Raman spectroscopy (≈10–100 mg/ml), most proteins are self-buffering and require little or no chemical buffering once the solution is at the desired pH. Alternatively, a dilute protein solution can be concentrated, without lyophilization and without concern for co-concentration of salts that may be present, by ultrafiltration on a membrane device (Centricon concentrator, Amicon, Inc.). The concentrated protein solution is transferred to the Raman cell (a glass or quartz capillary) by pipette, syringe, or direct suction and the cell is then sealed. Similar procedures can be employed for gels, homogeneous solids (including precipitates and colloids), crystals in mother liquor, and amorphous solids. In each step of the sample handling protocol the opportunity exists to introduce undesirable particulate matter, especially dust, into the sample. This can lead to an unacceptable level of noise in the Raman spectrum and should be carefully avoided. Glassware, plasticware, and utensils employed in handling of samples for Raman spectroscopy should be maintained free of dust.

Salts and chelating agents appropriate for specific Raman applications to proteins are described in the original literature. Generally, one employs monatomic cationic and anionic species whenever possible, since polyatomic species generate Raman bands that may interfere with those of the protein. Thus, NaCl would be preferred over NH_4Cl or $NaNO_3$ to increase the solution ionic strength. Low concentrations of chelating agents (<2 mM EDTA) are also acceptable. Although a low concentration of buffer is tolerable (for example, 5 mM Tris), use of such buffers is largely cosmetic since they would provide negligible buffering capacity in a protein solution of ≈100 mg/ml (equivalent to ≈10 mM of a 10 kDa protein and probably much greater than 10 mM in the acidic and basic side chains for which the buffering capacity is intended).

3. ANALYSIS OF SECONDARY STRUCTURE AND ENVIRONMENTS OF SIDE CHAINS

3.1. Main Chain Conformations

The 12 normal modes of vibration expected for the 6-atom trans-peptide group ($C^{\alpha}CONHC^{N}$), as modeled by *N*-methylacetamide, have been discussed recently by Bandekar (1992) and are reproduced in Table I and Fig. 2. Most useful for protein analysis by Raman spectroscopy are amide I, which is primarily a carbonyl stretching mode, and amide III, which combines both in-plane N-H bending and C-N stretching motions. In proteins, amide I and amide III are expected in the intervals 1640–1680 and 1230–1310 cm^{-1}, respectively. In synthetic polypeptides, the amide modes may occur as multiplets and may extend

Table I
Vibrational Modes of the Trans Peptide Group[a]

Description	Observed frequency	Calculated frequency	Potential energy distribution
NH stretch	3236	3254	NH s (100)
Amide I	1653	1646	CO a (83), CN s (15), C^{α}CN d (11)
Amide II	1567	1515	NH ib (49), CN s (33), CO ib (12)
Amide III	1299	1269	NH ib (52), C^{α}C s (18), CN s (14)
NC^{N} stretch	1096	1070	NC^{N} s (77), C^{α}C (17)
CN stretch	881	908	CN s (31), C^{α}C s (17), CO s (16)
Amide V	725	721	CN t (75), NH ob (38)
Amide IV	627	637	CO ib (44), C^{α}C s (34), CNC^{N} d (11)
Amide VI	600	655	CO ob (85), CN t (13)
C^{α}CN bend	436	498	C^{α}CN d (63), CO ib (11)
CNC^{N} bend	289	274	CNC^{N} (71), CO ib (19), C^{α}CN d (13)
Amide VII	206	226	NH ob (64), CN t (15), CO ob (12)

[a]Based on observed and calculated frequencies (cm^{-1} units) of *N*-methylacetamide ($C^{\alpha}CONHC^{N}$, where C^{α} and C^{N} refer to the acetyl and amide methyl groups, respectively) (Bandekar, 1992). The right column lists relative (unnormalized) contributions to the potential energy. Abbreviations: s, stretch, d, deformation, t, torsion, ib, in-plane bend, ob, out-of-plane bend.

outside these regions due to effects of coupling between identical residues in the polypeptide chain and various other factors (Bandekar, 1992). The precise Raman frequencies and band shapes of protein amide I and amide III bands within their respective intervals depend rather sensitively upon the protein secondary structure. Accordingly, the Raman amide bands are often exploited as indicators of protein secondary structure.

Table II lists the assigned Raman amide I and amide III bands of representative polypeptides and proteins of differing secondary structures. The data of Table II illustrate in a qualitative sense the conformational dependence of Raman amide I and amide III modes. Together with Raman data collected from a large number of other proteins of known three-dimensional structures, these results suggest the following generalizations: (1) α-Helical domains in proteins contribute amide I and amide III bands in the intervals 1645–1655 and 1260–1310 cm^{-1}, respectively; (2) for β-sheet structures, the corresponding intervals are roughly 1665–1680 and 1230–1245 cm^{-1}; and (3) for irregular or disordered domains, the respective amide I and amide III bands are anticipated in the intervals 1655–1665 and 1245–1270 cm^{-1}. Reverse turns and bends of defined structure also contribute variously to the spectra, and several specific examples have been discussed (Bandekar, 1992). Figure 3 shows the Raman amide I spectral region for representative peptide-chain conformations and illustrates the distinctive Raman amide I signatures of α-helix and β-sheet domains.

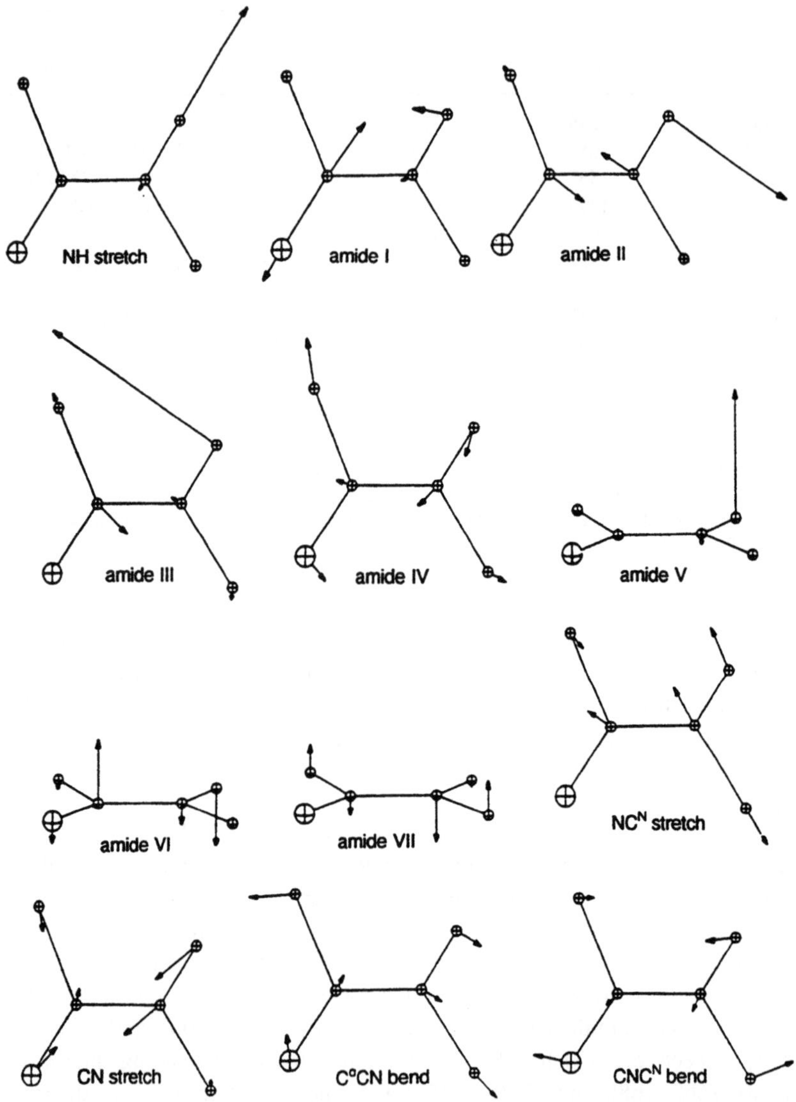

FIGURE 2. The 12 normal modes of vibration of planar, trans-*N*-methylacetamide, as defined in Table 1. *N*-methylacetamide is treated as a six-atom molecule in which the α-CH_3 group (upper left of each structure) and the N-CH_3 group (lower right of each structure) are considered as point masses of 15 a.m.u. (From Bandekar, 1992.)

Table II
Raman Amide I and Amide III Bands in Representative Polypeptides and Proteins

Structure	Amide I	Amide III	Conformation	Reference
α-Poly-L-alanine	1655	1265-1348	α-Helix	Carey (1982), Bandekar (1992)
α-Poly-L-glutamate	1652	1290	α-Helix	Carey (1982)
α-Poly-L-lysine	1645	1295-1311	α-Helix	Carey (1982), Bandekar (1992)
β-Poly-L-alanine	1669	1226-1243	β-Sheet	Carey (1982), Bandekar (1992)
β-Poly-L-glutamate	1672	1236	β-Sheet	Carey (1982)
β-Poly-L-lysine	1670	1240	β-Sheet	Carey (1982), Bandekar (1992)
Poly-L-lysine, pH4	1665	1243-1248	Irregular	Carey (1982)
Poly-L-glutamate, pH11	1656	1249	Irregular	Carey (1982)
fd subunit	1650	1270-1300	α-Helix	Thomas (1987)
Pfl subunit	1650	1270-1300	α-Helix	Thomas (1987)
cI repressor (1-102)	1650	1286-1297	α-Helix	Thomas *et al.* (1986)
	1675	1245	Turns and irregular	Thomas *et al.* (1986)
P22 subunit	1653	1273	α-Helix	Prevelige *et al.* (1990, 1993)
	1665	1235	β-Sheet and turns	Prevelige *et al.* (1990, 1993)
P22 tailspike	1668	1231	β-Sheet	Sargent *et al.* (1988)
	>1668	1249	Turns and irregular	Sargent *et al.* (1988)
BPMV subunit	1559	1236	β-Sheet	Li *et al.* (1990)

In principle, protein secondary structure may be estimated quantitatively from analysis of the conformation-sensitive amide I and amide III bands. In practice, however, a rigorous quantitative analysis is highly problematic and methods so far described have been of limited usefulness. A review of Raman (and IR) spectrophotometric methods employed for protein secondary structure analysis has been given (Krimm and Bandekar, 1986; Bandekar, 1992). Further discussion of this subject is given below.

In addition to amide I and amide III, other conformation-sensitive amide modes have been identified in UVRR spectra of peptide model compounds and proteins. For example, amide II (1500–1600 cm^{-1} interval), which involves substantial C-N stretching (Table I and Fig. 2), is promoted to prominence in UVRR spectra obtained with excitation near 200 nm, and its intensity depends on the protein α-helical content (Copeland and Spiro, 1986; Wang *et al.*, 1991). However, amide II is sometimes obscured by a strong band of tryptophan near 1550 cm^{-1} (normal mode W3). Additional amide-related modes near 1390 cm^{-1}

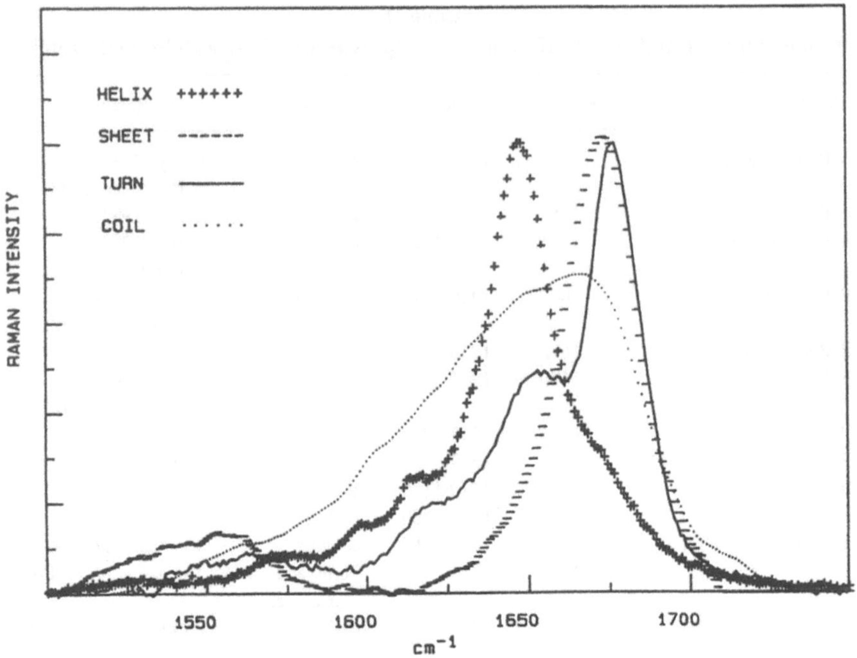

FIGURE 3. Raman amide I profiles of different polypeptide chain conformations. The α-helix (++++++) is represented by the coat protein of Pf1 virus; β-sheet (------------) by β-poly-L-serine; type II β-turn (————————) by cyclo-(VPGVG)$_3$; and irregular or random chain (············) by poly-L-lysine at low pH. (From Thomas *et al.*, 1987.)

have been identified in UVRR spectra, despite negligible intensities in off-resonance Raman spectra. The 1390 cm^{-1} band, assigned to the overtone of amide V (Table I and Fig. 2), is sensitive to the local conformation of the peptide linkage (Song and Asher, 1988, 1989). Spiro and co-workers observed that the intensity of the Raman band at 1390 cm^{-1}, like that of amide II, reflects the protein α-helical content (Wang *et al.*, 1991). Structural interpretations of these and related UVRR data from peptide models have been reviewed (Austin *et al.*, 1993).

3.2. Side-Chain Conformations and Local Environments

A number of correlations have been established between the data of Raman spectra (band intensities and/or frequencies) and the local environments or structures of specific side chains. Using such relationships, the side-chain conformations, amphipathicities, hydrogen bonding strengths, and so forth can be determined from the Raman spectra. In this section, we consider examples of indicator Raman bands of cysteine, tyrosine, tryptophan, and histidine and discuss representative applications. Structurally informative Raman bands of other side chains have been described in recent reviews (Thomas and Kyogoku, 1977; Carey, 1982; Harada and Takeuchi, 1986; Thomas, 1987; Austin *et al.*, 1993).

3.2.1. Cysteine

The S−H stretching vibration of the cysteinyl side chain is particularly well suited to Raman analysis. The normal mode is a nearly pure group frequency; the Raman intensity associated with the vibration is intrinsically high, and the frequency occurs in a region of the Raman spectrum (2500–2600 cm^{-1}), which is essentially devoid of interference from any other Raman bands of the protein or aqueous solvent. The structural value of the Raman S−H stretching band rests upon two key factors: First, the band frequency is highly sensitive to hydrogen bonding of the S−H donor and S acceptor groups (Li and Thomas, 1991). Second, the band intensity is a measure of the molecular concentration of thiol groups in the protein. The first of these factors may be exploited to determine the environment and hydrogen-bonding interactions of the S−H group(s) within the protein structure; the latter may be exploited to measure the pK_a of thiolate titration or the equilibrium constant governing oxidation of a cysteinyl thiol group (H. Li, *et al.*, 1993; Tuma *et al.*, 1993). The S−H stretching frequency is also sensitive, though to a lesser extent, upon the conformation of the cysteine side chain, i.e., upon torsions χ^1 (Cα-Cβ) and χ^2 (Cβ-S) of the Cα-Cβ-S-H side chain (Li and Thomas, 1991; Li *et al.*, 1992). The C−S stretching vibration of the cysteine side chain, which is also relatively easily measured in the region 640–725 cm^{-1} of the Raman spectrum, is much more highly sensitive to side-chain conformation (torsions χ^1 and χ^2) (Li *et al.*, 1992). Normal coordinate and empirical analyses of the cysteine side chain lead to the structural correlations summarized in Table III.

3.2.2. Tyrosine

A correlation between hydrogen bonding of the phenolic OH group of tyrosine and the ratio of intensities (I_{850}/I_{830}) of the tyrosyl Raman bands at 850 and 830 cm^{-1} was elucidated by Siamwiza *et al.* (1975). The correlation is now well established (Harada and Takeuchi, 1986) and is often used in the analysis of protein structures. More recently, several groups have investigated the effects of hydrogen bonding on other Raman bands of the tyrosine side chain, particularly for the bands at 1618, 1600, and 1180 cm^{-1}. These modes, which are designated Y8a, Y8b, and Y9a, respectively, are strongly enhanced in UVRR spectra of proteins.

Table III
Conformation-Sensitive Raman Bands of the Cysteine Side Chain[a]

A. Dependence of the Raman S-H Frequency and Band Width on Hydrogen Bonding

Hydrogen-bonding state of S-H group	S-H frequency (cm^{-1})	Band width (cm^{-1})	Examples
No hydrogen bond	2581–2589	12–17	Thiols in CCl_4 (dilute)
S acceptor	2590–2595	12–17	Thiols in $CHCl_3$
Weak S-H donor	2575–2580	20–25	Thiol neat liquids; thiols in thioethers
Moderate S-H donor	2560–2575	25–30	Thiols in acetone; crystal structures
Strong S-H donor	2525–2560	35–60	Thiols in dimethylacetamide; crystal structures
S-H donor and S acceptor	2565–2575	30–40	Thiols in H_2O

B. Dependence of the Raman S-H and C-S Frequencies on Side-Chain Conformation[b]

Mode	Rotamer	Frequency (cm^{-1})
S-H Stretch	Gauche χ^2	2581
	Trans χ^2	2589
C-S Stretch	Gauche χ^1	650
	Trans χ^1	704

[a]From results of Li and Thomas, 1991 and Li et al., 1992.
[b]Data for 1-propanethiol and 2-methyl-1-propanethiol in dilute CCl_4 solution.

3.2.2a. The Tyrosine Fermi Doublet at ca. 850 and 830 cm^{-1}. For the phenol ring of tyrosine only one Raman fundamental mode (designated Y1) is expected near 840 cm^{-1}. However, a pair of bands near 850 and 830 cm^{-1} is clearly evident in Raman spectra of tyrosine-containing proteins and model compounds. The doublet can be explained by the phenomenon of Fermi resonance, involving the expected Y1 fundamental and the overtone of another fundamental, Y16a, occurring near 420 cm^{-1} (Siamwiza *et al.*, 1975; Harada and Takeuchi, 1986). The accidental coincidence of the Y1 fundamental and the Y16a overtone (2 × 420) results in interaction of the corresponding vibrational states with consequent splitting of the vibrational energy levels, a phenomenon known as Fermi resonance. Both members of the Fermi doublet "share" the intensity originally resident in the unperturbed fundamental; and this partitioning of intensities is strongly dependent on the proximity of Y1 and 2 × Y16a. Since the doublet members are slightly perturbed by the local environment of the phenol ring (specifically by the hydrogen bonding state of the phenolic OH group) the ratio of intensities of the members of the Fermi doublet (I_{850}/I_{830}) is a sensitive indicator of the phenolic OH hydrogen bonding interactions of the tyrosine side chain. In a protein containing a single tyrosine, I_{850}/I_{830} is an unambiguous marker of the hydrogen bonding state. In a protein containing more than one tyrosine I_{850}/I_{830} reflects the average of the hydrogen bonding states of all the tyrosines.

With visible excitation of the Raman spectrum, the ratio I_{850}/I_{830} ranges from 0.3 (phenolic OH is a strong hydrogen bond donor) to 2.5 (strong hydrogen acceptor). With 200 nm excitation, the intensity ratio of the tyrosine doublet is also expected to reflect the state of phenolic OH hydrogen bonding (Rava and Spiro, 1985; Hildebrandt *et al.*, 1988). However, the precise quantitative interrelationship between I_{850}/I_{830} and hydrogen bonding has not yet been established for the case of UV excitation of the Raman spectrum.

3.2.2b. Normal Modes Y8a (ca. 1618 cm^{-1}) and Y8b (ca. 1600 cm^{-1}). Y8a and Y8b are ring skeletal stretching vibrations related to well-characterized benzene vibrations (Takeuchi *et al.* 1988). Here, *a* and *b* denote, respectively, that the vibrational modes are *symmetric* and *antisymmetric* with respect to the twofold rotation axis along the phenolic C–O bond. Y8b involves motion of the OH substituent (Harada and Takeuchi, 1986), and accordingly the frequency is sensitive to the environment (hydrogen bonding state) of the OH group: The stronger the hydrogen bonding of the phenolic OH donor, the lower the vibrational frequency of Y8b (Hildebrandt *et al.*, 1988; Rodgers *et al.*, 1992). A weak correlation exists between the frequency of Y8a and the strength of OH hydrogen bonding, as revealed in UVRR spectra of *p*-cresol solutions reported by Rodgers *et al.* (1992). In fact, Ames *et al.* (1992) reported no sensitivity of Y8a to *p*-cresol hydrogen bonding. This apparent discrepancy could be due to the difference in *p*-cresol concentrations employed in the two studies: 2-mM solute concentration in the former and 10 mM in the latter. At the more elevated cresol concentration and in the presence of apolar solvents, intermolecular hydrogen bonding of cresol is likely. Thus, Ames *et al.* (1992) may have failed to have observed a non-hydrogen-bonded reference state for *p*-cresol. In any case, the apparent frequency shift, if real, must be small, because Y8a involves rather little motion of the OH group (Harada and Takeuchi, 1986; Takeuchi *et al.*, 1989).

3.2.2c. Normal Mode Y7a′ (ca. 1260 cm^{-1}). Y7a′ includes a large contribution from the phenolic C–O stretching motion (Takeuchi *et al.*, 1988). The Raman frequency and bandwidth are both highly sensitive to hydrogen bonding (Takeuchi *et al.*, 1989). The observed range of Y7a′ in UVRR spectra is between 1272 (phenolic OH acts as a hydrogen bond donor) and 1232 cm^{-1} (phenolic O acts as a strong hydrogen bond acceptor). The width of the Raman band also increases as the degree of hydrogen bonding is increased. For example, the band is particularly broad for solutions of *p*-cresol in trifluoroacetic acid, a solvent that both donates and accepts strong hydrogen bonds.

3.2.2d. Normal Mode Y9a (ca. 1180 cm^{-1}). Y9a comprises exocyclic C–H and phenolic O–H bending motions in the plane of the ring (Takeuchi *et al.*, 1988). The Raman band has been correlated with out-of-ring-plane displacement of the phenolic OH group, which could be associated with hydrogen bonding effects (Takeuchi *et al.*, 1989). If the displacement is small, the repulsion between adjacent C–H and OH groups could elevate the frequency of Y9a. In the model compound studies of Takeuchi *et al.* (1989) and Ames *et al.* (1992), Y9a of *p*-cresol was observed at 1170 ± 1 in proton-acceptor solvents, 1173 ± 1 in inert solvents, and 1175 ± 2 cm^{-1} in proton-donor/acceptor solvents.

3.2.3. Tryptophan

3.2.3a. Normal Mode W3 (ca. 1550 cm^{-1}). Both off-resonance Raman and UVRR spectra of proteins give a strong band at 1550 ± 8 cm^{-1}, assignable to tryptophan. Normal coordinate calculations on 3-methylindole indicate that W3 involves C2=C3 stretching in the pyrrole nucleus (Takeuchi and Harada, 1986). A direct correlation between the Raman frequency of W3 and the absolute value of the torsional angle ($|\chi^{2,1}|$) of the C2-C3-Cβ-Cα network has been established by comparing Raman spectra of appropriate tryptophan derivatives with the molecular conformations revealed in X-ray crystal structures (Miura *et al.*, 1989). This relationship, shown in Fig. 4, is useful for determining by

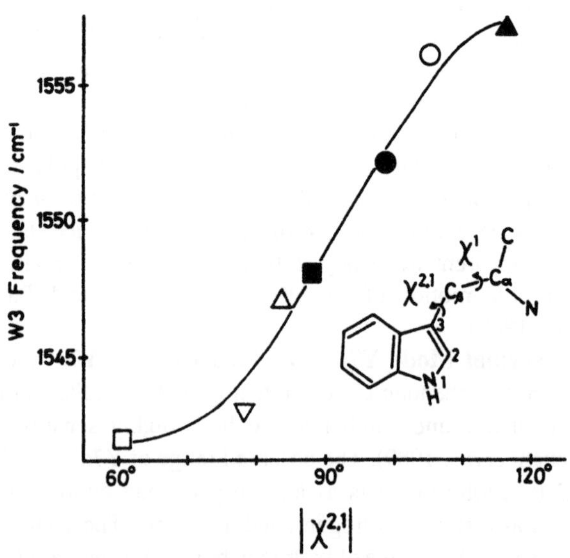

FIGURE 4. Relationship between the frequency of tryptophan normal mode W3 and the side-chain torsion $\chi^{2,1}$. (From Miura *et al.*, 1989.)

Raman spectroscopy the tryptophan side-chain conformation in proteins and their complexes. The correlation is applicable to both protein solutions and solids. In the case of proteins containing more than one tryptophan residue, the Raman peak position, bandwidth, and band shape give information on the distribution of side-chain conformations of the several tryptophan residues. An analysis of the W3 band envelope in the UVRR spectrum of hemoglobin, which contains three inequivalent tryptophans, is discussed in Section 4.3.2.

3.2.3b. Normal Mode W6 (ca. 1430 cm^{-1}). Because of a substantial contribution from indole N−H in-plane bending (Takeuchi and Harada, 1986), the Raman frequency of the tryptophan W6 mode exhibits a strong dependence upon the strength of hydrogen bonding of the NH donor (Miura *et al.*, 1989). Thus, W6 shifts from 1422 (no hydrogen bonding) to 1441 cm^{-1} (strong hydrogen bonding to an appropriate acceptor). The Wb Raman band is easily identified in UVRR spectra employing 266 nm excitation. However, the off-resonance Raman band is usually obscured by more intense Raman C−H bending modes of aliphatic side chains. (It should be noted that the indole N1 atom is not sufficiently electronegative to act as an acceptor of hydrogen bonding, owing to ring delocalization of the nitrogen lone-pair. On the other hand, the π-electrons of the ring can act as a weak proton acceptor.)

The W6 Raman band is observed to undergo a large N-deuteration shift—to 1382 cm^{-1}. The 1382 cm^{-1} band intensity thus provides a convenient means of monitoring hydrogen–deuterium exchange of the tryptophan side chain(s) in proteins, for both off-resonance Raman and UVRR spectra (Takesada *et al.*, 1976; Miura *et al.*, 1988). Unlike W6(NH), W6(ND) manifests no obvious dependency upon hydrogen bonding of the ND donor.

3.2.3c. The Tryptophan Fermi Doublet at ca. 1360 and 1340 cm^{-1}. A pair of Raman bands observed near 1360 and 1340 cm^{-1} in proteins containing tryptophan has been assigned to Fermi resonance involving a fundamental (W7) expected near 1350 cm^{-1} and one or more combination modes of out-of-plane vibrations (Harada *et al.*, 1986). The relative intensity ratio I_{1360}/I_{1340} is a useful and sensitive marker of the amphipathic environment of the ring. In model compounds, the intensity ratio I_{1360}/I_{1340} increases with increasing hydrophobicity, exhibiting values near 0.65–0.92 in hydrophilic environments and values within the range 1.23–1.32 in hydrophobic environments. This correlation has been confirmed for off-resonance spectra of proteins (Miura *et al.*, 1988). In off-resonance Raman spectra, the 1360 cm^{-1} band intensity is often used as the marker of hydrophobicity, because the 1340 cm^{-1} peak is usually overlapped by the C−H bending modes of aliphatic groups (Miura *et al.*, 1988). Upon deuteration of the N1 site, the environmental sensitivity of the doublet is lost. The tryptophan Fermi doublet is also observed in UVRR spectra of proteins (Rava and Spiro, 1985).

It should be kept in mind that Raman spectra of aqueous solutions of

L-tryptophan are of limited usefulness to assess spectra–structure correlations of the indole moiety. This is due to the fact that the indole normal modes are significantly affected by the states of protonation of the amino and carboxyl groups of the amino acid (Harada *et al.*, 1986). Aqueous solutions of simpler indole derivatives lacking dissociable groups, such as acetyltryptophan ethyl ester, serve as more useful model compounds (Uchida *et al.*, 1991).

3.2.3d. Normal Mode W17 (880 cm^{-1}). W17, which is intense in both off-resonance Raman and UVRR spectra of proteins, involves both deformation of the six-membered ring and displacement of the NH group along the direction of the N—H bond (Takeuchi and Harada, 1986). The frequency of W17 is sensitive to N—H hydrogen bond donation (Miura *et al.*, 1988), within the range from 883 cm^{-1} (no hydrogen bonding) to 871 cm^{-1} (strong hydrogen bonding). On the other hand, W17 exhibits a distinct intensity increase in very strong hydrophobic conditions, such as in the hydrophobic core regions of globular proteins (Kitagawa *et al.*, 1979; Miura *et al.*, 1991). N1 deuteration causes a shift of about -20 cm^{-1}. The sensitivity of W17(ND) to hydrogen bonding has not yet been assessed.

3.2.3e. Normal Mode W18 (760 cm^{-1}). W18, which is an indole ring-breathing vibration (Takeuchi and Harada, 1986), generates one of the intrinsically most intense Raman bands in off-resonance spectra of proteins (see Section 1.3.1). Therefore, this band is a potentially useful analytical indicator of tryptophan in proteins. Additionally, the band intensity is sensitive to the amphipathic environment of the indole ring (Miura *et al.*, 1991). In off-resonance spectra, the W18 Raman band increases in intensity if the hydrophobicity of the indole ring environment is decreased, as judged by normalization of the W18 band intensity to that of the W16 mode near 1010 cm^{-1} (Takeuchi and Harada, 1986).

3.2.4. Histidine

In Raman spectra of H_2O solutions of proteins, bands assignable to normal modes of histidine (or histidinium ion) are not clearly evident. This is attributable to both the intrinsically weak Raman cross sections associated with vibrations of the imidazole moiety in comparison to the more intense Raman bands of the aromatic side chains and the relatively small number of histidine residues in most proteins. However, a moderately prominent Raman band near 1408 cm^{-1} can be assigned unambiguously to the N-deuterated imidazolium group (His-d_2^+) in Raman spectra of D_2O solutions of proteins. This band is believed to be due to a vibration involving symmetric stretching of the N-C2-N network coupled with bending of the two N—D groups (Tasumi *et al.*, 1982). The band is particularly useful for monitoring histidine–histidinium ionization equilibria, as well as accessibility of the histidine ring through its hydrogen–deuterium exchange kinetics (Harada *et al.*, 1982).

A linear correlation between the frequency of the 1408 cm^{-1} Raman band and the histidine side-chain conformation has been proposed (Takeuchi *et al.* 1991). Changes in the absolute value of the torsion angle ($|\chi^{2,1}|$) of the Cα-Cβ-C4-C5 network were observed to shift the band from 1406 ($|\chi^{2,1}| = 106°$) to 1398 cm^{-1} ($|\chi^{2,1}| = 61°$). The frequency of the C4=C5 stretching mode of His-d_2^+ near 1590 cm^{-1} (Tasumi *et al.*, 1982) has also been correlated with $|\chi^{2,1}|$ (Takeuchi *et al.*, 1991). The latter is similar to the above-mentioned relationship between W3 of tryptophan and its $|\chi^{2,1}|$ torsion (Miura *et al.*, 1989).

4. APPLICATIONS TO PROTEIN STRUCTURES AND ASSEMBLIES

In this section we discuss representative Raman applications that illustrate the scope of the method and delineate many of the principles and practical considerations considered in the preceding sections.

4.1. Thioredoxin

The complete Raman spectrum of a typical monomeric protein, *Escherichia coli* thioredoxin in aqueous solution, is shown in Fig. 5 (H. Li, *et al.*, 1993). The sample concentration is 100 μg/μl. The abscissa is the Raman frequency in wave number units (cm^{-1}); the ordinate is the Raman intensity in arbitrary units. The majority of Raman bands, clustered between 500 and 1750 cm^{-1}, originate from vibrational modes of the protein main chain and various side chains. Detailed band assignments can be made by analogy with Raman data on model compounds.

The band at 2331 cm^{-1} is due to ambient nitrogen and serves as both a frequency (cm^{-1}) and relative intensity standard. The broad band at 2569 cm^{-1} is due to overlapping contributions from the two cysteinyl SH groups at the thioredoxin active site. (Further discussion of this complex Raman SH band is given below.) The bands near 2800–3000 cm^{-1} are due to the multitude of C−H stretching vibrations of all aliphatic groups (CH, CH_2, and CH_3) in the protein. The very intense and complex Raman bands between 3100 and 3600 cm^{-1} are due mainly to O−H stretching (symmetric and antisymmetric) vibrations of solvent water molecules.

The appearance in a protein Raman spectrum of bands assigned to specific types of side chains constitutes a basis for determining the number of such side chains in the protein. For example, the integrated intensity of the Raman SH

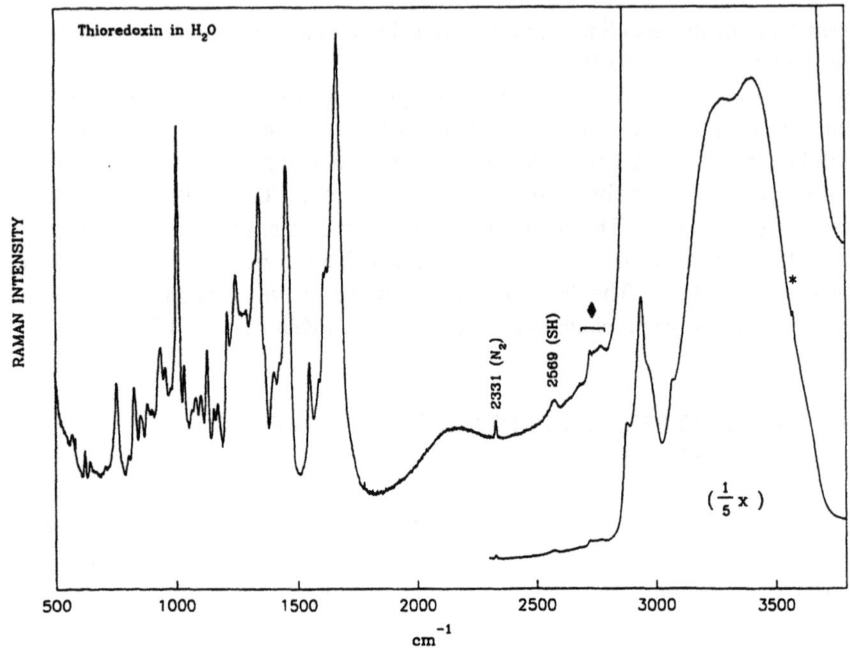

FIGURE 5. Raman spectrum in the region 500–3700 cm^{-1} of reduced thioredoxin at pH 4 and 6 °C (sequence: [1]SDKIIHLTDD[11]SFDTDVLKAD[21]GAILVDFWAE[31]WCGPCKMIAP[41]ILDEIADE-YQ[51]GKLTVAKLNI[61]DQNPGTAPLY[71]GIRGIPTLLL[81]FKNGEVAATK[91]VGALSKGQLK[101]EF-LDANLA). Protein concentration is ≈100 μg/μl. The Raman band of N_2 at 2331 cm^{-1} serves as an internal frequency and intensity reference (Tuma *et al.*, 1993). The complex Raman band at 2569 cm^{-1} is due to Cys32 and Cys35 sulfhydryl stretching modes. Raman bands just below 3000 cm^{-1}, which are due to aliphatic CH stretching modes of most side chains, and the very broad, complex band of H_2O solvent ~ 3100–3600 cm^{-1} are evident in the segment recorded at one-fifth amplification. Probable overtone or combination bands (♦) and a laser emission line (*) are also indicated. The spectrum shown was excited at 514.5 nm and represents the isotropic Raman scattering of thioredoxin, *i.e.*, the electric vector of the scattered light is collected for all (2π) orientations perpendicular to the direction of beam propagation. (From H. Li, *et al.*, 1993.)

band (2569 cm^{-1}) in the thioredoxin spectrum of Fig. 5 confirms the presence of two sulfhydryls in the protein at the conditions of this experiment (pH 4). Similar quantitative correlations exist for other types of amino acid residues.

The Raman spectrum is also diagnostic of both covalent modification and site-directed mutagenesis of side chains. One example is illustrated in Fig. 6, which compares the Raman spectra of oxidized [TRX$(S)_2$] and reduced [TRX$(SH)_2$] forms of thioredoxin. In this illustration, the 300–1750 cm^{-1} region of each Raman spectrum is shown, and the spectra are corrected for both the Raman scattering of the solvent and the gently sloping background underlying the Raman band intensities (cf. Fig. 5). The computed difference spectrum

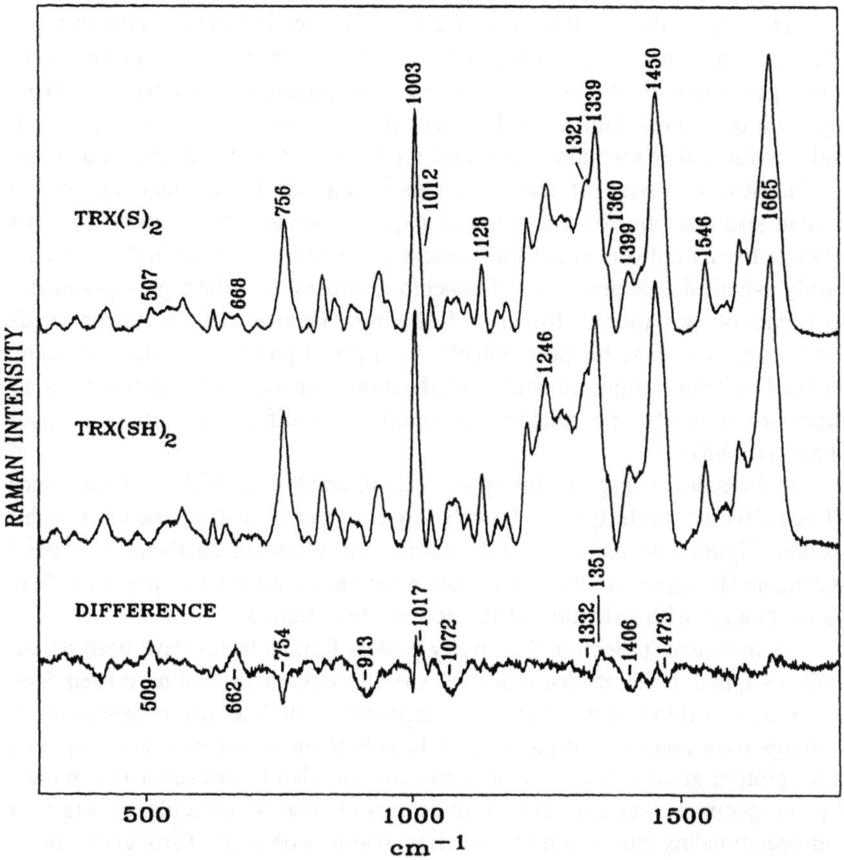

FIGURE 6. Raman spectra in the region 300–1750 cm^{-1} of oxidized (top) and reduced (middle) thioredoxin at pH 7.0 and 6 °C, and their difference spectrum (bottom). Protein concentrations are ≈100 μg/μl. Labels indicate frequencies of difference bands (peaks and troughs) which have been assigned to structural changes localized mainly near the active site. (From H. Li, *et al.*, 1993.)

(bottom trace of Fig. 6) shows clearly the Raman band centered near 509 cm^{-1}, which is assignable to the disulfide (S−S) stretching vibration of $TRX(S)_2$. Oxidized thioredoxin also lacks the sulfhydryl Raman band complex near 2569 cm^{-1} (data not shown).

Figure 5 shows further that the Raman spectrum of an aqueous protein contains prominent bands of both the protein solute and aqueous solvent, thus constituting a basis for determination of protein concentration. A roughly linear relationship exists between the protein concentration and the ratio ($I_{2800-3000}/I_{3100-3600}$) of Raman intensities of protein C−H and water O−H bands, over the concentration range ≈ 0.1–10 weight-percent protein.

The sensitivities of Raman amide I and amide III bands to protein main-chain conformation have been extensively documented and discussed. Many investigators have described quantitative or semiquantitative methods for extracting from the Raman amide band profiles the percentages of α-helix, β-strand, and possibly other secondary structure types present in the protein (Bandekar, 1992). However, none of these methods is universally accepted, and use of Raman spectroscopy for this purpose requires considerable caution. In cases where a highly uniform secondary structure is present, such as in the predominantly α-helical coat proteins of filamentous viruses or in the highly β-stranded structures of the tailspike trimer of P22 virus (Thomas, 1987), Raman amide band analysis tends to be more reliable. Changes of protein secondary structure, induced by temperature, covalent modification, mutation, or intermolecular association, may also be reliably measured using difference Raman methods (Thomas, 1987).

In thioredoxin (Fig. 6), the appearance of amide I at 1665 cm^{-1} and amide III at 1246 cm^{-1} reflects the relatively large content of β-sheet and turns in this protein. Figure 6 also shows, by the absence of difference bands in the amide I and amide III regions, that the thioredoxin secondary structure undergoes little or no net change with oxidation of the active-site cysteines.

As indicated in Section 3.2, many protein Raman bands have been definitively assigned to vibrational modes of specific side chains and have been demonstrated to exhibit either frequency dependence or intensity dependence, or both, upon the precise configuration of the side chain and/or its interactions with other protein groups. When such bands are confidently measured in a protein Raman spectrum, they can serve as indicators of detailed molecular geometry or hydrogen bonding interaction of the corresponding side chain. Particularly useful in the case of thioredoxin are bands of cysteine, cystine, tryptophan, and tyrosine.

Thus, the complex 2569 cm^{-1} band of thioredoxin (Fig. 5) is diagnostic of relatively strong S−H hydrogen bonding (Table III) (Li and Thomas, 1991; Li *et al.*, 1992) for both of the active-site cysteine residues. Detailed analysis of the band as a function of pH (Figs. 7 and 8) permits further characterization of the hydrogen bonding strengths of these sulfhydryls and determination of the pK_a values governing thiol/thiolate equilibria (H. Li, *et al.*, 1993). The 507 cm^{-1} band of $TRX(S)_2$ identifies the all gauche (g/g/g) rotamer of the disulfide bridge (Cα-Cβ-S-S-Cβ-Cα) (Sugeta *et al.*, 1972) in the oxidized protein. The difference bands at 754, 1332, and 1351 in Fig. 6 identify more hydrophobic environments for tryptophan residues (Miura *et al.*, 1988, 1991) of the oxidized protein than those of the reduced protein. The difference band near 1406 cm^{-1} identifies structural changes in the caroboxyl groups of acidic side chains (Asp/Glu). Finally, the tyrosine doublet near 850 and 830 cm^{-1} (Siamwiza *et al.*, 1975) exhibits intensities that identify the phenoxyl groups of both tyrosines, Tyr49 and

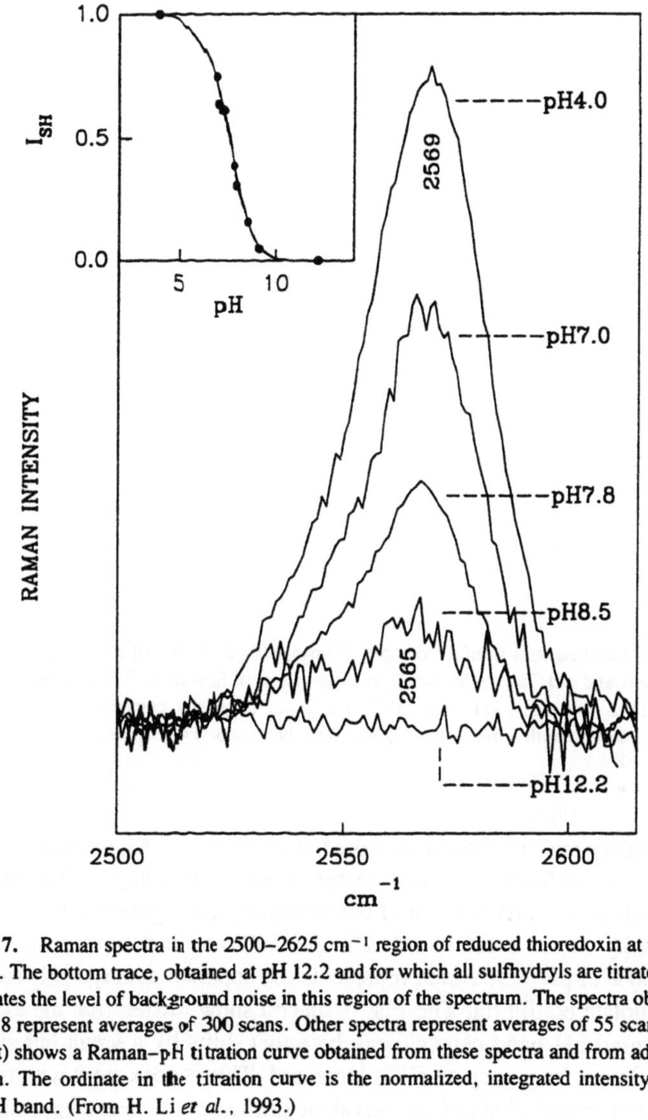

FIGURE 7. Raman spectra in the 2500–2625 cm^{-1} region of reduced thioredoxin at the indicated pH values. The bottom trace, obtained at pH 12.2 and for which all sulfhydryls are titrated to thiolate ion, indicates the level of background noise in this region of the spectrum. The spectra obtained at pH 4.0 and 7.8 represent averages of 300 scans. Other spectra represent averages of 55 scans. The inset (upper left) shows a Raman–pH titration curve obtained from these spectra and from additional data not shown. The ordinate in the titration curve is the normalized, integrated intensity of the total Raman SH band. (From H. Li *et al.*, 1993.)

Tyr70, as donors of strong hydrogen bonds in both the oxidized and reduced forms of thioredoxin. Thus, both tyrosines are sufficiently buried to prevent phenoxyl contacts with solvent molecules.

Figure 7 shows the complex and asymmetric Raman sulfhydryl band observed for reduced thioredoxin in solutions of different pH. At pH 4.0, where both Cys32 and Cys35 are fully protonated, the band exhibits maximum inte-

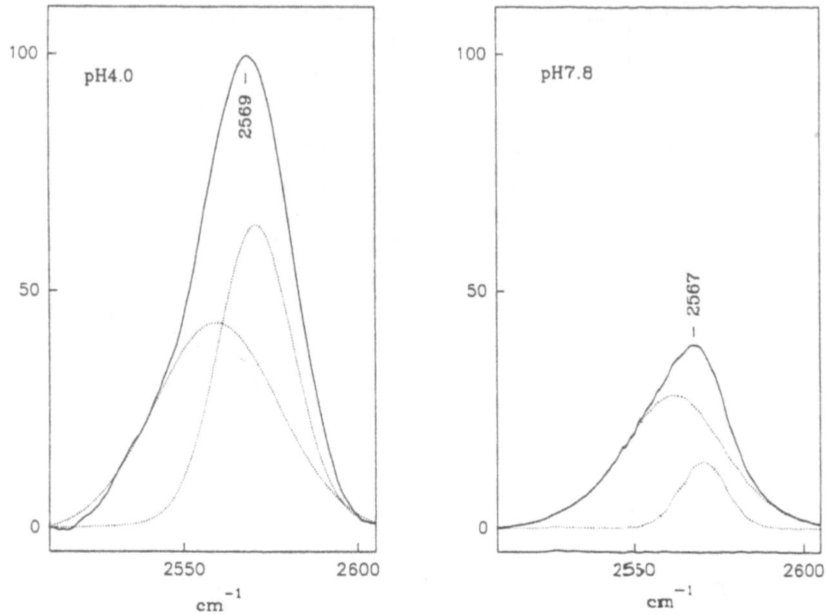

FIGURE 8. Least-squares fit of the complex Raman sulfhydryl band of reduced thioredoxin at pH 4.0 (left panel) and pH 7.8 (right panel) to Gauss-Lorentz functions. The component bands are centered at 2562 ± 1 and 2571 ± 1 cm^{-1}. The lower frequency component exhibits the larger bandwidth, consistent with a more strongly hydrogen-bonded S-H donor, assigned to Cys35. (From H. Li *et al.*, 1993.)

grated intensity and its peak is observed at 2569 cm^{-1}. As pH is increased, the intensity of the sulfhydryl band decreases, reaching zero at pH 12.2. Included in Fig. 7 (inset, upper left) is a plot of the normalized, integrated sulfhydryl Raman band intensity versus solution pH. The data indicate that the intensity is diminished by 50% at pH 7.5, which represents the apparent average pK_a for the two thiol → thiolate equilibria. The Fig. 7 spectra show further that the experimentally measured SH band center (peak bisector) shifts by a small but significant amount to lower frequency as pH is increased. This is consistent with the contribution of two populations of inequivalent sulfhydryls to the composite Raman band, and with the higher frequency component titrating preferentially to the thiolate form as pH is increased. In order to distinguish the two pK_a values corresponding to Cys32 and Cys35 thiolate formation, the thiol band complex can be decomposed into two components as a function of pH, as shown in Fig. 8. The data are interpreted to reveal $pK_1 = pK_{C32} = 7.1 \pm 0.2$ and $pK_2 = pK_{C35} = 7.9 \pm 0.2$ for the respective thiol → thiolate equilibria (H. Li, *et al.*, 1993).

4.2. Viruses

Raman spectroscopy has been applied extensively in structural studies of viruses and viral precursors. The methodology is unique in its ability to provide detailed structural information about both protein and nucleic acid constituents of native virions over a broad range of sampling conditions. A comprehensive review of the literature to around 1986 has been given previously (Thomas, 1987). Practical considerations in Raman spectroscopy of viruses and viral precursors have also been discussed (Hartman and Thomas, 1985). More recent applications have also been surveyed (Thomas and Tsuboi, 1993).

fd is a member of the Ff family of class I filamentous bacteriophages that includes also the members f1 and M13. The fd filament, $\approx 880 \times 7$ nm, comprises ≈ 2750 copies of a subunit (gpVIII) of 50 amino acids, which serves as the major coat protein, a few copies of each of four minor proteins at the filament ends, and a ssDNA genome of 6408 nucleotides packaged within the coat protein sheath. Since the gpVIII subunits constitute about 88% of the virion mass, the Raman bands associated with its sequence (^{1}AEGDDPAKAAFDSLQASATE-YIGYAWAMVVVIVGATIGIKLFKKFTSKAS^{50}) and conformation are expected to be the dominant features of the Raman spectrum of the virus. An extensive literature exists on the Raman spectroscopy of fd and other filamentous viruses (Thomas, 1987; Hartman and Thomas, 1985).

The Raman spectrum of the normal fd virion in the region 300–1800 cm^{-1} is shown in Fig. 9A. Included as Fig. 9C is the Raman spectrum of an isotopic fd derivative that incorporates deuterium in place of protium at the five hydrogenic ring sites of the single tryptophan residue (Trp26) in each gpVIII subunit. This fd isotopomer is referred to as fd(W_{d5}). Included as Fig. 9B is the Raman spectrum of the isotopomer fd($10A_{d3}$), which incorporates deuteriomethylalanine at each of the 10 alanine sites (Ala 1, 7, 9, 10, 16, 18, 25, 27, 35, 49) per gpVIII subunit. Also included as Fig. 9D is the Raman spectrum of the isotopomer fd(W_{d5} + $10A_{d3}$), which incorporates both deuterio-tryptophan and deuterio-alanine. The Raman frequency and intensity shifts accompanying these isotope substitutions, marked with arrows in the Fig. 9 spectra, identify unambiguously the bands assigned to the tryptophan and alanine residues. The isotope shifts in the spectral region 900–1800 cm^{-1} are depicted by difference Raman spectra in Fig. 10 and a detailed analysis has been given (Aubrey and Thomas, 1991). Extension of this approach to other side-chain types permits complete vibrational assignment of all bands in the Raman spectrum of the virus.

A number of interesting conclusions can be reached from the data of Figs. 9 and 10 and companion data obtained in other regions of the spectrum. First, the purity of heavy-isotope incorporation approaches 100% using standard microbiological methods. Second, it is clear that the Raman spectral contributions of a

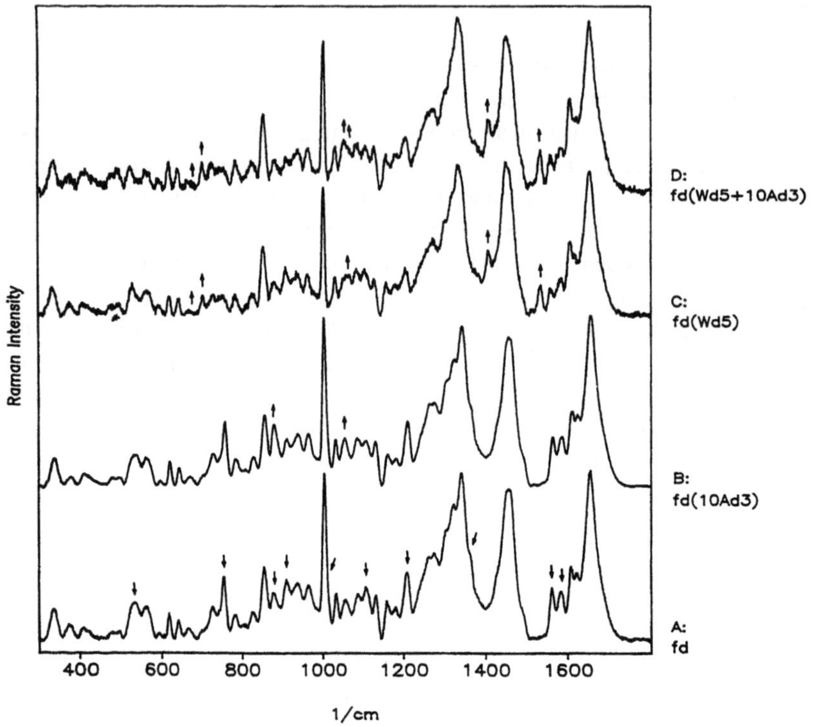

FIGURE 9. Raman spectra in the region 300–1800 cm^{-1} of (A) normal fd and its isotopomers, (B) fd(W_{d5}), (C) fd($10A_{d3}$), and (D) fd($W_{d5}+10A_{d3}$), as indicated. Phage are dissolved at 80 mg/ml in 10 mM Tris, pH 7.8, and 32 °C. Each spectrum is the average of 8 scans and is corrected for the solvent background. Downward-pointing arrows in the fd spectrum mark bands contributed by tryptophan and/or alanine residues which are eliminated, shifted in frequency, or diminished in intensity by incorporation of the labeled side chain(s). Upward-pointing arrows indicate bands in the isotopomer spectra assigned to W_{d5} and A_{d3}, which are generated by the labeled side chains. (From Aubrey and Thomas, 1991.)

single aromatic residue can greatly dominate over those of even a tenfold greater number of aliphatic side chains in the same protomer. Third, in regions of the spectrum where key conformational markers of the packaged ssDNA genome are expected to overlap with those of aromatic amino acid side chains, for example 600–900 cm^{-1}, the two can be effectively resolved from one another by appropriate isotope substitutions in coat protein side chains. Fourth, definitive assignments for the key bands of tryptophan (Trp26) permit structural conclusions to be reached about this residue in the native fd virus and in related filamentous viruses. This is illustrated by the examples given in the following two sections. Finally, the relatively sharp and intense amide I band at 1651 cm^{-1} and the

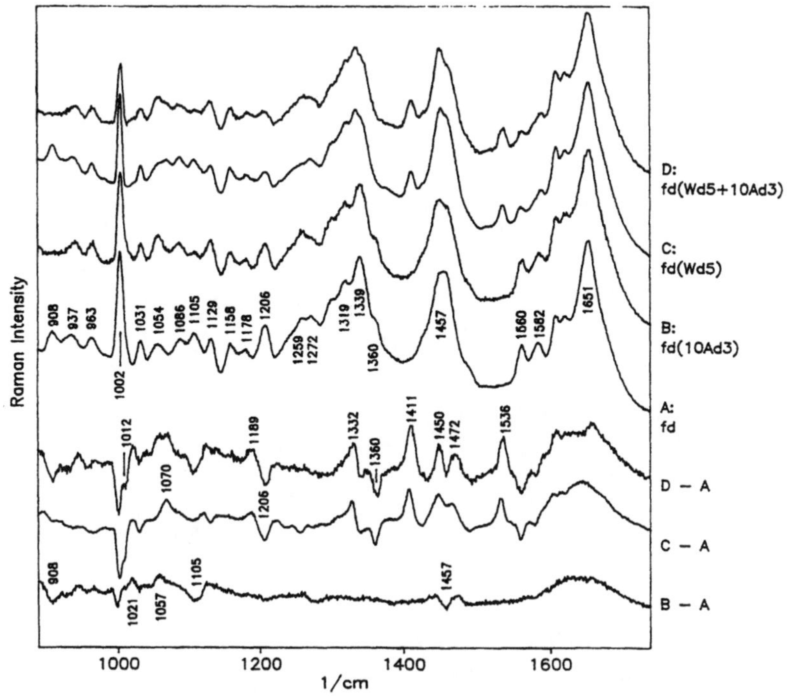

FIGURE 10. Raman spectra in the region 900–1750 cm^{-1} of: (A) normal fd, (B) fd($10A_{d3}$), (C) fd(W_{d5}), and (D) fd($W_{d5}+10A_{d3}$), obtained at the same conditions indicated in the legend to Fig. 9. In order to preserve natural band shapes, the spectra have not been corrected for solvent background or otherwise refined. Traces A, B, C, and D are averaged from as many as 70 multiple scans. The lower three traces show on an expanded ordinate scale the difference spectra computed with the deuterium isotopomer as minuend and normal fd as subtrahend. For computation of the difference spectra, intensities were scaled to compensate the amide I band at 1651 cm^{-1}, which is invariant to the isotope substitutions. The very broad difference bands in the region above 1600 cm^{-1} are due mainly to uncompensated H_2O. (From Aubrey and Thomas, 1991.)

companion amide III markers (1275–1300 cm^{-1}) identify the gpVIII subunit secondary structure in the virion as a virtually complete α-helix.

Figure 11 shows Raman spectra of fd in solutions of different ionic strength. The effect of salt is to alter a number of key Raman bands, including the conformational markers of the Trp26 side chain assigned on the basis of isotope substitution effects. Comparison of these results with standard correlations described in Section 3.2.3 leads to the following conclusions regarding Trp26. In solutions of low ionic strength (physiological conditions), the indole NH donor is engaged in a moderately strong hydrogen bond (876 cm^{-1} band), the indole ring environment is less hydrophobic (weak 1360 cm^{-1} band), and the side-chain

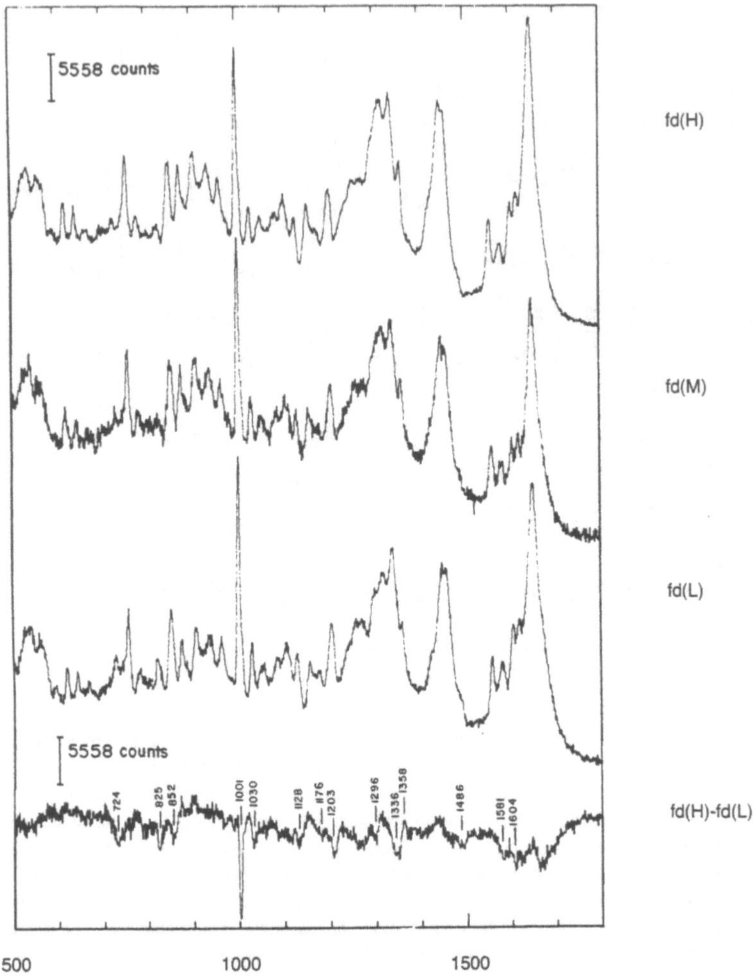

FIGURE 11. Raman spectra of fd virus in the region 500–1800 cm^{-1} for solutions containing different salt concentrations. From top to bottom: fd in high salt (H), fd in moderate salt (M), fd in low salt (L), and the difference spectrum between high and low salt. (From Thomas *et al.*, 1983.)

torsion $|\chi^{2,1}| = 120°$ (1560 cm^{-1} band). At high salt, NH is a weak hydrogen-bond donor, the indole environment is hydrophobic and $|\chi^{2,1}| = 100°$. Therefore, in existing fiber diffraction models of fd (D. A. Marvin, personal communication), the plane of the indole ring would be constrained roughly along the filament axis and adjacent to a cleft near the N-terminus of a neighboring subunit. Accordingly, the indole ring should be accessible to solvent. The Raman results

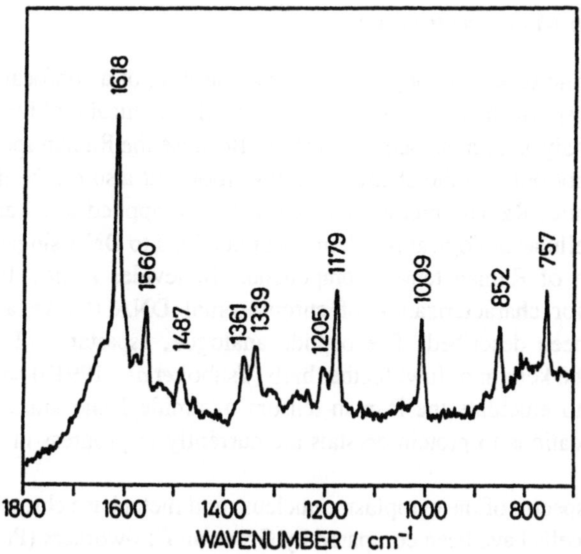

FIGURE 12. UVRR spectrum of fd virus in the region 700–1800 cm^{-1}, excited at 240 nm. (Unpublished results of A. Toyama, H. Takeuchi, and G.J. Thomas, Jr.)

suggest a mechanism for the salt-induced structure transition: At high salt, there occurs a displacement of hydrophilic agents (probably H_2O molecules) away from the cleft resulting in enhanced hydrophobic packing of the coat subunits.

As an example of the ultraviolet resonance Raman spectrum of a nucleoprotein complex, Fig. 12 shows the UVRR spectrum of fd obtained with 240 nm excitation (A. Toyama, H. Takeuchi, and G. J. Thomas, Jr., unpublished results). With this excitation wavelength, the most prominent bands expected in the fd spectrum are those of tryptophan (Trp26) and tyrosine (Tyr21 and Tyr24) residues. The data of Fig. 12 confirm and extend the fd assignments achieved by off-resonance Raman spectroscopy of fd and its isotopomers (Thomas *et al.*, 1983). Similar UVRR data obtained on oriented fd filaments permit the measurement of Raman anisotropies for specific aromatic residues, such as Trp26, thus facilitating determination of the preferred direction of orientation of the Trp26 indole ring in the virion assembly. To complement the results of Fig. 12, the UVRR method may also be implemented with 260 nm excitation, to enhance selectively the key Raman bands of the packaged ssDNA genome of fd.

More recently, Raman spectroscopy has been successfully exploited for analysis of protein structures and interactions in membrane-containing DNA and RNA viruses (Bamford *et al.*, 1993; T. Li, *et al.*, 1993a). Measurement of assembly-related subunit conformational changes affecting as few as 2% of the peptide groups can be detected (Prevelige *et al.*, 1993).

4.3. Raman Microspectroscopy

Raman microspectroscopy, which is the coupling of a confocal microscope to the Raman spectrometer, constitutes a powerful analytical tool for high resolution spatial analysis of microscopic samples. Because the Raman spectra provide information not only on the chemical composition but also on the detailed molecular structure, Raman microspectroscopy can be applied to a variety of molecular and cellular biological problems. Applications to DNA single crystals for determination of Raman tensor components (Benevides *et al.*, 1993) and to living cells for characterization of chromosomal DNA (Greve and Puppels, 1993) have been described. The peptide analogue, aspartame (Tsuboi *et al*,. 1991a), and the keratin of fowl feather barb (Tsuboi *et al.*, 1991b) have also been investigated to elucidate the Raman tensors of amide I and amide III modes. Similar applications to protein crystals are currently in progress in a number of laboratories.

Raman spectra of the cytoplasm, nucleus, and metaphase chromosome from single living cells have been compared by Greve and co-workers (Puppels *et al.*, 1990). The Raman spectrum obtained from a band of polytene chromosome of *Chironomus* was compared with that from an interband, and a significant difference was found in the respective DNA/protein ratios. Sureau *et al.* (1990) also succeeded in obtaining 257 nm-excited UVRR spectra of the nucleus of single living cells of a human tumor. These approaches have the potential of providing new insights into mechanisms of gene regulation in living cells.

The Raman microprobe approach has also been used for analysis of the structure of human gallstones. Generally, gallstones consist of several components, and the gallstone cross section shows a concentric layered and/or radiative structure that often contains islandlike small particles. The analysis of these microstructures is potentially useful for understanding mechanisms of gallstone growth. Ishida *et al.* (1987) obtained Raman spectra of the small white particles (20–100 μm diameter) in bilirubin and cholesterol–bilirubin gallstones. The spectra obtained were very similar to the Raman spectrum of stearate, suggesting that the principal molecular component of the white particles may be fatty acid.

4.4. Dynamic Processes

4.4.1. Hydrogen–Deuterium Exchange Kinetics

The deuteration reactions occurring at peptide NH sites, as well as those occurring at exchangeable side chain sites (tryptophanyl NH, tyrosyl OH, cysteinyl SH, etc.) in D_2O solutions of proteins, can be monitored independently by Raman spectroscopy. The rate of exchange is expected to depend strongly on the hydrogen bonding environment of the labile proton. In general, the exchange of a labile hydrogen located in the core of a protein structure or involved in a strong hydrogen bond at the protein surface is slow compared with that of protons exposed to solvent water molecules or not strongly hydrogen bonded. The retardation factor can amount to several orders of magnitude in the exchange rate constant, sometimes the equivalent of an exchange half-life of years. The hydrogen–deuterium exchange rate is thus a sensitive probe of local environment in proteins. [For more detailed discussion of hydrogen exchange phenomena in proteins, the reader is referred to the reviews of Englander and Kallenbach (1984) and Englander and Mayne (1992).]

Recent Raman applications have identified slowly exchanging fractions of cysteine sulfhydryl groups in the tailspike protein of the dsDNA bacteriophage P22 (Thomas *et al.*, 1990) and exchange-resistant peptide NH groups in the icosahedral capsid of bean pod mottle virus (T. Li, *et al.*, 1993b). Surprisingly, the capsid proteins of BPMV resist virtually completely the exchange of any peptide NH groups, despite the free flow of D_2O solvent molecules through the capsid shell.

Use of hydrogen–deuterium exchange kinetics in conjunction with Raman spectroscopy is also useful for the separation and assignment of component bands in a complex (overlapping) band envelope. Egg white lysozyme, for example, yields a composite W17 band near 875 cm^{-1} with contributions from each of the six tryptophan residues per subunit (Fig. 13). This composite band is actually broader in the native protein than in LiCl-denatured lysozyme (data not shown), indicating that a greater variety of tryptophan NH hydrogen bonding states (see Section 3.2.3.) prevails in the native structure (Miura *et al.*, 1988). However, the individual hydrogen bonding states cannot be deciphered from this single spectrum.

Figure 13b shows the Raman spectrum of lysozyme in D_2O solution, for which an estimated three tryptophans (Trp62, Trp63, and Trp123) are quickly deuterated. The percentages of deuteration of each tryptophan, evaluated from the intensities of ^{1}H-NMR signals, are 100% for Trp62 and Trp63, 85% for Trp123, 15% for Trp108, 5% for Trp111, and 0% for Trp28 (Miura *et al.*, 1991). Because the W17 Raman band shifts -20 cm^{-1} upon deuteration, the remaining W17 components at 879 and 871 cm^{-1} are assignable to the undeuterated three tryptophans. Upon the further deuterations of Trp108 (Fig. 13, b → c) and Trp111 (c → d), the higher frequency component loses intensity. The lower frequency component, at 871 cm^{-1}, disappears only upon deuteration of Trp28 (d → e), indicating that Trp28 is incorporated in a very strong hydrogen bond.

Raman spectra of stepwise deuterated lysozymes bound with a substrate-related inhibitor are shown in the right panel of Fig. 13. The W17 bands, due to Trp108, Trp111, and Trp28 exhibit no frequency shift upon inhibitor binding. However, an intensity increase of the 871 cm^{-1} band is seen clearly in the

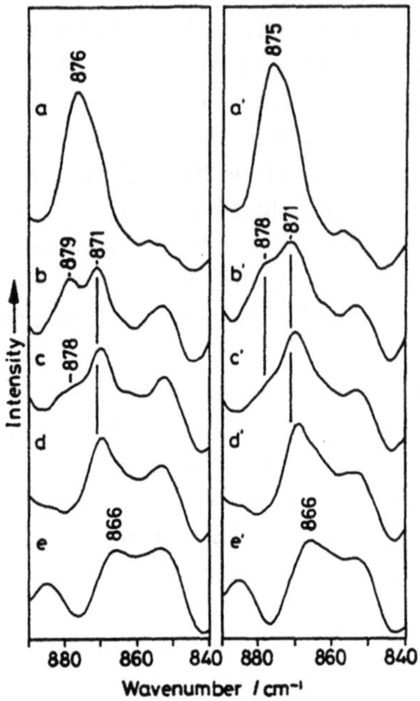

FIGURE 13. Comparisons of the Raman spectra (890–840 cm^{-1}) of egg white lysozyme (left) and lysozyme bound with N,N′,N″-triacetyl-chitotriose (right), following stepwise deuteration of tryptophan residues. (a) H_2O solution (no trytpophan deuteration); (b) D_2O solution in which Trp62, Trp63, and Trp123 are deuterated; (c) D_2O solution in which Trp62, Trp63, Trp123, and Trp108 are deuterated; (d) D_2O solution in which Trp62, Trp63, Trp123, Trp108, and Trp111 are deuterated; and (e) D_2O solution in which all tryptophans are deuterated. (From Miura *et al.*, 1991.)

spectra, reflecting stepwise deuterated lysozyme (cf. Figs. 13b and 13b′). The intensity of W17 is strong in a hydrophobic environment (Kitagawa *et al.*, 1979). Thus, the environment of Trp28 in the lysozyme-inhibitor complex is more hydrophobic than that in free lysozyme. Trp28 is located in the hydrophobic core of lysozyme and is in close contact with the side chain of Met105. Met105 is part of a flexible loop that is an important substrate recognition element. The Raman results of Fig. 13 may be considered indicative of a strong interaction between Trp28 and Met105 in the lysozyme-inhibitor complex. This study suggests that side chain interactions at the protein core may have a role in recognition and binding of substrates by the flexible loop.

4.4.2. Photoreactions of Proteins

4.4.2a. Bacteriorhodopsin. The purple membrane protein of *Halobacterium halobium* covalently binds a retinal chromophore to a lysine residue of the apoprotein via a Schiff base linkage. Photon absorption by retinal generates the photocycle, bR ($bR_{548} \leftrightarrow bR_{568}$) $\rightarrow K_{630} \rightarrow L_{550} \rightarrow M_{412} \rightarrow N_{550} \rightarrow 0_{640} \rightarrow$ bR. The protonation state of the Schiff base and the configuration of the conjugated polyene chain of retinal in the intermediates have been well established by using resonance Raman spectroscopy with visible excitation (summarized in Carey, 1982). Recently, several groups have reported UVRR studies of bacteriorhodopsin. UVRR offers the additional advantage over visible resonance Raman methods of probing the apoprotein structure and its interactions with retinal in the photocycle intermediates. Here, Raman spectroscopy offers clear-cut advantages over X-ray and NMR methods.

Recently, Ames *et al.* (1992) reported nanosecond- and millisecond-time-resolved UVRR spectra of bacteriorhodopsin. The nanosecond time scale spectra were obtained with a visible pump pulse followed by a time-delayed UV probe pulse (Fig. 14A). The 240 nm excitation is generated by frequency conversion of the 1064 nm emission of the Nd:YAG laser. After H_2 shifting, each pulse contains a 532-nm excitation line (second harmonic of 1064 nm) besides the 240-nm

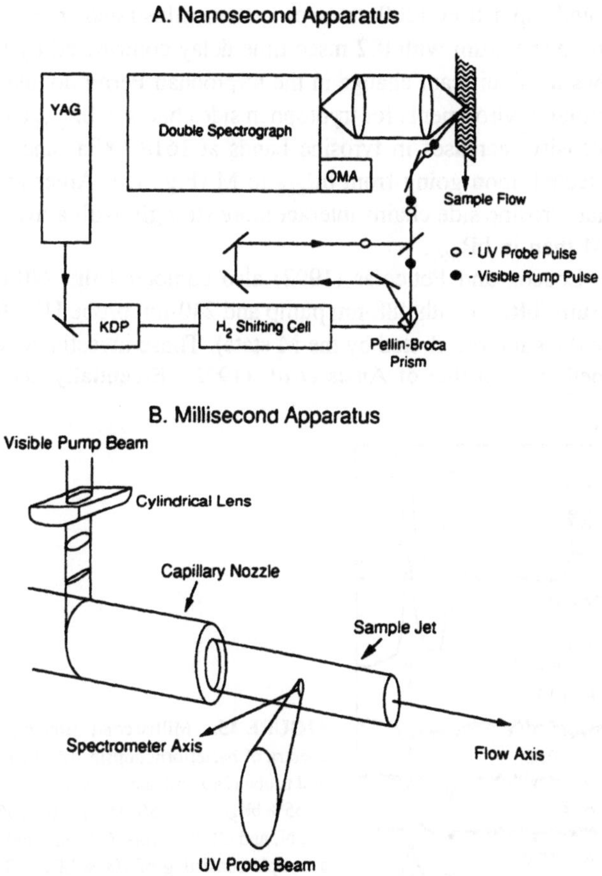

FIGURE 14. Scheme of a typical time-resoloved UVRR apparatus for (A) nanosecond and (B) millisecond experiments. Abbreviations: OMA, optical multichannel analyzer; KDP, potassium dihydrogen phosphate crystal (nonlinear frequency doubler). (From Ames *et al.*, 1992. Reprinted with permission from the American Chemical Society.)

line. The 240- and 532-nm lines are delineated with a prism and optically delayed. The delay time is varied by changing the difference in path length from the prism to the sample for the pump and probe beams. For example, a 3-m difference corresponds to a 10-nsec time delay. Ames and co-workers compared the pump and probe spectra of bacteriorhodopsin, which are contributed by the K state, with the probe only spectrum obtained from bR_{568}. Although FTIR studies have suggested a difference in ionization state of a tyrosine residue between the bR_{568} and K states (Braiman *et al.*, 1988), no significant changes of either tryptophan or tyrosine UVRR bands were observed (data not shown).

Time-resolved spectra on the millisecond time scale were obtained by combination of a dual-beam (514.5 or 568 nm CW pump laser and 240 nm pulse probe laser) and rapid-flow (200 cm/sec) system, illustrated in Fig. 14B. The pump and probe spectrum with 0.2 msec time delay contributed by the M state (≈30%) shows no significant change in the tryptophan Fermi doublet (Fig. 15), indicating similar environments for tryptophan side chains in bR_{568} and M states. However, intensity increases in tyrosine bands at 1618 (Y8a) and 1176 cm^{-1} (Y9a) are detected upon going from bR_{568} to M (Fig. 15). Ames *et al.* (1992) concluded that tyrosine side chains interact more strongly with a hydrogen-bond acceptor in M than in bR_{568}.

Moënne-Loccoz and Peticolas (1992) also compared the 240-nm excited UVRR spectrum (bR_{568}) with 488-nm pump and 240-nm probe UVRR spectrum of bacteriorhodopsin (contributed by the M state). These investigators employed a different method than that of Ames *et al.* (1992). Essentially, the pump and

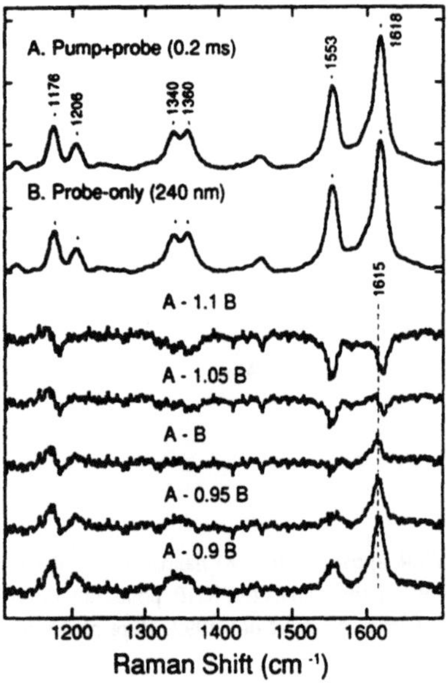

FIGURE 15. Millisecond time-resolved UVRR spectra of bacteriorhodopsin. (A) Pump (514.4 nm) and probe (240 nm) spectrum of a sample consisting of 55% bR_{568}, 30–35% M, and 10–15% other states (L, N, and O). (B) Probe (240 nm) only spectrum of a sample consisting of 100% bR_{568}. The difference spectra (A–B) indicate that intensity increases occur at 1615 and 1172 cm^{-1}. The 1553 cm^{-1} band (W3) is chosen as an internal intensity standard. (From Ames *et al.*, 1992. Reprinted with permission from the American Chemical Society.)

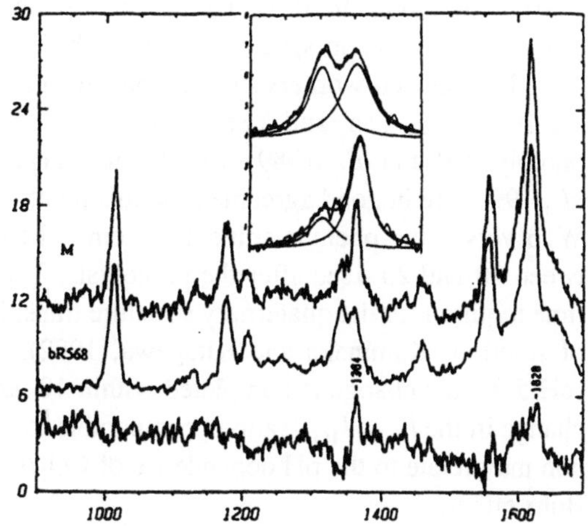

FIGURE 16. UVRR spectra (240 nm excitation) of bacteriorhodopsin with a UV probe only (bR_{568}, middle spectrum) and with a 488-nm pump and UV probe (contributed by the M state, top spectrum). The bottom trace shows the difference spectrum, M minus bR_{568}. (from Moënne-Loccoz and Peticolas, 1992. Reprinted with permission from the American Chemical Society.)

probe spectrum was obtained by defocusing a 488-nm argon laser beam (1.5-mm diameter) to surround a 240-nm probe beam (1-mm diameter). Because its lifetime is the longest among intermediates, the M state is dominant at this sampling condition. In contrast to the result obtained by Ames *et al.* (1992), Moënne-Loccoz and Peticolas (1992) observed a drastic change in the intensity ratio of the tryptophan doublet between bR_{568} and M states. Although the intensity ratio, I_{1360}/I_{1340}, is ≈ 1 in bR_{568} (probe only, Fig. 16, middle), it rises to ≈ 4 in the pump and probe spectrum (Fig. 16, top). This result indicates unusually strong hydrophobic environments for tryptophan side chains in intermediates of bacteriorhodopsin. Four tryptophan residues out of the total of eight in bacteriorhodopsin are located in the retinal pocket. The authors concluded that the change in hydrophobicity is due to deprotonation of the Schiff base with the bR $\rightarrow$ M reaction, which results in the elimination of charge in the retinal pocket.

4.4.2b. Heme Proteins. Hemoglobin (Hb) consists of four polypeptide chains, two α-chains, and two β-chains, each of which contains a heme group. The binding of oxygen to one heme group facilitates oxygen binding to the other heme groups, whereas there is no direct heme–heme interaction. Although the model of Monod *et al.* (1965) proposing a transition between R (high affinity) and T (low affinity) states has been supported by a number of studies, the detailed molecular mechanism of cooperative ligand binding is still not well understood.

The photolysis of carbon monoxide bound hemoglobin, HbCO $\rightarrow$ Hb + CO, is a feasible experimental model for the R $\rightarrow$ T transition, because this process takes place immediately after irradiation with visible light. Structural differences in the globin protein component of Hb in the R and T states are evidenced by changes in several Raman bands (Su *et al.*, 1989; Kaminaka *et al.*

1990). Therefore, the R → T quaternary structure change can be monitored by time-resolved Raman spectra after photolysis of HbCO.

Spiro and co-workers showed that changes in the Raman spectrum of HbCO involving W3, Y8a, and Y8b begin at ≈5 μsec and continue till 20 μsec after photolysis (Su *et al.*, 1989). Results of Kitagawa and co-workers (Kaminaka *et al.*, 1990) are in good agreement on this point; the change in relative intensity of W16 versus the phenylalanine 1004 cm^{-1} band is detected in the interval between 10 and 20 μsec after the photolysis. The latter group also found that the time required for the quaternary structure transition is strongly affected by the pH of solution (Kaminaka and Kitagawa, 1992). At moderately acidic conditions (pH 5.8), the change is completed within 10 μsec after photolysis, whereas the change in the I_{1010}/I_{1004} ratio is comparable to that at neutral pH. This phenomenon may relate to the pH dependence of CO_2/O_2 exchange by Hb, known as the Bohr effect.

Recently, Rodgers *et al.* (1992) reported the changes in local environments of individual tryptophan side chains accompanying the R → T transition. In the UVRR spectrum of Hb, the W3 Raman band is observed at 1558 cm^{-1} with a significant shoulder at 1550 cm^{-1} (Fig. 17). Since Hb contains three inequivalent

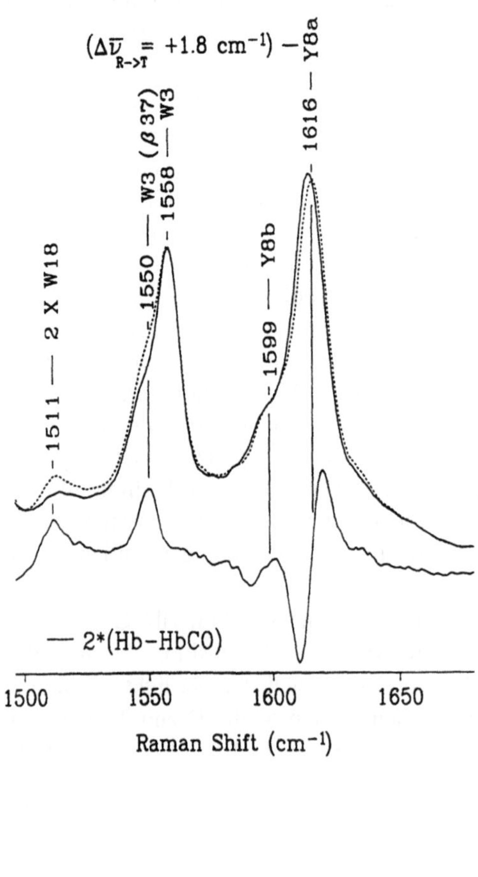

FIGURE 17. UVRR spectra (230-nm excitation) of deoxyHb (dashed line) and HbCO (solid line), and the difference spectrum, deoxyHb minus HbCO. (From Rodgers *et al.*, 1992. Reprinted with permission from the American Chemical Society.)

tryptophan residues—α14, β15, and β37—one of these must generate the shoulder and the remaining two the principal peak. Specific assignments were performed by two different methods. First, in the UVRR spectrum of a natural Hb mutant in which Trpβ37 was replaced by arginine, only a symmetric W3 band occurred at 1558 cm^{-1}, indicating that the shoulder in the spectrum of normal Hb is due to Trpβ37. Second, in the X-ray crystal structure of oxy-Hb (Shaanan, 1983), the side-chain conformation of β37 ($|\chi^{2,1}| = 92°$) is smaller than that of either α14 (122°) or β15 (115°). Using the correlation between W3 and $|\chi^{2,1}|$ (Fig. 4), these tryptophan residues are expected to give W3 at 1549 (β37), 1558 (α14), and 1557 cm^{-1} (β15), respectively. The observed Raman data are consistent with the X-ray structure.

According to the X-ray crystallographic studies (Shaanan, 1983; Perutz *et al.*, 1982), Trpβ37 is located at the interface between α and β subunits and forms an intrasubunit hydrogen bond with Asnβ102 in oxy-Hb and an intersubunit hydrogen bond with Aspα94 in deoxy-Hb. As clearly shown in Fig. 17, the intensity of the 1550 cm^{-1} shoulder [W3(β37)] in the UVRR spectrum of deoxy-Hb is greater than that of HbCO, whereas the 1558 cm^{-1} peak [W3(α14/β15)] exhibits no significant difference. Rodgers *et al.* (1992) found that most of the intensity increase of W3(β37) occurs between 1 and 10 μsec after photolysis of HbCO. Concomitant with this change, an intensity increase occurs at 882 cm^{-1} (tryptophan W17 mode), indicating a very weak hydrogen bond of the tryptophan NH donor during 1–10 μsec following photolysis. The time-resolved UVRR spectra detected a transitional non-hydrogen-bonded state along the pathway between the two hydrogen-bonded states of Trpβ37 in the R and T structures.

5. PERSPECTIVES

Methods of Raman and resonance-Raman spectroscopy provide versatile and sensitive probes of proteins, their biological assemblies and aggregates, and their complexes with other macromolecules. Often, the structural information obtained from Raman spectra of proteins is not obtainable with alternative structural probes. Protein applications of Raman spectroscopy have increased rapidly in number during the past decade and are expected to continue to increase with improvements in the sensitivity and reliability of Raman instrumentation.

The practice of Raman spectroscopy is technically more complex than that of many other spectroscopic probes applied to proteins, including FTIR absorption, UV absorption, and fluorescence and CD spectroscopies. Additionally, the requisite laser irradiation of proteins must be exercised very carefully, particularly for ultraviolet resonance Raman applications. *The investigator is well advised to commit the necessary efforts to establish protein purity and homogeneity before laser irradiation and to verify molecular integrity after the Raman spectra have been collected.* Finally, the Raman spectra of proteins must be interpreted in conjunction with well-established assignments and reliable correlations based upon analogous studies of appropriate model compounds of known structure.

ACKNOWLEDGMENTS. Research applications from the laboratory of GJT were supported by Grants AI11855, AI18758, and GM50776 from the U.S. National Institutes of Health. TM was supported by a postdoctoral fellowship from the Marion Merrell Dow Foundation and the University of Missouri—Kansas City. The authors also thank co-workers at UMKC and Tohoku University for their contributions to many of the applications discussed here.

6. REFERENCES

Ames, J. B., Ros, M., Raap, J., Lugtenburg, J., and Mathies, R. A., 1992, Time-resolved ultraviolet resonance Raman studies of protein structure: Application to bacteriorhodopsin, *Biochemistry* **31:**5328–5334.

Asher, S. A., 1993, UV resonance Raman spectroscopy for analytical, physical, and biophysical chemistry, Parts I and II, *Anal. Chem.* **65:**59A–66A, 201A–210A.

Aubrey, K. L., and Thomas, G. J., Jr., 1991, Raman spectroscopy of filamentous bacteriophage Ff (fd, M13, fl) incorporating specifically-deuterated alanine and tryptophan side chains, *Biophys. J.* **60:**1337–1349.

Austin, J. C., Jordan, T., and Spiro, T. G., 1993, Ultraviolet resonance Raman studies of proteins and related model compounds, in *Advances in Spectroscopy*, Vol. 20, Part A (R. J. H. Clark and R. E. Hester, eds.), pp. 55–127, Wiley, New York.

Bamford, J. K. H., Bamford, D. H., Li, T., and Thomas, G. J., Jr., 1993, Structural studies of the enveloped dsRNA bacteriophage φ6 of *Pseudomonas syringae* by Raman spectroscopy: II. Nucleocapsid structure and thermostability of the virion, nucleocapsid and polymerase complex, *J. Mol. Biol.* **230:**473–482.

Bandekar, J., 1992, Amide modes and protein conformation, *Biochim. Biophys. Acta* **1120:**123–143.

Benevides, J. M., Tsuboi, M., Wang, A. H.-J., and Thomas, G. J., Jr., 1993, Local Raman tensors of double-helical DNA in the crystal: A basis for determining DNA residue orientations, *J. Am. Chem. Soc.* **115:**5351–5359.

Braiman, M. S., Mogi, T., Stern, L. J., Khorana, H. G., and Rothschild, K. J., 1988, Vibrational spectroscopy of bacteriorhodopsin mutants, *Proteins: Struct. Funct. Genet.* **3:**219–229.

Carey, P. R., 1982, *Biochemical Applications of Raman and Resonance Raman Spectroscopies*, Academic Press, London.

Chiou, S-H., Lee, B-S., and Yu, N-T., 1992, Structural analysis and comparison of cobrotoxin and cardiotoxins by near-IR Fourier transform Raman spectroscopy, *Biochem. Int.* **26:**747–758.

Copeland, R. A., and Spiro, T. G., 1986, Ultraviolet Raman hypochromism of the tropomyosin amide modes: A new method for estimating α-helical content in proteins, *J. Am. Chem. Soc.* **108:**1281–1285.

Englander, S. W., and Kallenbach, N. R., 1984, Hydrogen exchange and structural dynamics of proteins and nucleic acids, *Q. Rev. Biophys.* **4:**521–655.

Englander, S. W., and Mayne, L., 1992, Protein folding studied using hydrogen-exchange labeling and two-dimensional NMR, *Annu. Rev. Biophys. Biomol. Struct.* **21**:243–265.

Greve, J., and Puppels, G. J., 1993, Raman microspectroscopy of single whole cells, in *Advances in Spectroscopy*, Vol. 20, Part A (R. J. H. Clark and R. E. Hester eds.), pp. 231–265, Wiley, New York.

Harada, I., and Takeuchi, H., 1986, Raman and ultraviolet resonance Raman spectra of proteins and related compounds, in *Advances in Spectroscopy*, Vol. 13 (R. J. H. Clark and R. E. Hester, eds.), pp. 113–175, Wiley, New York.

Harada, I., Takamatsu, T., Tasumi, M., and Lord, R. C., 1982, Raman spectroscopic study of the interaction between sulfate anion and an imidazolium ring in ribonuclease A, *Biochemistry* **21**:3674–3677.

Harada, I., Miura, T., and Takeuchi, H., 1986, Origin of the doublet at 1360 and 1340 cm^{-1} in the Raman spectra of tryptophan and related compounds, *Spectrochim. Acta* **42A**:307–312.

Hartman, K. A., and Thomas, G. J., Jr., 1985, The identification, interactions and structure of viruses by Raman spectroscopy, in *Instrumental Methods for Rapid Microbiological Analysis* (W. H. Nelson ed.), pp. 91–134, VCH Publishers, New York.

Hildebrandt, P. G., Copeland, R. A., Spiro, T. G., Otlewski, J., Laskowski, M,. Jr., and Prendergast, F. G., 1988, Tyrosine hydrogen-bonding and environmental effects in proteins probed by ultraviolet resonance Raman spectroscopy, *Biochemistry* **27**:5426–5433.

Ishida, H., Kamoto, R., Uchida, S., Ishitani, Y., Ozaki, Y., Iriyama, K., Tsukie, E., Shibata, F., Ishihara, F., and Kameda, H., 1987, Raman microprobe and Fourier transform-infrared microsampling studies of the microstructure of gallstones, *Appl. Spectrosc.* **41**:407–412.

Kaminaka, S., and Kitagawa, T., 1992, Time-resolved UV resonance Raman study on Bohr effects of hemoglobin, *J. Am. Chem. Soc.* **114**:3256–3260.

Kaminaka, S., Ogura, T., and Kitagawa, T., 1990, Time-resolved ultraviolet resonance Raman study of the photolysis of carbomonoxyhemoglobin. Relaxation of the globin structure, *J. Am. Chem. Soc.* **112**:23–27.

Kitagawa, T., Azuma, T., and Hamaguchi, K., 1979, The Raman spectra of Bence-Jones proteins. Disulfide stretching frequencies and dependence of Raman intensity of tryptophan residues on their environments, *Biopolymers* **18**:451–465.

Krimm, S., and Bandekar, J., 1986, Vibrational analysis of proteins, *Adv. Protein Chem.* **3**:181–364.

Li, H., and Thomas, G. J., Jr., 1991, Cysteine conformation and sulfhydryl interactions in proteins and viruses 1. Correlation of the Raman S–H band with hydrogen bonding and intramolecular geometry in model compounds, *J. Am. Chem. Soc.* **113**:456–462.

Li, H., Wurrey, C. J., and Thomas, G. J., Jr., 1992, Cysteine conformation and sulfhydryl interactions in proteins and viruses. 2. Normal coordinate analysis of the cysteine side chain in model compounds, *J. Am. Chem. Soc.* **114**:7463–7469.

Li, H., Hanson, C., Fuchs, J. A., Woodward, C., and Thomas, G. J., Jr., 1993, Determination of the pK_a values of active-center cysteines, cysteines-32 and -35, in *E. coli* thioredoxin by Raman spectroscopy, *Biochemistry* **32**:5800–5808.

Li, T., Chen, Z., Johnson, J. E., and Thomas, G. J., Jr., 1990, Structural studies of bean pod mottle virus, capsid and RNA in crystal and solution states by laser Raman spectroscopy, *Biochemistry* **29**:5018–5026.

Li, T., Bamford, D. H., Bamford, J. K. H., and Thomas, G. J., Jr., 1993a, Structural studies of the enveloped dsRNA bacteriophage φ6 of *Pseudomonas syringae* by Raman spectroscopy: I. The virion and its membrane envelope, *J. Mol. Biol.* **230**:461-472.

Li, T., Johnson, J. E., and Thomas, G. J., Jr., 1993b, Raman dynamic probe of hydrogen exchange in bean pod mottle virus: Base-specific retardation of exchange in packaged ssRNA, *Biophys. J.*, **65**:1963–1972.

Miura, T., Takeuchi, H., and Harada, I., 1988, Characterization of individual tryptophan side chains in proteins using Raman spectroscopy and hydrogen-deuterium exchange kinetics, *Biochemistry* **27**:88–94.

Miura, T., Takeuchi, H., and Harada, I., 1989, Tryptophan Raman bands sensitive to hydrogen bonding and side-chain conformation, *J. Raman Spectrosc.* **20**:667–671.

Miura, T., Takeuchi, H., and Harada, I., 1991, Raman spectroscopic characterization of tryptophan side chains in lysozyme bound to inhibitors: Role of the hydrophobic box in the enzymatic function, *Biochemistry* **30**:6074–6080.

Moënne-Loccoz, P., and Peticolas, W. L., 1992, Ultraviolet resonance Raman evidence for a change of hydrophobicity of the retinal pocket in the M state of bacteriorhodopsin, *J. Am. Chem. Soc.* **114**:5893–5894.

Monod, J., Wyman, J,. and Changeux, J. P., 1965, Nature of allosteric transitions, *J. Mol. Biol.* **12**:88–118.

Nishimura, Y., Hirakawa, A. Y., and Tsuboi, M., 1978, Resonance Raman spectroscopy of nucleic acids, in *Advances in Infrared and Raman Spectroscopy*, Vol. 5 (R. J. H. Clark and R. E. Hester, eds.), pp. 217–275, Heyden, London.

Perutz, M. F., Hasnain, S. S., Dukl, P. J., Sessler, J. L., and Hahn, J. E., 1982, Stereochemistry of iron in deoxyhemoglobin, *Nature* **295**:535–538.

Prevelige, P. E., Jr., Thomas, D., King, J., Towse, S. A., and Thomas, G. J., Jr., 1990, Conformational states of the bacteriophage P22 capsid subunit in relation to self-assembly, *Biochemistry* **29**:5626–5633.

Prevelige, P. E ., Thomas, D., Aubrey, K. L., Towse, S. A., and Thomas, G. J., Jr., 1993, Subunit conformational changes accompanying bacteriophage P22 capsid maturation, *Biochemistry* **32**:537–543.

Puppels, G. J., de Mul, F. F. M., Otto, C., Greve, J., Robert-Nicoud, M., Arndt-Jovin, D. J., and Jovin, T. M., 1990, Studying single living cells and chromosomes by confocal Raman microspectroscopy, *Nature* **347**:301–303.

Rath, P., Krebs, M. P., He, Y., Khorana, H. G., and Rothschild, K. J., 1993, Fourier transform Raman spectroscopy of the bacteriorhodopsin mutant Tyr-185 → Phe, *Biochemistry* **32**:2272–2281.

Rava, R. P., and Spiro, T. G., 1985, Ultraviolet resonance Raman spectra of insulin and α-lactalbumin with 218- and 200-nm laser excitation, *Biochemistry* **24**:1861–1865.

Rodgers, K. R., Su, C., Subramaniam, S., and Spiro, T. G., 1992, Hemoglobin R → T structural dynamics from simultaneous monitoring of tyrosine and tryptophan time-resolved UV resonance Raman signals, *J. Am. Chem. Soc.* **114**:3697–3709.

Sargent, D., Benevides, J. M., Yu, M.-H., King, J., and Thomas, G. J., Jr., 1988, Secondary structure and thermostability of the phage P22 tailspike: XX. Analysis by Raman spectroscopy of the wild-type protein and a temperature-sensitive folding mutant, *J. Mol. Biol.* **199**:491–502.

Shaanan, B., 1983, Structure of human oxyhemoglobin at 2.1 A resolution, *J. Mol. Biol.* **171**:31–59.

Siamwiza, M. N., Lord, R. C., Chen, M. C., Takamatsu, T., Harada, I., Matsuura, H., and Shimanouchi, T., 1975, Interpretation of the doublet at 850 and 830 cm^{-1} in the Raman spectra of tyrosyl residues in proteins and certain model compounds, *Biochemistry* **14**:4870–4876.

Song, S., and Asher, S. A., 1988, Assignment of a new conformation-sensitive UV resonance Raman band in peptides and proteins, *J. Am. Chem. Soc.* **110**:8547–8548.

Song, S., and Asher, S. A., 1989, UV resonance Raman studies of peptide conformation in poly(L-lysine), poly(L-glutamic acid), and model complexes: The basis for protein secondary structure determinations, *J. Am. Chem. Soc.* **111**:4295–4305.

Su, C., Park, Y. D., Liu, G.-Y., Spiro, T. G., 1989, Hemoglobin quaternary structure change monitored directly by transient UV resonance Raman spectroscopy, *J. Am. Chem. Soc.* **111**:3457–3459.

Sugeta, H., Go, A., and Miyazawa, T., 1972, S-S and C-S stretching vibrations and molecular conformations of dialkyl disulfides and cystine, *Chem. Lett.* 83–86.

Sureau, F., Chinsky, L., Amirand, C,. Ballini, J. P., Duquesne, M., Laigle, A., Turpin, P. Y., and Vigny, P., 1990, An ultraviolet micro-Raman spectrometer: Resonance Raman spectroscopy within single living cells, *Appl. Spectrosc.* **44:**1047–1051.

Takesada, H., Nakanishi, M., Hirakawa, A. Y., and Tsuboi, M., 1976, Hydrogen–deuterium exchange of the tryptophan residues in bovine α-lactalbumin, *Biopolymers* **15:**1929–1938.

Takeuchi, H., and Harada, I., 1986, Normal coordinate analysis of the indole ring, *Spectrochim. Acta* **42A:**1069–1078.

Takeuchi, H., Watanabe, N., and Harada, I., 1988, Vibrational spectra and normal coordinate analysis of p-cresol and its deuterated analogs, *Spectrochim. Acta* **44A:**749–761.

Takeuchi, H., Watanabe, N., Satoh, Y., and Harada, I., 1989, Effects of hydrogen bonding on the tyrosine Raman bands in the 1300-1150 cm^{-1} region, *J. Raman Spectrosc.* **20:**233–237.

Takeuchi, H., Kimura, Y., Koitabashi, I., and Harada, I., 1991, Raman bands of N-deuterated histidinium as markers of conformation and hydrogen bonding, *J. Raman Spectrosc.* **22:**233–236.

Tasumi, M., Harada, I., Takamatsu, T., and Takahashi, S., 1982, Raman studies on L-histidine and related compounds in aqueous solutions, *J. Raman Spectrosc.* **12:**149–151.

Thomas, G. J., Jr., 1987, Viruses and nucleoproteins, in *Biological Applications of Raman Spectroscopy,* Vol. 1, Raman Spectra and the Conformations of Biological Macromolecules (T. G. Spiro ed.), pp. 135–201, Wiley-Interscience, London.

Thomas, G. J., Jr., and Kyogoku, Y., 1977, Biological science, in *Infrared and Raman Spectroscopy,* Part C, *Practical Spectroscopy Series,* Vol. 1 (E. G. Brame, Jr. and J. G. Grasselli eds.), pp. 717–872, Dekker, New York.

Thomas, G. J., Jr. and Tsuboi, M., 1993, in *Advances in Biophysical Chemistry,* Vol. 3 (C. A. Bush ed.), pp. 1–70, JAI Press, Greenwich, CT.

Thomas, G. J., Jr., Prescott, B., and Day, L. A., 1983, Structure similarity, difference and variability in the filamentous viruses fd, Ifl, IKe, Pfl, Xf and Pf3, *J. Mol. Biol.* **165:**321–356.

Thomas, G. J., Jr., Prescott, B., Benevides, J. M., and Weiss, M. A., 1986, The N-terminal domain of the lambda repressor: Investigation of secondary structure and tyrosine hydrogen bonding in wild-type and mutant sequences by Raman spectroscopy, *Biochemistry* **25:**6768–6778.

Thomas, G. J., Jr., Prescott, B., and Urry, D. W., 1987, Raman amide bands of type-II β-turns in cyclo-$(VPGVG)_3$ and poly-(VPGVG) and implications for protein secondary structure analysis, *Biopolymers* **26:**921–934.

Thomas, G. J., Jr., Becka, R., Sargent, D., Yu, M.-H., and King, J., 1990, Conformational stability of P22 tailspike proteins carrying temperature-sensitive-folding mutations, *Biochemistry* **29:**4181–4187.

Tsuboi, M., Ikeda, T., and Ueda, T., 1991a, Raman microscopy of a small uniaxial crystal, *J. Raman Spectrosc.* **22:**619–626.

Tsuboi, M., Kaneuchi, F., Ikeda, T., and Akahane, K., 1991b, Infrared and Raman microscopy of fowl feather barb, *Can. J. Chem.* **69:**1752–1757.

Tuma, R., Vohnik, S., Li, H., and Thomas, G. J., Jr., 1993, Cysteine conformation and sulfhydryl interations in proteins and viruses. 3. Quantitative Measurement of the Raman S–H band intensity and frequency, *Biophys. J.,* **65:**1066–1072.

Uchida, K., Harada, I., Nakauchi, Y., and Maruyama, K., 1991, Structural properties of connectin studied by ultraviolet resonance Raman spectroscopy and infrared dichroism, *FEBS Lett.* **295:**35–38.

Wang, Y., Purrello, R., Jordan, T., and Spiro, T. G., 1991, UVRR spectroscopy of the peptide bond

Chapter 4

Fluorescence Spectroscopic Studies of Proteins

Siddhartha Roy and B. Bhattacharyya

1. INTRODUCTION

Fluorescence spectroscopy has emerged as one of the most important tools in the study of protein structure and function. Sensitivity of the fluorescence techniques is a major advantage. Although it cannot provide total structure like X-ray crystallography or nuclear magnetic resonance (NMR), the sensitivity of the technique allows one to work at or near the intracellular concentrations of macromolecules and ligands. This is particularly important for proteins that self-associate at higher concentrations. Although steady-state fluorescence spectra can now be easily obtained and steady-state fluorometers are inexpensive and widely available, they suffer from several disadvantages.

Recent advances in time-resolved fluorescence spectroscopy have solved many of the problems associated with the use of steady-state fluorescence spectroscopy. The reader may find a good description of basic theory and biological applications in Cantor and Schimmel (1980) and Lakowicz (1983). Methodology and applications of time-resolved fluorescence spectroscopy in proteins have been reviewed (Beecham and Brand, 1985) and will not be discussed here.

Siddhartha Roy Department of Biophysics, Bose Institute, Calcutta 700 054, India.
B. Bhattacharyya Department of Biochemistry, Bose Institute, Calcutta 700 054, India.

Subcellular Biochemistry, Volume 24. Proteins: Structure, Function, and Engineering, edited by B. B. Biswas and Siddhartha Roy. Plenum Press, New York, 1995.

Instead, we will focus our attention mainly on different types of information that are obtainable from fluorescence experiments in the areas of protein structure and function and illustrate with examples mainly from the areas of tubulin and λ-repressor structure and function, which are areas under active investigations in the authors' laboratories.

2. STUDY OF LIGAND BINDING

2.1. General Discussions on Method

Binding of ligands are an integral part of function of many proteins. Enzymes that catalyze chemical reactions bind ligands and are also regulated by ligand binding (e.g., allosteric effectors). Chemotherapeutic agents are often analogues of natural ligands that bind to the same site. The function of receptors is to bind ligands and transfer the information elsewhere. Thus, quantitative analysis of ligand binding to proteins plays a crucial role in the study of protein structure, function, as well as engineering.

The most popular methods of studying ligand binding to proteins either use a fluorescent ligand or monitor the effect of ligand binding on tryptophan fluorescence of proteins. Many biologically relevant ligands are either fluorescent or become fluorescent upon binding to proteins. A well-known example is NADH. It has very low fluorescence in the free state but becomes significantly fluorescent upon binding to many dehydrogenases. This property has been widely utilized to study NADH and NADPH binding to many dehydrogenases (Brochon *et al.*, 1976). In some cases fluorescent analogues of natural ligands have been synthesized to study its interaction with proteins. Many fluorescent ATP and GTP analogues have been synthesized for this purpose, e.g., ϵ-ATP (Secrist *et al.*, 1972).

Many fluorescent probes have been synthesized that are environment sensitive, e.g., ANS, PRODAN, and *bis*-ANS. The quantum yields and/or emission maxima of these probes change dramatically upon binding to hydrophobic sites of proteins and have been widely used to detect conformational changes (Stryer, 1965; Weber and Farris, 1979; Rosen and Weber, 1969).

The binding isotherms of these compounds can be derived from a "reverse titration" in which the ligand concentration is held constant and the protein concentration is varied to determine the enhancement factor. This is followed by a "direct titration" in which the protein is held constant and the ligand concentration is varied (Mas and Colman, 1985). For many fluorescent ligands, the fluorescence emission is not environment sensitive. Binding of such ligands to proteins does not lead to fluorescence enhancement or quenching. Hence, the methods described above are unsuitable for studying binding of such ligands.

Fluorescence polarization may be used, however, to obtain binding constants in such cases. The method is based on the fact that binding of ligand to a much larger protein leads to significant increase in polarization. A double reciprocal plot may be used to obtain binding constants (Yarbrough and Kirsch, 1981; Ghosh Choudhury *et al.*, 1983).

Binding of a ligand to a protein often leads to quenching or enhancement of tryptophan fluorescence. Information about both stoichiometry and binding constants are present in the data. Binding constants may be obtained by a double reciprocal plot when the protein concentrations are much below the dissociation constant. Information about stoichiometry may be obtained from titrations with protein concentrations at or above the dissociation constant (Lin *et al.*, 1988; Bhattacharyya *et al.*, 1991). Many other graphical methods are available to extract the same information (Klotz, 1986).

2.2. Tubulin–Colchicine Interaction

Promotion of colchicine fluorescence upon binding to tubulin has been applied extensively in the understanding of drug–protein interactions and is probably a unique example of potentiality and versatility of fluorescence spectroscopy. Colchicine and its analogues, which do not fluoresce in aqueous media, exhibit marked fluorescence when combined with tubulin. Fluorescence is a convenient method for determining the binding of colchicine to tubulin that does not require the separation of free colchicine from the protein-bound colchicine, and thus thermodynamic parameters are easily obtained under equilibrium conditions (Bhattacharyya and Wolff, 1974). Studies described below primarily used fluorescence spectroscopic techniques utilizing colchicine fluorescence.

The binding reaction of colchicine to tubulin has generally been studied by the use of labeled colchicine and subsequent separation of bound from free ligand by Sephadex gel filtration or ion exchange separation on DEAE cellulose impregnated filter disks (Weisenberg *et al.*, 1968; Wilson, 1970). Binding of colchicine to tubulin needs 90–120 min at 37 °C for the equilibration, and the binding reaction is essentially irreversible. Thus, these assay methods can be used quite successfully to quantitate binding of colchicine–tubulin interaction and determine its binding parameters at nonequilibrium conditions. However, to understand the mechanism of colchicine–tubulin interactions such as (1) specificity and necessity of different rings of colchicine in its interactions, (2) slow association rate and high activation energy, (3) reversibility, and (4) thermodynamic parameters of binding reaction, it is essential to study the binding properties of several other colchicine analogues to tubulin where fluorescence spectroscopic methods were applied successfully.

Colchicine and its analogues are composed of three rings: a trimethoxy benzene ring (ring A), a methoxy tropolone ring (ring C), and a seven-membered

B-ring, which anchors A- and C-rings. Among these three rings the structural requirements of the C-ring for the binding to the colchicine site of tubulin is relatively modest, since allocolchicine (containing a six-membered C-ring instead of the seven-membered ring) or thiocolchicine (containing SCH_3 group instead of the OCH_3 group in the C-ring) bind to tubulin (Andreu *et al.*, 1991; Bane-Hastie, 1989). Surprisingly, manipulations in the tropolone ring, as in isocolchicine, colchiceine, trimethyl colchicinic acid, and colchicinol, produced analogues that bind to tubulin with very very low affinity (Bhattacharyya and Wolff, 1974). In contrast, the structural requirement for the A-ring is relatively stringent since colchicoside, in which one methoxy group in the A-ring is replaced by a sugar and 4′-carbobenzoxy-4′-demethyl podophylotoxin which possesses bulky substituent in the 4′ position of the podophyllotoxin molecule, is totally ineffective in inhibiting the colchicine–tubulin interaction, when monitored by fluorescence (Flavin and Slaughter, 1974; Bhattacharyya and Wolff, 1974). On the other hand, colchicine analogues modified at or depleted of the B-ring are known to have potent antimitotic activity, which apparently rules out any major role of the B-ring in the antimitotic activity of colchicine (Fitzgerald, 1976). It was shown that even a very long side chain in the B-ring, e.g., such as *N*-deacetyl-*N*-retinoyl colchicine, does not affect the binding activity to a large extent (Bhattacharyya *et al.*, 1986). Nevertheless, a minor change in the B-ring substituents may significantly change the binding mechanism, kinetics, association rate, dissociation rate, temperature dependence, and thermodynamic properties of the binding reactions (Banerjee and Bhattacharyya, 1979; Ray *et al.*, 1981, 1984; Bane *et al.*, 1984; Andreu *et al.*, 1984, 1991; Menendez *et al.*, 1989; Medrano *et al.*, 1989; Pyles and Bane-Hastie, 1993).

The structural features of colchicine responsible for its slow association, high activation energy, and poor reversibility have not been clearly defined. Two hypothesis have been put forth to explain the differences in the kinetics of 2-methoxy-5(2,3,4 trimethoxy phenyl) tropone (AC), colchicine, and other B-ring analogs. An early study by Ray et al. (1981) suggested that not only the B-ring itself but also the substituents present in the B-ring are responsible for the slow rate and the high activation energy of colchicine binding to tubulin. According to the second hypothesis, the activation energy of AC binding to tubulin is decreased compared to colchicine binding to tubulin as a result of the relatively unhindered rotation of AC about the phenyl tropone bond. Recently, a kinetic study using the fluorescence of four B-ring analogues of colchicine—deacetamido colchicine, deacetyl colchicine, demecolchicine, and *N*-methyl-demecolchicine—have been carried out. In this study the association rate constants were determined as a function of temperature and the thermodynamic parameters for the transition states were also calculated. The important kinetic parameters such as global activation enthalpies of the formation of the deacetylcolchicine–

tubulin, demecolchicine–tubulin, and *N*-methyl-demecolchicine–tubulin complexes were very similar to each other, but were 3.5 kcal/mole greater than that of the colchicine–tubulin association (Pyles and Bane-Hastie, 1993). These results incidate that the electronic rather than the steric properties of the B-ring substituents are of greater importance in the activation enthalpy of colchicinoids binding to tubulin. Interestingly, the global activation enthalpy of deacetamidio colchicine, which lacks a substituent on the C-7 carbon, binding to tubulin was virtually identical to that previously found for the colchicine analogue, which lack the B-ring, i.e., 2-methoxy-5(2,3,4 trimethoxy phenyl) tropone binding to tubulin. This result demonstrates that the carbons of the B-ring are not involved in the transition state for the formation of colchicinoid–tubulin complex (Pyles and Bane-Hastie, 1993).

Another important property of colchicine–tubulin interaction that has been found to be related to the B-ring is the reversibility of the reaction. Colchicine and AC are two extreme drugs where AC–tubulin reaction highly reversible and colchicine–tubulin reaction is essentially irreversible (Bhattacharyya *et al.*, 1986; Ray *et al.*, 1981). Compounds such as DAAC, NH_2-DAAC, NH.Me-DAAC, and NMe_2-DAAC are reversible where rate constants varies from 7-3 $\times 10^{-3}$ sec^{-1} (Pyles and Bane-Hastie, 1993). The reversibility of colchicine–tubulin reactions were studied either by dilution of the complex or adding excess podophyllotoxin and then monitoring the lowering of drug fluorescence or quenching of tryptophan fluorescence of tubulin over time. Both methods gave similar results and can be monitored conveniently. Results of reversibility clearly indicates that it is the carbonyl group rather than the B-ring itself of the substituents present that is responsible for such an irreversibility of colchicine–tubulin interaction (Bhattacharyya *et al.*, 1986; Pyles and Bane-Hastie, 1993).

Fluorescence study of colchicine–tubulin interaction also provides information regarding the polarity of the binding site as well as the flexibility of the bound drug in the binding site. The basis of the promotion of colchicine fluorescence upon binding to tubulin is not well understood. Enhancement of fluorescence on binding of a small ligand to a protein is generally explained as due to the hydrophobic environment provided by the binding site, thus lowering the probability of mobile dipole dependent internal conversion reactions. Colchicine has a very low fluorescence in aqueous solution; however, binding to tubulin promotes a dramatic (approximately 400-fold) increase in colchicine fluorescence. It was established that the major contribution to colchicine fluorescence stems from immobilization of colchicine in the binding site of tubulin, since it has been established that colchicine solution fluoresces upon freezing, when present in glycerol, or upon inclusion into micelles (Letterier and Rieger, 1975; Bhattacharyya and Wolff, 1984; Shobha *et al.*, 1989). It is also established from this study that both viscosity and polarity determine the ability of colchicinoids to fluoresce in the absence of tubulin. However, the relative contribution of two factors in the promotion of fluorescence when drug is bound to tubulin is yet to be determined.

The role of the B-ring substituents of colchicine in both the binding and subsequent fluorescent properties of the complex is not well understood. Thus, the quantum yields of the colchicine analogues decrease with amine or alkyl amine substituents at the C-7 position, while colchicine analogues in which no substituents are present or the substituent is an acetylated amine show the characteristic fluorescence. Recently, this low fluorescence of the amino colchicinoids in solvents and bound to tubulin was explained by exciplex formation between the lone pair of the nitrogen and the allocolchicinoid π-system, which leads to the quenching of the allocolchicinoid fluorescence. Thus, here again we see how the fluorescence study provides excellent information of the intramolecular interactions in colchicinoids (Pyles and Bane-Hastie, 1992).

2.3. Magnesium Binding to Tubulin

Magnesium binding to tubulin illustrates many of the advantages of studying ligand binding by fluorescence spectroscopy where the ligand lacks intrinsic fluorescence. Magnesium ions are required for polymerization of tubulin. There are two tight binding sites on the tubulin molecule, which are required for structural integrity of tubulin molecule. When tubulin is passed through a gel filtration column, only two tightly associated magneisums per dimer are retained. Addition of exogenous magnesium ion causes quenching of tryptophan fluorescence, which is saturable (K_d = 180 μM). Acrylamide quenching revealed two classes of tryptophans: one quenchable by low concentrations of acrylamide and the other not. In the absence of added exogenous magnesium, the Lehrer plot cuts the *Y* axis at 1.6, whereas in the presence of magnesium it cuts at 1.15. In addition, there is a substantial change in K_{sv} upon magnesium binding for the quenchable tryptophans. This indicates substantial change in environment of more than one tryptophan and perhaps a global change in the tubulin molecule upon magnesium binding (Bhattacharyya *et al.*, 1994).

Removal of tubulin tail by limited subtilisin cleavage produces tubulin-S (Sackett *et al.*, 1985). Lehrer plot of tubulin-S is similar to gel-filtered tubulin without any added exogenous magnesium. Addition of exogenous magnesium has no effect on either tryptophan fluorescence or Lehrer plot of tubulin-S. Many other functional properties, such as colchicine off-rate and rate of loss of colchicine binding site upon incubation at 37 °C, which are magnesium dependent in tubulin become magnesium independent. It was suggested that the tubulin tail mediates many of the effects of magnesium. Thus, it was concluded that the tubulin tail interacts with the main body in the presence of added exogenous magnesium (Bhattacharyya *et al.*, 1994).

3. PROTEIN ASSOCIATION–DISSOCIATION

3.1. Methods

Heterologous and homologous protein association plays a key role in many important biological processes. Regulation of gene expression, for example, is controlled by many protein factors and protein–protein contact, and association is a vital part of the mechanism of regulation (Ptashne, 1992). Subunit association of multimeric proteins are important in many catalytic and regulatory processes. For example, protein associations between regulatory and catalytic subunits of cAMP-dependent protein kinase is of vital importance in the regulation of activity. Although many methods are available for studying protein association–dissociation, fluorescence spectroscopy is one of the very few techniques that can be used to study the process at equilibrium and at submicromolar concentrations. Two principal fluorescence methods are used to study protein–protein association: fluorescence anisotropy and fluorescence energy transfer.

Fluorescence anisotropy method is based on the fact that protein association leads to an increase in the molecular weight and hence increased rotational correlation time. Any fluorophore rigidly attached to the protein would also experience an increased rotational correlation time and hence an increased anisotropy value. The lifetime of the fluorophore chosen is of crucial importance and preferably should not be very much smaller than the correlation times of the species. Since steady-state anisotropy values are not a direct measure of molecular weight, independent molecular weight estimates are needed at both ends of association–dissociation equilibrium. Only then can the anisotropy values be fitted to a legitimate model (usually by a nonlinear least-square fit procedure).

The energy transfer method is based on labeling of one subunit with a fluorescence donor and another subunit with a fluorescence acceptor. A large R_o for the donor–acceptor pair is preferable. The dissociation of the hybrid dimer causes average distance between the subunits to increase beyond the range of energy transfer and hence the loss of energy transfer that existed in the hybrid dimer (Shore and Chakrabarti, 1976). The energy transfer method is suited for dimer–monomer dissociation but would be ordinarily inapplicable for tetramer or higher order structures because of subunit reshuffling.

3.2. Tetramer–Dimer Equilibrium of λ-Repressor

The dimer–dimer contact of λ-repressor is thought to be of crucial importance in binding cooperativity of λ-repressor to operator sites, and cooperative binding in turn plays a crucial role in the functioning of the λ "genetic switch."

Tetramer–dimer equilibrium of λ-repressor has been studied in our laboratory using the fluorescence anisotropy method. Since λ-repressor lacked a reactive sulfhydryl group we have labeled λ-repressor with dansyl chloride. The lifetime of dansyl chloride is sufficient (12–14 nsec) to detect protein association from dimer to higher-order structures. Using fluorescence anisotropy method we have been able to deduce thermodynamics parameters and the effect of operator fragments (O_R1) on the association equilibrium (Banik *et al.*, 1993).

3.3. Dimer–Monomer Equilibrium of Tubulin and Tubulin-S

Effect of C-terminal tails of tubulin on its dimer–monomer equilibrium has also been reported from this laboratory using fluorescence anisotropy and energy transfer study. In this study (Panda *et al.*, 1992), tubulin and tubulin-S were labeled with either fluorescein 5′-maleimide (FM) or 7-diethylamino-4-methyl coumarin 3-maleimide (CM). The advantage of labeling with FM and CM is that they are donor-acceptor pairs suitable for energy transfer study; in addition, their high quantum yields allow accurate measurement of anisotropy at nanomolar concentrations. Thermodynamic parameters, the effect of colchicine, and ionic strength on the equilibrium constants are substantially different for tubulin and tubulin-S. This study indicates that the removal of C-terminal tail of tubulin stabilizes dimer of tubulin-S by one order of magnitude compared to that of tubulin perhaps changing the interaction of α and β subunits in the dimer. A similar fluorescence anisotropy study has been reported for tubulin by Mejilano and Himes (1989).

4. CONFORMATIONAL CHANGES

4.1. General

Fluorescence spectroscopy is a very sensitive method for measuring conformational changes in proteins. Both quantum yield and emission maximum of many fluorophores are very sensitive to the microenvironment. A conformational change that changes the microenvironment also changes the fluorescence properties of the fluorophore. Tryptophans are often used as probes for conformational changes. Single tryptophan proteins are often ideal for such purposes since the localization of the conformational change is possible. With the advent of site-directed mutagenesis, it is now possible to construct single tryptophan proteins within the restrictions imposed by the protein structure. Another big boost came from increasing the availability of picosecond laser systems and phase-modulated spectrometers, which allows tryptophan signals to be resolved in the time domain. Such technologies have made possible detailed residue level analysis of conformational changes in multitryptophan proteins.

External environment-sensitive probes are also very useful tools for detection of conformational changes. Several types of environment-sensitive probes have been designed for protein structural studies. One class that has proved to be highly popular is anilino-napthalene sulfonic acid and its derivatives, particularly 1,8 ANS and *bis*-ANS. The quantum yields and to a moderate extent the emission maxima are sensitive to environmental polarity. Any conformational change that leads to change in polarity of the microenvironment leads to enhancement or quenching of fluorescence. Another class of fluorescent probes is PRODAN and its sufhydryl reactive analogue acrylodan (Weber and Farris, 1979; Prendergast *et al.*, 1983). The emission maximum of these probes is extraordinarily sensitive to polarity, and hence are sensitive to conformational changes. Recently a new class of probe has been introduced, exemplified by di-cyano vinyl julolidine (Kung and Reed, 1989), whose quantum yield is sensitive to the local flexibility of the binding site. Potentially, this class or probes may detect conformational changes that increase or reduce mobility restrictions.

4.2. Conformational Changes in λ-Repressor

The issue of DNA-induced global conformational change is an important one. In many DNA binding proteins, DNA binding and other functions often reside on different domains and are often functionally linked, e.g., *lac* repressor and *gal* repressor of *Escherichia coli*. Such functional linkages must be mediated by conformational changes, although satisfactory demonstration of such changes had been rare.

λ-Repressor is a two-domain protein whose N-terminal domain binds to the operator DNA and the C-terminal domain is involved in the intersubunit contacts necessary for protein dimerization and higher-order contacts necessary for cooperativity. All three tryptophans in the λ-repressor are located in the C-terminal part of the molecule. Binding of a 17-base pair (bp) operator fragment causes quenching and red shift of the tryptophan fluorescence. Using isolated domain and other experiments, the hydrophobic probe *bis*-ANS was shown to bind to the C-terminal domain of λ-repressor. The addition of 17-bp O_R1 causes approximately 40% enhancement of fluorescence. Use of two types of probes, tryptophan and *bis*-ANS, clearly demonstrated that binding of an operator to the N-terminal domain causes a conformational change in the C-terminal domain (Saha *et al.*, 1992). Although functional linkages of DNA binding and dimer–dimer contact is not well established, there are hints that this may be true, highlighting the importance of the observed conformational change (Banik *et al.*, 1993).

4.3. Substrate-Induced Conformational Changes in Glutaminyl-tRNA Synthetase

Glutaminyl-tRNA synthetase is a single-chain multidomain protein that acylates glutamine to $tRNA^{gln}$ with concomitant hydrolysis of ATP to AMP and pyrophosphate (Schimmel, 1987). Along with glutamyl and arginyl-tRNA synthetases, in contrast to others, glutaminyl-tRNA synthetase does not catalyze tRNA-independent ATP-PP_i exchange, implying a tRNA-induced conformational change (Ravel *et al.*, 1965). Crystal structure of glnRS/ATP/$tRNA^{gln}$ has been solved (Rould *et al.*, 1989, 1991). The protein is elongated and has five domains. The active site is located in the N-terminal domains and the $tRNA^{gln}$ virtually spans the whole molecule. The anticodon nucleotides bind to one of the C-terminal domains far away from the active site. It has long been recognized that anticodon nucleotides are important recognition elements of $tRNA^{gln}$. A recent study has shown that mutation of anticodon nucleotides cause a dramatic change V_{max} at a faraway active site, implying transmission of binding information either through tRNA or through protein (Jahn *et al.*, 1991).

Glutaminyl-tRNA synthetase has one very reactive nonessential sulfhydryl group that can be labeled predominantly by sulfhydryl reagents. Glutaminyl-tRNA synthetase that has been labeled at the reactive sulfhydryl by acrylodan was used to monitor conformational changes (Bhattacharyya and Roy, 1993). ATP has a dramatic effect on acrylodan emission maximum, shifting it to more than 10 nm to the blue, accompanied by almost 30% quenching of fluorescence. $tRNA^{gln}$ has only a modest effect, causing some enhancement of fluorescence accompanied by a 2-nm emission maximum shift. In the presence of ATP, the binding of $tRNA^{gln}$ leads to quenching of fluorescence, accompanied by an approximately 6-nm blue shift. Salt dependence of $tRNA^{gln}$ binding to glnRS in the presence and in the absence of ATP also differs significantly. This led to the conclusion that ATP induces a conformational change which significantly alters the binding of $tRNA^{gln}$ to glnRS.

5. FOLDING AND UNFOLDING OF PROTEINS

5.1. General

Denaturation–renaturation of proteins causes major change of microenvironment of side chains. Thus, fluorescent probes are ideal to probe structural changes during denaturation and renaturation of proteins. Environment-sensitive external fluorescent probes are widely used as well as tryptophans. In conjunction with circular dichroism, fluorescence has emerged as the most powerful tool for analyzing folding–unfolding of proteins (Creighton, 1992).

Many fluorescent techniques are useful in studying folding–unfolding reactions and characterization of intermediate states. Intermediate states (e.g., molten globules in protein folding) are often characterized by accessible hydrophobic areas of proteins and bind dyes such as 1,8 ANS to a great extent (Semiostnov *et al.*, 1987). Fluorescence energy transfer and polarization measurements are often used to estimate compactness of the intermediates (Semiostnov *et al.*, 1987).

5.2. Urea-Induced Equilibrium Denaturation of λ-Repressor

Urea-induced denaturation of λ-repressor occurs in several phases. At low urea concentrations, there is enhancement of tryptophan fluorescence and concomitant shift of emission maximum from 340 to 345 nm. There is only a small change in far ultraviolet circular dichroism (UV CD) spectrum around this concentration of urea. At around 3.5 M urea a second transition occurs that causes major loss of UV CD signal with little change in tryptophan fluorescence. Studies with isolated domains indicated that this phase of denaturation is the denaturation of the N-terminal domain. Studies of tryptophan fluorescence as well as other probes indicated that subunit dissociation and global denaturation of C-terminal domain occurs at much higher urea concentrations. Studies with isolated C-terminal domains indicated that the transition is still present but occurs at significantly lower urea concentrations. Thus, it was concluded that the transitions at low urea concentrations are the denaturation of a part of the C-terminal domain. The effect of N-terminal domain on the denaturation profile of this part of the C-terminal domain indicates the presence of N- and C-terminal domain contact in the native repressor (Banik *et al.*, 1992).

6. DISTANCE INFORMATION

6.1. General

There are two major techniques by which one can obtain distance estimates between two fluorophores. Forster energy transfer between two fluorophores occurs in the range of 10–80 Å and the magnitude of the energy transfer is inversely proportional to the sixth power of the distance, and thus fluorescence energy transfer may be used as a spectroscopic ruler. The main uncertainty is due to the orientation factor κ^2 and an estimate of uncertainty can be obtained by fluorescence polarization measurements. Recently, a time-resolved fluorescence method has been reported that reduces the uncertainty of distance estimate due to uncertainty of κ^2 (Wu and Brand, 1992).

6.2. Donor Quenching

The distances are even more difficult to estimate in multiple-donor single-acceptor systems, which often happen in multitryptophan proteins. If the energy donors differ in accessibility toward a collisional quencher and the acceptor fluorescence is not quenchable by such quenchers, then donor quenching can be used to resolve contributions to net energy transfer from a particular donor in favorable cases. In this technique, a collisional quencher is chosen such that it quenches the donor fluorescence, without affecting the acceptor fluorescence.

Quenching of energy transfer can thus be attributed to the accessible donors and Lehrer-type plot may be used to estimate the fraction of energy transfer from accessible and nonaccessible donors.

Tubulin has eight tryptophans, and in 0.5 mM magnesium about 17–20% of the fluorescence is not quenchable by acrylamide. Binding of colchicine causes severe quenching of tryptophan fluorescence. Donor quenching indicates that this quenching is mainly due to energy transfer to colchicine and this energy transfer occurs predominantly from the inaccessible tryptophans. An estimate of distance indicated that the tryptophans must be very close to the colchicine binding site (Bhattacharyya *et al.*, 1993).

7. REFERENCES

Andreu, J. M., Gorbunoff, M. J., Lee, J. C., and Timasheff, S. N., 1984, Interaction of tubulin with bi-functional colchicine analogues: An equilibrium study, *Biochemistry* **23:**1742–1752.

Andreu, J. M., Gorbunoff, M. J., Medrano, F. J., Rossi, M., and Timasheff, S. N., 1991, Mechanism of colchicine binding to tubulin: Tolerance of substituents in ring C′ of biphenyl analogues, *Biochemistry* **30:**3777–3785.

Bane, S., Puett, D., Macdonald, T. L., and Williams, R. C., 1984, Binding to tubulin of the colchicine analog 2-methoxy-5-(2′,3′,4′-trimethoxy phenyl) tropone, *J. Biol. Chem.* **259:**7391-7398.

Bane-Hastie, S., 1989, Spectroscopic and kinetic features of Allocolchicine binding to tubulin, *Biochemistry* **28:**7753–7760.

Banik, U., Saha, R., Mandal, N. C., Bhattacharyya, B., and Roy, S., 1992, Multiphasic denaturation of λ repressor by urea and its implications for the repressor structure, *Eur. J. Biochem.* **206:**15–21.

Banik, U., Mandal, N. C., Bhattacharyya, B., and Roy, S., 1993, A fluorescence anisotropy study of tetramer–dimer equilibrium of λ repressor and its implication for function, *J. Biol. Chem.* **268:**3938–3943.

Banerjee, A., and Bhattacharyya, B., 1979, Colcemid and colchicine binding to tubulin, *FEBS Lett.* **99:**333–336.

Beecham, J. M., and Brand, L., 1985, Time-resolved fluorescence of proteins, *Annu. Rev. Biochem.* **54:**43–71.

Bhattacharyya, T., and Roy, S., 1993, A fluorescence spectroscopic study of substrate induced conformational changes in glutaminyl-tRNA synthetase, *Biochemistry* **32:**9268–9273.

Bhattacharyya, B., and Wolff, J., 1974, Promotion of fluorescence upon binding of colchicine to tubulin, *Proc. Natl. Acad. Sci. USA* **71:**2627–2631.

Bhattacharyya, B., and Wolff, J., 1984, Immobilization dependent fluorescence of colchicine, *J. Biol. Chem.* **259:**11836–11843.

Bhattacharyya, B., Howard, R., Maity, S. N., Brossi, A., Sharma, P. N., and Wolff, J., 1986, B-ring regulation of colchicine binding kinetics and fluorescence, *Proc. Natl. Acad. Sci. USA* **83:**2052–2055.

Bhattacharyya, T., Bhattacharyya, A., and Roy, S., 1991, A fluorescence spectroscopic study of glutaminyl-tRNA synthetase from *Escherichia coli* and its implications for the enzyme mechanism. *Eur. J. Biochem.* **200:**739–745.

Bhattacharyya, A., Bhattacharyya, B., and Roy, S., 1993, A study of colchicine tubuline complex by donor quenching of fluorescence energy transfer, *Eur. J. Biochem.* **216:**757–761.

Bhattacharyya, A., Bhattacharyya, A., and Roy, S., 1994, Magnesium induced structural changes in tubulin, *J. Biol. Chem.*, in press.

Brochon, J. C., Wahl, P. H., Jallon, J. M., and Iwatsubo, M., 1976, Pulse fluorimetry study of beef liver glutamate dehydrogenase-reduced nicotinamide adeninie dinucleotide phosphate complex, *Biochemistry* **15**:3259–3265.

Cantor, C. R., and Schimmel, P., 1980, *Biophysical Chemistry*, W. H. Freeman, San Francisco.

Creighton, T., 1992, *Protein Folding*, W. H. Freeman, New York.

Fitzgerald, T. J., 1976, Molecular features of colchicine associated with antimitotic activity and inhibition of tubulin polymerization, *Biochem. Pharmacol.* **25**:1383–1387.

Flavin, M., and Slaughter, C., 1974, Microtubule assembly and function in chlamydomonas: Inhibition of growth and flagellar regeneration by antitubulin and other drugs and isolation of resistant mutants, *J. Bacteriol.* **118**:59–69.

Ghosh Choudhury, G., Bannerjee, A., Bhattacharyya, B., and Biswas, B. B., 1983, Interaction of colchicine analogues with purified tubulin, *FEBS Lett.* **161**:55–59.

Jahn, M., Rogers, M. J., and Soll, D., 1991, Anticodon and acceptor stem nucleotides in $tRNA^{gln}$ are major recognition elements for *E. coli* glutaminyl-tRNA synthetase, *Nature* **352**:268–260.

Klotz, I. M., 1986, *Introduction to Biomolecular Energetics*, Academic Press, Orlando, FL.

Kung, C. E., and Reed, J. R., 1989, Fluorescent molecular rotors: A new class of probes for tubulin structure and assembly, *Biochemistry* **28**:6678–6686.

Lakowicz, J. R., 1983, *Principles of Fluorescence Spectroscopy*, Plenum Press, New York.

Letterier, F., and Rieger, F., 1975, Immobilization of colchicine at low temperature, *CRSA Trav. Sci.* **5**:224–225.

Lin, S. X., Wang, Q., and Wang, Y. L., 1988, Interactions between *E. coli* arginyl-tRNA synthetase and its substrates, *Biochemistry* **27**:6348–6353.

Mas, M. T., and Colman, R. F., 1985, Spectroscopic studies of the interactions of coenzymes and coenzyme fragments with pig heart oxidised triphospho pyridine specific isocitrate dehydrogenase, *Biochemistry* **24**:1634–1646.

Medrano, F. J., Andreu, J. M., Gorbunoff, M. J., and Timasheff, S. N., 1989, Role of colchicine rings B and C in the binding process to tubulin, *Biochemistry* **28**:5589–5599.

Mejilano, M. R., and Himes, R. H., 1989, Tubulin dimer dissociation detected by fluorescence anisotropy, *Biochemistry* **28**:6518–6524.

Menendez, M., Laynez, J., Medrano, F. J., and Andreu, J. M., 1989, A thermodynamic study of the interaction of tubulin with colchicine site ligands, *J. Biol. Chem.* **264**:16367-16371.

Panda, D., Roy, S., and Bhattacharyya, B., 1992, Reversible dimer dissociation of tubulin-S and tubulin detected by fluorescence anistotropy, *Biochemistry* **31**:9709–9716.

Prendergast, F. G., Meyer, M., Carlson, G. L., Iida, S., and Potter, J. D., 1983, Synthesis, spectral properties and use of 6-acryloyl-2-dimethyl aminonapthalene (acrrylodan), *J. Biol. Chem.* **258**:7541–7544.

Ptashne, M., 1992, *A Genetic Switch*, Cell Press and Blackwell Scientific Publications, Cambridge, MA.

Pyles, E. A., and Bane-Hastie, S., 1992, Role of the B-ring substituent in the fluorescence of colchicinoid-tubulin and allocolchicinoid-tubulin complexes, *Biochemistry* **31**:7086–7093.

Pyles, E. A., and Bane-Hastie, S., 1993, Effect of the B-ring and C-7 substituent on the kinetics of colchicine-tubulin association, *Biochemistry* **32**:2329–2336.

Ravel, J. M., Wang, S., Heinemeyer, C., and Shive, W., 1965, Glutamyl and glutaminyl ribonucleic acid synthetase of *Escherichia coli* W, *J. Biol. Chem.* **240**:430–438.

Ray, K., Bhattacharyya, B., and Biswas, B. B., 1981, Role of B-ring of colchicine in its binding to tubulin, *J. Biol. Chem.* **256**: 6241–6244.

Ray, K., Bhattacharyya, B., and Biswas, B. B., 1984, Anion induced increases in the affinity of colcemid binding to tubulin, *Eur. J. Biochem.* **142**:577–581.

Rosen, C. G., and Weber, G., 1969, Dimer formation from 1-Anilino-8-napthalene sulfonate catalyzed by bovine serum albumin. A new fluorescent molecule with exceptional binding properties, *Biochemistry* **8:**3915–3920.

Rould, M. A., Perona, J. J., Soll, D., and Steitz, T. A., 1989, Structure of *E. coli* glutaminyl-tRNA synthetase complexed with $tRNA^{gln}$ and ATP at 2.8 Å resolution, *Science* **246:**1135–1142.

Rould, M. A., Perona, J. J., and Steitz, T. A., 1991, Structural basis of anticodon loop recognition by glutaminyl-tRNA synthetase, *Nature* **352:**213–218.

Sackett, D. L., Bhattacharyya, B., and Wolffe, J., 1985, Tubulin subunit carboxyl termini determine polymerization efficiency, *J. Biol. Chem.* **260:**43–45.

Saha, R., Banik, U., Bandyopadhyay, S., Mandal, N. C., Bhattacharyya, B., and Roy, S., 1992, An operator induced conformational change in the C-terminal domain of λ repressor, *J. Biol. Chem.* **267:**5862–5867.

Schimmel, P., 1987, Aminoacyl-tRNA synthetases: General scheme of structure–function relationships in the polypeptides and recognition of transfer RNAs, *Annu. Rev. Biochem.* **56:**125–158.

Secrist, J. A., III, Barrio, J. R., and Leonard, N. J., 1972, A fluorescent modification of adenosine tri-phosphate with activity in enzyme systems, *Science* **175:**646–647.

Semiostnov, G. V., Rodionova, N. A., Kutyshenko, V. P., Ebert, B., Blank, J., and Ptitsyn, O. B., 1987, Sequential mechanism of refolding of carbonic anhydrase B, *FEBS Lett.* **224:**9–13.

Shobha, J., Bhattacharyya, B., and Balasubramanium, D., 1989, Use of micelles in studying drug binding site: Simulation of the tubulin bound fluorescence of colchicine, *J. Biochem. Biophys. Methods* **18:**287–296.

Shore, J. D., and Chakrabarti, S. K., 1976, Subunit dissociation of malate dehydrogenase, *Biochemistry* **15:**875–879.

Stryer, L., 1965, The interaction of a Napthalene dye with Apomyoglobin and Apohemoglobin: A flourescent probe for non-polar binding sites, *J. Mol. Biol.* **13:**482–495.

Weber, G., and Farris, F. J., 1979, Synthesis and spectral properties of a hydrophobic fluorescent probe: 6-Propionyl-2-dimethyl amino napthalene, *Biochemistry* **18:**3075–3078.

Weisenberg, R. C., Borisy, G. G., and Taylor, E. W., 1968, The colchicine binding protein of mammalian brain and its relation to microtubule, *Biochemistry* **7:**4466–4478.

Wilson, L., 1970, Properties of colchicine binding protein from chick embryo brain. Interactions with vinca alkaloids and podophyllotoxin, *Biochemistry* **9:**4999–5007.

Wu, P., and Brand, L., 1992, Orientation factor in steady-state and time-resolved resonance energy transfer measurements, *Biochemistry* **31:**7939–7947.

Yarbrough, L. R., and Kirsch, M., 1981, Binding of fluorescent analogs of GTP to the exchangeable nucleotide binding site of tubulin, *J. Biol. Chem.* **256:**112–117.

Chapter 5

Pyrene Excimer Fluorescence as a Probe of Protein Conformational Change

Sherwin S. Lehrer

I. INTRODUCTION

Fluorescence probes are useful in studies of protein conformation and interactions because of the availability of fluorophors with a variety of fluorescence properties, of derivatives with different labeling specificities, and because of the accessibility of excellent fluorescence instruments (Haugland, 1992; Lakowicz, 1983). Derivatives of the fluorophor can be attached to specific amino acid residues of proteins while maintaining the spectroscopic properties of the parent chromophore. Information about the protein domain to which the probe is attached can be obtained because the sensitive fluorescence properties depend on the nature of the excited state. Due to the more polar electron distribution in the excited state than the ground state, the probe can interact strongly with the immediate environment, allowing for the use of a variety of fluorescence measurements: (1) the spectral characteristics depend on the polarity of the immediate environment; (2) the relative quantum yield or intensity depends on the proximity of quenching groups; (3) changes in relative probe exposure can be obtained by solute quenching techniques; (4) as fluorescence energy donors and

Sherwin S. Lehrer Boston Biomedical Research Institute, and Department of Neurology, Harvard Medical School, Boston, Massachusetts 02114.

Subcellular Biochemistry, Volume 24. Proteins: Structure, Function, and Engineering, edited by B. B. Biswas and Siddhartha Roy. Plenum Press, New York, 1995.

acceptors, probes can provide information on changes in distance in the 15–60 Å range; and (5) polarization of fluorescence measurements can give information about the flexibility of the domain to which the probes are attached.

A particular property of the excited state of some aromatic fluorophors—the ability to form an excited state dimer (excimer)—can be used to obtain information about proximity at close distances. The most useful fluorophors are derivatives of pyrene, whose excimer fluorescence was first reported by Kaspar and Förster in 1954 (reviewed in Förster, 1969) and whose properties have been extensively studied by Birks and co-workers (Birks, 1970). Excimer fluorescence has also been observed for benzene and toluene (Birks, 1970) and phenol and para-substituted phenols (Lehrer and Fasman, 1965), suggesting that phenylalanine and tyrosine could exhibit intrinsic excimer fluorescence in proteins if pairs of these groups can appropriately interact. The more efficient excimer fluorescence of pyrene derivatives attached to two protein sites that are close together can be used to probe conformational changes associated with a variety of protein interactions. With the use of site-directed mutagenesis, amino acid residues that selectively react with pyrene derivatives can be substituted for existing residues at regions of the protein that are suspected or known to be in close proximity in the structure. Although other spectroscopic properties of pyrene derivatives have been used in protein studies, only a limited number of studies involving protein excimer fluorescence have been reported since the initial studies of pyrene excimer fluorescence in proteins (Betcher-Lange and Lehrer, 1978; Zama *et al.*, 1978). A previous review in this series has emphasized the use of intermolecular excimer fluorescence to study properties of membranes (Wieb Van der Meer, 1988).

This chapter will emphasize intramolecular excimer fluorescence in aqueous solution relevant to protein studies. It will provide some background of the photophysical properties of pyrene, techniques for specific labeling of proteins, examples from published studies, and suggestions for experiments that could benefit from this approach.

2. PYRENE EXCIMER FORMATION IN "GOOD" SOLVENTS

In "good" (organic) solvents, pyrene molecules do not interact in the ground state, but at sufficiently high concentration ($\approx$ 50 μM) a diffusion-controlled collision between an excited and a ground-state molecule produces an excited-state dimer. Normally such collisions result in concentration quenching of the excited monomer, but due to the electronic structure of the excited state the excited monomer, M*, interacts with a ground state monomer, M, to produce an excimer, E*, as shown in the kinetic pathway of scheme A (Fig. 1).

The excimer produces a red-shifted unstructured spectrum (Fig. 2) because

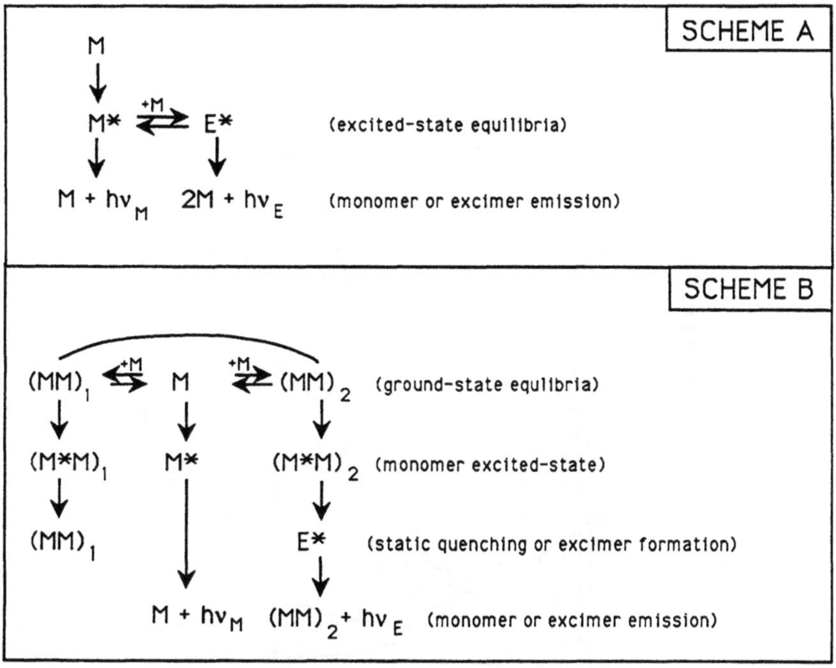

FIGURE 1. Schematic diagram of the formation of pyrene excimers, E*, upon excitation of pyrene monomers, M, for pyrene derivatives in "good" solvents (scheme A) and in aqueous solution (scheme B). (Scheme A) In good solvents the excimer fluorescence originates from pyrenes that do not interact in the ground state. (Scheme B) In aqueous solution the pyrenes equilibrate between those that do not interact, M, and those that interact hydrophobically in the ground state, $(MM)_1$ and $(MM)_2$. Upon excitation, M produces monomer fluorescence, $(MM)_1$ yields no fluorescence due to static quenching, and $(MM)_2$ produces excimer fluorescence due to the ability to attain the excimer configuration.

it originates from a lower energy excited state than the monomer and dissociates on returning to the ground state. The concentration dependence of the monomer fluorescence is described by an equation analogous to concentration quenching, with an inverse relationship for the excimer dependence (Birks, 1970). When pyrenes are attached to carrier molecules, intramolecular excimers can also form depending on steric accessibility of the pyrenes for each other (Hirayama, 1965). In good solvents, for both intra- and intermolecular excimer formation, the yield also depends on the temperature and the viscosity of the medium. In aqueous solution, the excimer yield is not as sensitive to these parameters, as will be discussed below. Information as to the structure of the excimer was obtained from pyrene crystals, which exclusively show excimer fluorescence (see Birks, 1970; Förster, 1969). The crystal structure indicates that the pyrenes interact as dimers, stacked in a precise manner with a distance of 3.5 Å between pyrene

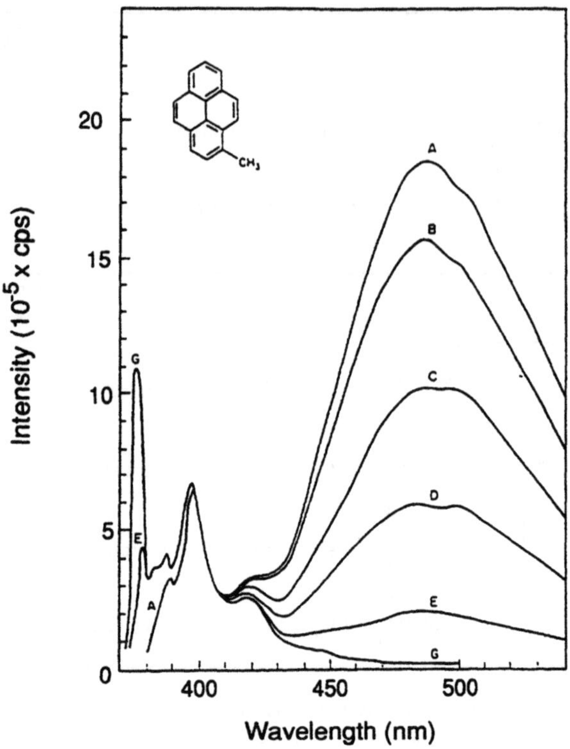

FIGURE 2. The monomer and excimer fluorescence of 1-methyl pyrene in cyclohexane at increasing concentrations. A = 10^{-2} M; B = 7.75×10^{-2} M; C = 5.5×10^{-3} M; D = 3.25×10^{-3} M; E = 10^{-3} M; G = 10^{-4} M. The spectra are normalized at 400 nm. Reprinted from Birks and Christophorou (1963), with kind permission from Pergamon Press Ltd.

planes. Thus, the formation of excimer in solution requires a specific interaction between pyrenes.

N-1 methyl pyrene, a compound substituted at the same position on the pyrene ring as in the labeling reagents used for proteins, does show efficient intermolecular excimer formation (Fig. 2), with similar spectra as pyrene. As the bulkiness and number of substituents on the pyrene ring increases, however, the efficiency of excimer formation decreases. Two additional properties of excimer fluorescence follow from scheme A: (1) the excitation spectrum of the excimer is identical to the monomer; (2) the time dependence of excimer fluorescence shows both a buildup and decrease. For labeled proteins in aqueous solution, however, the situation with respect to these two properties is somewhat altered, as discussed below. In aqueous solution, these compounds are not sufficiently soluble to reach the concentration necessary for excimer fluorescence to be observed. The problem of solubility can be solved by attaching the pyrenes to a water-soluble carrier via the techniques developed for the general labeling with probes. This has been done and information regarding quenching versus excimer formation has been obtained for some systems. It has been found that in aqueous solution there are nonspecific hydrophobic interactions between pyrenes in the ground state, which can result in concentration quenching that competes with excimer fluorescence if the excimeric configuration can not be reached.

3. PYRENE EXCIMER FORMATION IN AQUEOUS SOLUTION

3.1. Proteins

One of the first pyrene labeling reagents developed for proteins was N-1 pyrene maleimide (Weltman *et al.*, 1973) (note that the correct formula is N-1 not N-3), which due to its maleimide moiety was quite specific for the SH groups of exposed Cys side chains. It has two interesting properties as a general reagent: (1) a low quantum yield in aqueous solutions that markedly increases as a result of the labeling reaction (Weltman *et al.*, 1973); and (2) at high pH values (>pH 8), the resulting succinimide ring undergoes hydrolysis or aminolysis (with a neighboring Lys group) (Ishii and Lehrer, 1986; Wu *et al.*, 1976). N-1 pyrene iodoacetamide is now also available and both reagents are commonly used to label SH-groups with pyrene.

The first observation of excimer fluorescence in proteins employed pyrene maleimide labeling of tropomyosin (Betcher-Lange and Lehrer, 1978) and histone H3 in chromatin (Zama *et al.*, 1978). In both studies the SH-groups were on different peptide chains so the observation of an excimer immediately verified proximity. Both studies also showed that the excimer fluorescence was conformationally sensitive, i.e., the excimer fluorescence disappeared when the system was unfolded or dissociated. Conversion of type I (intact succinimide ring) to type II (cleaved succinimide ring) by incubation at high pH values (Wu *et al.*, 1976) red-shifted the monomer fluorescence and increased the excimer/monomer ratio (Fig. 3). The increased excimer is probably due to the greater rotational freedom around the extra C-C single bonds produced by the cleavage (see structure of Fig. 5). The monomer spectrum of type II pyrene maleimide is very similar to pyrene iodoacetamide.

Prior to these protein studies there was only one mention of pyrene excimer formation in aqueous solutions in the literature, for labeled butanediol, one of the systems tested in the original study of pyrene maleimide (Weltman *et al.*, 1973). It was therefore important to study the spectroscopic properties of pyrene in aqueous solution; e.g., does the excimer fluorescence in aqueous solution originate from noninteracting ground-state monomers as it does for excimers in good

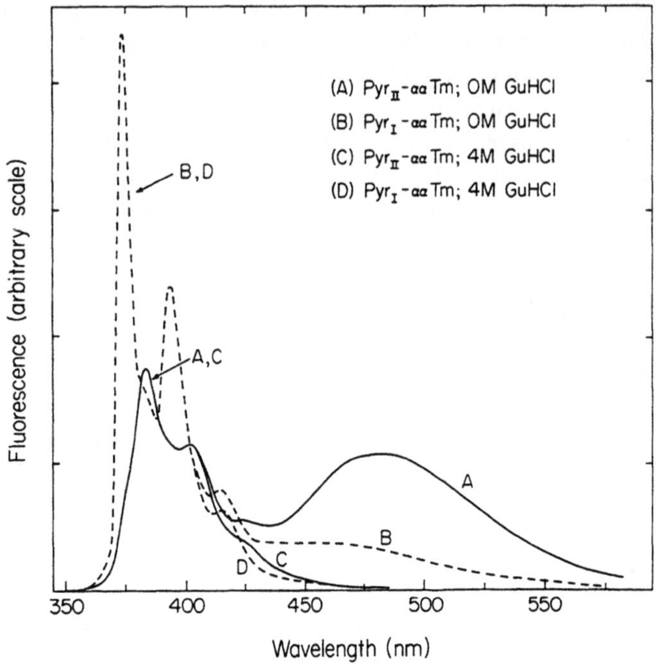

FIGURE 3. Fluorescence emission spectra of tropomyosin labeled with N-(1-pyrene)maleimide at pH 6.0 (Pyr$_{I}$ $-\alpha\alpha$ Tm) and pH 7.5 (Pyr$_{II}$ $-\alpha\alpha$Tm) in the absence and presence of GdmCl. Reprinted from Betcher-Lange and Lehrer (1978).

solvents. Although the excitation spectrum of the excimer was similar to the monomer, indicating that it did not originate from an impurity, the excimer excitation spectrum (Betcher-Lange and Lehrer, 1978; Zama *et al.*, 1978) and the absorption spectrum (Betcher-Lange and Lehrer, 1978) was broadened and slightly shifted to the red (Fig. 4). This indicated that the excimer fluorescence originated from pyrenes that were in a different environment than pyrenes that produced monomer fluorescence.

Further studies of the time dependence of the monomer and excimer emission in the nanosecond time region indicated that the excimer forms very quickly during monomer excitation (Graceffa and Lehrer, 1980); i.e., that the pyrenes are interacting in a pre-excimeric ground state, without requiring the time delay observed for intramolecular excimer formation of dipyrenyl compounds in organic solvents (Zachariasse *et al.*, 1984). It appeared that only a slight rotation was required to bring the pyrenes into the excimeric state after excitation. Thus, in contrast to pyrene excimer formation in good solvents, where the excimer is formed from noninteracting monomers by translational (intermolecular) or rotational (intramolecular) diffusion, for tropomyosin in aqueous solution the excimer is formed from interacting ground-state pyrenes.

Similar conclusions were reached from fluorescence studies of pyrene incorporated into water-soluble polymers (Herkstroeter *et al.*, 1983; Turro and Arora, 1986; Winnick *et al.*, 1987) and from studies with the protein gelsolin labeled with pyrene (Silva *et al.*, 1992). The question arose: Do all interacting ground-state pyrenes form excimer? Evidence for ground-state interaction that led to quenching rather than excimer was suggested from some studies with pyrene

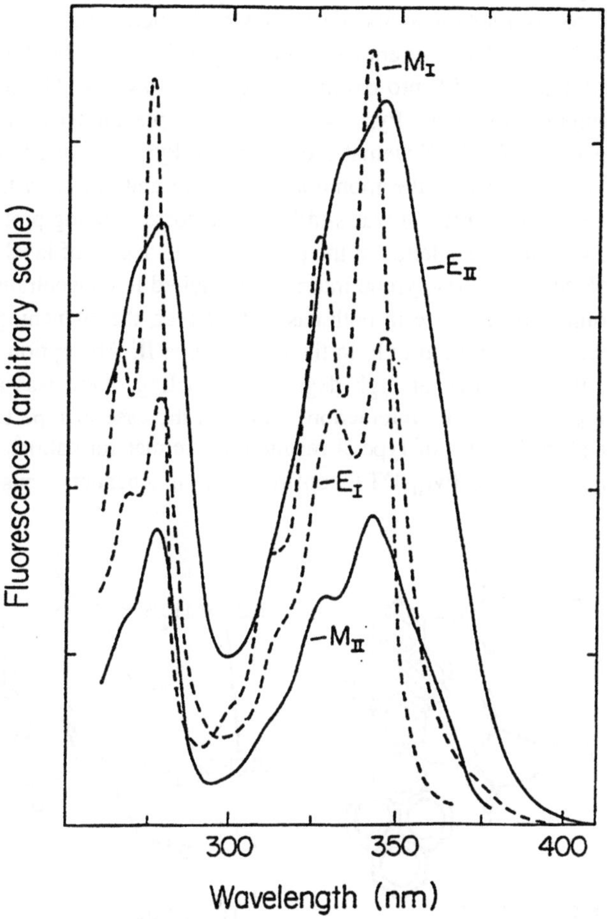

FIGURE 4. Fluorescence excitation spectra associated with monomer, M_I and M_{II}, and excimer, E_I and E_{II}, emission of Pyr_I-αα Tm and Pyr_{II}-αα Tm of Fig. 3, in the absence of GdmCl. Reprinted from Betcher-Lange and Lehrer (1978).

iodoacetamide-labeled tropomyosin (Ishii and Lehrer, 1990). To further explore this question, studies were performed with a model compound.

3.2. Pyrene–Dithiothreitol

In order to learn more of the factors that determine excimer formation in aqueous solution as opposed to organic solvents, studies were performed with pyrene-labeled dithiothreitol (Pyr-DTT) (Ishii and Lehrer, 1989), which contains two SH groups separated by four single-bonded C-atoms and is soluble in water as well as organic solvents. DTT in dimethyl formamide or dimethyl sulfoxide was reacted to completion at its SH-groups with increasing ratios of pyrene maleimide to form S-[N-(1-pyrene)succinimido]$_X$-DTT (Pyr$_I$-DTT), where $x = 0 \rightarrow 2$, and diluted to 1 μM into organic or aqueous solvents. The product with cleaved succinimido-groups (Pyr$_{II}$-DTT) was also prepared by incubation of Pyr$_I$-DTT in 1 mM NaOH. Schematic structures of Pyr$_I$-DTT and Pyr$_{II}$-DTT are shown in Fig. 5. In water, the monomer excitation and emission fluorescence spectrum of each type (Fig. 6) was similar to the corresponding protein adducts (Figs. 3 and 4). The dependence of the spectra on the degree of labeling showed that intramolecular pyrene–pyrene interaction resulted in concentration quenching of monomer fluorescence in both cases. However, the quenching is accompanied by excimer formation only in the case of type II. Absorption and excitation spectra indicated that at high degrees of labeling, there was appreciable ground-state pyrene–pyrene interaction, which in the case of type I resulted in quenching and in the case of type II resulted in excimer formation.

Lifetime studies of Pyr$_{II}$-DTT confirmed the independent emission of ex-

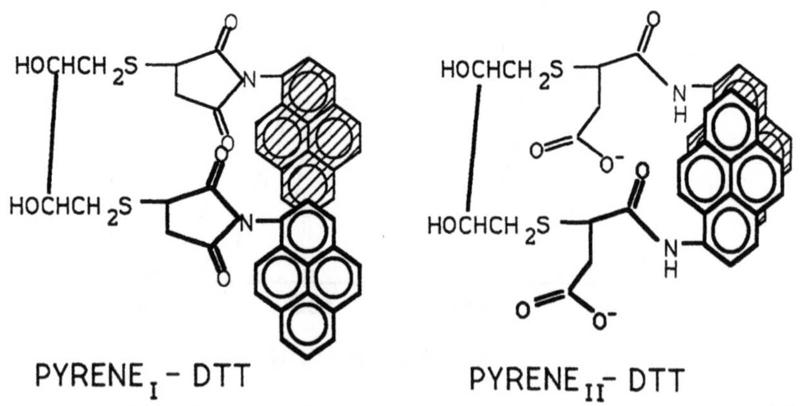

FIGURE 5. Schematic diagram of the two types of pyrene labeled dithiothreitol products obtained after reaction with pyrene maleimide. Pyrene$_I$-DTT, with intact succinimide rings, pyrene$_{II}$-DTT, cleaved succinimide rings. Type II was obtained after hydrolysis of type I at alkaline pH.

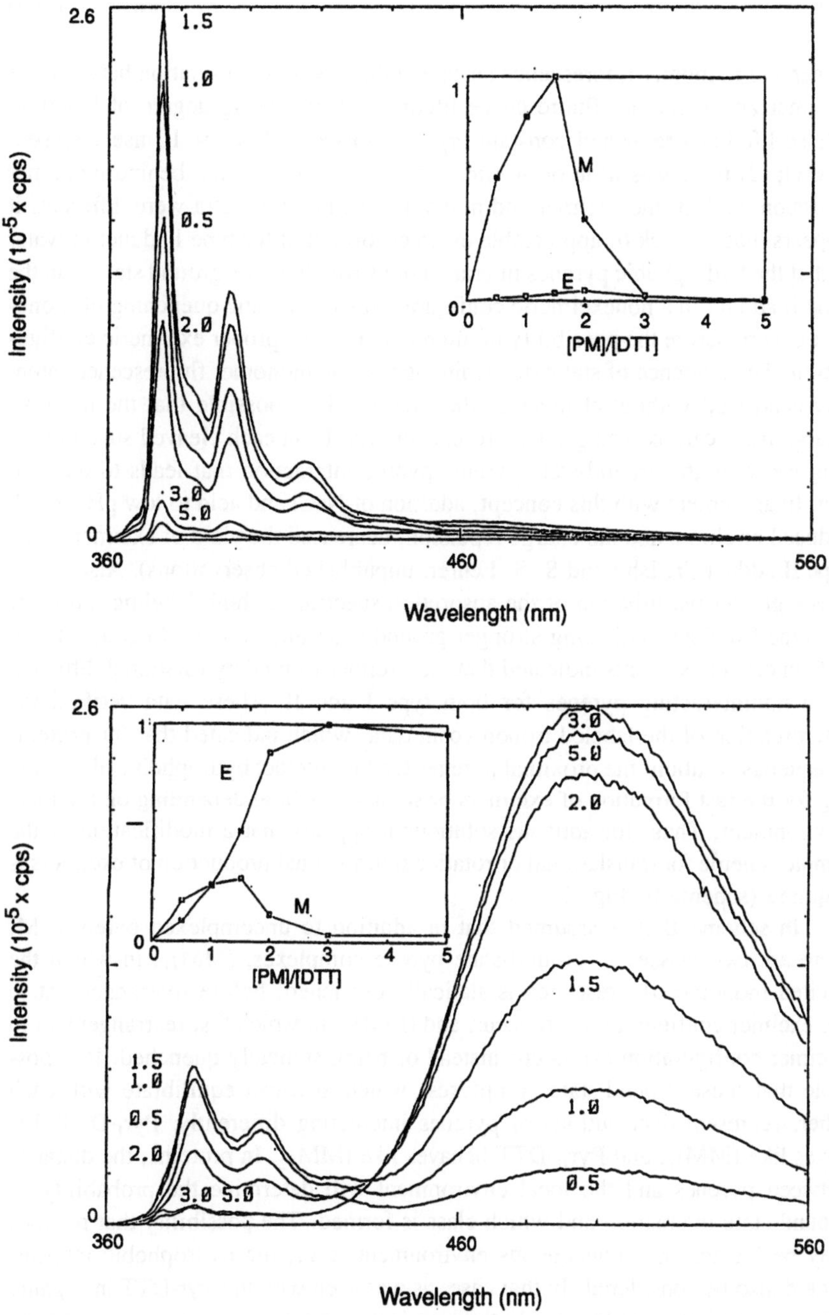

FIGURE 6. Fluorescence spectra of (A) pyrene$_I$-DTT and (B) pyrene$_{II}$-DTT, at the noted increasing ratios of reacted pyrene maleimide/DTT. Insert: The variation of monomer, M, and excimer, E, peak fluorescence with labeling ratio. Note that at increasing labeling ratios, the monomer fluorescence is quenched without formation of excimer for type I, but with formation of excimer for type II. Reprinted from Ishii and Lehrer (1989).

cimer and monomer fluorescence since: (1) there was no correlation between the excimer and monomer fluorescence lifetimes at increasing degree of labeling, whose lifetimes remained constant (τ_E = 40 nsec and τ_M = 10 nsec, respectively); (2) there was little or no time lag in excimer emission behind monomer emission; and (3) the excimer and monomer excitation spectra were different. It appears that the lack of appreciable excimer formation for type I adduct in water is that the hydrophobic pyrenes interact more strongly in the ground state than the type II adduct in a nonexcimeric configuration so that static quenching of monomer occurs before the possibility of alignment into the proper excimeric configuration. The evidence of static quenching is that the monomer fluorescence intensity decreased without changes in the lifetime. It is possible that the negative charge from carboxylate groups present for type II on each cleaved succinimide ring weakens the ground-state pyrene–pyrene interaction that leads to quenching. In agreement with this concept, addition of NaCl and acid to low pH, which reduced or eliminated the charge repulsion, decreased the yield of excimer in the type II adduct (Y. Ishii and S. S. Lehrer, unpublished observations). Also, there was a greater perturbation of the absorption spectrum at high labeling ratios for the type I adduct, indicating stronger ground-state interactions. In contrast, the data in organic solvents indicated that the excimer formed by rotational diffusion from noninteracting pyrenes for both type I and II. These data verified the interpretation of the labeled tropomyosin data, which indicated that for proteins in aqueous solution, the proximal pyrenes tend to interact hydrophobically allowing for the fast formation of excimers or static quenching depending on the local environment. Thus, for aqueous solutions it appears that a modification of the simple scheme for translational or rotational diffusional production of excimers is required (scheme B, Fig. 1).

In scheme B it is assumed that in addition to uncomplexed pyrenes, M, there are two classes of ground-state pyrene complexes: $(MM)_1$, in which the excited monomer fluorescence is statically quenched, before rearrangement to the excimer configuration can occur, and $(MM)_2$, in which fast rearrangement to excimer configuration can occur instead of being statically quenched. It is possible that these ground-state complexes, which also can equilibrate with each other, represent distributions of pyrenes interacting differently. Pyr_I-DTT, behaves like $(MM)_1$, and Pyr_{II}-DTT behaves like $(MM)_2$. In proteins, the distance between pyrenes and the local environment will determine the probability of ground-state interaction and which class is formed. The possibility that pyrenes may be located in a nonaqueous environment, e.g., the hydrophobic interior, should also be considered. In that case, as was seen with the Pyr-DTT in organic solvents, excimers could form in the classical manner, i.e., from noninteracting pyrenes (Ishii and Lehrer, 1989). In that case the excitation spectra of monomer and excimer should be quite similar.

4. LABELING OF PROTEINS

Cys groups are the prime choice for specific labeling of proteins for several reasons: (1) maleimide and iodoacetamide pyrene derivatives which preferentially label Cys are commercially available (e.g., Molecular Probes, Eugene, OR; Sigma, St. Louis, MO; Aldrich, Milwaukee, WI); (2) only a limited number of Cys are usually present in proteins; (3) the Cys that are present as the disulfide or can be oxidized to the disulfide will offer a high probability of forming excimer when labeled; and (4) Cys groups can be introduced into the amino acid sequence by site-directed mutation techniques. In the case of Cys present as disulfide, the disulfide cross-link must first be reduced. This can be accomplished with dithiothreitol (DTT), most readily in the unfolded state in the presence of denaturant such as GdmCl or urea. The DTT is then removed by dialysis or gel filtration, the labeling reaction is carried out in the presence of denaturant, and finally the denaturant is removed to renature the protein. Selective labeling of pairs of Cys is possible by varying the reaction conditions, e.g., at increased denaturant concentration to selectively reduce a pair which becomes exposed in a partially unfolded intermediate. In the case of the presence of Cys groups in the reduced form, labeling can be carried out in the native state if the Cys groups are accessible or in the unfolded state if they are not. The labeling times are controlled by quenching the reaction with excess DTT or by dropping the pH to low values ($<$ pH 4) where the reaction is inefficient. If the labeling reaction is carried out in the native state, the course of the reaction can be monitored fluorometrically, noting that excimer only forms when each molecule is doubly labeled and the reaction stopped when saturation begins to take place. The great wavelength separation between the excitation (340 nm) and the excimer emission (470–500 nm) will minimize effects of light scattering due to the insoluble pyrene maleimide or iodoacetamide, as will the use of low concentrations.

Mixed organic-water solutions can be explored to keep the reagent soluble and thereby facilitate reaction. Reaction with either pyrene maleimide or pyrene iodoacetamide is usually performed at pH 6.5–8.0 using a 5–10X excess of reagent over the SH concentration, diluted $\approx$20X from DMF, acetone, methanol, or DMSO. Reaction times will vary, but 1–5 hr at room temperature or below is often sufficient. Both reagents are not very soluble in aqueous solution so a precipitate usually appears, and therefore gentle shaking facilitates the reaction. After quenching the reaction, the solution is filtered or spun to remove excess undissolved reagent and dialyzed or gel filtered to remove unreacted dissolved reagent. Due to the hydrophobicity of pyrene, some noncovalent binding of the reagent may take place; therefore, exhaustive dialysis in the presence of DTT, which increases the solubility of the reagent on reaction, may be necessary. If maleimides are used, the different fluorescent properties of type I and type II labeled proteins can be explored (see above). Since the labeled protein may have altered properties, activity or other studies to determine if the labels have affected its properties should be performed.

Although both maleimides and iodoacetamides react preferentially with Cys, there is the possibility of reaction with other protein side chains such as Lys. Maleimides and iodoacetamides react with the basic form of both Cys and Lys, but Cys is preferentially labeled if present. To verify specific labeling at Cys, a control labeling reaction with the protein that has had its Cys groups reversibly blocked with 5,5′-dithiobis-2-nitrobenzoate can be performed (Lehrer and Ishii, 1988). Proteins with blocked Cys groups should have no fluorescence unless other groups reacted. Standard limited protein cleavage methods and peptide analysis can also be used to determine specificity and location of label.

5. EXAMPLES OF PROTEIN STUDIES

5.1. Tropomyosin

Tropomyosin (Tm) is a coiled-coil α-helix molecule, involved in the regulation of muscle contraction (Leavis and Gergely, 1984; Zot and Potter, 1987). Rabbit skeletal Tm mainly consists of αα and αβ isoforms with a Cys at position 190 in both α and β chains. The chains interact in register and in parallel as shown by the ability to produce a disulfide bond between the Cys 190s of each chain and the observation of excimer fluorescence from pyrene maleimide and pyrene-iodoacetamide-labeled Tm (Betcher-Lange and Lehrer, 1978; Lin, 1982). The labels did not perturb the structure appreciably since the thermal infolding profile of the α-helix, measured by CD, indicated that the label only slightly shifted the unfolding profile to somewhat lower temperature without decreasing the initial α-helicity. The temperature dependence of the excimer intensity showed an increased excimer fluorescence in the 30–40 °C region, associated with a small decrease in helix content, before decreasing to zero as the molecule cooperatively unfolded above 40 °C. This observation provided evidence for a partially unfolded intermediate. Presumably, the local unfolding provided greater flexibility for the pyrenes to more efficiently interact. That the interaction was a changed ground-state interaction was provided by excitation spectra and fluorescence lifetime studies. Thus, the increase in excimer was due to a shift in equilibrium between $(MM)_1$ and $(MM)_2$ ground-state classes (Fig. 1, scheme B) in going from the helix to the partially unfolded intermediate.

In another study, information about the composition of chicken gizzard smooth Tm (GTm) was obtained with pyrene labels on Cys. GTm consists of an equal mixture of two chains α and β, with a Cys at position 190 for α and position 36 for β (Lau *et al.*, 1985; Sanders and Smillie, 1985). Thus, excimer fluorescence is only possible for homodimers. The native molecule consists of αβ heterodimer (Sanders *et al.*, 1986), but after denaturation/renaturation a mixture of homodimers is produced (Burtnick *et al.*, 1988; Graceffa, 1989; Lehrer and Stafford, 1991). Thus, the native molecule showed little excimer fluorescence, which increased on denaturation/renaturation. Interestingly, the mixture of homodimers could exchange chains and form heterodimers when equilibrated at physiological temperature as shown by a loss of excimer fluorescence (Lehrer and Stafford, 1991) in agreement with the reversibility of $\alpha\beta = \alpha\alpha + \beta\beta$.

The greatest relative yield of pyrene excimer fluorescence was obtained with a nonmuscle Tm from platelets (Burtnick *et al.*, 1986). This appears to be due to the fact that the labeling site is the penultimate Cys, a region of the molecule that should be most flexible, allowing for efficient excimer formation, in contrast to the muscle Tm's where the Cys are located away from the ends. Interactions of Tm with actin (Burtnick *et al.*, 1986; Ishii and Lehrer, 1985) and with troponin (Ishii and Lehrer, 1991) produced a drop in the excimer yield, consistent with a decrease in tropomyosin flexibility. This change in excimer intensity has been used to obtain stoichiometry and binding constants. Long-range effects on excimer fluorescence of pyrene iodoacetamide-labeled Tm bound to actin were observed on binding myosin heads (S1) to actin by steady-state titrations and by kinetic methods (Ishii and Lehrer, 1990, 1993; Geeves and Lehrer, 1994).

5.2. Other Muscle Proteins

5.2.1. Troponin C and Troponin I

Troponin C (TnC) initiates the process of the regulation of striated muscle contraction when it undergoes a conformational change associated with binding of Ca^{2+} to specific sites. This causes reorganization of the other troponin subunits and Tm, allowing S1 to interact with actin (Leavis and Gergely, 1984; Zot and Potter, 1987). Cardiac troponin C, which contains two Cys groups about 20 Å apart, exhibits excimer fluorescence on labeling with pyrene iodoacetamide and pyrene maleimide (type II) (Liou and Fuchs, 1992; Verin and Gusev, 1988). Conformational changes on Ca^{2+} binding have been implicated by observing a decrease in excimer fluorescence. A recent study showed that the SH groups can be oxidized to the disulfide, indicating considerable conformational flexibility (Putkey *et al.*, 1993). In another study with skeletal TnC, two Cys were introduced by site-directed mutagenesis at Cys positions that were estimated from the X-ray structure to be 20 Å apart, which increased to 10–12 Å on binding Ca^{2+} by fluorescence-detected energy transfer measurements (Wang *et al.*, 1992). A small amount of excimer fluorescence was observed that was abolished in the presence of Ca^{2+}. Both of these experimental observations are consistent with the proposal of specific movements of domains on binding Ca^{2+} (Herzberg *et al.*, 1986). The observation of excimer fluorescence in these two cases also illustrates that the pyrenes can still stack properly at these large distances probably due to the flexibility of the structures and that the pyrene ring is attached to the S atom via at least three intervening single bonds.

Troponin I is involved in the Ca^{2+}-regulation of muscle contraction by changing its mode of interaction with TnC and actin when TnC binds Ca^{2+} (Leavis and Gergely, 1984; Zot and Potter, 1987). TnI contains three Cys groups, one of which is exposed and can be blocked with iodoacetamide and the other pair can be oxidized to disulfide. Instead of forming the disulfide, the other pair was labeled with pyrene maleimide and type II was formed by equilibration at alkaline pH (Strassburg and Leavis, 1985). The resulting labeled TnI showed excimer fluorescence that was Ca^{2+}-dependent in complexes with TnC. This provided information as to specific conformational changes within TnI involved in the transmission of information from TnC to TnI on Ca^{2+}-binding.

5.3. Other Proteins

5.3.1. Histone H3

As discussed above, conformationally dependent pyrene excimer fluorescence was seen between histone H3 molecules incorporated into erythrocyte nucleosomes (Zama *et al.*, 1978). In the native state both excimer and monomer fluorescence are observed. Under conditions that perturb the interaction between H3 molecules, only monomer fluorescence was observed. A visually interesting study was reported of the incorporation of pyrene-labeled H3 into nucleosomes in culture, which took advantage of the blue emission of the monomeric H3 compared to the green emission of the excimer when H3 molecules interact (Prior *et al.*, 1980). On incubation of histone containing specifically labeled H3, fluorescent micrographs clearly showed the presence of blue fluorescence in the cytoplasm, which decreased with time as green fluorescence continued to localize in the nuclei (isolated with a 470-nm cut-on filter). The presence of green fluorescence visually indicated that the histone molecules assembled properly in the nucleosome.

5.3.2. Proteins with Multiple Labeling Sites

Two Cys groups of gelsolin from horse plasma were specifically labeled with pyrene maleimide and pyrene iodoacetamide in the native state without appreciably affecting the structure (Silva *et al.*, 1992). The observation of excimer fluorescence immediately indicated proximity of the as yet unidentified groups. Studies of the excitation and absorption spectra indicated that the excimer fluorescence originated from pyrenes interacting in the ground state.

Several other studies of proteins that have multiple Cys and thereby multiple potential labeling sites for pyrene maleimide and pyrene iodoacetamide have shown pyrene excimer fluorescence. These include the Ca^{2+}–ATPase of sarcoplamic reticulum (Luedi and Hasselbach, 1983; Papp *et al.*, 1986), tubulin (Panda and Bhattacharyya, 1992) and the crystallins from lens (Sen and Chakrabarti, 1990; Walsh *et al.*, 1991). A recent study employed the excimer fluorescence of pyrene-labeled Cys mutants of lactose permease to show proximity of transmembrane helixes (Jung *et al.*, 1994).

6. PROSPECTS FOR FUTURE STUDIES

It is clear from the studies discussed above that pyrene excimer fluorescence is a sensitive probe of conformation change. In aqueous solution, excimer emission appears to originate from pyrenes that interact hydrophobically in the ground state, and on excitation, excimer emission competes with static quenching and monomer emission. Excimer fluorescence will predominate for those conformations that allow the interacting pyrenes to reorient to the excimeric configuration. This will depend on the flexibility of the pyrene ring at the attachment site, the distance between pyrenes and the local environment of the protein. For these reasons, excimer fluorescence has been observed from pairs of Cys as close as those that can form a disulfide bond up to about 20 Å. It is possible that pyrenes could locate in the hydrophobic interior of a protein without interacting in the ground state. In this situation, analogous to the lipid environment of a membrane, the probability of excimer formation would be determined by the local viscosity of the medium in addition to the other factors. Whatever the mechanism, the conformationally sensitive excimer fluorescence can be used to study the binding of substrates and inhibitors, complex formation, and unfolding/refolding.

An interesting application would involve studies of the equilibrium and kinetics of protein unfolding/refolding using pyrenes attached to Cys, which normally would form a disulfide bond, or from knowledge of the three-dimensional structure of Cys that are located near each other or that can be introduced by site-directed mutagenesis. In such cases it would be expected that the excimer fluorescence would increase from low or negligible values as the tertiary structure is formed. In the case of oligomeric proteins, where assembly involves quaternary interactions as well as tertiary interactions, information about the folding process may be obtained if pyrenes could be placed on the surface near the interface between two subunits. If the pyrenes do not overly affect the folding process, the appearance of excimer would monitor the later assembly processes. Comparison with other spectroscopic landmarks that monitor secondary and tertiary structure would provide information as to the sequence of folding events.

ACKNOWLEDGMENTS. Supported by NIH HL22461 and NIH AR41367.

7. REFERENCES

Betcher-Lange, S., and Lehrer, S. S., 1978, Pyrene excimer fluorescence in rabbit skeletal αα tropomyosin labeled with N-(1-pyrene) maleimide, *J. Biol. Chem.* **253**:3757–3760.

Birks, J. B., 1970, *Photophysics of Aromatic Molecules*, Wiley-Interscience, London.

Birks, J. B., and Christophorou, L. G., 1963, Excimer fluorescence spectra of pyrene derivatives, *Spectrochim. Acta.* **19**:301–410.

Burtnick, L. D., Stewart, D.I.H., and Smillie, L. B., 1986, Excimer fluorescence of equine platelet tropomyosin labeled with N-(1-pyrenyl)iodacetamide, *Biochemistry* **25**:3875–3880.

Burtnick, L. D., Sanders, C., and Smillie, L. B., 1988, Fluorescence from pyrene-labeled native and reconstituted chicken gizzard tropomyosins, *Arch. Biochem. Biophys.* **266**:622–627.

Förster, T., 1969, Excimers, *Angew. Chem. Internat. Edit.* **8**:333–343.

Geeves, M. A., and Lehrer, S. S., 1994, Dynamics of the muscle thin filament regulating switch, *Biophys. J.* **67**:273–282.

Graceffa, P., 1989, In-register homodimers of smooth muscle tropomyosin, *Biochemistry* **28**:1282–1287.

Graceffa, P., and Lehrer, S. S., 1980, The excimer fluorescence of pyrene-labeled tropomyosin, *J. Biol. Chem.* **255**:11296–11300.

Haugland, R. P., 1992, *Handbook of Fluorescent Probes and Research Chemicals*, Molecular Probes, Inc., Eugene, Oregon.

Herkstroeter, W. G., Martic, P. A., Hartman, S. E., Williams, J. L. R., and Farid, S., 1983, Unique hydrophobic interactions of pyrene in aqueous solution as effected by polyelectrolytes and surfactants, *J. Polym. Sci. Polym. Chem. Ed.* **21**:2473–2490.

Herzberg, O., Moult, J., and James, M. N. G., 1986, A model for the Ca^{2+}-induced conformational transition of troponin C, *J. Biol. Chem.* **261**:2638–2644.

Hirayama, F., 1965, Intramolecular excimer formation, *J. Chem. Phys.* **42**:3163–3171.

Ishii, Y., and Lehrer, S. S., 1985, Fluorescence studies of the conformation of pyrene-labeled tropomyosin: Effects of F-actin and myosin subfragment 1, *Biochemistry* **24**:6631–6638.

Ishii, Y., Lehrer, S. S., 1986, Effects of the state of the succinimido-ring on the fluorescence and structural properties of pyrene maleimide-labeled αα-tropomyosin, *Biophys. J.* **50**:75–80.

Ishii, Y., and Lehrer, S. S., 1989, Intramolecular excimer fluorescence of pyrene maleimide-labeled dithiothreitol, in *Fluorescent Biomolecules* (D. M. Jameson and G. D. Reinhart, eds.), pp. 423–425, Plenum Press, New York.

Ishii, Y., and Lehrer, S. S., 1990, Excimer fluorescence of pyrenyliodoacetamide-labeled tropomyosin, *Biochemistry* **29**:1160.

Ishii, Y., and Lehrer, S. S., 1991, 2-Site attachment of troponin to pyrene-labeled tropomyosin, *J. Biol. Chem.* **266**:6894–6903.

Ishii, Y., and Lehrer, S. S., 1993, Kinetics of the "on–off" change in regulatory state of the muscle thin filament, *Arch. Biochem. Biophys.* **305**:193–196.

Jung, K., Jung, H., Wu, J., Prive, G., and Kaback, H. R., 1993, Use of site-directed fluorescence labeling to study proximity relationships in the lactose permease of *E. coli*, *Biochemistry* **32:**12273–12278.

Lakowicz, J. R., 1983, *Principles of Fluorescence Spectroscopy*, Plenum Press, New York.

Lau, S.Y.M., Sanders, C., and Smillie, L. B., 1985, Amino acid sequence of chicken gizzard tropomyosin, *J. Biol. Chem.* **260:**7257–7263.

Leavis, P. C., and Gergely, J., 1984, Thin filament proteins, *Crit. Rev. Biochem.* **16:**235–305.

Lehrer, S. S., and Fasman, G. D., 1965, Excimer fluorescence in liquid phenol, p-ethyl phenol and anisole, *J. Am. Chem. Soc.* **87:**4678.

Lehrer, S. S., and Ishii, Y., 1988, Fluorescence properties of acrylodan-labeled tropomyosin and tropomyosin-actin, *Biochemistry* **27:**5899–5906.

Lehrer, S. S., and Stafford, W. F., 1991, Preferential assembly of the tropomyosin heterodimer, *Biochemistry* **30:**5682–5688.

Lin, T.-I, 1982, Excimer fluorescence of pyrene-tropomyosin adducts, *Biophys. Chem.* **15:**277–288.

Liou, Y.-M., and Fuchs, F., 1992, Pyrene labeled cardiac troponin C, *Biophys. J.* **61:**892–901.

Luedi, H., Hasselbach, W., 1983, Excimer formation of ATPase from sarcoplamic reticulum labeled with N-(3-pyrene)maleimide, *Eur. J. Biochem.* **130:**5–8.

Panda, D., and Bhattacharyya, B., 1992, Excimer fluorescence of pyrene-maleimide labeled-tubulin, *Eur. J. Biochem.* **204:**783–787.

Papp, S., Kracke, G., Joshi, N., and Martonosi, A., 1986, The reaction of N-(1-pyrene)maleimide with sarcoplasmic reticulum, *Biochem. J.* **49:**411–424.

Prior, C. P., Cantor, C. R., Johnson, E. M., and Allfrey, V. G., 1980, Incorporation of exogenous pyrene-labeled histone into physarum chromatin, *Cell* **20:**597–608.

Putkey, J. A., Cotson, D. G., and Mouawad, P., 1993, Formation of inter- and intramolecular disulfide bonds can activate cardiac troponin C, *J. Biol. Chem.* **268:**6827–6830.

Sanders, C., and Smillie, L. B., 1985, Amino acid sequence of chicken gizzard β-tropomyosin, *J. Biol. Chem.* **260:**7264–7275.

Sanders, C., Burtnick, L. D., and Smillie, L. B., 1986, Native chicken gizzard tropomyosin is predominently a βγ-heterodimer, *J. Biol. Chem.* **261:**12774–12778.

Sen, A. C., and Chakrabarti, B., 1990, Proximity of sulfhydryl groups in lens proteins, *J. Biol. Chem.* **265:**14277–14284.

Silva, B. E. R., Koef, E. K., Burtnick, L. D., and Turro, N. J., 1992, Monomer and excimer fluorescence of horse plasma gelsolin labeled with *N*-(1-pyrenyl) iodoacetamide, *Biochem. Cell Biol.* **70:**573–578.

Strasburg, G. M., and Leavis, P. L., 1985, Troponin C-mediated Ca-sensitive changes in the conformation of troponin I detected by pyrene excimer fluorescence, *J. Biol. Chem.* **260:**366–370.

Turro, N. J., and Arora, K. S., 1986, Pyrene as a photophysical probe for intermolecular interactions of water-soluble polymers in dilute solutions, *Polymer* **27:**783–796.

Verin, A. D., and Gusev, N. B., 1988, Ca^{2+}-induced conformational changes in cardiac troponin C as measured by N-(1-pyrene) maleimide fluorescence, *Biochim. Biophys. Acta.* **956:**197–208.

Walsh, M. T., Sen, A. C., Chakrabarti, B., 1991, Micellar subunit assembly in a three-layer model of oligomeric α-crystallin, *J. Biol. Chem.* **266:**20079–20084.

Wang, Z., Gergely, J., and Tao, T., 1992, Characterization of the Ca^{2+}-triggered conformational transition in troponin C, *Proc. Natl. Acad. Sci. USA* **89:**11814–11817.

Weltman, J. K., Szaro, R. P., Frackelton, A. R. J., Dowben, R. M., Bunting, J. R., and Cathou, R. E., 1973, N-(3-pyrene)maleimide: A long lifetime fluorescent sulfhydryl reagent, *J. Biol. Chem.* **248:**3173–3177.

Wieb Van der Meer, B., 1988, Biomembrane structure and dynamics viewed by fluorescence, in *Subcellular Biochemistry*, Vol. 13: *Fluorescence Studies on Biological Membranes* (H. J. Hilderson, ed.), pp. 1–53, Plenum Press, New York.

Winnick, F. M., Winnick, M. A., Tazuke, S., and Ober, C. K., 1987, Synthesis and characterization of pyrene-labeled (hydroxypropyl)cellulose and its fluorescence in solution, *Macromolecules* **20**:38–44.

Wu, C., Yarbrough, L. R., and Wu, F. Y., 1976, N-(1-pyrene)maleimide: A fluorescence crosslinking reagent, *Biochemistry* **15**:2863–2868.

Zachariasse, K. A., Duveneck, G., and Busse, R., 1984, Intramolecular excimer formation with 1,3-Di(1-pyrenyl)propane, *J. Am. Chem. Soc.* **106**:1045–1051.

Zama, M., Bryan, P. N., Harrington, R. E., Olins, A. L., and Olins, D. E., 1978, Conformational states of chromatin, *Cold Spring Harb. Symp. Quant. Biol.* **42**:31–41.

Zot, A. S., and Potter, J. D., 1987, Structural aspects of troponin-tropomyosin regulation of skeletal muscle contraction, *Annu. Rev. Biophys. Biophys. Chem.* **16**:535–559.

Chapter 6

Differential Scanning Calorimetry of Proteins

Jose M. Sanchez-Ruiz

1. INTRODUCTION

Differential scanning calorimetry (DSC) is a powerful technique to characterize temperature-induced conformational changes in proteins and other biological macromolecules. In fact, DSC studies on protein thermal denaturation have played a central role in the development of current views about the factors that determine protein stability. Reviews on the various aspects of this technique are available in the literature (see, for instance, Privalov, 1979, 1982, 1989; Mateo, 1984; Sturtevant, 1987; Sanchez-Ruiz and Mateo, 1987; Freire *et al.*, 1990) and, consequently, this chapter will mainly focus on recent developments in the field. The relevant features of the DSC experiment and the preliminary data analysis will be outlined in this section, and the well-known equilibrium thermodynamics procedures for the analysis of the DSC profiles will be briefly summarized in Section 2. Recent work on the thermodynamics of model systems, together with accurate DSC studies into the denaturation of small globular proteins, has allowed researchers to propose molecular interpretations for the relevant denaturational thermodynamic parameters and to address the biotechnologically impor-

Jose M. Sanchez-Ruiz Departamento de Quimica Fisica, Facultad de Ciencias, Universidad de Granada, 18071-Granada, Spain.

Subcellular Biochemistry, Volume 24. Proteins: Structure, Function, and Engineering, edited by B. B. Biswas and Siddhartha Roy. Plenum Press, New York, 1995.

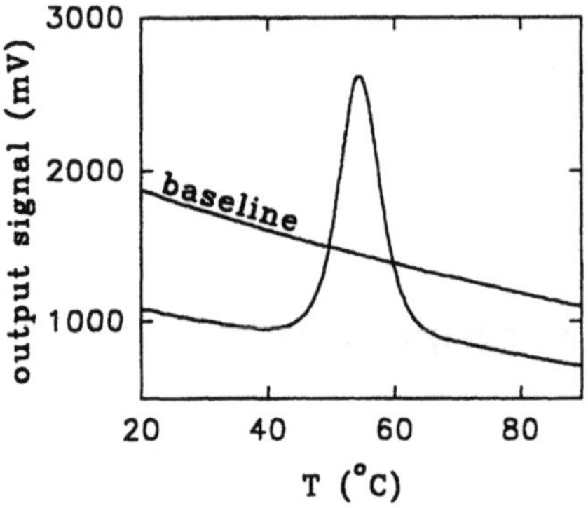

FIGURE 1. Original DSC data (output signal *versus* temperature) for an aqueous solution of ribonuclease a in 50 mM acetate buffer, pH 4 (protein concentration 4.9 mg/ml). The instrumental ("buffer–buffer") baseline is also shown. Recording from the author's laboratory using a DASM-1M calorimeter.

tant problem of the prediction of the folding–unfolding behavior of proteins; these developments will be described in Section 3. The controversial issue of the analysis of DSC data for protein irreversible denaturation has also received considerate attention during the last five years; the recently developed kinetic approaches to this analysis will be discussed in Section 4.

A differential scanning calorimeter basically consists of two cells: the reference cell and the sample cell (for a detailed technical description of this type of calorimeter, see Privalov *et al.*, 1975; Krishnan and Brandts, 1978; Privalov, 1980). This chapter is concerned with the case in which the sample is a diluted protein solution and the reference is the pure solvent (buffer). Both cells are simultaneously heated at a constant heating (i.e., scanning) rate and the instrument output is (proportional to) the difference in heat capacity between the two cells. A typical DSC profile (instrument output versus temperature) for a protein solution is shown in Fig. 1. Usually, the profile shows one peak (transition), which is attributed to the heat absorption associated with the denaturation of the protein (for "complex" proteins, several more or less overlapping transitions are often observed), while the pretransition and posttransition values of the output signal reflect the partial heat capacities of the native and denatured states of the protein, respectively. Also shown in Fig. 1 is the instrumental baseline obtained from a DSC scan with the two cells filled with buffer. This "buffer–buffer" baseline is commonly subtracted from the protein–buffer profile, in order to correct for instrumental distortions. Note (see Fig. 1) that, within most of the recorded temperature range, this subtraction would yield negative values. This indicates that the heat capacity of the reference is higher than that of the sample, which is the expected consequence of the lower water content in the latter (the protein "displaces" some water and the heat capacity of liquid water is higher than that of the protein).

Absolute values for the partial heat capacity of the protein can be obtained from the heat capacity difference between sample and reference and the partial

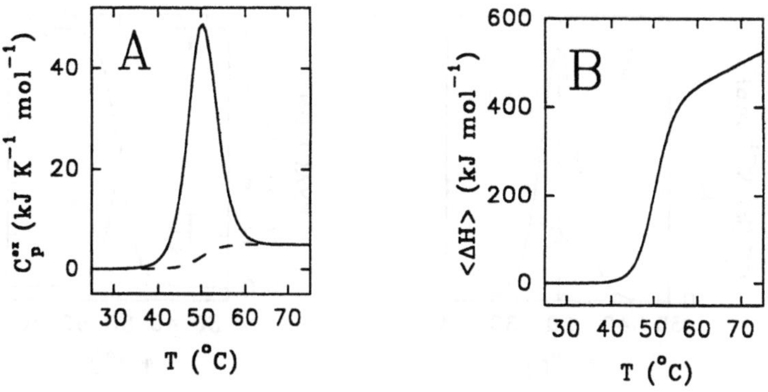

FIGURE 2. (A) Temperature-dependence of the excess heat capacity values predicted by the two-state equilibrium model. The heat capacity of the native state has been taken as the reference level and the dashed line represents the chemical baseline. The C_p^{ex} values were calculated by using Eq. (3) with the following parameters: T_m = 50 °C, $\Delta_N^D H(T_m)$ = 500 kJ/mole, and $\Delta_N^D C_p$ = 5 kJ·K^{-1}·mole^{-1}. In this simulation $\Delta_N^D C_p$ has been taken as constant. (B) Excess enthalpy *versus* temperature profile obtained by integration [Eq. (1)] of the C_p^{ex} *versus T* profile shown in A.

volume of the protein (obtained from densitometric measurements); see Privalov *et al.* (1989) for details. However, these absolute values are not required for many applications and, more often, the results of the DSC experiment are given as an excess heat capacity (C_p^{ex}) versus temperature profile, where C_p^{ex} is the partial heat capacity of the protein measured from an adequate reference level. The heat capacity of the native protein appears to be a reasonable choice for this reference level (this is in fact convenient, since many theoretical treatments of protein denaturation take the native protein as the reference state: see Section 2). Note, however, that the heat capacity of the native protein is only known at temperatures below those corresponding to the DSC transition; therefore, this reference level must be obtained by extrapolation from the (usually linear) temperature dependence of the pretransition values (see Fig. 1). The shape of C_p^{ex} versus temperature profile thus obtained is illustrated in Fig. 2A. By temperature integration of the C_p^{ex} data, the excess enthalpy (Fig. 2B) may be obtained:

$$\langle \Delta H \rangle = \int_{T_0}^{T} C_p^{ex} \cdot dT \tag{1}$$

where T_0 is a low temperature at which essentially all the protein is in the native state and the angle brackets are used to indicate that $<\Delta H>$ may be expressed as an average over protein states (see Section 2). According to theoretical treatments (Section 2), two terms contribute to the C_p^{ex} value obtained by taking the heat capacity of the native protein as reference level. The major contribution is related

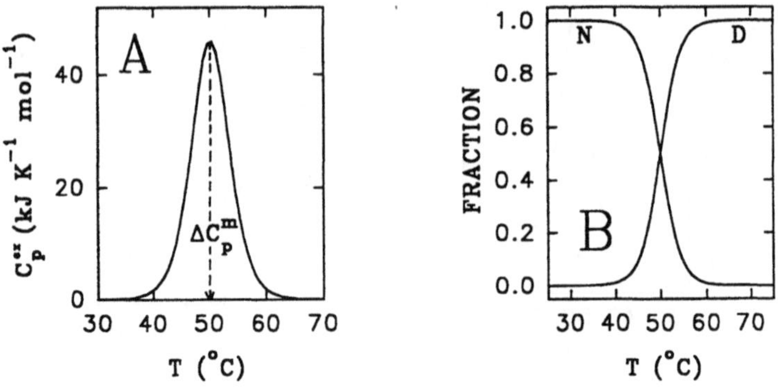

FIGURE 3. (A) Excess heat capacity *versus* temperature profile for the DSC transition shown in Fig. 2A, but, in this case, taking the chemical baseline as the reference level. (B) Population of native and denatured states *versus* temperature for the transition shown in A.

to the temperature dependence of the concentrations of the significantly populated protein states and is responsible for the presence of a "peak" (transition) in the thermogram. A second term can be identified with the average heat capacity of the protein states and is sometimes referred to as the "chemical baseline," given that it connects in a smooth way the pre- and posttransition values (see Fig. 2A). This chemical baseline could also be used as reference level for the calculation of excess heat capacity values. Of course, the average heat capacity of the protein states is not known beforehand; it is generally agreed, however, that any reasonable function that smoothly connects the pre- and posttransition values is an acceptable approximation to the true chemical baseline (see, for instance, Takahashi and Sturtevant, 1981). The shape of the C_p^{ex} versus temperature profiles obtained by using the chemical baseline as reference level is illustrated in Fig. 3A; obviously, these profiles only reflect the contribution from the temperature-induced changes in the populations of states. In this work, the reference level used in the C_p^{ex} calculations will always be indicated whenever confusion might otherwise arise.

In order to check the reversibility of the processes, DSC experiments on protein denaturation usually include a "reheating run"; that is, a second scan carried out after cooling the protein solution (inside the calorimetric cell) to room temperature. For several small globular proteins and some complex proteins (Privalov, 1979, 1982) the DSC thermograms corresponding to the first and reheating runs show identical (or almost identical) transitions, which indicates that the denaturation process is calorimetrically reversible. Protein reversible denaturation is thought to involve the unfolding of the protein structure and, consequently, the protein state obtained upon reversible denaturation is often described as the unfolded state. However, recent work (reviewed by Shortle, 1993) suggests that, in some cases, this state may not be completely unfolded and may still retain some residual structure. In this chapter, the reversibly denatured state of a protein will be simply referred to as the denatured state (D).

It must be recognized that protein thermal denaturation is often found to be calorimetrically irreversible, as shown by the lack of thermal effect (transitions) in the thermogram corresponding to the reheating run. Irreversible protein denaturation is usually attributed to alterations of the denatured state (aggregation, autolysis, chemical alteration of residues, etc.; see Klibanov and Ahern, 1987), which lead to a protein state (or states) that is unable to fold back to the native structure. This irreversibly denatured state will be referred to as the final state (F).

2. THE EQUILIBRIUM THERMODYNAMICS ANALYSIS OF DSC DATA

It is generally accepted that the analysis of DSC thermograms according to equilibrium thermodynamics is applicable to those cases in which calorimetric reversibility is found. It must be noted that, in some cases, reversible DSC profiles may be distorted due to the effect of the (possible) slow kinetics of the denaturation process. These and other distortions will be discussed in Section 4. In this section, we will assume that chemical equilibrium between the significantly populated protein states is always established and, also, that the experimental DSC profile accurately reflects the equilibrium denaturation process.

2.1. The Two-State Equilibrium Model

The simplest denaturation mechanism is that in which only the native (N) and denatured (D) states of the protein are significantly populated and their relative amounts at a given temperature are determined by the value of the denaturation equilibrium constant (K) at that temperature:

$$\mathrm{N} \rightleftharpoons \mathrm{D} \qquad K = (\mathrm{D})/(\mathrm{N})$$

Denaturational changes of thermodynamic quantities ($\Delta_{\mathrm{N}}^{\mathrm{D}}J$) are defined as the differences between the value of the quantity for the denatured state and that for the native one: $\Delta_{\mathrm{N}}^{\mathrm{D}}J = J(\mathrm{D}) - J(\mathrm{N})$, where J may be Gibbs energy (G), enthalpy (H), entropy (S), heat capacity at constant pressure (C_p), etc. Unless specifically noted, the thermodynamic quantities employed in this chapter refer to a mole of protein. Note that the above definition implies that we take the native state as the reference state (the usual convention in protein thermodynamics). Therefore, the excess enthalpy and the excess heat capacity are given by:

$$\langle \Delta H \rangle = x_D \, \Delta_N^D H = K \cdot \Delta_N^D H / (1 + K) \tag{2}$$

$$C_p^{ex} = \partial \langle \Delta H \rangle / \partial T = \frac{(\Delta_N^D H)^2}{RT^2} \times \frac{K}{(1+K)^2} + x_D \cdot \Delta_N^D C_p \tag{3}$$

where x_D [$= K/(1 + K)$] is the fraction of protein present as denatured state, and we have used the van't Hoff and Kirchoff equations to express the temperature dependencies of K and $\Delta_N^D H$: $\partial \ln K / \partial T = \Delta_N^D H / (RT^2)$ and $\partial \Delta_N^D H / \partial T = \Delta_N^D C_p$. The main features of the DSC transitions predicted by Eq. (3) are illustrated in Figs. 2 and 3 and are discussed below.

The denaturation temperature (T_m) is defined as the temperature at which $x_D = 1/2$ (and, hence, $K = 1$ and $\Delta_N^D G = 0$). The predicted transitions, however, are nearly symmetrical and the T_m value is close to the temperature corresponding to the maximum heat capacity. The second term in the right-hand side of Eq. (3) represents the average heat capacity of the protein and corresponds to the chemical baseline, while the first term arises from the temperature-induced shift of the denaturation equilibrium. The area enclosed by the transition and the chemical baseline (see Figs. 2 and 3) gives the total enthalpy change for the denaturation process, which is usually referred to as the *calorimetric enthalpy*. Clearly, the calorimetric enthalpy is just the $\Delta_N^D H$ value; note, however, that $\Delta_N^D H$ depends on temperature as predicted by the Kirchoff equation and the denaturational heat capacity increment [$\Delta_N^D C_p$]. It can be easily shown that little error is introduced if the calorimetric enthalpy is assigned to the denaturation temperature: $\Delta_N^D H(T_m)$. At $T = T_m$, the denaturational Gibbs energy change is zero; therefore, the denaturational entropy change at the denaturation temperature [$\Delta_N^D S(T_m)$] can be calculated as:

$$\Delta_N^D S(T_m) = \Delta_N^D H(T_m) / T_m \tag{4}$$

In principle, the denaturation heat capacity change can also be determined from the DSC profile (see Fig. 2) and, consequently, the $\Delta_N^D G$, $\Delta_N^D H$, and $\Delta_N^D S$ values at any temperature T can be calculated by using standard thermodynamic equations:

$$\Delta_N^D H(T) = \Delta_N^D H(T_m) + \int_{T_m}^{T} \Delta_N^D C_p \cdot dT \tag{5}$$

$$\Delta_N^D S(T) = \Delta_N^D H(T_m) / T_m + \int_{T_m}^{T} [\Delta_N^D C_p / T] \cdot dT \tag{6}$$

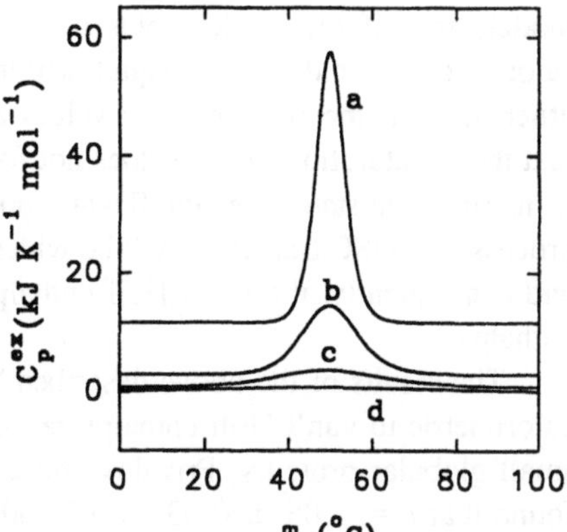

FIGURE 4. For an equilibrium, two-state transition, the calorimetric enthalpy, $\Delta_N^D H(T_m)$, determines both the area under the DSC transition and its width. This is illustrated by the excess heat capacity *versus* temperature profiles shown in this figure, which have been calculated by using Eq. (3) with $T_m = 50$ °C, $\Delta_N^D C_p = 0$ and the following values for the calorimetric enthalpy: a, 400 kJ/mole; b, 200 kJ/mole; c, 100 kJ; d, 50 kJ/mole. The profiles have been displaced in the *y* axis for display purposes. For a two-state, equilibrium denaturation with $\Delta_N^D H(T_m) < 100$ kJ/mole, the predicted C_p^{ex} values are very low (see profile d), and, in most cases, the "transition" will not be apparent in the DSC thermo gram. It must be noted that a process characterized by an enthalpy change below 100 kJ/mole takes place in such a wide temperature range that, in fact, we would hardly refer to it as a transition.

$$\Delta_N^D G(T) = \Delta_N^D H(T) - T \cdot \Delta_N^D S(T) \tag{7}$$

Therefore, if the values of T_m, $\Delta_N^D H(T_m)$, and $\Delta_N^D C_p$ can be reliably determined from the experimental DSC transitions, then a complete description of the temperature dependence of the relevant denaturation changes is available through the use of Eqs. (5–7).

Of course, the calculations described above assume that the two-state equilibrium model holds for the denaturation process under study. It is very important to note that the applicability of this model can be ascertained from a straightforward analysis of the DSC transition. Thus, as shown in Fig. 4, the denaturation enthalpy determines the area under the DSC transition, as well as its width. The denaturation enthalpy calculated from the shape (width) of the transition is usually known as the *apparent* or *van't Hoff enthalpy* [ΔH^{vH}]. A convenient expression for ΔH^{vH} can be derived by setting $T = T_m$ (and, hence, $K = 1$) in the first-term of the right-hand side of Eq. (3) (the term due to the shift of the denaturation equilibrium):

$$\Delta H^{vH} = 4RT_m^2 \cdot [\Delta C_p^m / \Delta_N^D H(T_m)] \tag{8}$$

where ΔC_p^m is the heat capacity at the temperature T_m and measured from the chemical baseline (see Fig. 3A). The ratio of this parameter to the calorimetric enthalpy [$\Delta C_p^m / \Delta_N^D H(T_m)$ in Eq. (8)] does not depend on the amount of protein present in the calorimetric cell, but is sensitive to the shape (width) of the transition (Fig. 4).

Of course, there cannot be two different enthalpy changes for the same process. Equation (8) has been derived on the basis of the two-state equilibrium model, and, if the model holds true, one must find that the van't Hoff and calorimetric enthalpies are equal within the experimental uncertainty; on the other hand, if the two enthalpy values are significantly different, we conclude that the denaturation process does not follow the two-state model. For instance, if intermediate states are significantly populated during the thermal denaturation process, the DSC transition will be wider than predicted by the two-state model, and consequently, the van't Hoff enthalpy will be smaller than the calorimetric enthalpy.

The results of the above-described "two-state test" are usually given as the calorimetric to van't Hoff enthalpy ratio: $r = \Delta_N^D H(T_m)/\Delta H^{vH}$. For a number of small globular proteins, Privalov and colleagues (reviewed by Privalov, 1979) found that $r = 1.05 \pm 0.03$, which indicates a low proportion of intermediate states and, hence, close adherence to the two-state model.

A two-state model in which a multimeric protein (N_μ) undergoes denaturation with concomitant dissociation into monomers has also been analyzed in the literature (Takahashi and Sturtevant, 1981):

$$N_\mu \rightleftharpoons \mu D$$

The DSC transitions predicted by this model are asymmetrical and shift to higher temperatures upon increasing protein concentration. Equations permitting the calculation of the van't Hoff enthalpy for this model are given by Manly *et al.* (1985); note that if, instead, Eq. (8) is used to calculate ΔH^{vH} and the multimeric character of the protein is ignored (that is, the calorimetric enthalpy is calculated per mole of monomer), then a calorimetric to van't Hoff enthalpy ratio smaller than unity will be obtained ($r < 1$), indicating the existence of intermolecular cooperativity (protein oligomerization) in the denaturation process.

The above two-state model has also been extended to include the effect of ligands on the DSC transitions (Manly *et al.*, 1985). If the ligand binds only to the native protein, then the denaturation temperature increases with increasing ligand concentration and the effect is predicted to be nonsaturating (that is, the T_m values do not reach a plateau). It is interesting that when the concentration of ligand is smaller than the concentration of binding sites and the binding is strong, the DSC profile is predicted to split into two transitions (Robert *et al.*, 1988; Brandts and Lin, 1990) due to the abrupt temperature dependence of the free ligand concentration (or ligand activity) during the scan. The possibility of using the ligand effects on DSC transitions to characterize strong to ultratight protein–ligand interactions has been recently discussed by Brandts and Lin (1990).

2.2. Multistate Equilibrium

When $r > 1$ (or when several transitions are evident in the DSC profile), the presence of intermediate states must be taken into account. Sixteen years ago,

Freire and Biltonen (1978) proposed a general mechanism for the equilibrium, thermal denaturation of proteins. According to this mechanism, the native state (here referred to as I_0) is converted to the denatured state (here referred to as I_n) through $n - 1$ intermediate states: $I_1, I_2, I_3, \ldots, I_{n-2}, I_{n-1}$.

The analysis of this mechanism is greatly simplified by the use of the partition function formalism. We take as reference the native state (I_0) and define the statistical weight (ω_i) of a given state I_i, as the ratio between its concentration that of the reference state:

$$\omega_i = [I_i]/[I_0] = K_i = \exp(-\Delta_N^i G/RT) \tag{9}$$

where we have used the fact that this ratio is equal to the constant (K_i) for the equilibrium $I_0 \Leftrightarrow I_i$ and that K_i can be expressed in terms of the Gibbs energy of the state I_i (taking the native state as reference; i.e., $\Delta_N^i G$). All the relevant information about this system is contained in the partition function (Q), which is defined as the sum of all statistical weights:

$$Q = \sum_{i=0}^{n} \omega_i = \sum_{i=0}^{n} \exp(-\Delta_N^i G/RT) = 1 + \sum_{i=1}^{n} \exp(-\Delta_N^i G/RT) \tag{10}$$

where we have taken into account that the statistical weight for the reference state is unity ($\Delta_N^N G = 0$, hence $K_0 = \omega_0 = 1$). The fraction of protein (x_i) present as a given state I_i, is given by the ratio between the statistical weight of that state and the partition function:

$$x_i = \omega_i/Q = (1/Q)\cdot\exp(-\Delta_N^i G/RT) \tag{11}$$

Some thermodynamic quantities (though not all of them) can be calculated as averages over all the protein states. For one such quantity, we write $\langle J\rangle = \Sigma\, x_i \cdot J_i$, where $\langle J\rangle$ is the average value and J_i is the value of the quantity for the state I_i. The excess enthalpy (taking the native as reference) is an average, hence:

$$\langle \Delta H\rangle = \sum_{i=0}^{n} x_i \cdot \Delta_N^i H = (1/Q) \cdot \sum_{i=0}^{n} \Delta_N^i H \cdot \exp(-\Delta_N^i G/RT) \tag{12}$$

The excess heat capacity is not an average over the protein states, but can be calculated as a derivative of the average $\langle \Delta H\rangle$; the result is:

$$C_p^{ex} = \partial\langle \Delta H\rangle/\partial T = (\langle \Delta H^2\rangle - \langle \Delta H\rangle^2)/(RT^2) + \langle \Delta C_p\rangle \tag{13}$$

where $\langle \Delta H^2\rangle$ is the average value of ΔH^2 [$= \Sigma\, x_i\cdot(\Delta_N^i H)^2$]. The first term in the right-hand side of Equation (13) is due to the temperature-induced shifts in denaturation equilibria, while the second term is the average heat capacity of the

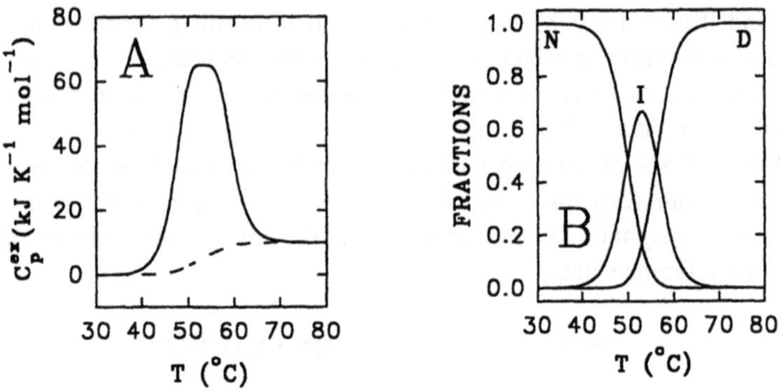

FIGURE 5. (A) DSC profile for a non–two-state denaturation process involving a significantly populated intermediate state (I). The heat capacity of the native state is taken as the reference level and the dashed curve represents the chemical baseline. The C_p^{ex} values were calculated by using Eq. (13) with the following parameters: $T_m^I = 50$ °C $\Delta_N^I H = 400$ kJ/mole, $\Delta_N^I C_p = 5$ kJ·K^{-1}·mole^{-1}, $T_m^D = 53$ °C, $\Delta_N^D H = 800$ kJ/mole, and $\Delta_N^D C_p = 10$ kJ·K^{-1}·mole^{-1}, where T_m^I and T_m^D are the temperatures at which $\Delta_N^I G = 0$ and $\Delta_N^D G = 0$, respectively. (B) Temperature dependence of the populations of native (N), denatured (D), and intermediate (I) states for the DSC profile shown in A.

protein [$\langle \Delta C_p \rangle = \Sigma\, \Delta_N^i C_p \cdot x_i$] and corresponds to the chemical baseline. For illustration, Fig. 5 shows a DSC profile predicted by Eq. (13) for the case in which one intermediate state is significantly populated during denaturation.

Another interesting relation can be obtained from Eq. (12) by integration:

$$\ln[Q(T)] = \int_{T_0}^{T} (1/RT^2) \cdot \langle \Delta H \rangle \cdot dT = \int_{T_0}^{T} (1/RT^2) \cdot \left(\int_{T_0}^{T} C_p^{ex} \cdot dT \right) \cdot dT \quad (14)$$

which shows that the partition function (as a function of temperature) can be obtained from the C_p^{ex}/T data by a double-integration procedure (T_0 is a low temperature at which essentially all the protein is present as native state). Hence, the experimental DSC profile contains (at least, in theory) all the information required to fully characterize the equilibrium denaturation process, a fact first pointed out by Freire and Biltonen (1978). These authors developed a recursive, stepwise procedure to determine the number and to develop a thermodynamic characterization of the intermediate states significantly populated during denaturation. Modifications of this procedure or alternative ways to "deconvolute" the DSC profiles have also been proposed by other authors (Filimonov *et al.*, 1982; Gill *et al.*, 1985), although recent approaches usually rely on the nonlinear, least-squares fitting to the excess heat capacity data. It must be noted that the partition function formalism described above can be generalized to include the effects of ligand binding and protein oligomerization (see Chapter 6 of Wyman and Gill, 1990).

The presence of significantly populated intermediate states in the equilibrium denaturation of complex proteins is generally attributed to the fact that different parts (thermodynamic domains) of the macromolecule may undergo denaturation in a more or less independent way (Privalov, 1982, 1989). Accordingly, complex DSC profiles are often interpreted as a sum of independent two-state transitions:

$$C_p^{\mathrm{ex}} = \sum_{j} \frac{(\Delta_{\mathrm{N}}^{\mathrm{D}} H_j)^2}{RT^2} \times \frac{K_j}{(1+K_j)^2} \tag{15}$$

where the excess heat capacity is calculated by taking the chemical baseline as reference level and $\Delta_{\mathrm{N}}^{\mathrm{D}} H_j$ and K_j stand for the enthalpy change and the equilibrium constant corresponding to the denaturation (unfolding) of the thermodynamic domain j [compare Eqs. (15) and (3)]. The validity of this approach is usually decided by comparing the DSC profiles for several protein fragments (obtained by proteolysis) with that corresponding to the intact macromolecule (see, for instance, Novokhatny *et al.*, 1984). Equation (15) assumes that the thermodynamic domains unfold in a strictly independent manner. Models that take into account the effect of domain–domain interactions have been recently developed and applied to the denaturation of two-domain proteins (Brandts *et al.*, 1989; Ramsay and Freire, 1990; Freire *et al.*, 1992).

According to the models described above, the intermediates in the Freire–Biltonen scheme are partially folded states; that is, they correspond to protein structures in which some parts of the macromolecule are folded, while other parts are unfolded. However, other types of intermediates are also possible; thus, Xie *et al.*, (1991) have analyzed the guanidine hydrochloride effect on the DSC profiles for the thermal denaturation of apo-α-lactalbumin in terms of the presence of a molten-globule intermediate.

3. THE THERMODYNAMIC STABILITY OF PROTEINS

The stability of a protein is often defined as the difference between the Gibbs energy of its denatured state and that of its native state and the $\Delta_{\mathrm{N}}^{\mathrm{D}} G$ versus temperature profile is sometimes referred to as the *protein stability curve* (Becktel and Schellman, 1987; Schellman, 1987). If the denaturation process follows the two-state equilibrium model, then the protein stability curve can be calculated from a DSC experiment by using Eqs. (5) through (7). The "standard first approximation of thermochemistry" (heat capacity change equal to zero) is not realistic for protein denaturation and, consequently, stability calculations require a reliable value for the denaturation heat capacity change. $\Delta_{\mathrm{N}}^{\mathrm{D}} C_p$ values can be obtained from the experimental DSC transitions (see Figs. 1 and 2) with an error that has been estimated to be not less than ± 4% by Becktel and Schellman (1987) and about ± 10% by Pace and Laurents (1989). It must be noted that $\Delta_N^D C_p$ values are often calculated from the slope of a $\Delta_N^D H(T_m)$ versus T_m plot, where the several DSC transitions (with different T_m values) needed to construct the plot are generated by changing the pH of the protein solution; this method assumes that $\Delta_N^D H$ and $\Delta_N^D C_p$ do not depend on pH and may require that the ionization heat effects on protein denaturation be compensated by an appropriate choice of buffers (Privalov, 1979; see, also, Privalov *et al.*, 1986). Regardless of the method employed, $\Delta_N^D C_p$ values are found to be large and positive, although accurate DSC measurements over a broad temperature range indicate that they depend on temperature, decreasing at high temperature and apparently extrapolating to zero at temperature well above 100 °C (Privalov and Gill, 1988; Privalov *et al.*, 1989; Privalov and Makhatadze, 1990 and 1992). Nevertheless, it appears that no serious error is introduced if the temperature dependence of $\Delta_N^D C_p$ is ignored, at least within the temperature range from 0 °C to 80 °C (Privalov and Gill, 1988; Privalov, 1989). Consequently, we will follow Becktel and Schellman (1987) and employ the "standard second approximation of thermochemistry"; that is, the constant (temperature-independent) heat capacity approximation.

It must be pointed out that protein stability can also be determined by noncalorimetric methods. In fact, solvent denaturation studies using urea or guanidine hydrochloride, together with extrapolation to zero denaturant concentration (Santoro and Bolen, 1988; Bolen and Santoro, 1988) are frequently used to obtain $\Delta_N^D G$ at a single temperature (usually 25 °C). It is surprising that until very recently (Hu *et al.*, 1992) the stability values obtained by using the two, more common methods had never been compared. Hu *et al.* (1992) have calculated $\Delta_N^D G$ at 25 °C, and as function or pH, for ribonuclease T1 denaturation on the basis of both DSC calorimetry and solvent denaturation studies; the agreement between the estimates obtained from the two approaches was described by these authors as "remarkably good."

3.1. The Protein Stability Curve

Under the constant heat capacity approximation, Eqs. (5) through (7) may be written as:

$$\Delta_N^D H(T) = \Delta_N^D H(T_m) + \Delta_N^D C_p \cdot (T - T_m) \qquad (16)$$

$$\Delta_N^D S(T) = \Delta_N^D H(T_m)/T_m + \Delta_N^D C_p \cdot \ln(T/T_m) \qquad (17)$$

$$\Delta_N^D G(T) = \Delta_N^D H(T_m) \cdot [1 - [(T/T_m)] + \Delta_N^D C_p \cdot [T - T_m - T \cdot \ln(T/T_m)] \qquad (18)$$

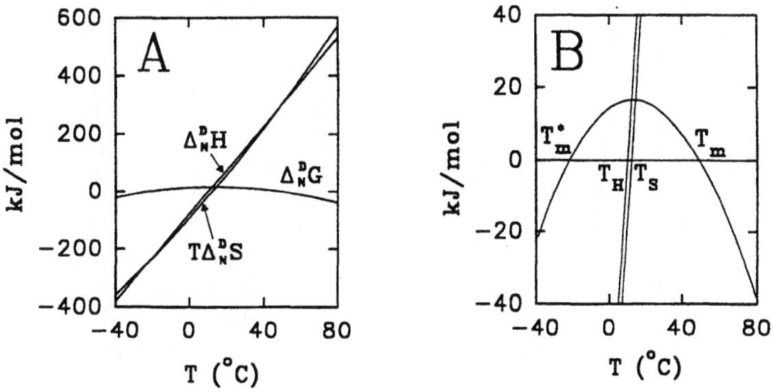

FIGURE 6. (A) Effect of temperature on the enthalpy, entropy, and Gibbs energy changes for protein denaturation. (B) Blowup showing the main features of the protein stability curve (the $\Delta_N^D G$ *versus* T profile); the heat (T_m) and cold (T_m^*) denaturation temperatures are shown, as well as the temperatures, T_H and T_S, at which the denaturational enthalpy and entropy changes become zero. The $\Delta_N^D H$, $\Delta_N^D S$, and $\Delta_N^D G$ values have been calculated by using Eqs. (16)–(18) with the parameters T_m = 48.3 °C, $\Delta_N^D H(T_m)$ = 290 kJ/mole, and $\Delta_N^D C_p$ = 7.6 kJ·K^{-1}·mole^{-1}, which correspond to the denaturation of staphylococcal nuclease at pH = 5 in 10 mM acetate buffer containing 100 mM NaCl (Griko *et al.*, 1988a); note that cold denaturation is predicted to occur at about −20 °C under these conditions. However, the stability curve is lowered upon pH-decreasing, thus resulting in a higher cold denaturation temperature (and a lower T_m) (see Fig. 7).

The elementary properties of the stability curve predicted by Eq. (18) have been discussed by Becktel and Schellman (1987) and Schellman (1987) and can be summarized as follows (Fig. 6). The curvature is given by $\partial^2\Delta_N^D G/\partial T^2 = -\Delta_N^D C_p/T$ and is always negative (the denaturation heat capacity change is positive). The slope of the curve is $-\partial\Delta_N^D G/\partial T = \Delta_N^D S$ and the curve shows only one extremum (which is a maximum) at the temperature, T_s, at which $\Delta_N^D S = 0$. The temperature, T_H, at which the denaturation enthalpy is zero, is slightly below T_s. The native protein is stable in some range of temperature ($\Delta_N^D G > 0$ within that range) and, therefore, the stability curve crosses the $\Delta_N^D G = 0$ axis at two points. Clearly, the higher point is the denaturation temperature value (T_m) used in the calculation of the stability curve. On the other hand, the low-temperature crossing is the result of the extrapolation of the curve, and suggests that the protein may also denature upon cooling (cold denaturation) (see Fig. 7). The early evidence suggesting low-temperature instability of proteins has been summarized by Chen and Schellman (1989) and Privalov (1990). In most cases, cold denaturation is predicted to occur at temperatures close to or below 0 °C, and therefore it is usually observed (Privalov, 1990) by supercooling the protein solution or by adding urea or guanidine hydrochloride to the protein solution (in order to raise the cold denaturation temperature). Several recent DSC studies (Privalov *et al.*, 1986; Griko *et al.*, 1988a, 1989; Tamura *et al.*, 1991; Griko Privalov, 1992; Azuaga *et al.*, 1992) suggest that cold denaturation is a general

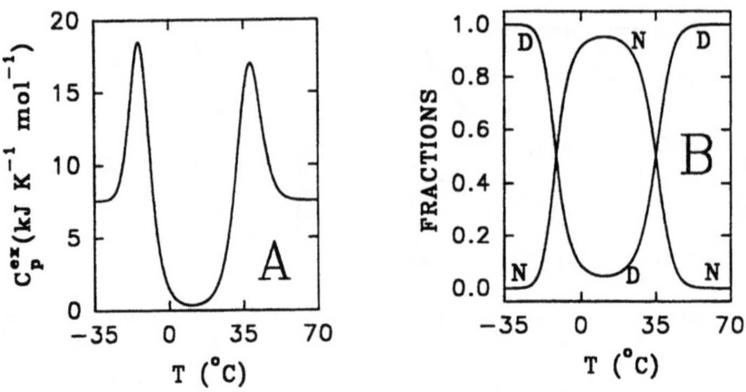

FIGURE 7. (A) Excess heat capacity *versus* temperature profile for a two-state denaturation process, showing both the heat and cold denaturation transitions. The heat capacity of the native state is taken as the reference level. The C_p^{ex} values were calculated by using Eq. (3) with the parameters $T_m = 35.0$ °C, $\Delta_N^D H(T_m) = 186$ kJ/mole, and $\Delta_N^D C_p = 7.6$ kJ·K^{-1}·mole^{-1}, which correspond to the thermal denaturation of staphylococcal nuclease at pH = 3.98 in 10 mM acetate buffer containing 100 mM NaCl (Griko *et al.*, 1988a). Under these conditions, the cold denaturation temperature is predicted to be about −11 °C and the high-temperature tail of the cold-denaturation transition may be observed in experimental DSC scans starting at about 0 °C (see Griko *et al.*, 1988a). (B) Effect of temperature on the population of the native (N) and denatured (D) states for the DSC profile shown in A.

property of globular proteins, thus confirming the essential features of the protein stability curve.

It is apparent in Fig. 6 that within most of the relevant temperature range, large values of $\Delta_N^D H$ and $T \cdot \Delta_N^D S$ compensate each other to yield comparatively low value for $\Delta_N^D G$. Enthalpy/entropy compensation is also observed when analyzing the thermodynamic changes for the denaturation of several compact proteins at a given temperature. Comparison between proteins that differ in size require that the thermodynamic quantities be normalized to a mole of residues (specific quantities). Figure 8 shows the specific changes in Gibbs energy, enthalpy, and entropy for the denaturation of several compact, globular proteins at 25 °C [data taken from Table I of Privalov and Gill (1988)]. For reasons that will be made clear in Section 3.3, these values are plotted versus the specific, denaturational heat capacity change. Due to efficient enthalpy/entropy compensation (see Fig. 8), all the compact proteins in this set show similar specific stabilities at 25 °C, the average value being 334 ± 93 J·mole − res^{-1}; this is a very low value (less than 0.1 $k_B T$ per residue, where k_B is the Boltzmann's constant) (see Privalov, 1979; Dill, 1990) and indicates that *native proteins are marginally stable at room temperature.*

It has been pointed out (Doig and Williams, 1992) that evolution could have generated more stable proteins if it were advantageous, as shown by the fact that,

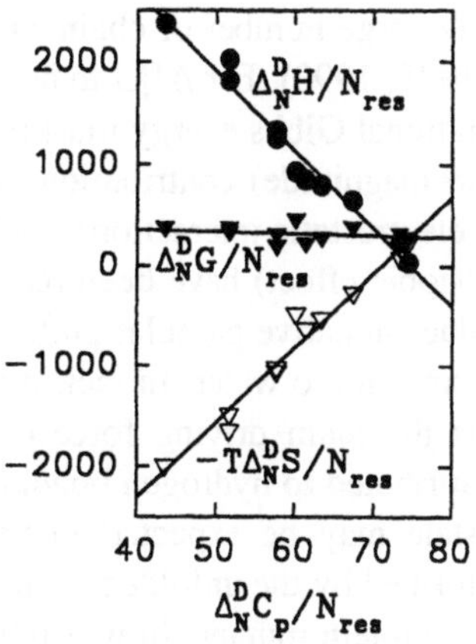

FIGURE 8. Plot of specific (per mole of residues) enthalpy, entropy, and Gibbs energy changes for the denaturation of several compact, globular proteins *versus* the corresponding specific, denaturational heat capacity change. Units are J·mole-res^{-1} for $\Delta_N^D H/N_{res}$, $\Delta_N^D G/N_{res}$, and $-T\cdot\Delta_N^D S/N_{res}$; and J·K^{-1}·mole − res^{-1} for $\Delta_N^D C_p/N_{res}$. All data correspond to a temperature of 25 °C and are taken from Table I of Privalov and Gill (1988). The proteins included in the plot are (by increasing specific heat capacity change): ribonuclease a, egg-white lysozyme, fragment K4 of plasminogen, β-trypsin, α-chymotrypsin, papain, staphylococcal nuclease, carbonic anhydrase, cytochrome *c*, pepsinogen, and myoglobin. Note that $\Delta_N^D H/N_{res}$ and $\Delta_N^D S/N_{res}$ change with $\Delta_N^D C_p/N_{res}$ in a linear fashion, as predicted by the convergence equations [Eqs. (21) and (22)]. Note also that due to efficient enthalpy/entropy compensation, the specific stability values of these proteins at 25 °C (i.e., the $\Delta_N^D G/N_{res}$ values shown in the figure) are small and do not significantly depend on the hydrophobicity of the protein (as measured by the specific, denaturational heat capacity change).

in several cases, modified proteins with enhanced stability have been designed and prepared in the laboratory (see, for instance, Stearman *et al.*, 1988; Pantoliano *et al.*, 1989). In fact, it appears that low stability may be advantageous, as suggested by the following arguments (recently summarized by Doig and Williams, 1992; see also Creighton, 1990; Fischer and Schmid, 1990): (1) low stability is likely to lead to a high degree of flexibility, which may be required for function; (2) proteolytic enzymes may have difficulties in degrading very stable proteins; hence low stability may be necessary to facilitate protein degradation; (3) low stability may also be required for transient unfolding for transport or assembly purposes; (4) if the native structure is very stable, folding intermediates may also be very stable, thus leading to a slow kinetic of folding; (5) low stability may help to ensure that the protein is not trapped in incorrectly folded structures, as these structures will easily yield the native, correctly folded protein via the denatured state (for details, see Doig and Williams, 1992). However, it is also clear that the stability cannot be too low either, since, in that case, denaturation could be induced by minor changes in external conditions. Furthermore, if stability is too low, the amount of denatured state in equilibrium with the native protein will be significant and denatured proteins are nonfunctional and highly susceptible to degradation. It appears then that a specific stability of about a few hundred J·mole − res^{-1} (low, but not too low) may have been determined by the effect of evolutionary pressure on protein stability (Doig and Williams, 1992).

From a molecular viewpoint, the low value for $\Delta_N^D G$ may be regarded as the result of a delicate balance between large driving and opposing forces (Dill, 1990). The main opposing force is the configurational entropy, which arises from the large number of chain configurations accessible in the denatured state (Dill, 1985, 1990). For $\Delta_N^D G$ to be positive at room temperature, the negative configurational Gibbs energy must be counterbalanced by a positive (and slightly larger in magnitude) contribution. The possible candidates for this driving-force role (electrostatic interactions, hydrogen bonding, van der Waals interactions, hydrophobic effect) have been recently reviewed by Dill (1990). Most nonpolar residues in native globular proteins are buried in "hydrophobic cores," thus avoiding exposure to water. In fact, it has been often assumed that the hydrophobic effect is the main driving force of folding, while little net effect has been usually attributed to hydrogen bonding, given that internal hydrogen bonds in the native state may be expected to have the same strength (Gibbs energy) than those formed by the unfolded chain to water. However, recent studies into the stability of protein mutants in which hydrogen bonds have been removed (Shirley *et al.*, 1992) and DSC analysis of the α-helix to coil transition of an alanine peptide in water (Scholtz *et al.*, 1991) appear to suggest that hydrogen bonding and hydrophobic effect make comparable contributions to the stability of globular proteins (see also Pace, 1992). It must be noted that, regardless of the issue of the driving force (or forces) of folding, the fact that the stability of globular proteins is low implies that no interaction is unimportant and even small interactions can make a significant contribution to the $\Delta_N^D G$ value (Dill, 1990).

3.2. The Molecular Interpretation of the Denaturational Heat Capacity Increment

Transfer of nonpolar solutes into water is accompanied by a large heat capacity increase. Consequently, it often has been taken for granted that the large and positive $\Delta_N^D C_p$ values found for protein denaturation are largely due to the exposure to the solvent of apolar groups previously buried in the native structure; therefore, the specific, denaturational heat capacity change (the values of the abscissa in Fig. 8) is usually considered to be an adequate measure of the hydrophobicity of the protein.

According to recent work on model transfer processes, however, the large positive contribution arising from the exposure of apolar groups is expected to be partially offset by a negative contribution from the exposure of polar (hydrogen bonding) groups. Thus, Murphy and Gill (1990, 1991) have studied the thermodynamics of dissolution (into water) of a set of solid cyclic dipeptides with apolar side chains and have calculated the contributions of various groups to the dissolution thermodynamic changes. They find that, while the exposure of apolar groups leads to the expected increase in heat capacity, the contribution of the peptide backbone and the hydroxyl group to the dissolution heat capacity is negative. The rationale behind the use of organic solids as model compounds is that several properties suggest that the interior of a protein resembles a crystalline solid (Murphy and Gill, 1991; Murphy *et al.*, 1990). It is interesting nevertheless that the same conclusion (negative ΔC_p contribution from polar groups) has been reached by Spolar *et al.* (1992) from the analysis of thermodynamic data for the transfer of hydrocarbons and organic amides from the pure liquid phase to water. Furthermore, Privalov and Makhatadze (Makhatadze and Privalov, 1990; Privalov and Makhatadze, 1990, 1992) have analyzed the heat capacities in aqueous solution for the various constituent groups of proteins and have calculated the corresponding heat capacity effects of hydration, defined as the difference between the heat capacity in aqueous solution and the heat capacity in a compact gaseous state (a hypothetical gaseous state with the same density as the liquid state of the substance). These authors find that the hydration heat capacity is positive (and decreases as temperature increases) for apolar groups, while it is negative (and decreases in magnitude as temperature increases) for polar groups [other possible contributions to $\Delta_N^D C_p$, such as the disruption of noncovalent bonds and the increase of configurational freedom, are also discussed by Privalov and Makhatadze (1992)].

It is important to recognize that the thermodynamic stability of proteins is determined by the *difference* between the Gibbs energies of the denatured and native states. According to recent work (reviewed by Shortle, 1993; see also Dill and Shortle, 1991), denatured states of proteins often exhibit significant amounts or residual structure. Theoretical (Shortle *et al.*, 1992) and experimental (Shortle *et al.*, 1990; Green *et al.*, 1992) studies suggest that, in some cases, the interactions that give rise to this residual structure may contribute to the stabilization of the folded structure through their effect on the Gibbs energy of the denatured state. In connection with this, it is interesting that Privalov and Makhatadze (1990) (see also Privalov *et al.*, 1989; Yang *et al.*, 1992) have shown that the experimental values for heat capacities of the denatured states of several small globular proteins are close to those obtained by adding the heat capacities of the constituent amino acid residues. Hence, although the denatured states of these proteins cannot be considered as random coils, they appear to behave as fully solvated chains in aqueous solution, at least with regard to their heat capacity. On the other hand, Griko and Privalov (1992) have proposed that heat-denatured β-lackoglobulin is not completely unfolded, as judged from its heat capacity, which is significantly lower than that calculated for a fully solvated chain. A similar situation has been described by Tamura *et al.* (1991) for the cold-denatured state of the *Streptomyces* subtilisin inhibitor.

If the denatured state is assumed to behave as a fully solvated chain, then the heat capacity change for the denaturation of a given protein can be estimated from its native structure. The simplest approach that allows one to do so is based upon the two following assumptions: (1) the heat capacity is mainly contributed to by the exposure to water of apolar and polar groups; and (2) these contributions are proportional to the corresponding water-accessible surface areas (ASA), usually calculated by using one of the several available algorithms [see, for instance, Lee and Richards (1971), Shrake and Rupley (1973)] with a water probe radius of 1.4 Å. Accordingly (see Murphy *et al.*, 1992) we may write:

$$\Delta_N^D C_p = \Delta\bar{C}_{ap} \cdot \Delta_N^D A_{ap} + \Delta\bar{C}_{pol} \cdot \Delta_N^D A_{pol} \tag{19}$$

where $\Delta_N^D A_{ap}$ and $\Delta_N^D A_{pol}$ are the denaturational changes in apolar (ap) and polar (pol) ASA; these changes are given by the difference between the (apolar or polar) ASA exposed in the denatured state (calculated, assuming a fully solvated chain, as the sum of the ASA values of the constituent amonoacids) and that exposed in the native state (calculated from the known X-ray structure). $\Delta\bar{C}_{ap}$ and $\Delta\bar{C}_{pol}$ stand for the heat capacity changes associated with the exposure to water of 1 mole − Å² of apolar and polar ASA. Recent estimates (Murphy *et al.*, 1992) of these parameters are: $\Delta\bar{C}_{ap} = 1.88\ \mathrm{J \cdot K^{-1} \cdot (mole - Å^2)^{-1}}$ and $\Delta\bar{C}_{pol} = -1.09\ \mathrm{J \cdot K^{-1} \cdot (mole - Å^2)^{-1}}$. This approach [Eq. (19)] assumes the constant heat capacity approximation and assigns a single $\Delta\bar{C}$ value to all groups of a given type (polar or apolar). A more detailed analysis may be carried out on the basis of the data provided by Privalov and Makhatadze (Makhatadze and Privalov, 1990; Privalov and Makhatadze, 1990; see also Yang *et al.*, 1992), which include the heat capacities of all the constituent groups of proteins as a function of temperature. Nevertheless, the simpler approach embodied in Eq. (19) (or an equivalent approach couched in terms of the numbers of hydrogen bonds and apolar hydrogens, rather than in terms of ASA values) (Murphy and Gill, 1991) has been found to acceptably reproduce the experimental heat capacity changes for the denaturation of several globular proteins (Murphy and Gill, 1991; Murphy and Freire, 1992) and has been used in several recent studies of the molecular basis of protein stability and cooperativity (Murphy and Gill, 1991; Freire and Murphy, 1991; Freire *et al.*, 1992; Murphy *et al.*, 1992; Murphy and Freire, 1992). The application of Eq. (19) is illustrated in Fig. 9.

Several recent interpretations of some prominent features of the thermodynamics of protein denaturation (Section 3.3) are based upon Chothia's observation (Chothia, 1976) that native globular proteins bury a constant proportion of their polar surface area and a variable proportion of their apolar surface area. Equation (19) can be written in terms of specific quantities (per mole of residues) as:

$$\Delta_N^D C_p/N_{res} = \Delta\bar{C}_{ap} \cdot (\Delta_N^D A_{ap}/N_{res}) + \Delta\bar{C}_{pol} \cdot (\Delta_N^D A_{pol}/N_{res}) \tag{20}$$

where N_{res} is the number of residues. If we assume that the denatured state behaves as a fully solvated chain, then Chothia's observation appears to suggest that different globular proteins have similar values of $\Delta_N^D A_{pol}/N_{res}$ and different values of $\Delta_N^D A_{ap}/N_{res}$. According to this interpretation (Murphy and Gill, 1991;

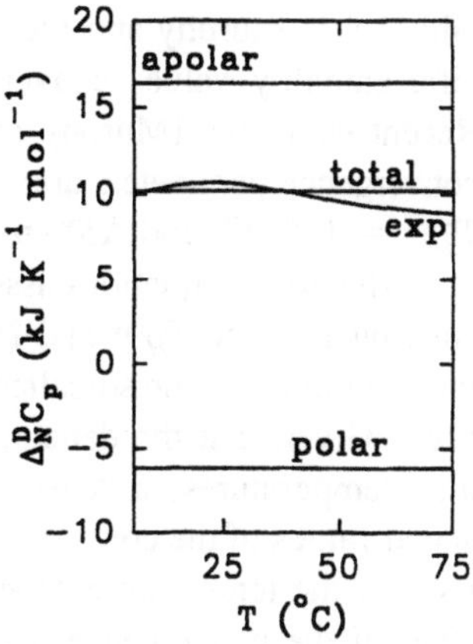

FIGURE 9. Calculation of the denaturation heat capacity change of myoglobin on the basis of Eq. (19) and structural features of the native protein. The denaturational changes in apolar and polar water-accessible surface area used in the calculation are: $\Delta_N^D A_{ap} = 8720$ Å² and $\Delta_N^D A_{pol} = 5660$ Å². These values are taken from Murphy *et al.*, (1992) and are equal to the difference between the (polar or apolar) ASA in the denatured and native states, where the ASA of the native protein was calculated from its tridimensional structure and that of the denatured state was obtained as the sum of the ASA values for the individual residues in an extended Ala-X-Ala tripeptide (Murphy *et al.*, 1992). The predicted apolar contribution (positive), polar contribution (negative), and total heat capacity change (polar + apolar) are shown in the figure. Note that these calculations assume the constant (temperature independent) heat capacity approximation, and therefore cannot reproduce the temperature dependence of the experimental $\Delta_N^D C_p$ values [taken from Privalov and Makhatadze (1990) and labeled "exp" in the figure]. This dependence is not large within the temperature range of this figure; however, extrapolation of the linear temperature dependence of the native heat capacity suggests that $\Delta_N^D C_p$ decreases at higher temperatures, extrapolating to zero at a temperature above 100 °C (see, for instance, Privalov and Gill, 1988).

Murphy and Freire, 1992), the polar contribution to $\Delta_N^D C_p/N_{res}$ is approximately constant, and therefore the specific heat capacity changes found for globular proteins (the values of the abscissa in Fig. 8) correlate with the specific apolar contribution; that is, with the hydrophobicity of the protein (as measured by the proportion of apolar ASA buried upon folding).

3.3. The Convergence Behavior of the Denaturational Enthalpy and Entropy Changes

Some 20 years ago, Privalov and co-workers (Privalov and Khechinashvili, 1974; Privalov, 1979) observed that the specific entropy and enthalpy changes for the denaturation of several globular compact proteins attain common entropy and enthalpy values when extrapolated to a temperature around 100 °C under the constant heat capacity approximation. This behavior can be formulated as:

$$\Delta_N^D H/N_{res} = \Delta H^* + (\Delta_N^D C_p/N_{res})\cdot(T - T_H^{conv}) \tag{21}$$

$$\Delta_N^D S/N_{res} = \Delta S^* + (\Delta_N^D C_p/N_{res})\cdot\ln(T/T_S^{conv}) \tag{22}$$

where ΔH^* and ΔS^* stand for the specific enthalpy and entropy changes at the convergence temperatures: T_H^{conv} T_S^{conv}. It was initially thought that T_H^{conv} and T_S^{conv} were equal and had a value about 110 °C. It appears clear now (Murphy and

Gill, 1991; Murphy and Freire, 1992; Baldwin and Muller, 1992) that the specific enthalpy values converge at a temperature a few degrees below $T_S^{\rm conv}$. Recent estimates (Murphy and Gill, 1991; Murphy and Freire, 1992) of the convergence parameters are: $T_H^{\rm conv} = 100.5$ °C, $\Delta H^* = 5.65$ kJ·(mole − res)$^{-1}$, $T_S^{\rm conv} = 112$ °C, and $\Delta S^* = 18.0$ J·K^{-1}·(mole − res)$^{-1}$.

The above equations assume the constant heat capacity approximation, and consequently the $T_H^{\rm conv}$ and $T_S^{\rm conv}$ values given above are hypothetical temperatures obtained by the so-called "linear extrapolation" [i.e., by using Eq. (21) and (22)]. Given that the denaturation heat capacity change appears to decrease at high temperatures, a "curved" extrapolation would be required to obtain the actual values of the convergence temperatures (see Privalov and Gill, 1988). The issue of the temperature-dependence of $\Delta_N^D C_p$ can be bypassed (Murphy *et al.*, 1990) if the convergence behavior is demonstrated by plotting the specific enthalpy or entropy changes versus the specific heat capacity changes for a set of proteins at a given temperature, usually $T = 25$ °C. The slopes of these "MPG plots" (Lee, 1991) are $298.15 - T_H^{\rm conv}$ for the enthalpy plot and $\ln(298.15/T_S^{\rm conv})$ for the entropy plot, while the intercepts are ΔH^* and ΔS^*, respectively. Clearly, the $T_H^{\rm conv}$ and $T_S^{\rm conv}$ values calculated from these slopes correspond to the hypothetical convergence temperatures that would be obtained by linear extrapolation from $T = 25$ °C. The MPG plots for the globular proteins of the Privalov–Gill data set (Privalov and Gill, 1988) have already been shown in Fig. 8, although the entropy plot is given as a $-T\cdot(\Delta_N^D S/N_{\rm res})$ versus $\Delta_N^D C_p/N_{\rm res}$ plot with a slope equal to $-298.15\cdot\ln(298.15/T_S^{\rm conv})$.

A likely explanation for the convergence behavior was advanced by Murphy and Gill (1991) on the basis of the group additivity scheme previously used to analyze the thermodynamics of solid cyclic dipeptide dissolution (Murphy and Gill, 1990). The following argument is equivalent to that given by these authors. We assume that the variation in specific enthalpy change found for the denaturation of globular proteins arises mainly from the apolar contribution, while other contributions [mainly those arising from the exposure of polar (hydrogen bonding) groups] are approximately constant per mole of residues:

$$\Delta_N^D H/N_{\rm res} \approx \text{constant} + \Delta\bar{H}_{\rm ap}\cdot(\Delta_N^D A_{\rm ap}/N_{\rm res}) \tag{23}$$

where the apolar contribution has been expressed in terms of the apolar ASA exposed to water upon denaturation and $\Delta\bar{H}_{\rm ap}$ is the enthalpy change associated with the exposure of 1 mole − Å^2 of apolar surface. The temperature dependence of $\Delta\bar{H}_{\rm ap}$ can be formulated with reference to a temperature, $T_H^{\rm ap}$, at which the apolar contribution to the enthalpy change is zero:

$$\Delta\bar{H}_{\rm ap} = \Delta\bar{C}_{\rm ap}\cdot(T - T_H^{\rm ap}) \tag{24}$$

Clearly, the apolar contribution to the denaturational enthalpy change ($\Delta_N^D H_{ap}$) is written in terms of T_H^{ap} as:

$$\Delta_N^D H_{ap} = \Delta_N^D A_{ap} \cdot \Delta \bar{C}_{ap} \cdot (T - T_H^{ap}) \tag{25}$$

The convergence temperature is related to the slope of the MPG enthalpy plot, which is given by the ratio between the derivatives of the specific enthalpy and heat capacity changes with respect to $\Delta_N^D A_{ap}/N_{res}$. Then, using Eq. (23) and (20) [note that the polar contribution in Eq. (20) is taken as constant], we arrive at:

$$T - T_H^{conv} = \frac{\partial(\Delta_N^D H/N_{res}0)/\partial(\Delta_N^D A_{ap}/N_{res})}{\partial(\Delta_N^D C_p/N_{res})/\partial(\Delta_N^D A_{ap}/N_{res})} = \Delta^{-} H_{ap}/\Delta \bar{C}_{ap} \tag{26}$$

which leads to $T_H^{conv} = T_H^{ap}$ [compare Eqs. (24) and (26)]. According to this interpretation, the ΔH^* term in Eq. (21) is the polar contribution to the specific enthalpy change at the temperature T_H^{conv} ($= T_H^{ap}$). In general, the polar contribution to the denaturational enthalpy change ($\Delta_N^D H_{pol}$) may be written as:

$$\Delta_N^D H_{pol} = \Delta_N^D A_{pol} \cdot [\Delta \bar{H}_{pol}^* + \Delta \bar{C}_{pol} \cdot (T - T_H^{ap})] \tag{27}$$

where $\Delta \bar{H}_{pol}^*$ [calculated from the ΔH^* value; see Murphy and Freire (1992) and Murphy *et al.* (1992)] is the enthalpy change associated with the exposure to water of 1 mole − Å^2 of polar surface at the temperature T_H^{ap} (assumed to be equal to the T_H^{conv} value for protein denaturation). It is to be noted that this polar contribution is defined in an operational way and it includes the effect of the van der Walls interactions of the polar groups.

The analogy of Eq. (23) for the specific denaturation entropy is:

$$\Delta_N^D S/N_{res} \approx \text{constant}' + \Delta \bar{S}_{ap} \cdot (\Delta_N^D A_{ap}/N_{res}) \tag{28}$$

where $\Delta \bar{S}_{ap}$ is the entropy change associated with the exposure to water of 1 mole − Å^2 of apolar surface and the constant′ in Eq. (28) includes contributions from the polar (hydrogen bonding) groups as well as from the configurational entropy. Using the same type of argument as above, it is shown that the T_S^{conv} value of Eq. (22) may be interpreted as the temperature, T_S^{ap}, at which the apolar contribution to the entropy change is zero:

$$\Delta \bar{S}_{ap} = \Delta \bar{C}_{ap} \cdot \ln(T/T_S^{ap}) \tag{29}$$

Then, the apolar contribution to the denaturational entropy change ($\Delta_N^D S_{ap}$) can be expressed as:

$$\Delta_N^D S_{ap} = \Delta_N^D A_{ap} \cdot \Delta \bar{C}_{ap} \cdot \ln(T/T_S^{ap}) \tag{30}$$

and the contribution from polar groups plus configurational entropy ($\Delta_N^D S_{conf/pol}$) is easily found to be:

$$\Delta_N^D S_{conf/pol} = N_{res} \cdot \Delta S^* + \Delta_N^D A_{pol} \cdot \Delta \bar{C}_{pol} \cdot \ln((T/T_S^{ap}) \tag{31}$$

Note, however, that the configurational entropy is expected to be the major contribution to $\Delta_N^D S_{conf/pol}$. Equations (25), (27), (30), and (31) are essentially those given by Murphy *et al.* (1992), although an equivalent approach couched in terms of the numbers of hydrogen bonds and apolar hydrogens has also been employed (Murphy and Gill, 1991; Murphy and Freire, 1992; Freire *et al.*, 1992). Recent estimates (Murphy *et al.*, 1992) of the parameters in these equations are:

$$\Delta \bar{C}_{ap} = 1.88 \pm 0.08 \text{ J} \cdot \text{K}^{-1} \cdot (\text{mole} - \text{Å}^2)^{-1}$$

$$\Delta \bar{C}_{pol} = -1.09 \pm 0.13 \text{ J} \cdot \text{K}^{-1} \cdot (\text{mole} - \text{Å}^2)^{-1}$$

$$T_H^{ap} = 100 \pm 6\ ^\circ\text{C}$$

$$T_S^{ap} = 112 \pm 1\ ^\circ\text{C}$$

$$\Delta \bar{H}^*_{pol} = 146 \pm 13 \text{ J} \cdot (\text{mole} - \text{Å}^2)^{-1}$$

$$\Delta S^* = 18.0 \pm 0.4 \text{ J} \cdot \text{K}^{-1} \cdot \text{mole} - \text{res}^{-1}$$

Equations (25), (27), (30), and (31), together with the above values, allow the thermodynamic properties of proteins to be calculated from their structural features (number of residues and amounts of polar and apolar ASA exposed to the solvent upon denaturation calculated assuming a fully solvated denatured state). This type of calculation is illustrated in Figs. 10–12. The thermodynamic properties of intermediate states can also be calculated if their structure is modeled using the native protein as a template [for instance (Freire *et al.*, 1992), if the native protein shows two differentiated structural domains, then a possible intermediate may be constructed assuming that one of the domains retains the same structure as in the native protein, while the other is unfolded]. On this basis, the above approach has been recently applied to the prediction of the effect of domain–domain interactions on the folding–unfolding behavior of two-domain proteins (Freire *et al.*, 1992; Murphy and Freire, 1992) and to the structural characterization of the intramolecular interactions responsible for the

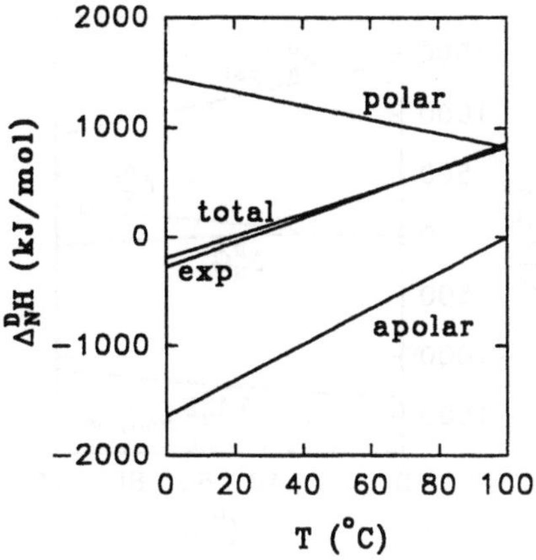

FGURE 10. Calculation of the enthalpy change for myoglobin denaturation of the basis of Eqs. (25) and (27) and structural features of the native protein: $\Delta_N^D A_{ap} = 8720$ Å² and $\Delta_N^D A_{pol} = 5660$ Å² (see legend to Fig. 9). The predicted apolar contribution [Eq. (25)], polar contribution [Eq. (27)], and total enthalpy change (polar + apolar) are shown in the figure. The line labeled "exp" has been calculated, using the constant heat capacity approximation, from the $\Delta_N^D H$ and $\Delta_N^D C_p$ values at 25 °C given by Privalov and Gill (1988) which were derived from experimental DSC studies.

cooperative character of protein denaturation (Freire and Murphy, 1991; Murphy and Freire, 1992; Murphy *et al.*, 1992).

It must be noted, nevertheless, that Eqs. (25), (27), (30), and (31) assume the constant heat capacity approximation and that they only consider the (presumed) major driving and opposing forces and in a simplified way; accordingly, they are only expected to provide a "first-order prediction" (Murphy and Gill, 1991) of the major thermodynamic features. Another important point (already referred to in Section 3.1) is that "minor" effects (such as effects of pH, ligands, ionic strength) are not unimportant, because they make a significant contribution to the small $\Delta_N^D G$ value, and therefore they strongly affect the denaturation temperature value. Agreement between the predicted and experimental T_m values is usually achieved by small adjustments of the ΔS^* value (Murphy *et al.*, 1992).

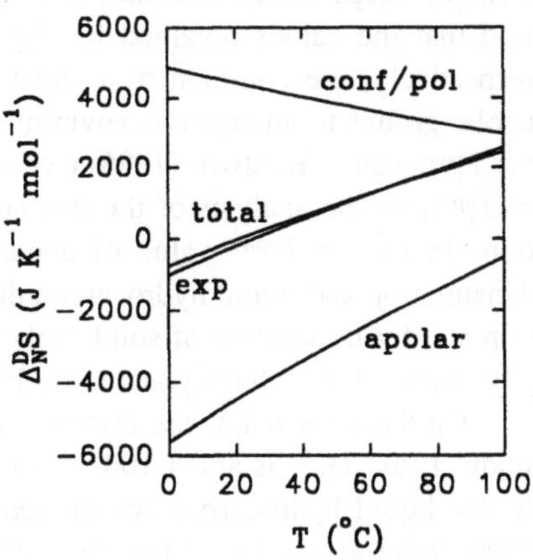

FIGURE 11. Calculation of the entropy change for myoglobin denaturation on the basis of Eqs. (30) and (31) and structural features of the native protein: $\Delta_N^D A_{ap} = 8720$ Å², $\Delta_N^D A_{pol} = 5660$ Å² (see legend to Fig. 9), and number of residues $N_{res} = 153$. The predicted apolar contribution [Eq. (30)], configurational *plus* polar contribution [Eq. (31)], and total entropy change (conf/pol + apolar) are shown in the figure. The line labeled "exp" has been calculated, using the constant heat capacity approximation, from the $\Delta_N^D S$ and $\Delta_N^D C_p$ values at 25 °C given by Privalov and Gill (1988), which were derived from experimental DSC studies.

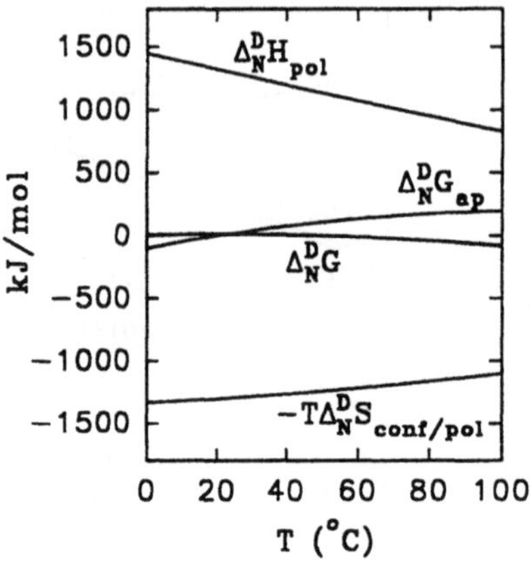

FIGURE 12. Balance between opposing and driving forces in myoglobin denaturation, as inferred from the calculations shown in Figs. 10 and 11. The main opposing force is the configurational Gibbs energy (the major contribution to the term labeled $-T\cdot\Delta_N^D S_{conf/pol}$ in this figure). There is a strong driving force arising from the enthalpic, polar contribution ($\Delta_N^D H_{pol}$ in the figure), in qualitative agreement with recent studies (Scholtz *et al.*, 1991; Shirley *et al.*, 1992), which suggests that hydrogen bonding makes a significant contribution to protein stability. On the other hand, the apolar contribution ($\Delta_N^D G_{ap} = \Delta_N^D H_{ap} - T\cdot\Delta_N^D S_{ap}$) is not large and in fact is close to zero at room temperature, due to compensation between the large enthalpic and entropic apolar terms (see Figs. 10 and 11). This is a result of having taken 100 °C (the value of the enthalpy convergence temperature for protein denaturation: T_H^{conv}) as the temperature (T_H^{ap}) at which the apolar contribution to the enthalpy change is zero (see Section 3.3); note, however, that the molecular interpretation of the T_H^{conv} value found for protein denaturation is still a matter of debate (see Section 3.4). [See note added in proof]. Note also that the calculated (as well as the experimental) protein stability values are very low as a result of the delicate balance between driving and opposing forces ($\Delta_N^D G = \Delta_N^D H_{pol} + \Delta_N^D G_{ap} - T\cdot\Delta_N^D S_{conf/pol}$).

3.4. The Convergence Temperature Puzzle

As described in the preceding section, recent attempts to predict the thermodynamic features of protein denaturation are based on the assumption that the T_H^{conv} and T_S^{conv} values found for protein denaturation correspond to the temperatures (T_H^{ap} and T_S^{ap}) at which the apolar contribution to the enthalpy and entropy changes, respectively, becomes zero. If this interpretation is correct, we should find that the values obtained for T_H^{conv} and T_S^{conv} (about 100 °C and 112 °C, respectively) are common to a variety of processes involving the exposure of apolar groups to an aqueous environment. This condition appears to be met by the T_S^{conv} value. Baldwin (1986) (see also Sturtevant, 1977) proposed that $T_S^{conv} = T_S^{ap}$ from the analysis of the thermodynamics of liquid hydrocarbon dissolution. In fact, a T_S^{conv} value of about 112 °C is found, not only for protein denaturation and liquid hydrocarbon dissolution, but also for apolar gas dissolution and for dissolution of solid cyclic dipeptides with apolar side chains (Murphy *et al.*, 1990; Murphy and Gill, 1991).

On the other hand, the T_H^{conv} value appears to depend on the process under study. Thus, T_H^{conv} is about 100 °C for protein denaturation, while it is about 25 °C for liquid hydrocarbon dissolution. It has been suggested (Murphy *et al.*, 1990; Yang *et al.*, 1992) that the T_H^{conv} value is sensitive to the van der Walls interactions in the initial state of the process (protein interior versus liquid hydrocarbon). Several properties suggest that the interior of a protein resembles a crystalline solid and, in fact, the T_H^{conv} value for the dissolution of solid cyclic dipeptides is about 80 °C (Murphy and Gill, 1991), closer to the protein value. However, this is a dangerous line of reasoning because the T_H^{conv} value for apolar gas dissolution is about 100 °C (Yang *et al.*, 1992) in good agreement with the protein value (but, of course, we cannot accept a gaslike protein interior). Thus, while it appears that the T_s^{ap} value is about 112 °C (the T_s^{conv} value for protein denaturation and other processes), it is not clear which value must be assigned to the temperature T_H^{ap}, and the interpretation offered in the preceding section (T_H^{ap} equal to T_H^{conv} for protein denaturation) must be considered with due caution.

It is important to note that the calculated apolar (i.e., hydrophobic) contribution to protein stability ($\Delta_N^D G_{\text{ap}}$) depends on the value chosen for T_H^{ap}. Thus,

$$\Delta_N^D G_{\text{ap}} = \Delta_N^D H_{\text{ap}} - T \cdot \Delta_N^D S_{\text{ap}} \tag{32}$$

where $\Delta_N^D H_{\text{ap}}$ and $\Delta_N^D S_{\text{ap}}$ are given by Eqs. (25) and (30). If we assume $T_H^{\text{ap}} =$ 100 °C (the T_H^{conv} value for protein denaturation), then the enthalpy and entropy terms in Eq. (32) cancel each other to yield a very low $\Delta_N^D G_{\text{ap}}$ value, which is the result shown in Fig. 12; this low value appears to suggest that the hydrophobic effect contributes little to protein stability and is consistent with the fact that the specific stability of compact globular proteins does not significantly depend on their hydrophobicity (as measured by the specific denaturational heat capacity change) (see Fig. 8). On the other hand, if we assume T_H^{ap} = 25 °C (the T_H^{conv} value for liquid hydrocarbon dissolution), the enthalpy term in Eq. (32) is zero for T = 25 °C [see Eq. (25) and the apolar contribution to the denaturational Gibbs energy change is stabilizing and mainly entropic. This is consistent with the generally accepted view of the hydrophobic effect as a driving force for folding, which, it is important to note, is supported by recent studies of the stability of mutant proteins with altered aliphatic residues (see Pace, 1992).

The above paragraphs help to understand the current interest on the origin of the T_H^{conv} value of about 100 °C found for protein denaturation (see, for instance, Lee, 1991; Baldwin and Muller, 1992; Doig and Williams, 1992; Yang *et al.*, 1992; Murphy *et al.*, 1992; Fu and Freire, 1992). Baldwin and Muller (1992) and Doig and Williams (1992) have shown that, provided that T_s^{conv} is about 112 °C, then the general features of globular protein denaturation (denaturation temperatures around 330 K or specific denaturation Gibbs energies around 300 J·mole − res^{-1}) imply that T_H^{conv} must be around 100 °C for protein denaturation; this result is interesting in itself, but does not provide an answer to the question on the molecular origin of the T_H^{conv} value. Yang *et al.* (1992) (see, also, Honig *et al.*, 1993) have carried out electrostatic calculations that support the idea that the burying (and desolvation) of polar groups in the protein interior brings about a large destabilizing (primarily enthalpic) contribution to the thermodynamics of protein denaturation; this polar contribution may explain the apparently anomalous T_H^{conv} value found for protein denaturation, but only if it is assumed to contain a term that correlates with the apolar contribution to the heat capacity change (see Yang *et al.*, 1992).

In this author's opinion, the molecular origins of the T_H^{conv} value found for protein denaturation have not been conclusively explained yet. [See, however, the note added in proof].

4. KINETIC ANALYSIS OF DSC THERMOGRAMS

Equilibrium thermodynamics analysis of DSC data, as described in Section 2, applies to those cases in which chemical equilibrium is maintained during the DSC scan. However, if the folding–unfolding kinetics is slow (Section 4.2) or if denaturation is overall irreversible (Section 4.3), equilibrium may not be attained and, in that case, the concentrations of the significantly populated states at a given temperature will depend on the time required to reach that temperature. The immediate consequence of this situation is that the calorimetric transitions will depend on the scanning rate at which the DSC experiment is carried out. Clearly, DSC transitions that show a significant scanning rate dependence must be analyzed on the basis of kinetic models (to be described in Sections 4.2 and 4.3). However, prior to this analysis (in fact, prior to any type of analysis), the usually small (but omnipresent) scanning-rate-dependent distortion introduced by the calorimeter time response should be corrected for (Section 4.1).

4.1. The Effect of the Calorimeter Time Response

All DSC thermograms are to some extent distorted, due to the slow time response of the calorimeter. The response times (τ) of commercial instruments range between 10 and 100 sec, which indicates that distortions may become significant for sharp transitions and for thermograms obtained at high scanning rates (Lopez-Mayorga and Freire, 1987; Freire *et al.*, 1990). Lopez-Mayorga and Freire (1987) have shown how these distortions can be corrected for. Most commercial calorimeters can be adequately characterized by a single response time; in this case, the input signal [$Y(t)$, the actual thermal effect] can be recovered from the distorted calorimeter response [$Y^*(t)$] by using:

$$Y = \frac{1}{A}\,[Y^* + \tau\cdot(dY^*/dt)] = \frac{1}{A}\,[Y^* + \tau v\cdot(dY^*/dT)] \tag{33}$$

where τ is the calorimeter response time, A is the calorimeter calibration constant and v stands for the scanning rate ($= dT/dt$). Practical aspects of the application of Eq. (33) and the determination of the response time for a given instrument are discussed in detail by Lopez-Mayorga and Freire (1987). These authors also give the correction equation for a second-order calorimeter exhibiting two response time constants.

4.2. The Effect of Slow Equilibrium

In some cases, reversible DSC transitions show significant scanning-rate dependence, even after correcting for the effect of the finite time response of the instrument (Lopez-Mayorga and Freire, 1987; Freire *et al.*, 1990; Plaza del Pino *et al.*, 1992). This "residual" scanning-rate dependence is attributed to the fact that the populations of the relevant protein states deviate from their equilibrium values, due to the slow kinetics of the thermally induced denaturation process. Lopez-Mayorga and Freire (1987) approached this problem by analogy with the formalism employed to analyze and correct the instrumental distortions (Section 4.1). Thus, the equilibrium excess heat capacity [C_p^{ex} (eq)] and the experimental excess heat capacity [C_p^{ex}, measured at a given scanning rate (v) and corrected for the instrumental distortions] are related by:

$$C_p^{ex}\,(\mathrm{eq}) = C_p^{ex} + \tau_c \cdot (dC_p^{ex}/dt) = C_p^{ex} + v\tau_c \cdot (dC_p^{ex}/dT) \tag{34}$$

where τ_c is the relaxation time of the heat capacity function [the analogous of the instrument response time in Eq. (33)]. The equilibrium C_p^{ex} (eq) versus T profile can be calculated from the experimental C_p^{ex} and dC_p^{ex}/dT data corresponding to two or more thermograms (obtained at different scanning rates) by application of Eq. (34), as illustrated by Lopez-Mayorga and Freire (1987). Note that C_p^{ex} (eq) is obtained as an extrapolation to zero scanning rate, and that this procedure also yields the τ_c versus T profile.

Later (Freire *et al.*, 1990), it was realized that a relaxation time (τ_H) can also be defined for the enthalpy function:

$$<\Delta H>_{eq} = <\Delta H> + \tau_H \cdot (d<H>/dt) = <\Delta H> + v\tau_H \cdot C_p^{ex} \tag{35}$$

where $<\Delta H>_{eq}$ is the equilibrium excess enthalpy and $<\Delta H>$ stands for the excess enthalpy calculated from the experimental profiles (obtained at a given scanning rate, v, and corrected for the instrumental distortion). Note that τ_c [Eq. (34)] and τ_H> are not equal and the Eq. (34) is not the temperature-derivative of Eq. (35), because the relaxation times depend on temperature.

An important issue, which has not been addressed yet, is the relation between these relaxation times and the rate constants for the kinetically relevant steps in the denaturation mechanism. I wish to show here that this relation should be based on the relaxation time of the enthalpy function: τ_H. The simplest kinetic mechanism consistent with a two-state reversible denaturation can be depicted as:

$$N \underset{k_r}{\overset{k_f}{\rightleftarrows}} D$$

where k_f and k_r are the first-order rate constants for the forward and reverse processes. Note that the denaturation equilibrium constant is given by $K = k_f/k_r$. The rate equation for this mechanism is:

$$dx_D/dt = k_f{\cdot}x_N - k_r{\cdot}x_D \tag{36}$$

which, taking into account that $x_d + x_N = 1$ and that $K = k_f/k_r$, can be written as:

$$dx_D/dt = (k_f + k_r){\cdot}(x_D^{eq} - x_D) \tag{37}$$

where $x_D^{eq} = K/(K + 1)$ is the fraction of denatured protein when equilibrium is reached at a given temperature (for instance, the x_D^{eq} versus T profile would be observed in a hypothetical DSC experiment carried out at an infinitely slow scanning rate). Equation (37) can be integrated (with constant k_f, k_r, and x_D^{eq}) to yield the time-dependence of x_D for a kinetic experiment carried out at constant temperature; the well-known result is an exponential relaxation characterized by a relaxation time equal to $(k_f + k_r)^{-1}$. In a DSC experiment, however, temperature changes with time according to a constant (to a good degree of approximation) scanning rate, $v = dT/dt$. Hence, the differential equation for a DSC experiment is*:

*Note that the differential equation corresponding to a temperature-scanning experiment is obtained by substituting dT/v for dt. This substitution must be carried out in the original rate equations that describe the kinetic process under consideration; that is, in equations of the form:

$$\text{rate} = f(\text{concentrations, rate constants}) \tag{a}$$

Rate equations are often integrated for constant temperature (see any standard physical chemistry textbook) to yield expressions of the type:

$$\text{concentrations} = f(\text{time, rate constants}) \tag{b}$$

which can be used to fit concentration/time experimental data determined at a given temperature. The time-derivative of Eq. (b) leads to the rate as a function of time:

$$\text{rate} = f(\text{time, rate constants}) \tag{c}$$

However, it is wrong to substitute dT/v for dt (and a linear function of temperature for time) in Eq. (c), since this equation was obtained from Eq. (b) and, therefore, only applies to a kinetic experiment at constant temperature.

$$dx_D/dT = (1/v)\cdot(k_f + k_r)\cdot(x_d^{eq} - x_D) \tag{38}$$

where now k_f, k_r, and x_D^{eq} are functions of the independent variable (T). If the excess heat capacity is measured from the chemical baseline, then we have that $x_D = \langle\Delta H\rangle/\Delta_N^D H$, $x_D^{eq} = \langle\Delta H\rangle_{eq}/\Delta_N^D H$ and $dx_D/dT = C_p^{ex}/\Delta_N^D H$. Consequently, Eq. (38) may be easily written as:

$$\langle\Delta H\rangle_{eq} = \langle\Delta H\rangle + \frac{v}{k_f + k_r} C_p^{ex} \tag{39}$$

Equations (39) and (34) are identical, provided that,

$$\tau_H = (k_f + k_r)^{-1} \tag{40}$$

Thus, the τ_H value is in fact the actual relaxation time of the chemical process under consideration. It can be easily shown that this simple relation holds for τ_H, but not for τ_c. Of course, the model we have analyzed is extremely simple; however, it may provide an adequate description of the kinetics in some cases, as is illustrated by the work of Chen *et al.* (1989, 1992) on the folding–unfolding kinetics of T4 lysozyme and several of its mutants.

4.3. The Effect of Irreversibility

The thermal denaturation of many proteins is irreversible, as shown by the lack of thermal effect in the thermogram corresponding to the reheating run. Irreversible protein denaturation is generally attributed to alterations (such as autolysis, aggregation, chemical alteration of residues) (see Klibanov and Ahern, 1987) that lock the protein in a state (the final state: F) that is unable to fold back to the native structure. These irreversible alterations are fundamentally kinetic processes that must be described by rate equations. It has been often assumed in the past that irreversible alterations are slow and do not take place significantly during the comparatively short time the protein spends in the temperature range of the DSC transition, but occur (with little thermal effect) at somewhat higher temperatures; consequently, irreversible DSC transitions have been often assumed to reflect the equilibrium denaturation of the protein and have been analyzed according to the well-known equilibrium thermodynamics procedures (described in Section 2). However, more recent work indicates that very often irreversible DSC transitions are highly distorted by the kinetics of the irreversible protein alterations, as shown by the strong scanning rate effects found in many cases (Sanchez-Ruiz *et al.*, 1988a,b; Guzman-Casado *et al.*, 1990; Morin *et al.*, 1990; Lepock *et al.*, 1990; Galisteo *et al.*, 1991; Conejero-Lara *et al.*, 1991a,b; Galisteo and Sanchez-Ruiz, 1993) and by the correlation between the DSC-detected thermal effect and the extent of irreversible denaturation measured by independent methods (Morin *et al.*, 1990; Lepock *et al.*, 1990; Galisteo *et al.*, 1991). It appears clear now that most irreversible DSC transitions must be analyzed on the basis of kinetic models (described in Sections 4.3.1 and 4.3.2). In addition, these results have general consequences concerning the factors that determine protein stability, as is explained below.

Protein stability is often defined as the Gibbs energy difference between the denatured and native states (the definition we have employed in Section 3), but, for some practical purposes, the denaturation temperature (T_m) may be a more useful measure of protein stability. For two-state reversible denaturation cases, the denaturation temperature is simply determined by the stability curve (Section 3.1); however, it can be easily shown (Section 4.3.2) that irreversible alterations shift the DSC transitions to temperatures significantly lower than that corresponding to $\Delta_N^D G = 0$. Therefore, in many cases of irreversible denaturation, operational thermal stability (as measured by the denaturation temperature under given conditions) may be subject to kinetic constraints. As a result (Conejero-Lara *et al.*, 1991a; Sanchez-Ruiz, 1992), equilibrium approaches to the design of proteins with enhanced thermal stability (such as introducing mutations that are expected to increase $\Delta_N^D G$) may not be adequate in some cases of irreversible denaturation, while changes (in the protein or in the solvent conditions) that decrease the rate of irreversible denaturation might be found to be more successful.

4.3.1. The Two-State Irreversible Model

The simplest model accounting for irreversible denaturation is a two-state, first-order kinetic mechanism in which only the native state (N) and the irreversibly denatured state (i.e., the final state: F) of the protein are significantly populated:

$$\mathrm{N} \xrightarrow{k_{\mathrm{ap}}} \mathrm{F}$$

where k_{ap} is the first-order rate constant for the irreversible denaturation process (Sanchez-Ruiz *et al.*, 1988a; Sanchez-Ruiz, 1992). No equilibrium step is present in this model and the excess heat capacity function is entirely determined by the rate equation for the irreversible formation of the final state:

$$-\,dx_{\mathrm{N}}/dt = dx_{\mathrm{F}}/dt = k_{\mathrm{ap}} \cdot x_{\mathrm{N}} \tag{41}$$

where x_{N} and x_{F} are the mole fractions of molecules in states N and F. For a DSC experiment carried out at constant scanning rate the relevant differential equation is:

$$dx_N/dT = -[k_{ap}(T)/v]\cdot x_N \tag{42}$$

where the rate constant is a function of the independent variable (T). Separation of variables in Eq. (42), followed by integration from a low temperature (T_0), at which the reaction rate is negligible and $x_N \cong 1$, leads to the temperature dependence of x_N and x_F:

$$x_N = e^{-F(T)/v} \qquad x_F = 1 - e^{-F(T)/v} \tag{43}$$

where F(T) is the following integral:

$$F(T) = \int_{T_0}^{T} k_{ap}(T) \cdot dT \tag{44}$$

The excess heat capacity (measured from the chemical baseline) is given by,

$$C_p^{ex} = \Delta_N^F H \cdot (dx_F/dT) = -\Delta_N^F H \cdot (dx_N/dT) \tag{45}$$

where $\Delta_N^F H$ is the denaturation enthalpy (the enthalpy difference between the final and native states). If the temperature dependence of the rate constant is known (or assumed), then the DSC profiles can be simulated by using Eqs. (41) to (43). In the illustrative example shown in Fig. 13, we have assumed that the rate constant changes with temperature as given by the Arrhenius equation, $k_{ap} = A \cdot e^{-E/RT}$, where E is the energy of activation and A is the frequency factor. Note (see Fig. 13) that the DSC transitions predicted by the two-state irreversible model are strongly scanning rate dependent and asymmetrical, being steeper in the high-temperature side.

Equilibrium thermodynamics analysis of DSC transitions that follow the two-state irreversible model is not permissible. This statement means that

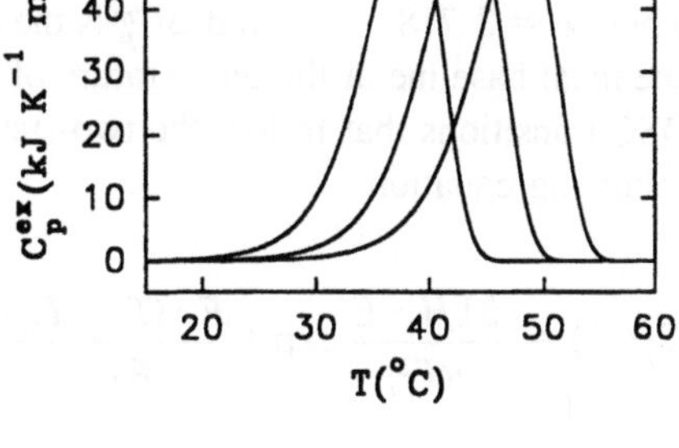

FIGURE 13. Excess heat capacity *versus* temperature profiles predicted by the two-state irreversible model [Eqs. (41)–(45)] with the parameters $\Delta_N^F H$ = 500 kJ/mole (assumed constant), E = 250 kJ/mole, and T^* = 50 °C, where T^* is the temperature at which the apparent first-order rate constant for irreversible denaturation equals 1 min^{-1} (T^* is related to the frequency factor of the Arrhenius equation by $A = e^{-E/RT^*}$). The numbers alongside the transitions stand for the scanning rate in K/min.

(Sanchez-Ruiz, 1992): (1) the denaturation entropy change (and, hence, the Gibbs energy change) cannot be obtained from the transitions, given that entropy calculations from experimental heat capacity data are based upon the Clausius equality, which does not hold for an irreversible, rate-limited process; (2) analysis of the transitions based on the assumption that two or more protein states coexist in equilibrium during the scan are obviously incorrect. Nevertheless, according to the first law, the total heat absorbed (calculated by integration of the C_p^{ex} versus T profile) is equal to the denaturation enthalpy change, even if the denaturation process is irreversible. Clearly, this denaturation enthalpy is the enthalpy difference between the final and native states: $\Delta_{\text{N}}^{\text{F}}H$. In addition to $\Delta_{\text{N}}^{\text{F}}H$, kinetic parameters for the irreversible process can be obtained by nonlinear, least-squares fitting procedures (Morin *et al.*, 1990; Conejereo-Lara *et al.*, 1991a) or by using the following simple expressions (Sanchez-Ruiz *et al.*, 1988a; Sanchez-Ruiz, 1992). Thus, the first-order rate constant is related to the calorimetric data by:

$$k_{\text{ap}} = v \cdot C_p^{\text{ex}} / (\Delta_{\text{N}}^{\text{F}}H - \langle \Delta H \rangle) \tag{46}$$

Note that there must be agreement between the k_{ap} values calculated [by using Eq. (46)] from transitions obtained at different scanning rates, a property that can be used to test the validity of the model. The scanning rate effect on the DSC transitions is described by:

$$\ln(v/T_m^2) = \text{constant} - E/RT_m \tag{47}$$

which indicates that the energy of activation can be obtained from the slope of a plot of $\ln(v/T_m^2)$ versus $1/T_m$. In this equation, as well as in the following Eqs. (48) and (49), T_m stands for the temperature of the maximum excess heat capacity (which, in this case, differs by several degrees from the temperature at which $x_{\text{F}} = 1/2$). The activation energy can also be calculated from the parameters of the maximum of the transitions by using:

$$E = eRT_m^2 \cdot \Delta C_p^m / \Delta_{\text{N}}^{\text{F}}H \tag{48}$$

where $e = 2.718\ldots$, and ΔC_p^m is the excess heat capacity measured from the chemical baseline at the temperature of the maximum. Finally, the shape of the DSC transitions that follow the two-state irreversible model is described by the following equation

$$C_p^{\text{ex}}(T) = \frac{\Delta_{\text{N}}^{\text{F}}H \cdot E}{RT_m^2} \exp\left(\frac{E \cdot (T - T_m)}{RT_m^2}\right) \times \left\{\exp\left[-\exp\left(\frac{E \cdot (T - T_m)}{RT_m^2}\right)\right]\right\} \tag{49}$$

which may be useful for fitting purposes (see, for instance, Conejero-Lara *et al.*, 1991a).

The two-state irreversible model may appear excessively simple. However, recent work has shown that the thermal denaturation of many proteins follow this model, at least under some solvent conditions. The examples include: *Bacillus thermoproteollyticus rokko* thermolysin (Sanchez-Ruiz *et al.*, 1988a), pro- and carboxypeptidase A from porcine pancreas (Sanchez-Ruiz *et al.*, 1988b), lobster hemocyanin (Guzman-Casado *et al.*, 1990), cytochrome *c* oxidase from yeast (Morin *et al.*, 1990), Ca^{2+} -ATPase of sacroplasmic reticulum (Lepock *et al.*, 1990) pro- and carboxypeptidase B from porcine pancreas (Conejero-Lara *et al.*, 1991b), *Escherichia coli* 5-enolpyruvoil shikimate-3-phosphate synthase (Merabet *et al.*, 1993), and *Halobacterium halobium* bacteriorhodopsin (Galisteo and Sanchez-Ruiz, 1993).

4.3.2. The Lumry–Eyring Model

There are, of course, irreversible, scanning-rate dependent DSC transitions that do not follow the two-state irreversible model (see, for instance, Galisteo *et al.*, 1991). It is also clear that although the two-state irreversible model provides an adequate phenomenological description of the thermal denaturation in many cases, it is too simple to be physically realistic. Thus, this model would appear to suggest that the final state is formed by direct alteration of the native protein, while it appears likely that unfolded or partially unfolded states are more susceptible to irreversible modifications. Accordingly, a still very simple but more realistic model should include two steps: (1) reversible denaturation of the native protein to yield and unfolded, or partially unfolded state (D); and (2) irreversible alteration of the denatured state to produce the final state (F), which is unable to fold back to the native protein. This model can be depicted as:

$$\mathrm{N} \rightleftharpoons \mathrm{D} \longrightarrow \mathrm{F}$$

and is usually known as the Lumry–Eyring model (Lumry and Eyring, 1954). The several processes responsible for the irreversible step (aggregation, autolysis, chemical alteration of residues) have been reviewed by Klibanov and Ahern (1987).

This author (Sanchez-Ruiz, 1992) has recently carried out a theoretical analysis of the DSC transitions predicted by the Lumry–Eyring model on the basis of the following assumptions: (1) equilibrium is always established for the reversible step; (2) the kinetics of the irreversible process is of first order with a rate constant that changes with temperature as given by the Arrhenius equation; and (3) the irreversible step shows no thermal effect ($\Delta_D^F H = 0$). Depending on the approximate temperature at which the irreversible step proceeds with

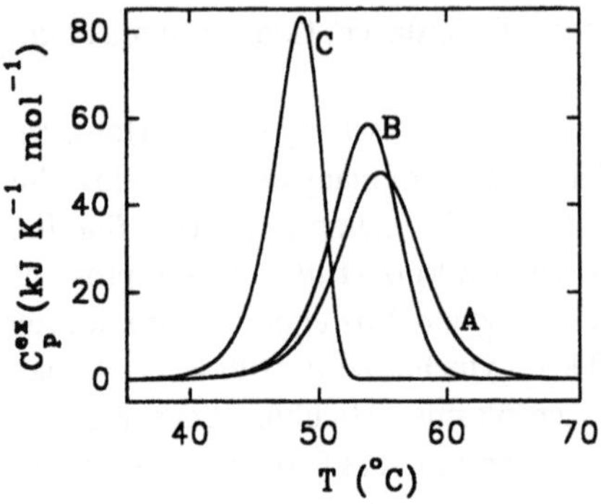

FIGURE 14. Excess heat capacity *versus* temperature profiles predicted by the Lumry–Eyring model. The parameters for the reversible step (assumed at equilibrium) are: $\Delta_N^D H$ = 400 kJ/mole (assumed constant) and T_m = 55 °C. The irreversible step is characterized by an activation energy of E = 100 kJ/mole and shows no thermal effect ($\Delta_D^F H$ = 0). The transitions shown correspond to v = 2 K/min and differ in the value of the temperature, T^*, at which the rate constant for irreversible denaturation is 1 min^{-1}. (A) T^* = 75 °C; the irreversible step proceeds at significant rate at temperatures above T_m and the DSC transition differs little from that corresponding to the reversible denaturation. (B) T^* = 55 °C; the irreversible step is fast in the temperature range of the reversible denaturation and the DSC transition is distorted, although there exists a significant amount of denatured state in equilibrium with the native one within the temperature range of the transition (see Sanchez-Ruiz, 1992). (C) T^* = 25 °C; the irreversible step becomes fast at temperatures well below T_m, the population of denatured state is always very low and the DSC transition follows the two-state equilibrium model.

significant rate, the simulated DSC transitions are broadly classified into several groups (situations A, B, and C), which are illustrated in Fig. 14. (A) The transition is calorimetrically irreversible, but the irreversible step takes place with significant rate only at temperatures somewhat above than those corresponding to the transition. Equilibrium thermodynamics analysis of the transition is permissible. The irreversible thermal denaturation of the B subunit of cholera toxin (Goins and Freire, 1988) appears to be an example of this situation. (B) The DSC transition is distorted by the occurrence of the irreversible step; there exists, however, a significant amount of denatured state in equilibrium with the native protein during the thermally induced denaturation. That is, the DSC transition contains thermodynamic information about the reversible step, which (at least in theory) could be obtained upon the appropriate data treatment. (C) The amount of denatured state is always very low and only the native and final states are significantly populated during denaturation; thus, the thermal denaturation process is phenomenologically described by the two-state irreversible model with an apparent rate constant (k_{ap}) equal to the denaturation equilibrium constant (K) times the rate constant for the irreversible step (k):

$$\mathrm{N} \xrightarrow{kK} \mathrm{F}$$

Clearly, the two-state irreversible model can be considered as a limiting case of the Lumry–Eyring model.

The simulations described above (Fig. 14) were carried out assuming that chemical equilibrium between the states N and D is always established, that is, assuming that the rate-limiting step is the irreversible process itself (D→F). It was noted (Sanchez-Ruiz, 1992) that a more general analysis of the Lumry–Eyring model would require that the kinetics of the denaturation step be taken into account, according to this scheme:

$$\mathrm{N} \underset{k_r}{\overset{k_f}{\rightleftharpoons}} \mathrm{D} \xrightarrow{\;k\;} \mathrm{F}$$

where, for the sake of simplicity, we will assume that all processes are of first order. A detailed theoretical analysis of this "fully kinetic" Lumry–Eyring model can be found in Lepock *et al.* (1992). Here, we will discuss a central issue over which some confusion appears to persist in the literature, and which may be posed in the following way: Provided that the amount of denatured protein is very low (that is, the state D is an unstable intermediate), what are the factors that determine the rate-limiting step? The answer (see Chapter 10 of Jencks, 1987) is that the k_f value has nothing to do with it, and that the rate-limiting step is solely determined by the relative values of the constants k and k_r. This is illustrated by the Gibbs energy versus reaction coordinate diagrams shown in Fig. 15. For the simple model we are analyzing, these diagrams show one intermediate (D) and two transition states (1 and 2 in Fig. 15). If $k_r >> k$, the highest point in the diagram (i.e., the transition state for the overall process) occurs in the irreversible step (transition state 2 in the diagram); therefore, the irreversible step is rate-limiting and the activation Gibbs energy for the overall process is given by $\Delta_{\mathrm{N}}^{\neq}G = \Delta_{\mathrm{N}}^{\mathrm{D}}G + \Delta_{\mathrm{N}}^{2}G$, which leads to $k_{\mathrm{ap}} = kK$ (see Fig. 15 for details). Obviously, this is the situation C described above. If $k >> k_r$, the highest point in the diagram occurs in the reversible step (transition state 1 in the diagram); then the formation of D becomes rate-limiting and the activation Gibbs energy for the overall process is given by $\Delta_{\mathrm{N}}^{\neq}G = \Delta_{\mathrm{N}}^{1}G$ (Fig. 15), which leads to $k_{\mathrm{ap}} = k_f$. This case has been previously referred to (Sanchez-Ruiz, 1992) as situation C′:

$$\mathrm{N} \xrightarrow{\;k_f\;} \mathrm{D}$$

It must be noted that the reaction coordinate diagrams of Fig. 15 are only meant to reflect the main features of the denaturation process. In the real case (see, for instance, Fig. 1 in Matouschek *et al.*, 1989), we should expect many unstable intermediates (i.e., many local minima along the Gibbs energy versus reaction coordinate profile) and, consequently, many transition states (i.e., many

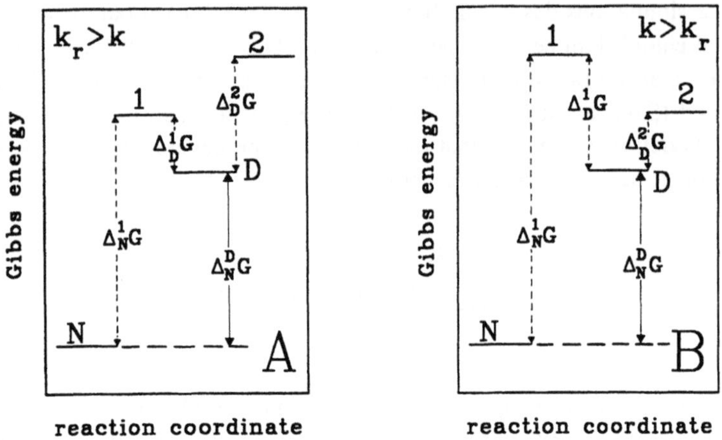

FIGURE 15. Gibbs energy *versus* reaction coordinate profiles for the Lumry–Eyring model, illustrating the fact that the rate-limiting step is determined by the relative values of k and k_r. These diagrams depict the Gibbs energies of the native (N) and denatured states (D), as well as those of the transition states for the reversible (1) and irreversible (2) steps. The Gibbs energy of the final state (not shown) is much lower than those of the native and denatured states (the overall process is irreversible). These profiles correspond to the cases (situations C or C′) in which the denatured state is an unstable intermediate ($\Delta_N^D G >> 0$). Activation Gibbs energies ($\Delta G^\#$) and rate constants (k_{ap}) are related by transition state theory: $k_{ap} = (k_B T/h) \cdot \exp(-\Delta G^\# / RT)$. (A) $k_r > k$ and $\Delta_D^1 G < \Delta_D^2 G$; the irreversible step is rate-limiting and the transition state for the overall process is 2: $\Delta G^\# = \Delta_N^\# G = \Delta_N^D G + \Delta_D^2 G$ and $k_{ap} = kK$. (B) $k > k_r$ and $\Delta_D^2 G < \Delta_D^1 G$; the reversible step is rate-limiting and the transition state for the overall process is 1: $\Delta G^\# = \Delta_N^\# G = \Delta_N^1 G$ and $k_{ap} = \Delta k_f$.

local maxima along the Gibbs energy profile). Clearly, the position of the transition state for the overall process (the global maximum in the Gibbs energy profile) cannot be known from the sole fact that the thermal denaturation follows the two-state irreversible model; in other words, it is not possible to decide whether the rate-limiting step is the irreversible process itself (situation C) or an earlier event in the denaturation pathway (situation C′).* Nevertheless, a recent comparative analysis (see Conejero-Lara *et al.*, 1991b) of the kinetic parameters for reversible and irreversible thermal denaturation of several proteins appears to suggest that the irreversible step is not rate-limiting in many cases of irreversible denaturation.

*Strictly speaking, the above reasonings assume that all kinetically relevant processes are of first order. Some information about the rate-limiting step will be obvious if one of the processes is expected to be non–first order. For instance, the process responsible for the irreversibility in the thermal denaturation of thermolysin is the autolysis of the enzyme (Sanchez-Ruiz *et al.*, 1988a), which is expected to be a second-order process; however, the overall irreversible denaturation follows first-order kinetics, suggesting that autolysis is not rate-limiting in this case.

Regardless of the nature of the rate-limiting step, it is clear that, for a two-state irreversible denaturation, only events that take place prior to this step may affect the rate of the overall process. Thus, for a protein that has a binding site for a ligand, the concentration of ligand will affect the rate of irreversible denaturation if ligand dissociation takes place prior to the rate-limiting step (the ligand does not bind to the transition state); in this case, the rate will decrease upon increasing ligand concentration and, consequently, the DSC transitions will be shifted to higher temperatures (Conejero-Lara *et al.*, 1991a; Sanchez-Ruiz, 1992). On the other hand, the DSC transitions will not be affected by ligand concentration if ligand dissociation takes place after the rate-limiting step. The same type of reasoning can be used to explain other effects (for instance, the protein concentration effect arising from the dissociation of a multimeric protein into monomers) or the lack of them (if the dissociation takes place after the rate-limiting step; for details, see Sanchez-Ruiz, 1992).

So far, we have mainly discussed cases of two-state irreversible denaturation. It is conceivable, however, to find DSC transitions that are distorted by the occurrence of irreversible processes, but that contain thermodynamic information about the reversible denaturation of the protein (this is the situation B described above). Recently, the possibility of extracting this information from the distorted DSC profiles has been theoretically discussed (Sanchez-Ruiz, 1992; see also Freire *et al.*, 1990) on the basis of the following straightforward generalization of the Lumry–Eyring model (for an experimental illustration of this approach, see Arriaga *et al.*, 1992; Hernandez-Arana *et al.*, 1993). We assume that the protein may exist in a number of different states [I_0 (the native state, N), I_1, I_2, . . .] and that the rate-limiting step is the reaction of the state I_n (which is not necessarily the denatured state, D) to yield the final, irreversibly denatured state (if the irreversible process itself is rate-limiting) or another protein state (if a previous step in the denaturation pathway is rate-limiting). In both cases, we phenomenologically describe the formation of the final state from I_n as:

$$I_n \xrightarrow{k} F$$

where, for simplicity, we assume first-order kinetics (for non–first-order kinetics, see Sanchez-Ruiz, 1992). We further assume that all processes that take place prior to the rate-limiting step are fast enough to be at equilibrium during the scan; this means that chemical equilibrium between I_n and all the previous protein states (I_0, I_1, I_2, . . . , I_{n-1}) is established and that their relative amounts can be expressed in terms of the Freire–Biltonen, partition–function formalism (see Section 2.2). Thus, the fraction of states I in equilibrium that exists as a given state, I_i, is:

$$y_i = [\mathrm{I}_i]/\sum_{i=0}^{n}[\mathrm{I}_i] = (1/Q)\cdot\exp(-\Delta_{\mathrm{N}}^{i}G/RT) \qquad (50)$$

and the average enthalpy of these states is given by,

$$\langle\Delta H\rangle_{\mathrm{eq}} = \sum_{i=0}^{n} y_i\cdot\Delta_{\mathrm{N}}^{i}H \qquad (51)$$

Of course, these are just the Eqs. (11) and (12) of Section 2.2; we have used y_i and $<\Delta H>_{\mathrm{eq}}$ (instead of x_i and $<\Delta H>$) to emphasize the fact that the above fractions and excess enthalpy are calculated excluding the contribution from the final state. The corresponding quantities, including the contribution from F, are:

$$x_{\mathrm{F}} = [\mathrm{F}]/\left([\mathrm{F}] + \sum_{i=0}^{n}[\mathrm{I}_i]\right) \qquad (52)$$

$$x_i = [\mathrm{I}_i]/\left([\mathrm{F}] + \sum_{i=0}^{n}[\mathrm{I}_i]\right) = y_i\cdot(1 - x_{\mathrm{F}}) \qquad (53)$$

$$\langle\Delta H\rangle = x_{\mathrm{F}}\cdot\Delta_{\mathrm{N}}^{\mathrm{F}}H + \sum_{i=0}^{n} x_i\cdot\Delta_{\mathrm{N}}^{i}H \qquad (54)$$

where $<\Delta H>$ of Eq. (54) (an average over *all* protein states) is the quantity experimentally accessible from the DSC profiles, while $<\Delta H>_{\mathrm{eq}}$ of Eq. (51) (an average over the states in equilibrium only) contains the thermodynamic information about the "equilibrium part" of the overall irreversible process. Using Eq. (53), the relation between both excess enthalpies is found to be:

$$<\Delta H> = (1 - x_{\mathrm{F}})\cdot<\Delta H>_{\mathrm{eq}} + x_{\mathrm{F}}\cdot\Delta_{\mathrm{N}}^{\mathrm{F}}H \qquad (55)$$

Taking into account Eqs. (50) and (53), the rate equation for the irreversible formation of F may be written as:

$$dx_{\mathrm{F}}/dt = k\cdot x_n = k\cdot y_n\cdot(1 - x_{\mathrm{F}}) = (1 - x_{\mathrm{F}})\cdot(k/Q)\cdot\exp(-\Delta_{\mathrm{N}}^{n}G/RT) \qquad (56)$$

where the expression $(k/Q)\cdot\exp(-\Delta_{\mathrm{N}}^{n}G/RT)$ depends only on temperature and plays the role of an apparent, first-order rate constant: k_{ap}. For a DSC experiment at constant scanning rate, the relevant differential equation is:

$$dx_F/dT = (k_{ap}/v)\cdot(1 - x_F) \tag{57}$$

which can be integrated (from a low temperature, T_0, at which the rate is negligible and $x_F \cong 0$) to yield the temperature dependence of x_F:

$$x_F = 1 - e^{-F(T)/v} \tag{58}$$

$$F(T) = \int_{T_0}^{T} k_{ap} \cdot dT \tag{59}$$

which, upon substitution in Eq. (55), leads to:

$$\Delta_N^F H - <\Delta H> = (\Delta_N^F H - <\Delta H>_{eq})\cdot e^{-F(T)/v} \tag{60}$$

It can easily be shown that Eq. (60) remains unaffected if we assume that all the I states may react to yield the final state, provided that the apparent rate constant is defined as an average over the states I in equilibrium:

$$k_{ap} = \langle k \rangle = (1/Q) \cdot \sum_{i=0}^{n} k_i \cdot \exp(-\Delta_N^i G/RT) \tag{61}$$

where k_i is the first-order rate constant for the process $I_i \rightarrow F$.

According to Eq. (60), $<\Delta H>_{eq}$ and F(T) can be calculated from a series of DSC transitions obtained at different scanning rates. Thus, a plot of $\ln(\Delta_N^F H - <\Delta H>/)$ versus $1/v$ should result in a straight line; $<\Delta H>_{eq}$ and F(T) could then be obtained from the slope and intercept of that line [of course, a nonlinear, least-squares fitting of Eq. (6) to the experimental $<\Delta H>$ versus v data would also be possible]. This procedure should be applied to several temperatures (within the denaturational range) to produce the $<\Delta H>_{eq}/T$ profile (which contains the thermodynamic information associated with the DSC transitions) and the F(T)/T profile (which contains kinetic information about the irreversible formation of F). Clearly, the thermodynamic information ($<\Delta H>_{eq}$) is obtained as an extrapolation to infinite scanning rate [$1/v = 0$ in Eq. (60)]. It must be noted, nevertheless, that the above approach [Eqs. (50)–(60)] is based on a very simple kinetic model, since the collection of states I in equilibrium behaves as a single species from a kinetic point of view. Given the limited range of scanning rates usually available, it does not seem possible to discern, on the basis of DSC data alone, between this simple "two-species" model and other, more complex kinetic situations. Consequently, the equilibrium information obtained from the extrapolation to infinite scanning rate [Eq. (60)] must be interpreted with caution.

Note Added in Proof

After this chapter had been submitted, Privalov and Makhatadze (Makhatadze and Privalov, 1993; Privalov and Makhatadze, 1993) reported an analysis of the thermodynamic data for the denaturation of four globular proteins (cytochrome *c*, ribonuclease a, lysozyme, and myoglobin) that is more detailed than that used for illustration in this chapter. These authors consider contributions arising from the hydration of polar and apolar groups (both modeled on the basis of gas dissolution data), configurational entropy, hydrogen bonding, and van der Walls interactions of the buried apolar groups (modeled on the basis of sublimation data for organic crystals). In view of their analysis, it appears that hydrophobic contributions calculated on the basis of $T_H^{ap} = 100\ °C$ *(see Sections 3.3 and 3.4) do not include the van der Waalls interactions of the apolar groups in the protein interior.*

5. REFERENCES

Arriaga, P., Menendez, M., Villacorta, J. M., and Laynez, J., 1992, Differential scanning calorimetric study of the thermal unfolding of β-lactamase I from *Bacillus cereus*, *Biochemistry* **31:**6603–6607.

Azuaga, A. I., Galisteo, M. L., Mayorga, O. L., Cortijo, M., and Mateo, P. L., 1992, Heat and cold denaturation of β-lactoglobulin B, *FEBS Lett.* **309:**258–260.

Baldwin, R. L., 1986, Temperature dependence of the hydrophobic interaction in protein folding, *Proc. Natl. Acad. Sci. USA* **83:**8069–8072.

Baldwin, R. L., and Muller, N., 1992, Relation between the convergence temperatures T_h^* and T_s^* in protein unfolding, *Proc. Natl. Acad. Sci. USA* **89:**7110–7113.

Becktel, W. J., and Schellman, J. A., 1987, Protein stability curves, *Biopolymers* **26:**1859–1877.

Bolen, D. W., and Santoro, M. M., 1988, Unfolding free energy determined by the linear extrapolation method. 2. Incorporation of $\Delta G°_{N\text{-}U}$ values in a thermodynamic cycle, *Biochemistry* **27:**8069–8074.

Brandts, J. F., and Lin, L-N., 1990, Study of strong to ultratight protein interactions using differential scanning calorimetry, *Biochemistry* **29:**6927–6940.

Brandts, J. F., Hu, C. Q., and Lin, L-N., 1989, A simple model for proteins with interacting domains. Applications to scanning calorimetry data, *Biochemistry* **28:**8588–8596.

Chen, B., and Schellman, J. A., 1989, Low-temperature unfolding of a mutant of phage T4 lysozyme. 1. Equilibrium studies, *Biochemistry* **28:**685–691.

Chen, B., Baase, W. A., and Schellman, J. A., 1989, Low temperature unfolding of a mutant of phage T4 lysozyme. 2. Kinetic investigations, *Biochemistry* **28:**691–699.

Chen, B., Baase, W. A., Nicholson, H., and Schellman, J. A., 1992, Folding kinetics of T4 lysozyme and nine mutants at 12°C, *Biochemistry* **31:**1464–1476.

Chothia, C., 1976, The nature of the accessible and buried surfaces in proteins, *J. Mol. Biol.* **105:**1–14.

Conejero-Lara, F., Mateo, P. L., Aviles, F. X., and Sanchez-Ruiz, J. M. 1991a, Effect of Zn^{2+} on the thermal denaturation of carboxypeptidase B, *Biochemistry* **30:**2067–2072.

Conejero-Lara, F., Sanchez-Ruiz, J. M., Mateo, P. L., Burgos, F. J., Vendrell, J., and Aviles, F. X.,

1991b, Differential scanning calorimetric study of carboxypeptidase B, procarboxypeptidase B and its globular activation domain, *Eur. J. Biochem.* **220:**663–670.

Creighton, T. E., 1990, Protein folding, *Biochem. J.* **270:**1–16.

Dill, K. A., 1985, Theory for the folding and stability of globular proteins, *Biochemistry* **24:**1501–1509.

Dill, K. A., 1990, Dominant forces in protein folding, *Biochemistry* **29:**7133–7155.

Dill, K. A., and Shortle, D., 1991, Denatured states of proteins, *Annu. Rev. Biochem.* **60:**795–825.

Doig, A. J., and Williams, D. H., 1992, Why water-soluble, compact, globular proteins have similar specific enthalpies of unfolding at 110°C, *Biochemistry* **31:**9371–9375.

Filimonov, V. V., Matveyev, S. V., Potekhin, S. A., and Privalov, P. L., 1982, Thermodynamic analysis of scanning microcalorimetric data, *Mol. Biol.* **16:**551–562.

Fischer, G., and Schmid, F. X., 1990, The mechanism of protein folding. Implications of *in vitro* refolding models for *de novo* protein folding and translocation in the cell, *Biochemistry* **29:**2205–2212.

Freire, E., and Biltonen, R. L., 1978, Statistical mechanical deconvolution of thermal transitions in macromolecules. I. Theory and applications to homogeneous systems, *Biopolymers* **17:**463–479.

Freire, E., and Murphy, K. P., 1991, The molecular basis of co-operativity in protein folding, *J. Mol. Biol.* **222:**687–698.

Freire, E., van Osdol, W. W., Mayorga, O. L., and Sanchez-Ruiz, J. M., 1990, Calorimetrically determined dynamics of complex unfolding transitions in proteins, *Annu. Rev. Biophys. Biophys. Chem.* **19:**159–188.

Freire, E., Murphy, K. P., Sanchez-Ruiz, J. M., Galisteo, M. L., and Privalov, P. L., 1992, The molecular basis of co-operativity in protein folding. Thermodynamic dissection of interdomain interactions in phosphoglycerate kinase, *Biochemistry* **31:**250–256.

Fu, L., and Freire, E., 1992, On the origin of the enthalpy and entropy convergence temperatures in protein folding, *Proc. Natl. Acad. Sci. USA* **89:**9335–9338.

Galisteo, M. L., and Sanchez-Ruiz, J. M., 1993, Kinetic study into the irreversible thermal denaturation of bacteriorhodopsin, *Eur. Biophys. J.* **22:**25–30.

Galisteo, M. L., Mateo, P. L., and Sanchez-Ruiz, J. M., 1991, Kinetic study on the irreversible thermal denaturation of yeast phosphoglycerate kinase, *Biochemistry* **30:**2061–2066.

Gill, S. J., Richey, B., Bishop, G., and Wyman, J., 1985, Generalized binding phenomena in an allosteric macromolecule, *Biophys. Chem.* **21:**1–14.

Goins, B., and Freire, E., 1988, Thermal stability and intersubunit interactions of cholera toxin in solution and in association with its cell-surface receptor ganglioside G_{M1}, *Biochemistry* **27:**2046–2052.

Green, S. M., Meeker, A. K., and Shortle, D., 1992, Contributions of the polar, uncharged amino acids to the stability of staphylococcal nuclease: Evidence for mutational effects on the free energy of the denatured state, *Biochemistry* **31:**5717–5728.

Griko, Yu. V., and Privalov, P. L., 1992, Calorimetric study of the heat and cold denaturation of β-lactoglobulin, *Biochemistry* **31:**8810–8815.

Griko, Yu. V., Privalov, P. L., Sturtevant, J. M., and Venyamov, S. Yu., 1988a, Cold denaturation of staphylococcal nuclease, *Proc. Natl. Acad. Sci. USA* **85:**3343–3347.

Griko, Yu. V., Privalov, P. L., Venyamov, S.Yu., and Kutishenko, V. P., 1988b, Thermodynamic study of the apomyoglobin structure, *J. Mol. Biol.* **202:**128–138.

Griko, Yu. V., Venyamov, S. Yu., and Privalov, P. L., 1989, Heat and cold denaturation of phosphoglycerate kinase (interaction of domains), *FEBS Lett.* **244:**276–278.

Guzman-Casado, M., Parody-Morreale, A., Mateo, P. L., and Sanchez-Ruiz, J. M., 1990, Differential scanning calorimetry of lobster hemocyanin, *Eur. J. Biochem.* **188:**181–185.

Hernandez-Arana, A., Rojo-Dominguez, A., Altamirano, M. M., and Calcagno, M. L., 1993,

Differential scanning calorimetry of the irreversible denaturation of *Escherichia coli* glucosamine-6-phosphate deaminase, *Biochemistry* **32**:3644–3648.
Honig, B., Sharp, K., and Yang, A-S., 1993, Macroscopic models of aqueous solutions: Biological and chemical applications, *J. Phys. Chem.* **97**:1101–1109.
Hu, C-Q., Sturtevant, J. M., Thomson, J. A., Erickson, R. E., and Pace, C. N., 1992, Thermodynamics of ribonuclease T1 denaturation, *Biochemistry* **31**:4876–4882.
Jencks, W. P., 1987, *Catalysis in Chemistry and Enzymology*, Dover Publications Inc., New York.
Klibanov, A. M., and Ahern, T. J., 1987, Thermal stability of proteins, in *Protein Engineering* (D. L. Oxender and C. F. Fox, eds.), pp. 213–218, Alan R. Liss, New York.
Krishnan, K. S., and Brandts, J. F., 1978, Scanning calorimetry, *Methods Enzymol.* **49**:3–14.
Lee, B., 1991, Isoenthalpic and isoentropic temperatures and the thermodynamics of protein denaturaiton, *Proc. Natl. Acad. Sci. USA* **88**:5154–5158.
Lee, B., and Richards, F. M., 1971, The interpretation of protein structures: estimation of static accessibility, *J. Mol. Biol.* **55**:379–400.
Lepock, J. R., Rodhal, A. M., Zhang, M. L., Heynen, M. L., Waters, B., and Cheng, K. H., 1990, Thermal denaturation of the Ca^{2+}-ATPase of sarcoplasmic reticulum reveals two thermodynamically independent domains, *Biochemistry* **29**:681–689.
Lepock, J. R., Ritchie, K. P., Kolios, M. C., Rodahl, A. M., Heinz, K. A., and Kruuv, J., 1992, Influence of transition rates and scan rate on kinetic simulations of differential scanning calorimetry profiles of reversible and irreversible protein denaturation, *Biochemistry* **31**:12706–12712.
Lopez-Mayorga, O., and Freire, E., 1987, Dynamic analysis of differential scanning calorimetry data, *Biophys. Chem.* **87**:87–96.
Lumry, R., and Eyring, E., 1954, Conformation changes of proteins, *J. Phys. Chem.* **58**:110–120.
Makhatadze, G. I., and Privalov, P. L., 1990, Heat capacity of proteins. I. Partial molar heat capacity of individual amino acid residues in aqueous solution: Hydration effect, *J. Mol. Biol.* **213**:375–384.
Makhatadze, G. I., and Privalov, P. L., 1993, Contribution of hydration to protein folding thermodynamics. I. The enthalpy of hydration, *J. Mol. Biol.* **232**:639–659.
Manly, S. P., Matthews, K. S., and Sturtevant, J. M., 1985, Thermal denaturation of the core protein of the *lac* repressor, *Biochemistry* **24**:3842–3846.
Mateo, P. L., 1984, Differential scanning calorimetry of protein solutions, in *Thermochemistry and Its Applications to Chemical and Biochemical Systems* (R. da Silva, ed.), pp. 541–568, Reidel, Dordrecht.
Matouschek, A., Kellis, J. T., Serrano, L., and Fersht, A. R., 1989, Mapping the transition state and pathway of protein folding by protein engineering, *Nature* **340**:122–126.
Merabet, E. K., Walker, M. C., Yuen, H. K., and Sikorski, J. A., 1993, Differential scanning calorimetric study of 5-enolpyruvoil shikimate-3-phosphate synthase and its complexes with shikimate-3-phosphate and glyphosate: Irreversible thermal transitions, *Biochim. Biophys. Acta* **1161**:272–278.
Morin, P. E., Diggs, D., and Freire, E., 1990, Thermal stability of membrane-reconstituted yeast cytochrome *c* oxidase, *Biochemistry* **29**:781–788.
Murphy K. P., and Freire, E., 1992, Thermodynamics of structural stability and cooperative folding behavior in proteins, *Adv. Prot. Chem.* **43**:313–361.
Murphy, K. P., and Gill, S. J., 1990, Group additivity thermodynamics for dissolution of solid cyclic dipeptides in water, *Thermochim. Acta* **172**:11–20.
Murphy, K. P., and Gill, S. J., 1991, Solid model compounds and the thermodynamics of protein unfolding, *J. Mol. Biol.* **222**:699–709.
Murphy, K. P., Privalov, P. L., and Gill, S. J., 1990, Common features of protein unfolding and dissolution of hydrophobic compounds, *Science* **247**:559–561.
Murphy, K. P., Bhakuni, V., Xie, D., and Freire, E., 1992, Molecular basis of co-operativity in

protein folding. III. Structural identification of cooperative folding units and folding intermediates, *J. Mol. Biol.* **227**:293–306.

Novokhatny, V. V., Kudinov, S. A., and Privalov, P. L., 1984, Domains in human plasminogen, *J. Mol. Biol.* **179**:215–232.

Pace, C. N., 1992, Contribution of the hydrophobic effect to globular protein stability, *J. Mol. Biol.* **226**:29–35.

Pace, C. N., and Laurents, D. V., 1989, A new method for determining the heat capacity change for protein folding, *Biochemistry* **28**:2520–2525.

Pantoliano, M. W., Withlow, M., Wood, J. F., Dodd, S. W., Hardman, K. D., Rollence, M. L., and Bryan, P. N., 1989, Large increases in general stability for subtilisin BPN′ through incremental changes in the free energy of unfolding, *Biochemistry* **28**:7205–7213.

Plaza del Pino, I. M., Pace, C. N., and Freire, E., 1992, Temperature and guanidine hydrochloride dependence of the structural stability of ribonuclease T_1, *Biochemistry* **31**:11196–11202.

Privalov, P. L., 1979, Stability of proteins. Small globular proteins, *Adv. Prot. Chem.* **33**:167–241.

Privalov, P. L., 1980, Scanning microcalorimeters for studying macromolecules, *Pure Appl. Chem.* **52**:479–497.

Privalov, P. L., 1982, Stability of proteins. Proteins which do not present a single cooperative system, *Adv. Prot. Chem.* **35**:1–104.

Privalov, P. L., 1989, Thermodynamic problems of protein structure, *Annu. Rev. Biophys. Biophys. Chem.* **18**:47–69.

Privalov, P. L., 1990, Cold denaturation of proteins, *Crit. Rev. Biochem. Mol. Biol.* **25**:281–306.

Privalov, P. L., and Gill, S. J., 1988, Stability of protein structure and hydrophobic interaction, *Adv. Prot. Chem.* **39**:191–234.

Privalov, P. L., and Khechinashvili, N. N., 1974, A thermodynamic approach to the problem of stabilization of globular protein structure: a calorimetric study, *J. Mol. Biol.* **86**:665–684.

Privalov, P. L., and Makhatadze, G. I., 1990, Heat capacity of proteins. II. Partial molar heat capacity of the unfolded polypeptide chain of proteins: protein unfolding effects, *J. Mol. Biol.* **213**:385–391.

Privalov, P. L., and Makhatadze, G. I., 1992, Contribution of hydration and non-covalent interactions to the heat capacity effect on protein unfolding, *J. Mol. Biol.* **224**:715–723.

Privalov, P. L., and Makhatadze, G. I., 1993, Contribution of hydration to protein folding thermodynamics. II. The entropy of Gibbs energy of hydration. *J. Mol. Biol.* **232**:660–679.

Privalov, P. L., Plotnikov, V. V., and Filimonov, V. V., 1975, Precision scanning microcalorimeter for the study of liquids, *J. Chem. Thermodyn.* **7**:41–47.

Privalov, P. L., Griko, Yu. V., Venyamov, S. Yu., and Kutyshenko, V. P., 1986, Cold denaturation of myoglobin, *J. Mol. Biol.* **190**:487–498.

Privalov, P. L., Tiktopulo, E. I., Venyaminov, S. Y., Griko, Y. V., Makhatadze, G. I., and Khechinashvili, N. N., 1989, Heat capacity and conformation of proteins in the denatured state, *J. Mol. Biol.* **205**:737–750.

Ramsay, G., and Freire, E., 1990, Linked thermal and solute perturbation analysis of cooperative domains interactions in proteins. Structural stability of diphteria toxin. *Biochemistry* **29**:8677–8683.

Robert, C. H., Gill, S. J., and Wyman, J., 1988, Quantitative analysis of linkage in macromolecules when one ligand is present in limited total quantity, *Biochemistry* **27**:6829–6835.

Sanchez-Ruiz, J. M., 1992, Theoretical analysis of Lumry–Eyring models in differential scanning calorimetry, *Biophys. J.* **61**:921–935.

Sanchez-Ruiz, J. M., and Mateo, P. L., 1987, Differential scanning calorimetry of membrane proteins, *Cell Biol. Rev.* **11**:15–45.

Sanchez-Ruiz, J. M., Lopez-Lacomba, J. L., Cortijo, M., and Mateo, P. L., 1988a, Differential scanning calorimetry of the irreversible thermal denaturation of thermolysin, *Biochemistry* **27**:1648–1652.

Sanchez-Ruiz, J. M., Lopez-Lacomba, J. L., Mateo, P. L., Vilanova, M., Serra, M. A., and Aviles, F. X., 1988b, Analysis of the thermal unfolding of porcine procarboxypeptidase A and its functional pieces by differential scanning calorimetry, *Eur. J. Biochem.* **176**:225–230.

Santoro, M. M., and Bolen, D. W., 1988, Unfolding free energy changes determined by the linear extrapolation method. 1. Unfolding of phenylmethanesulfonyl α-chymotrypsin using different denaturants, *Biochemistry* **27**:8063–8068.

Schellman, J. A., 1987, The thermodynamic stability of proteins, *Annu. Rev. Biophys. Chem.* **16**:115–137.

Scholtz, J. M., Marqusee, S., Baldwin, R. L., York, E. J., Stewart, J. M., Santoro, M., and Bolen, D. W., 1991, Calorimetric determination of the enthalpy change for the α-helix to coil transition of an alanine peptide in water, *Proc. Natl. Acad. Sci. USA* **88**:2854–2858.

Shirley, B. A., Stanssens, P., Hahn, U., and Pace, C. N., 1992, Contribution of hydrogen bonding to the conformational stability of ribonuclease T1, *Biochemistry* **31**:725–732.

Shortle, D., 1993, Denatured states of proteins and their roles in folding and stability, *Curr. Opin. Struc. Biol.* **3**:66–74.

Shortle, D., Stites, W. E., and Meeker, A. K., 1990, Contributions of large hydrophobic amino acids to the stability of staphylococal nuclease, *Biochemistry* **29**:8033–8041.

Shortle, D., Chan, H. S., and Dill, K., 1992, Modeling the effects of mutations on the denatured states of proteins, *Protein Sci.* **1**:201–215.

Shrake, A., and Rupley, J. A., 1973, Environment and exposure to solvent of protein atoms. Lysozyme and insuline, *J. Mol. Biol.* **79**:351–372.

Spolar, R. S., Livingstone, J. R., and Record, M. T., 1992, Use of liquid hydrocarbon and amide transfer data to estimate contributions to thermodynamic functions of protein folding from the removal of non-polar and polar surface from water, *Biochemistry* **31**:3947–3955.

Stearman, R. S., Frankel, A. D., Freire, E., Liu, B., and Pabo, C. O., 1988, Combining thermostable mutations increases the stability of λ repressor, *Biochemistry* **27**:7571–7574.

Sturtevant, J. M., 1977, Heat capacity and entropy changes in processes involving proteins, *Proc. Natl. Acad. Sci. USA* **74**:2236–2240.

Sturtevant, J. M., 1987, Biochemical applications of differential scanning calorimetry, *Annu. Rev. Phys. Chem.* **38**:463–488.

Takahashi, K., and Sturtevant, J. M., 1981, Thermal denaturation of *Streptomyces* subtilisin inhibitor, subtilisin BPN′, and the inhibitor-subtilisin complex, *Biochemistry* **20**:6185–6190.

Tamura, A., Kimura, K., Takahara, H., and Akasaka, K., 1991, Cold denaturation and heat denaturation of *Streptomyces* subtilisin inhibitor. 1. CD and DSC studies, *Biochemistry* **30**:11307–11313.

Wyman, J., and Gill, S. J., 1990, *Binding and Linkage. Functional Chemistry of Biological Macromolecules*, University Science Books, Mill Valley, California.

Xie, D., Bhakuni, V., and Freire, E., 1991, Calorimetric determination of the energetics of the molten globule intermediate in protein folding: Apo-α-lactalbumin, *Biochemistry* **30**:10673–10678.

Yang, A-S., Sharp, K. A., and Honig, B., 1992, Analysis of the heat capacity dependence of protein folding, *J. Mol. Biol.* **227**:889–900.

Chapter 7

Affinity Labeling and Related Approaches to Targeting Specific Enzyme Sites

Roberta F. Colman

1. GENERAL CONSIDERATIONS

1.1. Approach of Affinity Labeling

Identification and mapping of the active and allosteric sites of enzymes is important to the enzymologist who seeks understanding of the structural basis of catalysis and regulation, to the medicinal chemist who aims at designing rationally new drugs based on their ability to inhibit or inactivate particular enzymes, and to protein engineers who propose the construction of improved enzymes or novel proteins that can catalyze desirable new reactions. One approach to locating the active site of an enzyme is to chemically modify an amino acid residue and then to correlate the extent of modification of that amino acid with the degree of inactivation of the enzyme. However, limitation of chemical modification of a protein to one or a few residues is difficult to accomplish. The strategy of affinity labeling restricts the extent of chemical modification to the region of the catalytic

Roberta F. Colman Department of Chemistry and Biochemistry, University of Delaware, Newark, Delaware 19716.

Subcellular Biochemistry, Volume 24. Proteins: Structure, Function, and Engineering, edited by B. B. Biswas and Siddhartha Roy. Plenum Press, New York, 1995.

site by exploiting the specificity of the enzyme for its natural substrate: a reagent is designed that is related structurally to the normal substrate or regulatory compound of the enzyme but that also features a functional group capable of reacting covalently with various types of amino acid. Such a compound is intended to mimic the substrate in forming a reversible enzyme–reagent complex analogous to the enzyme–substrate complex and, once bound at the specific ligand site, to react irreversibly with an amino acid residue within that site. This model can be described by

$$E + I \underset{k_{-1}}{\overset{k_1}{\rightleftharpoons}} EI \xrightarrow{k_{max}} \mathrm{EI}' \tag{1}$$

where E represents free enzyme, I is the affinity label, EI the reversible enzyme–reagent complex, and EI′ the covalently modified enzyme. The existence of a reversible enzyme–reagent complex is indicated by a "rate saturation effect" in which the rate constant for modification increases with increasing reagent concentration until the enzyme site is saturated with reagent; thereafter, the rate constant is not changed by further increases in reagent concentration. The observed rate constant at a given reagent concentration (k_{obs}) can be described by the equation

$$k_{obs} = \frac{k_{\max}}{1 + \frac{K_I}{[I]}} \tag{2}$$

where the apparent dissociation constant for the enzyme–reagent complex, K_I, equals $\frac{(k_{-1} + k_{\max})}{k_1}$ and k_{max} is the maximum rate of modification at saturating concentrations of reagent. These kinetic parameters are frequently evaluated from a double reciprocal plot of $1/k_{obs}$ versus $1/[I]$ on the basis of the equation:

$$\frac{1}{k_{obs}} = \frac{1}{k_{max}} + \frac{K_I}{k_{max}}\frac{1}{[I]} \tag{3}$$

Another characteristic of an affinity label is the small number of moles of reagent incorporated (ideally, 1 mole per mole of enzyme active sites) concomitant with complete inactivation. Additionally, it is expected that a natural ligand, which binds reversibly to the site targeted by the reagent, will markedly decrease the rate constant for modification. Most of these features are illustrated by the specific examples of affinity labeling presented in Section 2 of this chapter.

Affinity labeling can be used to identify a single enzyme or receptor within a complex mixture of proteins. For a purified enzyme in solution, affinity labeling may be used to identify specific catalytic or regulatory sites, allowing comparison to be made between the structure of such sites for the enzyme in solution and those sites located by X-ray diffraction in a protein crystal. This chemical technique allows an experimental evaluation of the location of a particular type of ligand site which is predicted on the basis of amino acid consensus sequences. By irreversibly binding to the active or allosteric site, an affinity label can yield information on the effect of occupying that site on the conformation of the enzyme, on the reactivity of other sites, or on subunit interactions. Affinity labels with chromophores can be used to introduce spectral probes into specific sites on enzymes with the goal of monitoring conformational changes or of making estimates of distances between known functional sites on enzymes. Finally, affinity labeling can provide a rational basis for selection of targets for site-directed mutagenesis. The general aspects of affinity labeling have been reviewed (Jakoby and Wilchek, 1977; Colman, 1990). This chapter will therefore emphasize recent studies from my laboratory in which chemical approaches have been used to target specific sites in enzymes.

1.2. Representative Types of Reactive Compounds

1.2.1. Ring Substituted Bromoodioxobutyl Nucleotides

Examples of compounds designed in my laboratory as affinity labels are the bromodioxobutyl nucleotides shown in Fig. 1 (Colman, 1989, 1990). These nucleotide analogues all have the same reactive functional group located at distinctive positions of the purine ring. Figure 1a shows the structure of 2-(4-bromo-2,3-dioxobutylthio)adenosine 5′-diphosphate (2-BDB-TA 5′-DP),* which is closely related to natural adenosine 5′-diphosphate; accordingly, it is water soluble and negatively charged at neutral pH. The compound illustrated in Fig. 1b, 2-(4-bromo-2,3-dioxobutylthio)-1,N^6-ethenoadenosine 5′-diphosphate (2-BDB-TϵA 5′-DP), is similar except that it features the fluorescent eth-

*Abbreviations used in this chapter: 2-BDB-TA 5′-DP, 2-(4-bromo-2,3-dioxobutylthio)adenosine 5′-diphosphate; 2-BDB-TϵA 5′-D,2P-(4-bromo-2,3-dioxobutylthio)-1,N^6-ethenoadenosine 5′-diphosphate; 6-BDB-TA 5′-DP, 6-(4-bromo-2,3-dioxobutylthio)adenosine 5′-diphosphate; 8-BDB-TA 5′-DP, 8-(4-bromo-2,3-dioxobutylthio)adenosine 5′-diphosphate; GMPS-BOP, guanosine 5′-0-[S-(3-bromo-2-oxopropyl)]thiothosphate; AMPS-BOP, adenosine 5′-O-[S-(3-bromo-2-oxopropyl)]thiophosphate; NMNS-BOP, nicotinamide ribose 5′-O-[S-(3-bromo-2-oxopropyl)]thiophosphate; 5′-FSBA, 5′-p-fluorosulfonylbenzoyl adenosine; 5′-FSBAzA, 5′-p-fluorosulfonylbenzoyl-8-azido-adenosine; 5′-FSBG, 5′-p-fluorosulfonylbenzoyl guanosine; 5′-FSBϵA, 5′-p-fluorosulfonylbenzoyl-1,N^6-ethenoadenosine; PEP, phosphoenol pyruvate; 2-BOP-TϵA 5′-DP 2-(3-bromo-2-oxopropylthio)-1,N^6-ethenoadenosine 5′-diphosphate; S-BDB-G, S-(4-bromo-2,3-dioxobutyl)glutathione; 2-BDB-TϵA 2′,5′-DP, 2-(4-bromo-2,3-dioxobutylthio)-1,N^6-ethenoadenosine 2′,5′-diphosphate.

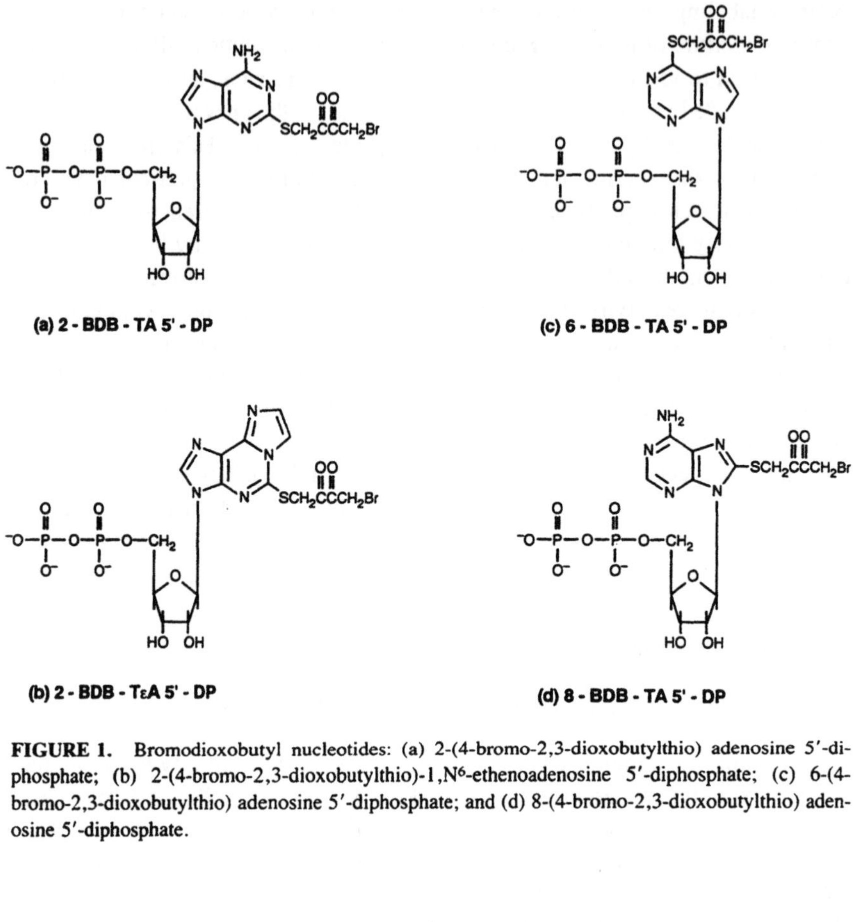

FIGURE 1. Bromodioxobutyl nucleotides: (a) 2-(4-bromo-2,3-dioxobutylthio) adenosine 5′-diphosphate; (b) 2-(4-bromo-2,3-dioxobutylthio)-1,N^6-ethenoadenosine 5′-diphosphate; (c) 6-(4-bromo-2,3-dioxobutylthio) adenosine 5′-diphosphate; and (d) 8-(4-bromo-2,3-dioxobutylthio) adenosine 5′-diphosphate.

enoadenosine moiety: when excited at 302 nm, it has a characteristic emission peak at 428 nm. Figures 1c and 1d show, respectively, the 6-, and 8-bromodioxobutyl derivatives of adenosine 5′-diphosphate, which permit a systematic probing of the amino acid residues around the purine region of the ADP binding site of enzymes. The bromoketo group is potentially reactive with several nucleophiles found in proteins, including cysteine, histidine, tyrosine, lysine, glutamic, and aspartic acid; and the dioxo group lends the possibility of reaction with arginine residues.

1.2.2. Bromooxopropyl Derivatives of Nucleoside Thiophosphates

Figure 2 illustrates another class of compounds in which the reactive bromoketo moiety is linked to the sulfur of a purine nucleotide thiophosphate. The GMPS-BOP shown in Fig. 2a preserves the guanine and ribose and adds an

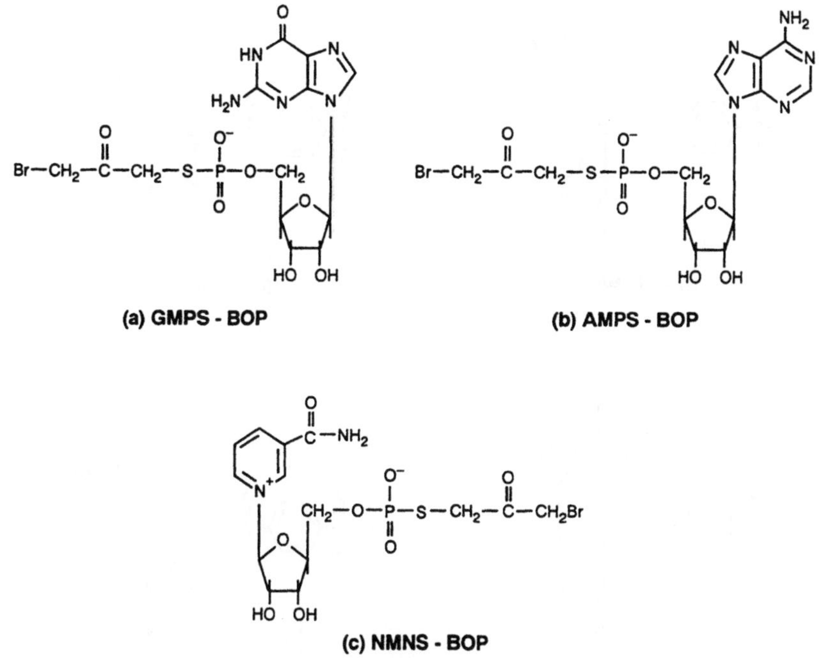

FIGURE 2. Bromooxopropyl derivatives of nucleoside thiophosphates: (a) guanosine 5′-0-[S-(3-bromo-2-oxopropyl)]thiophosphate; (b) adenosine 5′-0-[S-(3-bromo-2-oxopropyl)]thiophosphate; and (c) nicotinamide ribose 5′-0-[S-(3-bromo-2-oxopropyl)]thiophosphate.

alkylating group at a position equivalent to the pyrophosphoryl region of the natural nucleotide (Ozturk *et al.*, 1992). This new analogue is negatively charged at neutral pH and molecular modeling indicates that its size is equivalent to that of GTP. The corresponding ATP analogue (Walner and Colman, 1990) is shown in Figure 2b. A reactive NMN analogue is pictured in Fig. 2c, which might be expected to function as an affinity label of coenzyme binding sites in NAD/NADP-dependent dehydrogenases in which it will probe residues in the subdomain normally occupied by the coenzyme's pyrophosphate (Park *et al.*, 1993).

1.2.3. Fluorosulfonylbenzoyl Nucleosides

Figure 3 illustrates examples of a different type of purine analogue that we have synthesized, the fluorosulfonylbenzoyl derivatives of nucleosides (Colman, 1983a, 1990). Compound (a) is 5′-p-fluorosulfonylbenzoyl adenosine (5′-FSBA), which might reasonably be considered as an analogue of ADP, ATP, or NADH. In addition to the adenine and ribose moieties, it has a carbonyl group

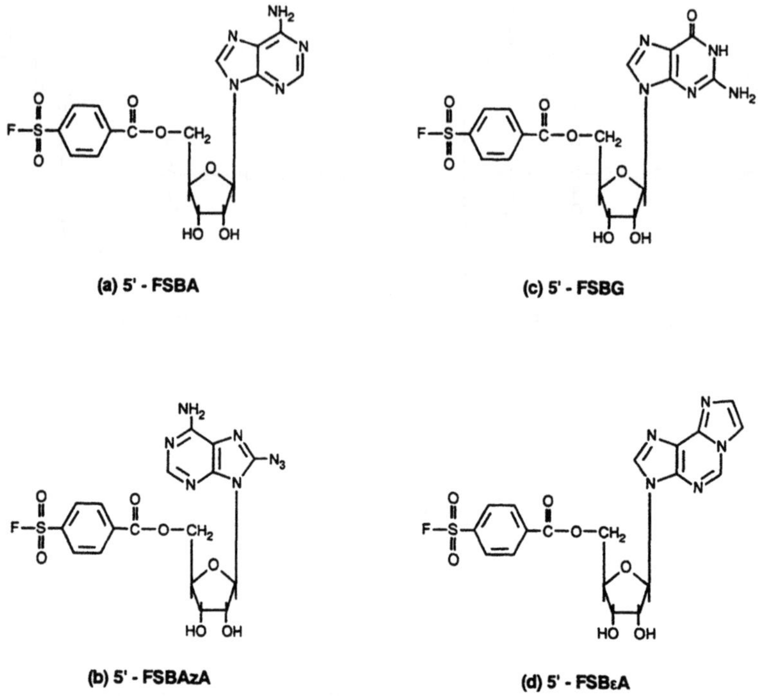

FIGURE 3. Fluorosulfonylbenzoyl nucleosides: (1) 5′-p-fluorosulfonylbenzoyl adenosine; (b) 5′-p-fluorosulfonylbenzoyl-8-azidoadenosine; (c) 5′-p-fluorosulfonylbenzoyl guanosine; and (d) 5′-p-fluorosulfonylbenzoyl-1,N^6-ethenoadenosine.

adjacent to the 5′-position, which is structurally similar to the first phosphoryl group of the naturally occurring purine nucleotides. If the molecule is arranged in an extended conformation, the sulfonyl fluoride moiety may be located in a position analogous to the terminal phosphate of ATP or to the ribose proximal to the nicotinamide ring of NADH. This sulfonyl fluoride moiety can act as an electrophilic agent in covalent reactions with several classes of amino acids, including tyrosine, lysine, histidine, serine, and cysteine.

Compound (b) is a bifunctional affinity label, 5′-p-fluorosulfonyl-benzoyl--8-azidoadenosine, 5′-FSBAzA, which contains both an electrophilic fluorosulfonyl moiety and a photoactivatable azido group (Dombrowski and Colman, 1989). Following stoichiometric incorporation of reagent through the fluorosulfonyl at a specific site, photolysis of the tethered molecule can lead to reaction with amino acids adjacent to the residue that is initially labeled. This two-step reaction can help to elucidate the tertiary structure of the enzyme in the region of the nucleotide site.

Structure (c) of Fig. 3 is 5′-p-fluorosulfonylbenzoyl guanosine (5′-FSBG) in which guanine replaces the adenine moiety in the first derivative. It might be anticipated that this purine nucleotide alkylating agent would be specifically directed toward GTP sites in proteins. Figure 3d is the fluorescent compound 5′-p-fluorosulfonylbenzoyl-1,N^6-ethenoadenosine. This nucleotide analogue, with a fluorescence emission maximum at 412 nm, provides a means of introducing a covalently bound fluorescent probe into nucleotide sites in proteins.

Each of the nucleotide derivatives in Figs. 1–3 contains a reactive functional group that is relatively indiscriminate in reactions with amino acids. In affinity labeling studies, the types of amino acid participants in the ligand binding site are frequently unknown. Using a nonspecific functional group improves the likelihood of covalent reaction after the binding specificity is determined by the remainder of the structure of the affinity label.

In this chapter, several bromodioxobutyl derivatives are used to illustrate affinity labeling of three different types of enzymes: pyruvate kinase, glutathione S-transferase, and glutamate dehydrogenase. For pyruvate kinase and glutathione S-transferase, these studies were used to identify the active site of the enzymes. For bovine liver glutamate dehydrogenase, both bromoketo- and fluorosulfonylbenzoyl nucleotides were directed to allosteric sites. For bacterial glutamate dehydrogenase, affinity labeling was used to indicate the region of the coenzyme binding site of the enzyme; we were then positioned to rationally select target sites for mutagenesis so that we could subsequently examine the properties of mutant glutamate dehydrogenase with particular amino acids replace. For isocitrate dehydrogenase, we chose a divalent metal ion or the metal–isocitrate complex to target the active site of the enzyme. By using a redox-active metal ion, such as Fe^{2+}, in the presence of oxygen and a reducing agent, oxidative cleavage of the polypeptide backbone of the enzyme is promoted at the substrate binding site. This process, which is termed *affinity cleavage*, can locate the amino acid sequence adjacent to the metal ion and metal–isocitrate sites of a metal-dependent enzyme.

2. SPECIFIC EXAMPLES

2.1. Rabbit Muscle Pyruvate Kinase: Affinity Labeling of the Active Site

Pyruvate kinase is the critical glycolytic enzyme that catalyzes the transfer of a phosphoryl group from phosphoenol pyruvate to ADP, yielding pyruvate and ATP as the products. The structure of this enzyme has received considerable attention. Complete amino acid sequences are known for several isozymes, and the amino acid sequence of the cat muscle pyruvate kinase has been fitted to the 2.6 Å resolution electron density map (Muirhead *et al.*, 1986). Affinity labeling

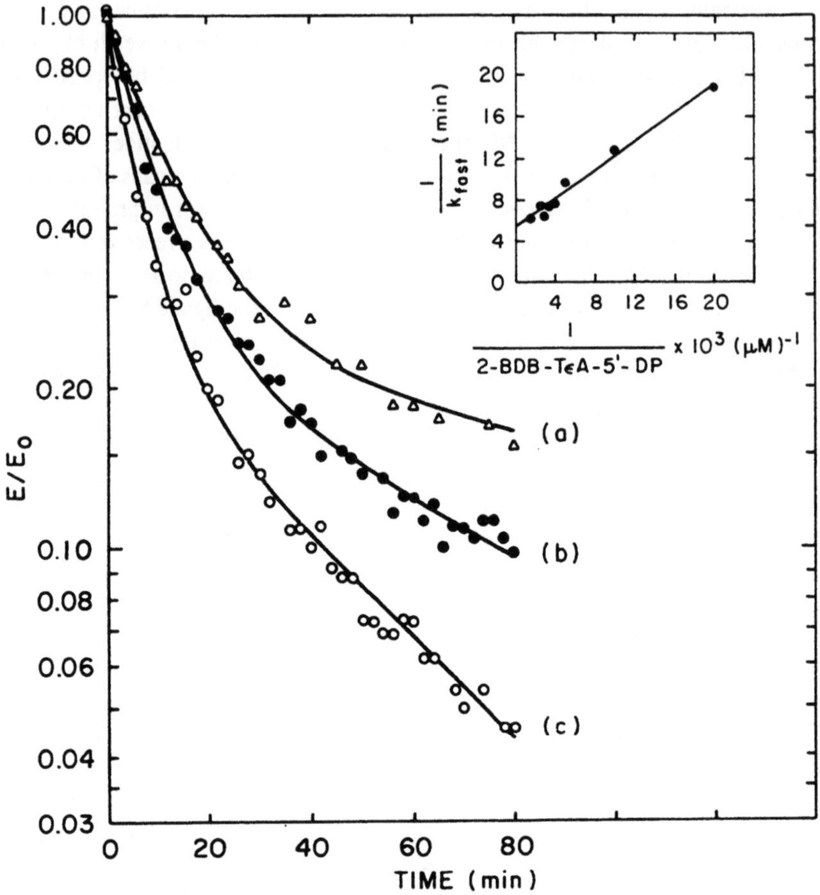

FIGURE 4. Inactivation of pyruvate kinase by 2-BDB-TεA-5′-DP. Rabbit muscle pyruvate kinase was incubated with (a) 100 μM (△), (b) 200 μM (θ), and (c) 600 μM 2-BDB-TεA 5′-DP (O) at 25 °C in 0.05 M HEPES buffer, pH 7.0. Residual activity, E/E_0, was measured at the indicated times. The solid lines are theoretical lines generated from a computer fit to Eq. (1) of DeCamp and Colman (1989). Values for k_{fast} and k_{slow} were found to be (a) 0.077 min^{-1}, 0.005 min^{-1}, (b) 0.101 min^{-1}, 0.012 min^{-1}, and (c) 0.159 min^{-1}, 0.022 min^{-1}. Inset: double-reciprocal plot in accordance with Eq. (3) for the fast phase of inactivation by 2-BDB-TεA 5′-DP. Reproduced from DeCamp and Colman (1989), with permission.

with reactive nucleotide analogues offers a complementary technique for investigating the active site as it exists when the enzyme is in solution.

The first analogue that we used is 2-[4-bromo-2,3-dioxobutylthio]-1,N^6-ethenoadenosine 5′-diphosphate (Fig. 1b). Incubation of rabbit muscle pyruvate kinase with 100–600 μM 2-BDB-TεA 5′-DP resulted in a time-dependent inactivation of the enzyme, as shown in Fig. 4. Biphasic inactivation kinetics were observed that could be described in terms of a fast initial phase of inactivation yielding a partially active enzyme with 25% residual activity, followed by a slower phase leading to totally inactive enzyme (DeCamp and Colman, 1989). Both phases exhibit a "rate saturation effect" in the dependence of the rate constant for inactivation on the reagent concentration, indicative of the initial formation of a reversible enzyme–reagent complex prior to irreversible modification. For the fast phase, a double reciprocal plot of $1/k_{obs}$ versus 1/[reagent] yields values of K_I = 133 μM and k_{max} = 0.193 min^{-1} at saturating concentrations of reagent (see Fig. 4, inset). The slow phase has a k_{max} that is ten times lower (DeCamp and Colman, 1989). Pyruvate kinase was incubated with 200 μM 2-BDB-TεA 5′-DP and the fluorescence intensity of the protein-bound organic phosphorus was used to quantitate the incorporation of reagent into the enzyme at various times. At 80 min, when the enzyme was 90% inactivated, about 1.7 moles of reagent/mole of enzyme subunit were incorporated. Including either ADP + Mn^{2+} or phosphoenol pyruvate + K^+ + Mn^{2+} in the reaction mixture caused a substantial reduction in the percent inactivation at 80 min, as well as a decrease in the reagent incorporation to about 1 mole/mole enzyme subunit (DeCamp and Colman, 1989). These results indicate that 2-BDB-TεA 5′-DP reacts with two groups on the enzyme, one of which is at or near the active site. It is likely that reaction occurs near the phosphoenol pyruvate (PEP) binding site of pyruvate kinase since PEP is the most effective in protecting against inactivation.

Nuclear magnetic resonance (NMR) spectra of the BDB-nucleotides indicate that the diketo groups of bromodioxobutyl-nucleotides exist predominantly in enolized form when they are dissolved in aqueous solution, as shown schematically in Fig. 5. The reagent 2-BDB-TεA 5′-DP thus resembles phosphoenolypyruvate. In contrast, a monoketo compound, such as 2-(3-bromo-2-oxopropylthio)-1,N^6-ethenoadenosine 5′-diphosphate (2-BOP-TεA 5′-DP), does

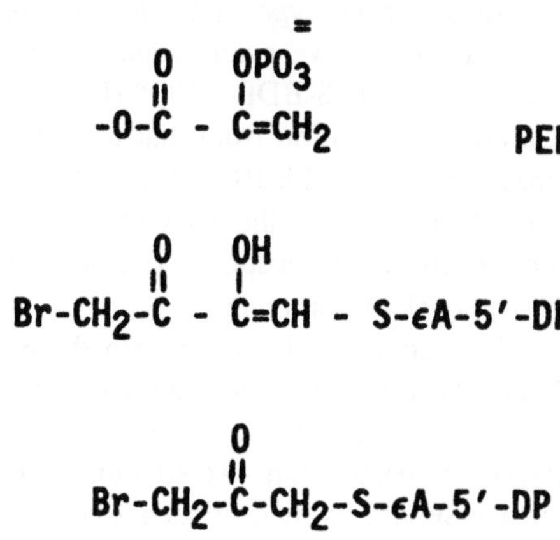

FIGURE 5. Comparison of the structure of phosphoenolpyruvate (upper) with those of side chains of 2-BDB-TεA 5′-DP (middle) and 2-BOP-TεA 5′-DP (lower). Reproduced from DeCamp and Colman (1989), with permission.

not enolize in aqueous solution. If the enolate directs the reagent to the PEP binding site, the bromooxopropyl derivative should be unreactive. As expected, incubation of the enzyme with 200 μM 2-BOP-TεA 5′-DP for 80 min resulted in only 14% loss of enzyme activity, whereas the bromo*dioxo* derivative caused 90% inactivation under the same conditions (DeCamp and Colman, 1989).

With the goal of identifying the amino acids that were reacting, modified enzyme was prepared by incubating pyruvate kinase with 2-BDB-TεA 5′-DP either in the absence or in the presence of phosphoenol pyruvate, K^+ and Mn^{2+}. The isolated modified enzymes were reduced with [^{3}H]-$NaBH_4$ to introduce the radioactive tag and were then digested with trypsin. The digests were subsequently purified by chromatography on a phenylboronate agarose column followed by reverse phase HPLC using a trifluoroacetic acid solvent system.

Tryptic digests of the inactive modified enzyme yielded two labeled peptides with the following sequences:

I. Asn-Ile-Cys^{164}-Lys
II. Ile-Thr-Leu-Asp-Asn-Ala-Tyr^{147}-Met-Glu-Lys

Only peptide I was isolated from digests of pyruvate kinase that had been modified in the presence of phosphoenol pyruvate and therefore retained most of its activity. Since Cys^{164} is labeled equally whether the enzyme is active or inactive, it cannot be essential for activity. In contrast, modification of Tyr^{147} by 2-BDB-TεA 5′-DP is associated with inactivation (DeCamp and Colman, 1989).

The three-dimensional structure of pyruvate kinase has been determined in the crystalline state by Muirhead *et al.* (1986). The active site is thought to lie in a pocket between two domains with Tyr^{147} being located at the entrance to the active site; specific reaction of Tyr^{147} with 2-BDB-TεA 5′-DP causes loss in activity. Cysteine^{164} is an internal residue in domain B outside the active site; this location is consistent with its modification having little effect on catalysis.

As a second type of nucleotide analogue for probing the active site of pyruvate kinase, we synthesized 8-(4-bromo-2,3-dioxobutylthio)adenosine di- and triphosphate (8-BDB-TATP) (DeCamp *et al.*, 1988), with the reactive group at a different position from that of 2-BDB-TεA 5′-DP. Incubation of pyruvate kinase with 175 μM 8-BDB-TATP at pH 7.0 and 25 °C caused biphasic inactivation. The reactions of the two eight-substituted analogues are similar, but the rate constant for the triphosphate derivative is about three times greater than that of the diphosphate compound.

Incorporation of reagent into the enzyme was measured at various times during the incubation by quantitation of the organic phosphorus. The triphosphate derivative is more specific, exhibiting lower incorporation for a higher degree of inactivation. At 80 min, 2.05 mole of 8-BDB-TATP per mole of enzyme subunit is incorporated when the enzyme has lost 97% of its original activity. Protection against inactivation by 8-BDB-TATP is provided by various combinations of substrates. In the presence of substrates, the enzyme loses only 15–20% of its activity, and the incorporation of reagent is reduced to about 1 mole/mole enzyme subunit (DeCamp *et al.*, 1988). These results suggest that 8-BDB-TATP also reacts with two groups on the enzyme, one of which is at or near the PEP site.

The residues modified by 8-BDB-TATP have been identified by reduction of the modified enzyme with [^{3}H]-sodium borohydride, digestion with trypsin, and purification of the resultant peptides on a phenylboronate agarose column followed by HPLC (Vollmer and Colman, 1990). Two cysteine-modified peptides were isolated from the digest of inactive enzyme:

I. Asn-Ile-Cys^{164}-Lys
II. Cys^{151}-Asp-Glu-Asn-Ile-Leu-Trp-Leu-Asp-Tyr-Lys

Reaction in the presence of the protectants phosphoenol pyruvate, K + Mn^{2+} yielded Asn-Ile-Cys-Lys as the only labeled peptide, with Cys^{164} as the modified residue. This is the same residue that reacts with the two-substituted nucleotide analogue under protecting conditions. Although Cys^{164} is clearly not required for activity and is internal, it appears to be favorably positioned to react with nucleotide analogues. The major residue responsible for inactivation of pyruvate kinase by 8-BDB-TATP seems to be Cys^{151}, in contrast to the finding that 2-BDB-TεA 5′-DP causes inactivation by modifying Tyr^{147}. A distance of 8.9 Å can be measured between the α-carbons of Cys^{151} and Tyr^{147} in the crystal structure of pyruvate kinase. Since the target residues of the two affinity labels are approximately 9 Å apart, the question arises as to whether, if the nucleotide moieties were superimposed, the reactive -CH_2Br groups would be situated 9 Å apart.

Figure 6 compares the structures of 2-BDB-TεA 5′-DP, shown in (a) in the anti conformation about the purine–ribose bond, and 8-BDB-TATP, pictured in (b) in the syn conformation about the purine–ribose bond. The two structures are superimposed in the middle so that the riboses and phosphates coincide. Although a range of distances between the two -CH_2Br groups is possible, it *is* energetically permissible to arrange the compounds on the enzyme such that the distance between the reactive groups coincides with the 9-Å distance between Cys^{151} and Tyr^{147} (Vollmer and Colman, 1990). Thus, if the purines, riboses, and phosphates of the two compounds bind similarly to the enzyme, the experimental results from affinity labeling are consistent with the atomic coordinates of the enzyme in the crystalline state.

2.2. Rat Liver Glutathione S-Transferases: Affinity Labeling of the Hydrophobic Substrate Site

A different type of enzyme that will be used to illustrate the approach of affinity labeling is glutathione S-transferase. The glutathione S-transferases con-

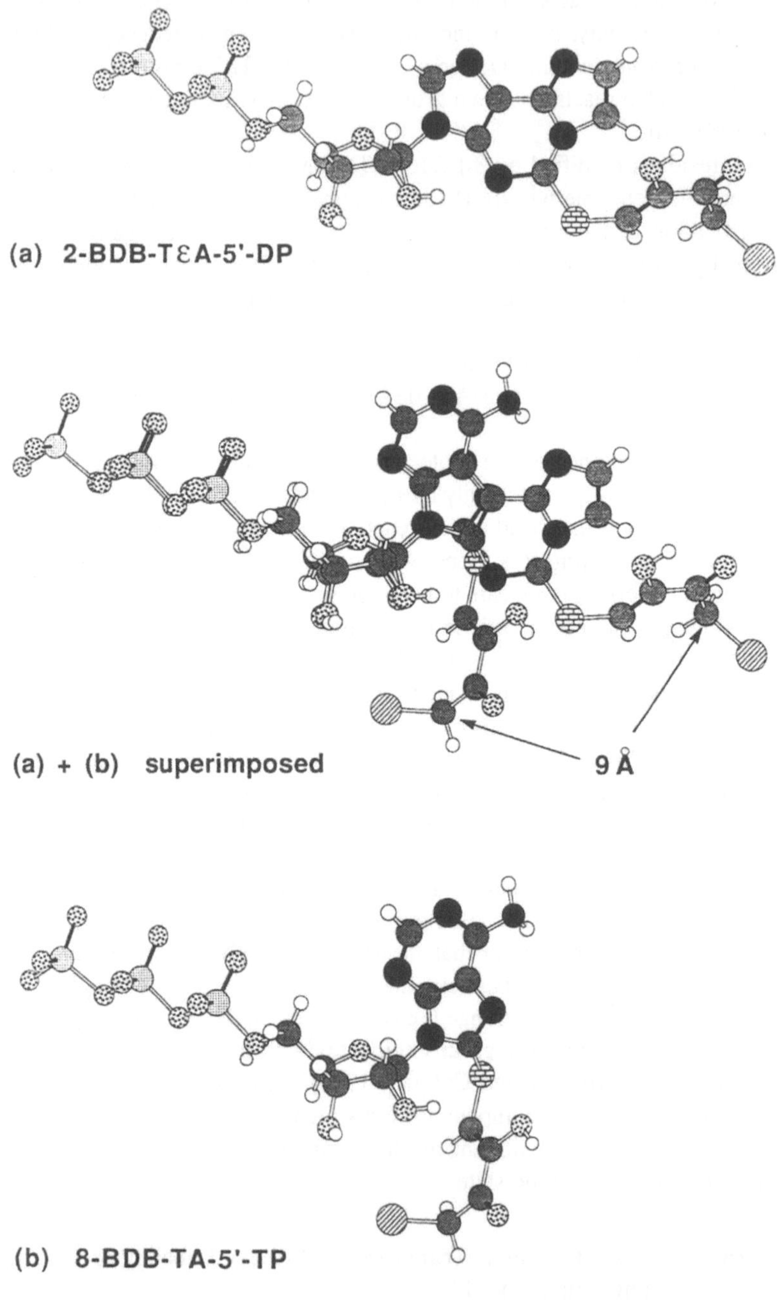

FIGURE 6. Comparison of the structures of (a) 2-BDB-TεA 5′-DP and (b) 8-BDB-TATP. Reproduced from Vollmer and Colman (1990), with permission.

stitute a family of enzymes important in the detoxification of xenobiotics or foreign chemicals. These enzymes catalyze reactions in which the thiol of glutathione undergoes nucleophilic addition to electrophilic substrates; the product is a glutathione conjugate of the xenobiotic compound which can then be more readily excreted. Glutathione S-transferases can use a wide variety of electrophilic acceptors as substrates. Since they catalyze the inactivation of several known carcinogens, such as aflatoxin B_1, these enzymes are considered to provide a defense against carcinogenesis. On the other hand, glutathione S-transferase can also metabolize as substrates various alkylating agents, such as chlorambucil or melphalan, which are commonly used as anticancer drugs. Since these reactions cause melphalan and chlorambucil to lose their cytotoxic effect, the activity of glutathione S-transferases may contribute to resistance of tumors to various anticancer drugs. The development of specific inhibitors of glutathione S-transferase may yield novel protocols for combination chemotherapy that will enhance the efficacy of currently used alkylating anticancer drugs and will prolong the life of oncology patients.

The glutathione S-transferases are dimers with subunit molecular weights of 25,000–28,000, which, on the basis of isoelectric point, substrate specificity, and primary sequence similarity, can be divided into at least three classes: α, μ, and π (Rushmore and Pickett, 1993, Mannervik and Danielson, 1988; Armstrong, 1991). Three-dimensional structures at high resolution have now been published for representatives of the π and μ classes (Reinemer *et al.*, 1991; Ji *et al.*, 1992). Our aim in initiating affinity labeling studies was to obtain information on the active site of the enzyme in solution that would complement the crystallographic studies.

We have synthesized the bromodioxobutyl derivative of glutathione in which the reactive side chain is linked to the glutathione backbone through the original -SH of cysteine (Katusz and Colman, 1991). The expectation was that the compound S-(4-bromo-2,3-dioxobutyl)glutathione (S-BDB-G) would bind to glutathione S-transferases since it is known that S-alkyl derivatives of glutathione, such as S-hexylglutathione, are effective competitive inhibitors of the enzyme.

Rat liver isozyme 4-4 is a member of the μ class of glutathione transferases. Incubation of the 3-3 isoenzyme at pH 6.5 with 200 μM S-BDB-G results in a time-dependent total inactivation (Katusz and Colman, 1991). The rate constant for inactivation exhibits a nonlinear dependence on reagent concentration as measured from 50 to 1000 μM. This result suggests that prior to irreversible modification, a reversible enzyme–reagent complex is formed, as expected for an affinity label, and the existence of this complex accounts for the relative specificity of such reagents. A K_I value of 66 μM can be calculated with a maximum rate constant of 0.078 min^{-1} at saturating concentrations of reagent.

As shown in Table I, complete protection against inactivation by S-BDB-G

Table I
Effect of Substrate Analogues on Rate Constant for Inactivation of Glutathione S-Transferase, 4-4 Isozyme, by S-BDB-G[a]

Ligands added to reaction mixture	$k_{obs} \times 10^3$ (min^{-1})
None	58.0
S-Hexylglutathione (5 mM)	0
S-Butylglutathione (5 mM)	0
S-Propylglutathione (5 mM)	1.7
S-Ethylglutathione (5 mM)	3.3
S-Methylglutathione (5 mM)	10.4

[a]Reproduced from Katusz and Colman (1991), with permission.

was provided by glutathione analogues with hydrophobic side chains of at least four carbons (i.e., S-butyl- and S-hexylglutathione). As the size of the alkyl side chain is decreased, the effect of the S-alkylglutathione in lowering the inactivation rate constant is decreased (e.g., S-ethyl- and S-methylglutathione). These results indicate that the reaction target of S-BDB-G is the active site of the enzyme, probably the subsite normally occupied by the electrophilic (hydrophobic) substrate.

Incubation of glutathione S-transferase with radioactive S-BDB-G leads to a time-dependent incorporation, with 1.3 moles of reagent incorporated when the enzyme is completely inactivated. In the presence of the protecting substrate analogue, S-hexylglutathione, the highest incorporation measured was 0.48 mole of reagent per mole of subunit, and the enzyme retained all of its activity. These results suggest that an amino acid at or near the active site of the enzyme is modified by the reagent, but that there is some additional reaction at an amino acid not essential for enzymatic activity.

Enzyme modified with radioactive reagent in the absence or the presence of S-hexylglutathione was digested with trypsin and the resulting peptides were separated by high-pressure liquid chromatography (HPLC). Two major radioactive peaks (I and II) were observed in the digest from the inactive enzyme, with peak II predominating. Upon gas phase sequencing, the peptides were identified as:

I. Lys^{82}-His-Asn-Leu-X-Gly-Glu-Thr-Glu-Glu-Glu-Arg^{93}
II. Leu^{109}-Gln-Leu-Ala-Met-CmCys-Y-Ser-Pro-Asp-Phe-Glu-Arg^{121}

The X is modified Cys^{86}, while the Y is modified Tyr^{115}. Only peptide I was modified in the presence of S-hexylglutathione when the enzyme retained full this amino acid is not essential for enzymatic activity, while Tyr^{115} is considered to be in the region of the active site (Katusz and Colman, 1991).

Among the glutathione S-transferases within a given family of isozymes, there is a high degree of sequence homology, but between families there is relatively little sequence similarity. The extent of the differences between isozyme sequences also appears to correlate with differences in substrate specificity. In addition to our study of the modification of the 4-4 isozyme of rat liver glutathione S-transferase by S-BDB-G, we have examined the reaction of the same affinity label with two other isoenzymes of glutathione S-transferase: the 3-3 and 1-1 isozymes (Katusz *et al.*, 1992a,b). The 3-3 enzyme is a member of the same μ family and is highly homologous to the 4-4 enzyme. Not surprisingly, S-BDB-G inactivates the 3-3 isoenzyme concomitant with labeling of the same residue, Tyr^{115} (Katusz *et al.*, 1992a) as in the case of the 4-4 isozyme. This result indicates that the active site of the 3-3 and 4-4 isoenzymes of glutathione S-transferase are very similar indeed, at least in the region occupied by the reactive BDB group of the reagent, which is likely that of the hydrophobic substrate.

The structure of crystalline rat liver glutathione S-transferase, isoenzyme 3-3, has recently been determined by X-ray diffraction (Ji *et al.*, 1992). The distance between the sulfur of glutathione (bound in the active site cleft) and the oxygen of Tyr^{115}, which we have modified, has been measured as 7.5 Å (R. M. Armstrong, personal communication). This distance is close to that expected to be occupied by the bromodioxobutyl side chain of S-BDB-G (6.8 Å is the distance between the sulfur and the bromo of $-CH_2Br$). This analysis supports our conclusion that the critical Tyr^{115} of the 3-3 isozyme is located in the hydrophobic substrate binding portion of the enzyme's active site where its major function is to facilitate binding of the nonglutathione substrate through hydrophobic interactions. In the case of 1-1 isozyme, which is a member of a different family of glutathione S-transferases, the α family, S-BDB-G also inactivates the enzyme in a specific manner at the active site (Katusz *et al.*, 1992b). However, the mode of binding of S-BDB-G and the critical amino acid labeled is distinct: Cys^{111} is modified. Clearly the active site of the α family is distinguishable from that of the μ family.

2.3. Pig Heart NADP-Dependent Isocitrate Dehydrogenase: Affinity Cleavage at the Metal–Isocitrate Site

Our recent study of the active site of the metal-dependent, isocitrate dehydrogenase by affinity cleavage exemplifies an alternate approach to affinity labeling for locating the metal-substrate site of an enzyme (Soundar and Colman, 1993). The NADP-specific isocitrate dehydrogenase from pig heart mitochondria catalyzes the oxidative decarboxylation of isocitrate to yield α-ketoglutarate and carbon dioxide. This is a nonallosteric enzyme, which is a dimer of identical subunits, each of which has a molecular weight of 46,600. The entire sequence of this 413 amino acid protein has recently been determined (Haselbeck *et al.*, 1992).

Isocitrate dehydrogenase requires a divalent metal ion for activity. Several metals have been shown to fulfill this function, including Mn^{2+}, Mg^{2+}, Zn^{2+}, and Cd^{2+} (Colman, 1972). Direct binding experiments indicate that isocitrate dehydrogenase binds 1 mole of Mn^{2+}/mole of enzyme subunit under all conditions (Villafranca and Colman, 1972). However, analysis of kinetics and binding experiments suggests that the metal ion occupies different sites in the absence and presence of the substrate isocitrate (Ehrlich and Colman, 1976; Colman, 1983b).

There is considerable literature on the action of "chemical nucleases" in which ferrous-EDTA is tethered to a DNA-binding drug to provide sequence specificity for the cleavage of nucleic acids (Sigman and Chen, 1990). It occurred to us that the Fe^{2+}–isocitrate complex might occupy the Mn^{2+}–isocitrate site and, in the presence of oxygen, oxidatively cleave the polypeptide backbone at the substrate site in analogy to the chemical nucleases; this is the process that is termed *affinity cleavage*.

Addition of Fe^{2+} in the presence of the reducing agent ascorbate causes an irreversible inactivation of NADP-dependent isocitrate dehydrogenase at 25 °C and pH 7.7, to give 61% of the original activity after 3 hr (Table II). The rate of inactivation by Fe^{2+} is much more rapid when isocitrate is included in the

Table II
Inactivation of Isocitrate Dehydrogenase by Fe^{2+} or Fe^{2+}-Isocitrate[a,b]

Addition	Residual activity (%)
None	100
Fe^{2+} (20 μM)	103
Fe^{2+} (20 μM) + ascorbate (20 mM)	61
Mn^{2+} (2 mM) + ascorbate (20 mM)	98
Fe^{2+} (20 μM) + Mn^{2+} (2 mM) + ascorbate (20 mM)	101
Isocitrate (4 mM) + Fe^{2+} (20 μM)	97
Isocitrate (4 mM) + Fe^{2+} (20 μM) + ascorbate (20 mM)	2
Isocitrate (4 mM) + Mn^{2+} (2 mM) + ascorbate (20 mM)	98
Isocitrate (4 mM) + Fe^{2+} (20 μM) + Mn^{2+} (2 mM) + ascorbate (20 mM)	104

[a]Isocitrate dehydrogenase (22 μM) was incubated at 25 °C for 3 hr in 0.1 M triethanolamine chloride buffer, pH 7.7, containing 10% glycerol and 0.3 M Na_2SO_4, with additions as indicated. After 3 hr, aliquots were removed, diluted about 5,000-fold, and assayed for enzymatic activity.

[b]Reproduced from Soundar and Colman (1993), with permission.

Table III
Fragments of Isocitrate Dehydrogenase Produced by Affinity Cleavage

	By Fe^{2+}	By Fe^{2+}-isocitrate
Band designation	M_r	M_r
Enzyme	46,000	46,000
F_1		35,000
F_2	32,000	
F_3		30,000
F_4		17,000
F_5	15,000	
F_6		11,000

F_2 (32,000) + F_5 (15,000) ≐ Enzyme 46,000

F_1 (35,000) + F_6 (11,000) ≐ Enzyme 46,000
or
F_3 (30,000) + F_4 (17,000) ≐ 46,000

incubation mixture, reaching 2% of the initial activity after 3 hr. These results are consistent with a distinct mode of binding by the enzyme of free metal ion compared with metal–isocitrate complex (Soundar and Colman, 1993).

Ascorbate is required for inactivation by added Fe^{2+} when isocitrate is either absent or present, as shown in Table II. An electron donor (such as ascorbate) has previously been shown to be required, along with oxygen and the ferrous–ferric ion pair to promote metal-catalyzed oxidation of proteins. Fenton chemistry is involved in which reactive oxygen species are generated locally at the metal binding site as the metal is oxidized and a reductant, such as ascorbate, is used to recycle the oxidized metal ion. Inclusion of Mn^{2+} totally prevents inactivation of isocitrate dehydrogenase by Fe^{2+} or by Fe^{2+}–isocitrate. These results suggest that inactivation by Fe^{2+} involves specific interaction at the natural metal site of the enzyme. Furthermore, other tri- or dicarboxylates (such as citrate, tricarballylate, or malate), which are not effective substrates of the enzyme, fail to promote inactivation of the enzyme by ferrous ion. These results support the specificity of interaction of metal-isocitrate at the active site (Soundar and Colman, 1993).

In addition to inactivation, incubation of isocitrate dehydrogenase with Fe^{2+} or Fe^{2+}–isocitrate causes cleavage of the enzyme subunit. The fragments were separated and their molecular weights estimated by polyacrylamide gel electrophoresis in the presence of sodium dodecyl sulfate (Soundar and Colman, 1993). The molecular weights of the fragments resulting from enzyme cleavage after 3 hr are summarized in Table III. About 60–70% of the enzyme inactivated by Fe^{2+} or Fe^{2+}–isocitrate is cleaved into smaller fragments. After incubation of the enzyme for 3 hr in the presence of Fe^{2+} alone, one pair of predominant fragments is observed, in addition to the intact enzyme subunit: these are F_2 + F_5. The molecular weights of these two fragments, produced in the presence of free Fe^{2+}, indicate that they derive from a single cleavage of the intact enzyme subunit: F_2 (32,000) + F_5 (15,000 ≅ 46,000.

In contrast, after incubation of the enzyme for 3 hr in the presence of Fe^{2+}–isocitrate, two distinguishable pairs of major fragments are seen: F_1 + F_6 and F_3 + F_4 (Table III). The molecular weights of the four major bands are best explained by two mutually exclusive cleavages of the intact subunit: F_1 (35,000) + F_6 (11,000) ≅ 46,000; or F_3 (30,000) + F_4 (17,000) ≅ 46,000. Cleavage to produce the F_1 + F_6 fragments occurs more frequently. The addition of Mn^{2+} prevents cleavage by Fe^{2+} or by Fe^{2+}–isocitrate. Since Mn^{2+} protects against both the inactivation and cleavage reactions promoted by Fe^{2+}, it seems reasonable to conclude that Fe^{2+} binds to the same sites as Mn^{2+} and causes specific cleavage at the enzyme's metal binding site (Soundar and Colman, 1993).

The identification of all of these peptides and of the cleavage sites has been made by gas phase sequencing. The distinctive fragmentation patterns produced by affinity cleavage of isocitrate dehydrogenase by Fe^{2+}–isocitrate and by Fe^{2+} alone are shown schematically in Fig. 7, as deduced from the molecular weight and amino acid sequence data. The darkened rectangles represent the peptide sections that have actually been sequenced. Fragment F_1 is the N-terminal portion and F_6 is the C-terminal portion of an enzyme subunit cleaved at the His^{309}-Gly^{310} peptide bond of the 413 amino acid protein. Fragment F_3 is the N-terminal portion and F_4 is the C-terminal portion of an enzyme subunit cleaved at the Asp^{253}-Met^{254} peptide bond. The results of cleavage by Fe^{2+} in the presence of isocitrate are best understood in terms of a major cleavage of the His^{309}-Gly^{310} peptide bond to yield F_1 + F_6, with a minor cleavage at the Asp^{253}-Met^{254} peptide bond to produce F_3 + F_4. It seems likely that His^{309} and Asp^{253} are coordination sites for metal ion in Mn^{2+}–isocitrate as pictured in the top part of Fig. 7. The NADP-dependent isocitrate dehydrogenase is a dimer of identical subunits. Each metal ion may either be shared between two subunits with one ligand contributed by each subunit or both coordination sites may be located within a single subunit. In either case, after an oxidative cleavage occurs in the vicinity of one of the ligand sites of Fe^{2+}-isocitrate, binding of metal–isocitrate to that subunit will be disrupted and no further cleavage will take place. Thus, cleavage at Asp^{253}-Met^{254} or His^{309}-Glu^{310} would be expected to be alternative reactions within a given subunit (Soundar and Colman, 1993).

The lower part of Fig. 7 indicates that Fe^{2+} ion alone produces peptides F_2 and F_5 by specific cleavage of the Tyr^{272}-Asp^{273} peptide bond. The enzyme ligand to the metal ion in the binary enzyme–metal complex is likely to be the carboxylate of Asp^{273}, although coordination to the -OH of Tyr^{272} cannot be excluded. Clearly, the enzyme binding site for free metal ion is close to (but is

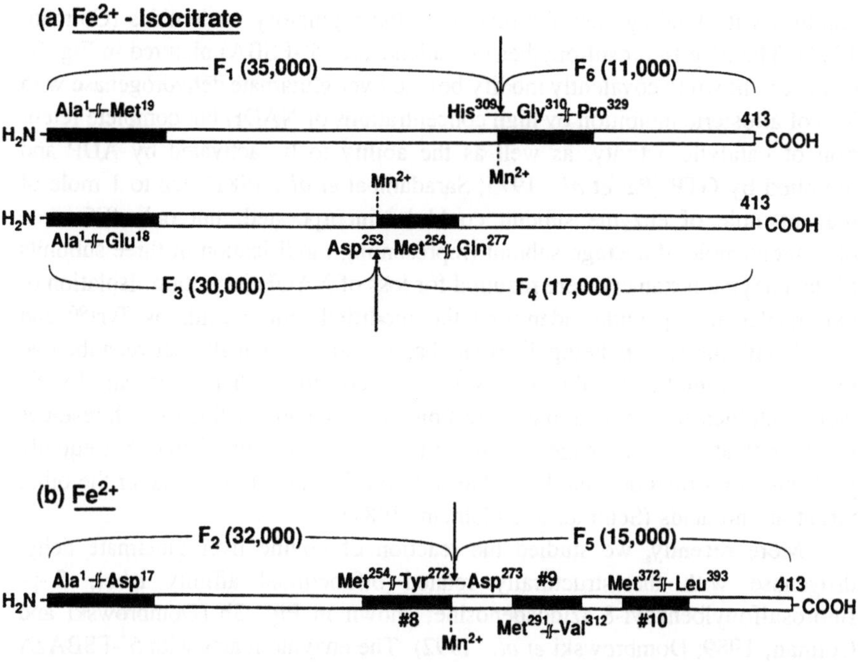

FIGURE 7. Scheme of affinity cleavage of pig heart NADP-dependent isocitrate dehydrogenase by (a) Fe^{2+}-isocitrate and (b) Fe^{2+}. The sequences that have been determined are shaded. Reproduced from Soundar and Colman (1993), with permission.

distinguishable from) that for metal–isocitrate, which is consistent with previous studies of the kinetics and binding characteristics of this enzyme. The potential of the method of affinity cleavage to probe the vicinity of a metal or metal–substrate site is apparent from this study; this approach may be generally useful in locating such sites in other enzymes.

2.4. Bovine Liver Glutamate Dehydrogenase: Affinity Labeling of Regulatory Sites with Mono- and Bifunctional Reactive Nucleotides

Bovine liver glutamate dehydrogenase is an allosteric enzyme that is activated by ADP and inhibited by GTP, as well as by high concentrations of NADH. All of these allosteric effectors bind to regulatory sites distinct from the catalytic coenzyme site (Colman, 1991). The enzyme in its smallest active form is a hexamer of identical subunits, each of which binds 2 moles of NADH (at a catalytic and at a regulatory site), 2 moles of ADP, and 2 moles of GTP (one at a high-affinity site and one at a low-affinity site) (Colman, 1991). Affinity labeling experiments with monofunctional purine nucleotides have associated amino acid residues with binding sites for several of the regulatory nucleotides (Colman, 1991). The 5′-p-fluorosulfonylbenzoyl adenosine (5′-FSBA) pictured in Fig. 3a has been shown to covalently modify bovine liver glutamate dehydrogenase with loss of allosteric inhibition by high concentrations of NADH but complete retention of catalytic activity, as well as the ability to be activated by ADP and inhibited by GTP (Pal *et al.*, 1975; Saradambal *et al.*, 1981). Up to 1 mole of reagent/mole of enzyme subunit could be incorporated, but only 0.5 mole of reagent/mole of average subunit (representing modification of three subunits of the enzyme hexamer) were required for loss of NADH inhibition. Isolation of two nucleosidyl peptides identified the modified amino acids as Tyr^{190} and Lys^{420} with the reagent being distributed approximately equally between the two residues (Schmidt and Colman, 1984). It was considered that Tyr^{190} and Lys^{420} were both located in or near the NADH binding site and labeling of both residues might indicate that the reagent binds to the enzyme in two distinct but equally probable conformations that bring the sulfonyl fluoride close to one or the other target amino acids (Schmidt and Colman, 1984).

More recently, we studied the reaction of bovine liver glutamate dehydrogenase with the structurally related bifunctional affinity label, 5′-p-fluorosulfonylbenzoyl-8-azidoadenosine, shown in Fig. 3b (Dombrowski and Colman, 1989; Dombrowski *et al.*, 1992). The enzyme reacts with 5′-FSBAzA in a two-step process: (1) an initial dark reaction occurring through the fluorosulfonyl group yielding about 0.5 mole of the -SBAzA group bound/mole enzyme subunit, (2) followed by photoactivation of the azido group leading to 25% of the covalently bound -SBAzA becoming cross-linked to another part of the enzyme. The 5′-FSBAzA-modified enzyme, both before and after photolysis, retains full catalytic activity but is strikingly less sensitive to inhibition by high concentrations of NADH, characteristics that are similar to those of 5′-FSBA-modified enzyme (Pal *et al.*, 1975; Saradambal *et al.*, 1981).

Table IV summarizes the effect of natural nucleotides on the rate of the initial dark reaction of 5′-FSBAzA with glutamate dehydrogenase. Neither the activator ADP nor the inhibitor GTP decreases the reaction rate, indicating that these sites are not the target of 5′-FSBAzA. At 0.2 mM NADH, binding occurs predominantly at the catalytic site, and k_{obs} remains unchanged. At 2 mM, NADH binds to both the catalytic and NADH regulatory site; the sevenfold decrease in k_{obs} produced by 2 mM NADH suggests that 5′-FSBAzA modifies the NADH regulatory site. The affinity of NADH for its regulatory site is enhanced by GTP, and the additional fivefold decrease in k_{obs} (Table IV) is consistent with reaction at or near the GTP-dependent NADH regulatory site. These results are similar to those obtained with the monofunctional reagent 5′-FSBA (Pal *et al.*, 1975; Saradambal *et al.*, 1981), implying that 5′-FSBA and 5′-FSBAzA modify the same operational site.

We have recently identified the amino acid residues modified in each phase

Table IV
Effect of Nucleotides on Rate Constant for Dark Reaction of Bovine Liver Glutamate Dehydrogenase with 5′-FSBAzA[a,b]

Addition to reaction mixture	k_{obs}(min^{-1}) × 10^3
None	41.6
ADP (1 mM)	43.3
GTP (200 μM)	37.5
NADH (0.2 mM)	38.9
NADH (2 mM)	6.0
GTP (2 μM) + NADH (2 mM)	4.6
GTP (20 μM) + NADH (2 mM)	5.7
GTP (200 μM) + NADH (2 mM)	1.2

[a]Glutamate dehydrogenase (2 mg/ml) was incubated with 0.3 mM 5′-FSBAzA at 30 °C and pH 8.0 and the rate constant was measured as described by Dombrowski *et al.* (1992).
[b]Reproduced from Dombrowski *et al.* (1992), with permission.

of the reaction of glutamate dehydrogenase with 5′-FSBAzA (Dombrowski *et al.*, 1992). Tyr^{190} is the major amino acid that reacts with the fluorosulfonyl group of 5′-FSBAzA, in contrast to the equal amounts of Tyr^{190} and Lys^{420} labeled by 5′-FSBA. It is likely that this difference in reaction products is attributable to a distinction in the predominant conformation of the reactive nucleotide analogues. When adenine nucleotides are free in solution, the orientation of the purine ring about the glycosidic bond is equally distributed between the syn and the anti conformation. It seems reasonable that the equal labeling of Tyr^{190} and Lys^{420} is due to reaction of each of these residues with, respectively, the syn or anti conformation of 5′-FSBA. In contrast, substitution of the eight-position of the adenine ring with a bulky group, such as the azido group of 5′-FSBAzA, causes a shift in the conformational equilibrium such that the syn form predominates. The major reaction of 5′-FSBAzA with Tyr^{190} may thus correspond to reaction with the fluorosulfonyl group of 5′-FSBAzA bound in the syn conformation.

The predominant cross-link formed during photolysis of the 5′-FSBAzA-modified enzyme is between modified Tyr^{190} and the peptide Leu^{475}-Asp^{476}-Leu^{477}-Arg^{478}, which is located near the C-terminus of glutamate dehydrogenase. Figure 8 shows the bifunctional nucleotide analogue 5′-FSBAzA in the syn (Fig. 8a) or in the anti (Fig. 8b) conformation. The maximum distance between the nitrogen of the nitrene and the sulfur of the fluorosulfonyl group is 14.2 Å with 5′-FSBAzA in the syn conformation and 11.0 Å in the anti conformation. Although a range of distances between the two functional groups is possible, a value of approximately 14 Å is most likely if the syn conformation, as expected, predominates in this eight-substituted nucleotide. This study illustrates

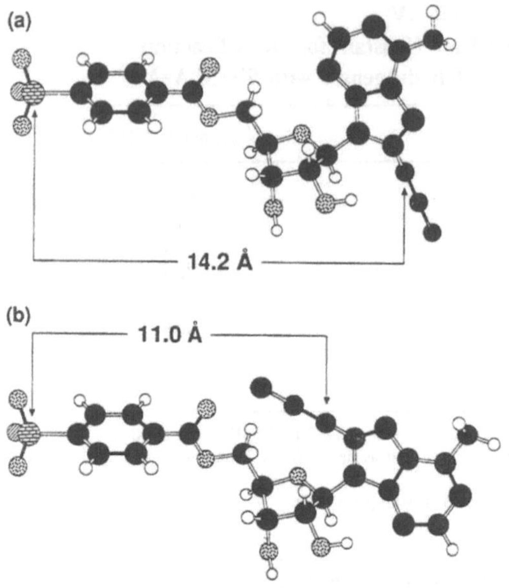

FIGURE 8. Distance relationships between the azido and fluorosulfonyl groups of 5′-FSBAzA. Distance measurements were made using the Chem 3D computer program (Cambridge Scientific Corp.) with the purine ring of 5′-FSBAzA in the (a) syn and (b) anti conformations about the glycosidic bond. The distances labeled are those between the nitrogen of the nitrene and the sulfur of the sulfonyl fluoride. Reproduced from Dombrowski *et al.* (1992), with permission.

the usefulness of a bifunctional affinity label such as 5′-FSBAzA in identifying pairs of amino acids and yielding estimates of the distances between these residues in the folded structure of the enzyme.

Bovine liver glutamate dehydrogenase has also been used in the initial evaluations of the specificity and effectiveness as affinity labels of the bromooxopropyl derivatives of nucleoside thiophosphates illustrated in Fig. 2. The NMNS-BOP inactivates glutamate dehydrogenase and appears to target the catalytic coenzyme site (Park *et al.*, 1993). In contrast, GMPS-BOP does not cause loss of activity, but rather decreases the sensitivity of the enzyme to regulation by NADH, GTP, and ADP (Ozturk *et al.*, 1992); and AMPS-BOP reacts with glutamate dehydrogenase to yield an irreversibly activated enzyme, mimicking the reversible effect of ADP (Walner and Colman, 1990). These three affinity labels have the same reactive bromooxopropyl moiety (Fig. 2); clearly, the specificity for their reaction with bovine liver glutamate dehydrogenase is determined by the purine or nicotinamide portion of their structures.

2.5. *Salmonella typhimurium* NADP-Dependent Glutamate Dehydrogenase: Affinity Labeling and Site-Directed Mutagenesis at the Catalytic Coenzyme Site

Glutamate dehydrogenase is widespread in nature, exhibiting a coenzyme specificity that depends on the source of the enzyme. As distinguished from the mammalian enzyme that can utilize both NAD and NADP, the enzyme from microorganisms is specific for either NAD or NADP; for example, the enzyme from *Salmonella typhimurium* exhibits a great preference for NADP. Furthermore, the glutamate dehydrogenases from microorganisms are not allosteric enzymes.

The gene from *S. typhimurium* encoding glutamate dehydrogenase has been cloned and sequenced (Bansal *et al.*, 1989). The enzyme has been overexpressed in *Escherichia coli*, making it a convenient source of bacterial glutamate dehydrogenase and raising the possibility of engineering new forms of the enzyme by site-directed mutagenesis. Furthermore, a crystallographic group in Sheffield, England has been studying the structure of crystals of a related bacterial glutamate dehydrogenase from *Clostridium symbiosum* (Baker *et al.*, 1992). Affinity labeling can indicate, for an enzyme in solution, that there is proximity between the modified amino acid and the natural ligand that is mimicked by the affinity reagent; it does not prove that the target amino acid makes a direct contribution to catalysis, regulation, or binding. However, the possibility of a direct role for a particular amino acid, which is suggested by either crystallographic data or affinity labeling, can be explored further by site-directed mutagenesis. The bacterial glutamate dehydrogenase system offers the possibility of using the complementary techniques of crystallography, affinity labeling, and site-directed mutagenesis to examine structure–function relationships. We initiated our exploration of the *Salmonella* glutamate dehydrogenase by affinity labeling studies to provide the knowledge base for rational choices of target sites for mutagenesis experiments. Since the enzyme requires NADP, we selected a reactive nucleotide analogue with a 2′-phosphate: 2-(4-bromo-2,3-dioxobutylthio)-1, N^6-ethenoadenosine-2′,5′-diphosphate (Bansal *et al.*, 1989), related to the compound shown in Fig. 1b, although we later carried out similar studies with the adenosine analogue related to the compound shown in Fig. 1a (Haeffner-Gormley *et al.*, 1992a).

Incubation of *S. typhimurium* glutamate dehydrogenase with 200 μM 2-BDB-TεA 2′,5′-DP causes a time-dependent inactivation to yield a partially active enzyme of 40% residual activity. Neither additional reagent nor increases in time led to further inactivation, demonstrating that this was the limit of the reaction for this enzyme. The rate of inactivation exhibits a nonlinear dependence on the reagent concentration, indicating that 2-BDB-TεA 2′,5′-DP binds reversibly to the enzyme prior to the irreversible reaction (Bansal *et al.*, 1989).

Complete protection against inactivation was provided by the specific coenzymes NADPH or NADP or the coenzyme fragment 2′-phospho-adenosine diphosphoribose, but not by NADH or the substrate α-ketoglutarate (Bansal *et al.*, 1989). Thus, reaction appeared to be directed to the region of the coenzyme binding site.

The amount of reagent incorporated into glutamate dehydrogenase was measured from the fluorescence of enzyme-bound 2-BDB-TεA 2′,5′-DP as a function of time of incubation. Up to about 1 mole of reagent was incorporated per peptide chain. The incorporation of reagent was decreased to 0.5 mole per peptide chain in the presence of NADPH, when the enzyme retained full activity. These results indicate that 2-BDB-TεA 2′,5′-DP functions as an affinity label of the coenzyme binding site and that specific reaction occurs at only about 0.5 sites per enzyme subunit or 3 sites/enzyme hexamer (Bansal *et al.*, 1989). Glutamate dehydrogenase from several species has been shown to be organized as a dimer of trimers. In this case, modification and inactivation of one trimer must indirectly diminish the activity of the second trimer and decrease below the level of detection the reaction rate of that second trimer with 2-BDB-TεA 2′,5′-DP.

Two major tryptic peptides modified by the reagent rate were isolated and identified: the pentapeptide Leu^{282}-Cys-Glu-Ile-Lys^{286} and an 11-membered peptide containing a modified Glu^{153} (Bansal *et al.*, 1989; Haeffner-Gormley *et al.*, 1991). In the presence of NADPH, which completely prevents inactivation, only the 11-membered peptide was labeled.These results indicate that modification of the pentapeptide causes loss of activity (Bansal *et al.*, 1989). However, we did not isolate directly the modified residue and initially there was some uncertainty as to whether the cysteine or the glutamate was the actual target of 2-BDB-TεA 2′,5′-DP.

In order to examine these possibilities, site-directed mutagenesis was used to engineer and then to express in *E. coli* three mutant *Salmonella* glutamate dehydrogenases: one in which glutamate was changed to the nonnucleophilic glutamine at position 284, a second in which the cysteine was changed to the nonreactive isoleucine at position 283, and a double mutant in which both amino acids 283 and 284 were changed (Haeffner-Gormley *et al.*, 1991, 1992a). The substitution of isoleucine for cysteine was chosen because isoleucine is not a nucleophile and is found in the corresponding position in the *E. coli* enzyme; while replacement of glutamic acid by glutamine was selected to eliminate the nucleophilic character of the residue but retain its size, in order to avoid altering the conformation of the enzyme. The glutamine mutant enzyme reacts with 2-BDB-TεA 2′-5′-DP at about the same rate as the original wild-type enzyme, demonstrating that Glu^{284} could not be the target of 2-BDB-TεA 2′-5′-DP. On the other hand, for the two mutant *Salmonella* glutamate dehydrogenases in which cysteine has been converted to isoleucine, there was no inactivation by added 2-BDB-TεADP. These results identify Cys^{283} as the reaction site of this nucleotide analogue (Haeffner-Gormley *et al.*, 1991, 1992a).

We have purified to homogeneity these three mutant enzymes. The striking result is that they all have about the same specificity activity as the wild-type enzyme and exhibit the same K_m values for the three substrates NADPH, NH_4Cl, and α-ketoglutarate (Haeffner-Gormley *et al.*, 1991). Since the mutant enzymes are fully active, Cys^{283} cannot be required for catalysis. However, we consider that the results of the affinity labeling experiments locate Cys^{283} within the coenzyme binding site. The 2-(4-bromo-2,3-dioxobutylthio)-1,N^6-ethenoadenosine 2′,5′-diphosphate probably binds to the enzyme at the coenzyme binding site. If a nucleophilic amino acid is available, it then reacts covalently, thereby permanently occupying the coenzyme binding site and preventing further catalysis and inactivating the enzyme. Examination of the crystal structure of the glutamate dehydrogenase of *Clostridium symbiosum* (Baker *et al.*, 1992) supports this conclusion. The two bacterial enzymes exhibit 64% sequence homology, which makes feasible an alignment of the amino acid sequence of the *S. typhimurium* enzyme with the atomic coordinates of the *C. symbiosum* enzyme. The loop 282-286 is located close to the bound coenzyme in the general region of the adenine ribose and pyrophosphate moieties.

Further examination of the crystal structure of glutamate dehydrogenase suggested that the Lys at position 286 might be closer to the phosphate binding region of the coenzyme binding site. Accordingly, we engineered a set of mutants with substitutions at this position. Lysine^{286} was replaced by the positively charged arginine, by neutral glutamine, and by negatively charged glutamate in order to assess the effect of the amino acid charge on coenzyme binding, catalysis, and specificity. The mutant enzymes were constructed by site-directed mutagenesis, expressed in *E. coli*, and then purified to homogeneity and characterized (Haeffner-Gormley *et al.*, 1992b).

All three mutant enzymes exhibited elevated K_m values for the coenzyme NADPH, as shown in Table V. The arginine mutant enzyme exhibits the smallest change, as might have been expected since the positively charged arginine was substituted for the positively charged lysine. Substitution of the neutral amino acid glutamine caused a greater increase in K_m for NADPH. However, replacement of the lysyl residue by the negatively charged glutamate at position 286 resulted in a 30-fold increase in the K_m for NADPH (Haeffner-Gormley *et al.*, 1992). This observation is consistent with our postulate that Lys^{286} is normally involved in coenzyme binding. In contrast, the K_m values for the other substrates, ammonium chloride and α-ketoglutarate, do not differ appreciably between the wild-type and mutant enzymes.

Table V
Determination of K_m (NADPH) and k_{cat} for Wild-Type and Mutant Glutamate Dehydrogenases of *Salmonella typhimurium*

Enzyme	K_m (NADPH) (μM)	k_{cat} (NADPH) min^{-1} $\times 10^{-3}$	Efficiency k_{cat}/K_m ($\mu M^{-1}\ min^{-1}$)
Wild type	9.8	11.3	1200
Lys 286 Arg	30	18.1	610
Lys 286 Gln	66	13.8	210
Lys 286 Glu	280	11.8	42

Table VI
Comparison of NADPH and NADH as Coenzymes for Wild-Type and Mutant Glutamate Dehydrogenases

Enzyme	K_m (NADH) (μM)	K_{cat}/K_m(min^{-1}μM^{-1})		
		NADPH	NADH	NADPH/NADH
Wild type	2000	1200	0.62	1900
Lys 286 Arg	1800	610	1.60	380
Lys 286 Gln	2100	210	1.10	200
Lys 286 Glu	2100	42	0.46	91

The maximum velocity for each enzyme, expressed in Table V as k_{cat} (μmoles of NADPH per min per μmole of enzyme subunit), was obtained at high, saturating concentrations of the coenzyme and substrates. The wild-type and mutant enzymes have similar values of k_{cat} (Haeffner-Gormley *et al.*, 1992b). The last column in Table V assesses the efficiency of the enzyme, which is defined as k_{cat}/K_m. The k_{cat}/K_m values show that the decreasing efficiency in the mutant enzymes, particularly that of the glutamate mutant, is attributable to the increase in K_m for NADPH.

The only structural feature that distinguishes NADPH from NADH is its negatively charged phosphate at the 2′-position. Although the *Salmonella* glutamate dehydrogenase is generally considered to be NADPH-specific, we were able to detect the activity of the wild–type enzyme using NADH and to determine the K_m for NADH. With NADH as coenzyme, the k_{cat} is tenfold lower and K_m is 200-fold higher than for NADPH as the coenzyme (Haeffner-Gormley *et al.*, 1992b). Table VI shows that, in contrast to the major differences in K_m for NADPH, wild-type and mutant enzymes have similar K_m values for NADH. Since the efficiency of the mutant enzymes decreases for NADPH as coenzyme but does not change appreciably when NADH is used as coenzyme, there is a decrease in the relative efficiency of the mutant enzymes for NADPH compared to NADH and the mutants become less specific for coenzyme. These results indicate that the positively charged residue Lys^{286} normally contributes to the great preference of the enzyme for this coenzyme by participating in the binding of the 2′-phosphate group of NADPH.

3. CONCLUDING REMARKS

The examples cited in this chapter are representative only: the techniques of affinity labeling and affinity cleavage are broadly applicable to the identification of specific sites and to the evaluation of the function of these sites in numerous enzymes. The 5′-p-fluorosulfonylbenzoyl adenosine, which was the first of the reactive nucleotide analogues described by my laboratory, has been found to yield specific labeling of NAD sites in dehydrogenase and reductases (Colman, 1983a, 1990). It has also labeled ATP or ADP binding sites in a large number of kinases and synthetases, in addition to providing an effective handle for examining an ADP receptor protein of platelet membranes. Furthermore, it has modified specific nucleotide sites in such diverse proteins as the ATPases, actin, myosin, luciferase, and oxoprolinase. In all, more than 50 proteins have been studied using this compound (Colman, 1990). It is anticipated that the other mono- and bifunctional fluorosulfonylbenzoyl nucleosides, the bromo-oxopropyl and bromodioxobutyl derivatives, as well as metal-dependent affinity cleavage reactions, will similarly have widespread applications in targeting specific sites in diverse enzymes and receptors. Affinity labeling is a powerful technique, particularly when it is applied in combination with spectroscopic measurements, site directed mutagenesis, and X-ray crystallography.

ACKNOWLEDGMENTS. This work was supported by grants DMB-9105116 from the National Science Foundation, BE-111 from the American Cancer Society, and 1 RO1 DK17522 from the United States Public Health Service.

4. REFERENCES

Armstrong, R. N., 1991, Glutathione S-transferases: Reaction mechanism, structure and function, *Chem. Res. Toxicol.* **4:**131–140.

Baker, P. J., Britton, K. L., Engel, P. C., Farrants, G. S., Lilley, K. S., Rice, D. W., and Stillmen, T. J., 1992, Subunit assembly and active site location in the structure of glutamate dehydrogenase, *Proteins* **12:**75–86.

Bansal, A., Dayton, M. A., Zalkin, H., and Colman, R. F., 1989, Affinity labeling of a glutamyl peptide in the coenzyme binding site of NADP+-specific glutamate dehydrogenase of *Salmonella typhimurium* by 2-](4-bromo-2,3-dioxobutyl)]-1,N6-ethenoadenosine-2′,5′-bisphosphate, *J. Biol. Chem.* **264:**9827–9835.

Colman, R. F., 1972, Role of metal ions in reactions catalyzed by pig heart TPN-dependent isocitrate dehydrogenase. II. Effect on catalytic properties and reactivity of amino acid residues. *J. Biol. Chem.* **247:**215–223.

Colman, R. F., 1983a, Affinity labeling of purine nucleotide sites in proteins, *Annu. Rev. Biochem.* **52:**67–91.

Colman, R. F., 1983b, Aspects of the structure and function of the isocitrate dehydrogenases, *Pept. Protein Rev.* **1:**41–69.

Colman, R. F., 1989, Affinity labelling, in *Protein Function: A Practical Approach*, (T. E. Creighton, ed.), pp. 77–99, IRL Press, New York.

Colman, R. F., 1991, Glutamate dehydrogenase (bovine), in *A Study of Enzymes*, vol. II, *Mechanism of Enzyme Action* (S. A. Kuby, ed.), pp. 173–192, CRC Press, Boca Raton, Florida.

DeCamp, D. L., and Colman, R. F., 1989, 2-[(4-Bromo-2,3-dioxobutyl)thio]-1,N^6-ethenoadenosine 5′-diphosphate: A new fluorescent affinity label of a tyrosyl residue in the active site of rabbit muscle pyruvate kinase, *J. Biol. Chem.* **264:**8430–8441.

DeCamp, D. L., Lim, S., and Colman, R. F., 1988, Reaction of pyruvate kinase with the new nucleotide affinity labels 8-](4-bromo-2,3-dioxobutyl)-thio]adenosine 5′-diphosphate and triphosphate, *Biochemistry* **27:**7651–7658.

Dombrowski, K. E., and Colman, R. F., 1989, 5′-p-Fluorosulfonylbenzoyl-8-azidoadenosine: A new bifunctional affinity label for nucleotide binding sites in proteins, *Arch. Biochem. Biophys.* **275:**302–308.

Dombrowski, K. E., Huang, Y.-C., and Colman, R. F., 1992, Identification of amino acids modified by the bifunctional affinity label 5′-p-(fluoro-sulfonylbenzoyl)-8-azidoadenosine in the reduced coenzyme site of bovine liver glutamate dehydrogenase, *Biochemistry* **31:**3785–3793.

Ehrlich, R. S., and Colman, R. F., 1976, Influence of substrates and coenzymes on the role of manganous ion in reactions catalyzed by pig heart TPN-dependent isocitrate dehydrogenase, *Biochemistry* **15:**4034–4041.

Haeffner-Gormley, L., Chen, Z., Zalkin, H., and Colman, R. F., 1991, Evaluation of cysteine-283 and glutamic acid-284 in the coenzyme binding site of *Salmonella typhimurium* glutamate dehydrogenase by site directed mutagenesis and reaction with the nucleotide analogue 2-[(4-bromo-2,3-dioxobutyl)thio]-1,N^6-ethenoadenosine 2′,5′-diphosphate, *J. Biol. Chem.* **266:**5388–5394.

Haeffner-Gormley, L., Chen, Z., Zalkin, H., and Colman, R. F., 1992a, Reaction of the nucleotide analogue 2-[(4-bromo-2,3-dioxobutyl)thio]adenosine 2′,5′-bisphosphate at the coenzyme site of wild type and mutant $NADP^+$-specific glutamate dehydrogenases from *Salmonella typhimurium*, *Arch. Biochem. Biophys.* **292:**179–189.

Haeffner-Gormley, L., Chen, Z., Zalkin, H., and Colman, R. F., 1992b, Importance of lysine-286 at the NADP site of glutamate dehydrogenase from *Salmonella typhimurium*, *Biochemistry* **31:**7807–7814.

Haselbeck, R. J., Colman, R. F., and McAlister-Henn, L., 1992, Isolation and sequence of a cDNA encoding porcine mitochondrial NADP-specific isocitrate dehydrogenase, *Biochemistry* **31:**6219–6223.

Jakoby, W. B., and Wilchek, M., eds., 1977, *Methods in Enzymology*, Vol. 46: *Affinity Labeling*, Academic Press, New York.

Ji, X., Zhang, P., Armstrong, R. H., and Gilliland, G. L., 1992, The 3-dimensional structure of a glutathione S-transferase from the Mμ class (isoenzyme 3-3), *Biochemistry* **31:**10169–10184.

Katusz, R. M., and Colman, R. F., 1991, S-(4-bromo-2,3-dioxobutyl)glutathione: A new affinity label for the 4-4 isoenzyme of rat liver glutathione S-transferase, *Biochemistry* **30:**11230–11238.

Katusz, R. M., Bono, B., and Colman, R. F., 1992a, Identification of Tyr^{115} labeled by S-(4-bromo-2,3-dioxobutyl)glutathione in the hydrophobic substrate binding site of glutathione S-transferase, isoenzyme 3-3, *Arch. Biochem. Biophys.* **298:**667–677.

Katusz, R. M., Bono, B., and Colman, R. F., 1992b, Affinity labeling of Cys^{111} of glutathione S-transferase, isoenzyme 1-1, by S-(4-bromo-2,3-dioxobutyl)glutathione, *Biochemistry* **31:**8984–8990.

Mannervik, B., and Danielson, U. H., 1988, Glutathione S-transferases—structure and catalytic activity, *Crit. Rev. Biochem.* **23:**283–337.

Muirhead, H., Clayden, D. A., Barford, D., Lorimer, C. G., Fothergill-Gilmore, L. A., Schlitz, E., and Schmitt, E., 1986, The structure of cat muscle pyruvate kinase, *EMBO J.* **5:**475–481.

Ozturk, D. H., Park, I., and Colman, R. F., 1992, Guanosine 5′-O-[S-(3-bromo-2-oxopropyl)]-

thiophosphate: A new reactive purine nucleotide analog labeling Met-169 and Tyr-262 in bovine liver glutamate dehydrogenase, *Biochemistry* **31:**10544–10555.

Pal, P. K., Wechter, W. J., and Colman, R. F., 1975, Affinity labeling of the DPNH inhibitory site of bovine liver glutamate dehydrogenase by 5′-p-fluorosulfonylbenzoyl adenosine, *J. Biol. Chem.* **250:**8140–8147.

Park, I., Ozturk, D. H., Soundar, S., and Colman, R. F., 1993, Nicotinamide ribose 5′-O-[S-(3-bromo-2-oxopropyl)thiophosphate: A new affinity label for NMN sites in enzymes, *Arch. Biochem. Biophys.* **303:**483–488.

Reinemer, P., Dirr, H. W., Landenstein, R., Schaffer, J., Gallay, O., and Huber, R., 1991, The 3-dimensional structure of class π glutathione S-transferase in complex with glutathione sulfonate at 2.3 Å resolution, *EMBO J.* **10:**1997–2005.

Rushmore, T. H., and Pickett, C. B., 1993, Glutathione S-transferases, structure, regulation and therepeutic implications, *J. Biol. Chem.* **268:**11475–11478.

Saradambal, K. V., Bednar, R. A., and Colman, R. F., 1981, Lysine and tyrosine in the NADH inhibitory site of bovine liver glutamate dehydrogenase, *J. Biol. Chem.* **256:**11866–11872.

Schmidt, J. A., and Colman, R. F., 1984, Identification of the lysine and tyrosine peptides labeled by 5′-p-fluorosulfonylbenzoyl adenosine in the NADH inhibitory site of glutamate dehydrogenase, *J. Biol. Chem.* **259:**14515–14519.

Sigman, D. S., and Chen, C. B., 1990, Chemical nucleases: New reagents in molecular biology, *Annu. Rev. Biochem.* **59:**207–236.

Soundar, S., and Colman, R. F., 1993, Identification of metal–isocitrate binding site of pig heart NADP-specific dehydrogenase by affinity cleavage of the enzyme by Fe^{2+}-isocitrate, *J. Biol. Chem.* **268:**5264–5271.

Villafranca, J. J., and Colman, R. F., 1972, Role of metal ions in reactions catalyzed by pig heart TPN-dependent isocitrate dehydrogenase. I. Magnetic resonance and binding studies of the complexes of enzyme, manganous ion and substrates, *J. Biol. Chem.* **247:**209–214.

Vollmer, S. H., and Colman, R. F., 1990, Cysteinyl peptides of rabbit muscle pyruvate kinase labeled by the affinity label 8-[(4-bromo-2,3-dioxobutyl)thio]adenosine 5′-triphosphate, *Biochemistry* **29:**2495–2501.

Walner, M. B., and Colman, R. F., 1990, Adenosine 5′-O-[(3-bromo-2-oxopropyl)thiophosphate]: A new affinity label of an activity site of glutamate dehydrogenase, *Biochemistry* **29:**2202.

Chapter 8

Tryptophan Synthase
Structure, Function, and Protein Engineering

Edith Wilson Miles

1. INTRODUCTION

Two important questions in modern biochemistry are how different proteins interact and how protein–protein interaction regulates enzymatic activity. The bacterial tryptophan synthase $\alpha_2\beta_2$ complex (EC 4.2.1.20) is an excellent model system for seeking answers to these questions.

Biochemical and genetic studies in the 1950s and 1960s established that bacterial tryptophan synthase is an $\alpha_2\beta_2$ complex composed of nonidentical and dissociable α and β subunits. Investigations of tryptophan synthase during the same period played an important role in determining gene–protein colinearity. Several reviews and commentaries cover these early studies (Yanofsky, 1987, 1989; Yanofsky and Crawford, 1972). Other reviews have emphasized evolutionary aspects (Crawford, 1989), enzyme mechanism (Miles, 1979, 1986; Miles *et al.*, 1987), the structural basis for catalysis (Miles, 1991a; Miles *et al.*, 1994),

Edith Wilson Miles Enzyme Structure and Function Section, Laboratory of Biochemical Pharmacology, National Institute of Diabetes and Digestive and Kidney Diseases, National Institutes of Health, Bethesda, Maryland 20892.

Subcellular Biochemistry, Volume 24. Proteins: Structure, Function, and Engineering, edited by B. B. Biswas and Siddhartha Roy. Plenum Press, New York, 1995.

and the use of tryptophan synthase as a model for protein–protein interaction (Swift and Stewart, 1991). The present chapter will concentrate on recent studies, especially studies that have used genetically engineered mutants of tryptophan synthase to probe structure, function, and protein–protein interaction.

The α and β subunits of tryptophan synthase catalyze the last two reactions in L-tryptophan biosynthesis (scheme I). The separate α subunit catalyzes the *α reaction,* the reversible cleavage of indole-3-glycerol phosphate to form indole and D-glyceraldehyde 3-phosphate. The β subunit normally contains the cofactor pyridoxal phosphate. The separate β subunit, which exists as a β_2 dimer in solution, catalyzes the *β reaction,* the irreversible formation of L-tryptophan from L-serine and indole. Association of the α and β subunits results in mutual activation of the individual subunits and renders each subunit sensitive to effects of ligands bound to the neighboring subunit. The $\alpha_2\beta_2$ complex also catalyzes the αβ reaction, which is essentially the sum of the α and β reactions (see Section 4.7).

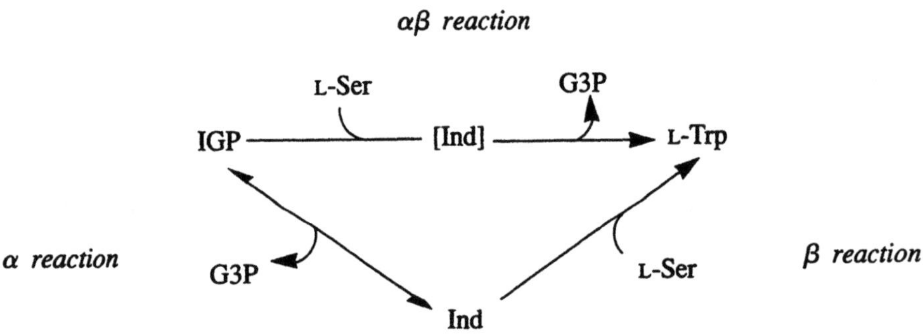

Scheme 1

Crystallographic studies of the tryptophan synthase $\alpha_2\beta_2$ complex from *Salmonella typhimurium* reveal that the complex has an extended αββα subunit arrangement with an overall length of ~150 Å (Hyde *et al.*, 1988). The β_2 dimer is located at the center and the two α subunits are separated from each other at opposite ends of the complex. The active sites of the neighboring α and β subunits are ~25 Å apart and are connected by a tunnel having a diameter and length sufficient to accommodate four molecules of indole. The tunnel permits the internal diffusion of indole between the heterologous active sites and prevents escape of indole into the solution.

Figure 1 shows our current picture of the mechanism of the αβ reaction based on the crystal structure. The indole intermediate, which is produced by cleavage of indole-3-glycerol phosphate at the active site of the α subunit,

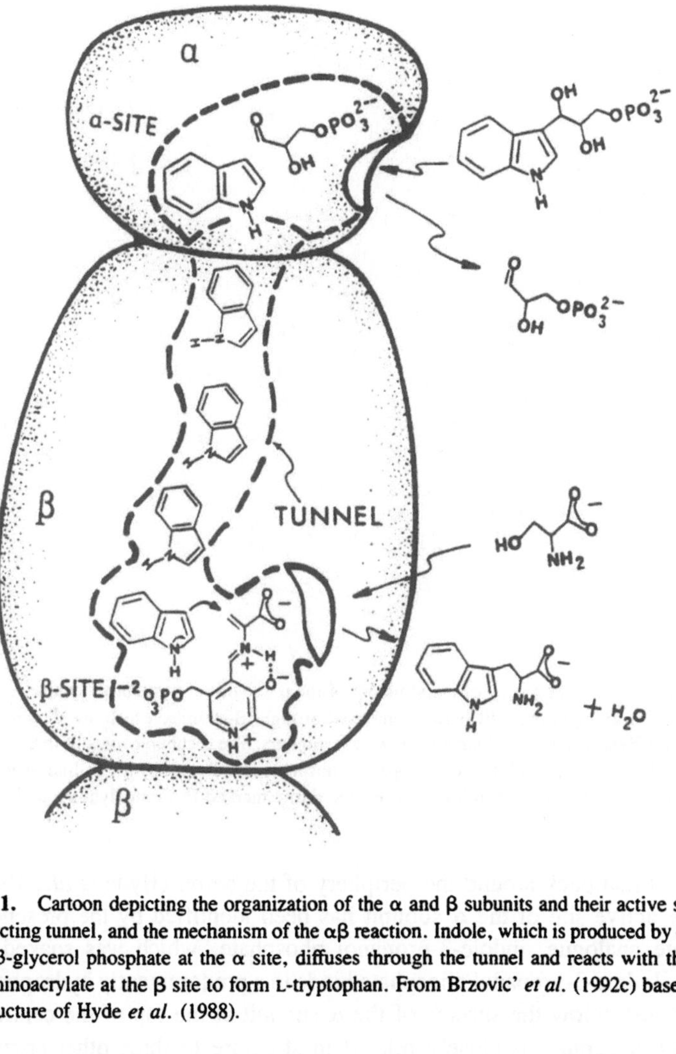

FIGURE 1. Cartoon depicting the organization of the α and β subunits and their active sites, the interconnecting tunnel, and the mechanism of the αβ reaction. Indole, which is produced by cleavage of indole-3-glycerol phosphate at the α site, diffuses through the tunnel and reacts with the Schiff base of aminoacrylate at the β site to form L-tryptophan. From Brzovic' *et al.* (1992c) based on the crystal structure of Hyde *et al.* (1988).

diffuses through an intramolecular tunnel to the active site of the β subunit where it undergoes a pyridoxal phosphate-dependent β-replacement reaction with L-serine to form L-tryptophan (see Section 4.7).

2. STRUCTURE AND FUNCTION OF THE α SUBUNIT

2.1. Three-Dimensional Structure

The tryptophan synthase α subunit has an eightfold α/β barrel polypeptide fold composed of a central core of eight parallel β-strands with eight parallel

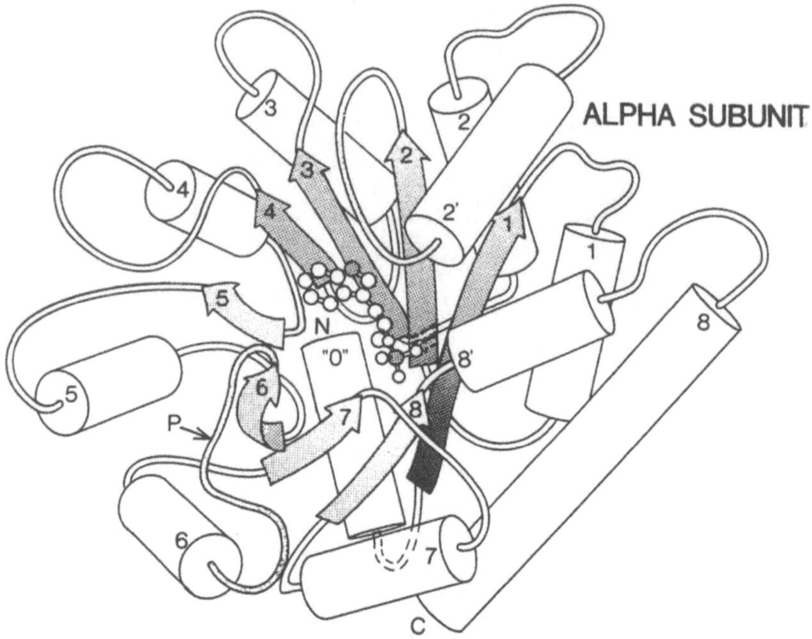

FIGURE 2. Schematic view of the structure of the α subunit of tryptophan synthase. The overall fold is that of an eightfold α/β barrel composed of eight β-strands, which are shown as flattened arrows, and eight α-helices, which are shown as cylinders. The α subunit contains three other helices labeled 0, 2′, and 8′. Bound indole-3-propanol phosphate is indicated by the ball-and-stick model. A site of proteolysis at Arg-188 in loop 6 is by the arrow marked P. From Hyde *et al.* (1988).

α-helices that pack around the periphery of the barrel (Hyde *et al.*, 1988) (Fig. 2). The active site of the α subunit has been identified by the presence of the substrate analogue, indole-3-propanol phosphate, which was soaked into the crystal. Indole-3-propanol phosphate binds in a predominantly hydrophobic "pit" formed just below the surface of the α subunit at the top of the central barrel.

The α subunit is closely related in structure to three other enzymes: triosephosphate isomerase and both parts of the bifunctional enzyme (phosphoribosyl)anthranilate isomerase: indole-3-glycerol phosphate synthetase (Farber and Petsko, 1990; Wilmanns *et al.*, 1991, 1992). The active site of each of these enzymes is in a largely hydrophobic region located at the carboxy terminal ends of the β-strands that compose the central barrel. In each enzyme the phosphate of the substrate analogue binds to a site created by residues in the loop between strand 7 and helix 7 and the N-terminus of an additional helix 8′.

The tryptophan synthase α subunit, triosephosphate isomerase, and (phosphoribosyl)anthranilate isomerase each contains an extended loop (loop 6) that connects strand 6 and helix 6 (Wilmanns *et al.*, 1991). The weak electron density of the α subunit in this region suggests that loop 6 has a high degree of conformational mobility or flexibility. Crystallographic studies of triosephosphate isomerase show that loop 6 moves nearly 7 Å from an "open" position to a "closed" position upon binding a substrate analogue, effectively covering the active site (Davenport *et al.*, 1991; Lolis and Petsko, 1990; Wierenga *et al.*, 1991). Modeling of the triosephosphate isomerase flexible loop 6 onto the analogous position of the tryptophan synthase α subunits (Fig. 3) suggests that closure of loop 6 may

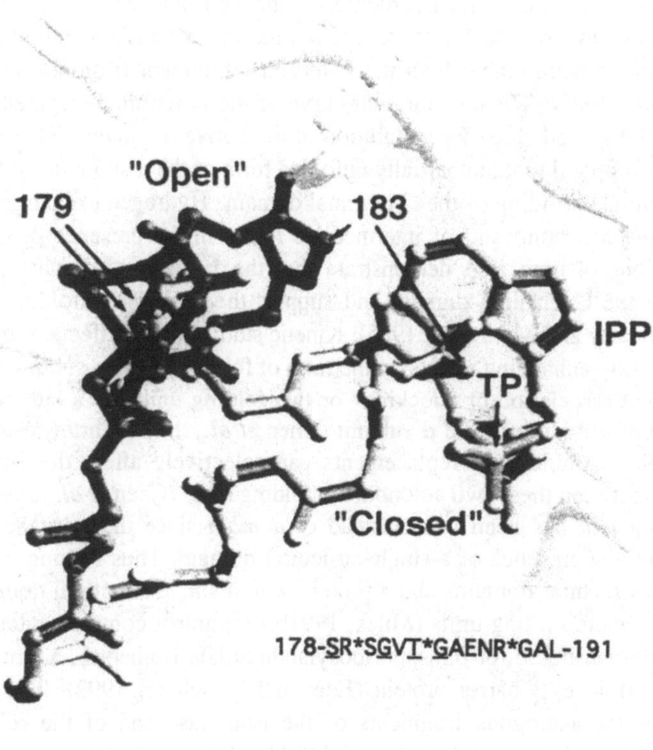

FIGURE 3. Model for the "open" and "closed" conformation of the flexible loop 6 of the α subunit based on structural homology to triosephosphate isomerase (see Section 2.1). The model shows the backbone atoms of loop 6 from triosephosphate isomerase in the "open" conformation and in the "closed" conformation, which is observed after binding of a substrate analogue (TP) (Davenport *et al.*, 1991; Wierenga *et al.*, 1991). The phosphate of indole-3-propanol phosphate (IPP) is modeled at the position of the phosphate in the triosephosphate isomerase substrate analogue (TP). In the closed position of loop 6, the residue that comes closest to the bound phosphate corresponds to position 183 in the α subunit sequence (see text). The sequence of the α subunit loop 6 (below) shows the sites of mutation at positions 179* and 183* and a site of proteolysis at position 188*. Conserved residues are underlined. (From Brzovic' *et al.*, 1993).

interact both with the bound substrate and with residues in loop 2 (Brzovic' *et al.*, 1992c, 1993). Roles of loop 6 in substrate binding, intersubunit communication, and metabolite channeling are discussed in Section 4.

2.2. Folding and Conformational Stability

Although the crystal structure of the $\alpha_2\beta_2$ complex shows that the α subunit has a single structural domain (Hyde *et al.*, 1988), urea or guanidine hydrochloride-induced unfolding of the α subunit is biphasic or stepwise (Matthews and Crisanti, 1981; Matthews *et al.*, 1983; Yutani *et al.*, 1979, 1980). A comparison of the guanidine hydrochloride-induced unfolding of the intact α subunit and of two tryptic fragments shows that the N-terminal fragment (residues 1–188) is more stable than the C-terminal fragment (residues 189–268) (Miles *et al.*, 1982). The denaturation curve of the N-terminal fragment corresponds to the second step of denaturation of the native α subunit. These results support the proposal that the partially unfolded form of the α subunit results from the preferential unfolding of the C-terminal domain. Hydrogen exchange studies of the native α subunit and of intermediate forms in the presence of different concentrations of urea also demonstrate that the N-terminal domain is more stable than the C-terminal domain and support the stepwise unfolding of two domains (Beasty and Matthews, 1985). Kinetic studies of the effects of mutation and of viscosity-enhancing agents on the rates of folding of the α subunit provide evidence that association or "docking" or two folding units are a late and rate-limiting step in folding of the α subunit (Chen *et al.*, 1992; Chrunyk and Matthews, 1990). Amino acid replacements can selectively affect the energy of interaction between these two autonomous folding units (Chen *et al.*, 1992). The term *folding unit* has been used instead of *domain* since the crystallographic results show the presence of a single structural domain. Thus folding units can differ from structural domains and a protein with a single structural domain can have two or more folding units (Miles, 1991b). Fragment complementation has also been demonstrated for phosphoribosylanthranilate isomerase, a structurally related eightfold α/β barrel protein (Eder and Kirschner, 1992). The similar behavior of the analogous fragments of the isomerase and of the α subunit supports the hypothesis that these two eightfold α/β barrel proteins have evolved from a common ancestor.

Recent characterization of the α subunit folding intermediate by differential scanning calorimetry, circular dichroism spectroscopy, and nuclear magnetic resonance demonstrates that the folding intermediate has very little tertiary structure, differs greatly from the native protein in conformation, and may correspond to a molten globule (K. Yutani, personal communication). Thus folding units may attain their final tertiary structures only after association.

The effects of single amino acid substitution on urea and guanidine-induced unfolding of the α subunit have been extensively investigated by Yutani's group (Yutani *et al.*, 1979, 1980, 1984b, 1987b) and by Matthews' group (Beasty *et al.*, 1986; Chen *et al.*, 1992; Hurle *et al.*, 1986; Matthews *et al.*, 1983; Tweedy *et al.*, 1990). These studies provide important information on factors that influence protein stability and folding. Studies of the thermal stability of the α subunit are described in Section 4.3.

2.3. Active Site and Mutagenesis Studies

Mutagenesis has played a key role in elucidating the structure and function of tryptophan synthase. Three general approaches have been used: (1) penicillin selection and reversion analysis; (2) site-directed mutagenesis; and (3) random mutagenesis. This section describes results from some of these studies and correlates them with the crystal structure of the α subunit in the $\alpha_2\beta_2$ complex (Hyde *et al.*, 1988).

Genetic and biochemical studies in the 1960s led to the identification of a set of missense amino acid substitutions that inactivate the tryptophan synthase α subunit from *Escherichia coli* (Allen and Yanofsky, 1963; Helinski and Yanofsky, 1963; Yanofsky *et al.*, 1961; Yanofsky and Horn, 1972). The positions of seven of eight of these substitutions (Phe-22, Glu-49, Tyr-175, Leu-177, Gly-211, Gly-34, and Ser-235) are located near a bound substrate analogue in the active site of the α subunit in the crystal structure of the $\alpha_2\beta_2$ complex from *S. typhimurium* (Hyde *et al.*, 1988). The eighth site of substitution, Thr-183, is in loop 6, a region of the structure that is highly disordered. Modeling of the structure of loop 6 on the basis of the structure of the analogous loop in triosephosphate isomerase (see Fig. 3) implies that Thr-183 may interact with the phosphate of the bound substrate analogue (Brzovic' *et al.*, 1993). Gly-213, which is changed in some second site revertants (Yanofsky, 1971), is also located close to the bound substrate analogue. Later mutational studies of Asp-60, another residue located close to the bound substrate analogue, provide evidence that Asp-60 serves a critical role in catalysis (Milton *et al.*, 1986; Nagata *et al.*, 1989). Mutations altering Asp-60 were not identified in the many *trpA* missense mutants isolated in the 1960s. However, there was one genetic class of these mutants, represented by mutant *trpA34*, in which the amino acid replacement responsible for α subunit inactivation was not identified. Recent DNA sequence analysis has shown that the *trpA34* mutation is in codon 60 and that the mutation results in replacement of Asp-60 by Asn (Shirvanee *et al.*, 1990). Two distinguishable partial revertants of the Asp-60→ Asn mutation have been recently characterized by sequence analysis (Yanofsky *et al.*, 1993). Two slow-growing partial revertants have second-site changes in which Tyr-102 is converted to Asp or Phe. A faster-growing partial revertant lacks three consecutive base pairs, resulting in the replacement of Ala-59 and Asn-60 of the mutant polypeptide by

Asp. Thus each of the second-site mutations affects a residue located in the vicinity of the active-site residue, which is altered by the primary mutation.

Interestingly, random saturation mutagenesis studies examining the importance of the N-terminal 121 amino acid residues in the α subunit find the most marked defects in catalysis by the α subunit at positions 22, 49, and 60, all residues described above (Lim *et al.*, 1991a, b; Milton *et al.*, 1986). These results provide additional support for the conclusion that the tryptophan synthase α subunit contains only a small number of absolutely essential residues. An examination of the enzymatic properties of 39 purified mutant α subunits reveals marked defects in α subunits with single amino acid replacements at other residues that are involved in substrate binding (Thr-63, Gln-65, Tyr-102, and Leu-107) and in subunit interaction (see Section 4.2) (Lim *et al.*, 1991a).

2.4. Reaction Mechanism

The mechanism of the α reaction, the reversible cleavage of indole-3-glycerol phosphate, has been investigated by steady-state and rapid kinetic techniques (Anderson *et al.*, 1991; Creighton, 1970; Faeder and Hammes, 1971; Kirschner *et al.*, 1992; Lane and Kirschner, 1991; Weischet and Kirschner, 1976a, b). The pH dependence of the α reaction shows that at least one protonated group with a pK between 7 and 8 is required for catalysis (Kirschner *et al.*, 1991). Mutagenesis studies designed to identify catalytic residues in the α site first focused on Glu-49, one of the sites of missense mutation that inactivate the α subunit (see Section 2.3). Because Glu-49 has an unusually high pK of about 7.5 (Yutani *et al.*, 1984a), Glu-49 is a potential proton acceptor at neutral pH. Studies of 19 mutants of the α subunit from *E. coli* at position 49 provide strong evidence that Glu-49 is essential for activity and may serve as a catalytic residue in the α reaction (Yutani *et al.*, 1987a). Additional studies of five of the mutant α subunits at position 49 demonstrate that the inactive proteins bind α subunit ligands and transmit ligand-dependent effects to the β subunit (Miles *et al.*, 1988). These results support a catalytic role for Glu-49. Subsequent crystallographic results show that Glu-49 is located in the interior of the α subunit near the binding site of a substrate analogue, indole-3-propanol phosphate (Hyde *et al.*, 1988).

These X-ray crystallographic studies also reveal that the carboxylate of Asp-60 is near the bound inhibitor and seems to pair with the indole nitrogen atom of the inhibitor. Substitution of Asp-60 by Asn (Milton *et al.*, 1986; Shirvanee *et al.*, 1990) or by Asn, Ala, or Tyr (Nagata *et al.*, 1989) results in loss of α subunit activity. These results support a catalytic role for Asp-60. Since the mutant form with the conservative substitution of Glu for Asp at position 60 has partial activity, Glu-60 may serve as an alternative catalytic base. This result is consistent with the occurrence of Glu in the homologous position in the α subunit

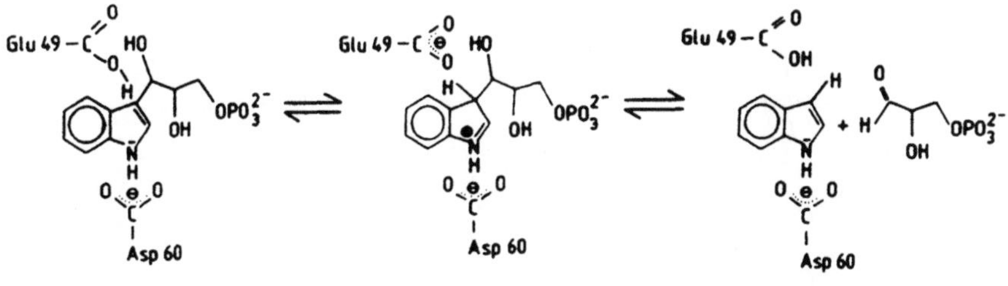

FIGURE 4. Proposed mechanism of the proton-transfer steps during cleavage and synthesis of indole-3-glycerol phosphate in the active site of the α subunit (see Section 2.3). (From Kirschner *et al.*, 1991. Reprinted with permission from American Chemical Society.)

from *Caulobacter crescentus* (Ross and Winkler, 1988). The proposed mechanism for proton transfer steps in the α reaction catalyzed by Glu-49 and Asp-60 (Fig. 4) is consistent with the pH dependence of the reaction (Kirschner *et al.*, 1991). Cleavage of the C3-C3′ carbon–carbon bond of indole-3-glycerol phosphate (I) is activated by tautomerization of the indole ring to yield the indolenine tautomer (II). Asp-60 could function by forming a strong hydrogen bond to polarize the indole nitrogen atom of indole-3-glycerol phosphate with concomitant protonation of the indole ring at C3′. The proton donor could be either an unidentified residue (Nagata *et al.*, 1989) or Glu-49 (Kirschner *et al.*, 1991). In the latter case, Glu-49 would perform two functions, first protonating the C-3 position of the indole moiety and subsequently accepting the hydroxyl proton of the glycerolphosphate moiety in the cleavage reaction to form III (Kirschner *et al.*, 1991).

3. STRUCTURE AND FUNCTION OF THE β SUBUNIT

3.1. Three-Dimensional Structure

The three-dimensional structure of the tryptophan synthase $\alpha_2\beta_2$ complex shows that the β subunit is composed of two structural domains, termed the N- and C-domains, that correspond approximately to the N- and C-terminal halves of the polypeptide chain (Fig. 5) (Hyde *et al.*, 1988). One stretch of the N-terminal sequence (residues 53-85) "crosses over" into the C-domain, forming the first two strands of a β-sheet at the center of the C-domain. The pyridoxal phosphate coenzyme is "sandwiched" between these two domains and is shown on the surface of each domain in Fig. 5. The central core regions of the two domains (see hatched regions in Fig. 5) have very similar folding topologies. Each core contains four parallel strands with three helices packed on the interior side of the

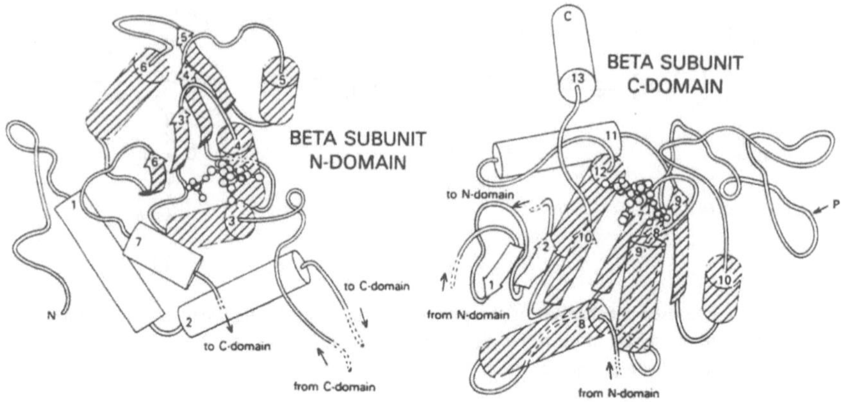

FIGURE 5. Folding patterns of the two domains of the β subunit. The "core" regions of N-domain and C-domain, which are shown with diagonal stripes, have very similar topologies and are nearly superimposable. The pyridoxal phosphate–Lys 87 Schiff base complex, shown in a ball-and-stick model, lies on the surface of each domain. A region susceptible to limited proteolysis is indicated by the arrow marked P. (From Miles *et al.*, 1991.)

sheet and a fourth helix packed on its exterior side. Because the cores of the two domains are nearly superimposable, the β subunit may have evolved via gene duplication and fusion.

3.2. Proteolysis, Folding, and Assembly

Mild proteolytic treatment of the β subunit results in cleavage of the polypeptide chain of the β subunit in a protease-sensitive region between residues 272 and 296 (see arrow marked "P" in Fig. 5) (Ahmed *et al.*, 1986a; Högberg-Raibaud and Goldberg, 1977; Kaufmann *et al.*, 1991). The two isolated tryptic fragments, termed F1 and F2, refold independently into globular polypeptides and reassemble to form the original native nicked β subunit (Högberg-Raibaud and Goldberg, 1977). The nicked β subunit, although enzymatically inactive, binds pyridoxal phosphate and forms a fluorescent enzyme–substrate intermediate with L-serine. The dual findings that the nicked enzyme does not associate with α subunits (Högberg-Raibaud and Goldberg, 1977) and that trypsin does not cleave the β subunit in the $\alpha_2\beta_2$ complex (Higgins *et al.*, 1979) led to the suggestion that the sites of trypsinolysis are in or near the interface between the α and β subunits (Higgins *et al.*, 1979). This suggestion is confirmed by the crystal structure of the $\alpha_2\beta_2$ complex (Hyde *et al.*, 1988) which shows that the sites of trypsinolysis are near the interface between the α and β subunits. Endoproteinase Glu-C cleaves the β subunit at a site (Glu-296) in the loop that is more distant from the interface between the α and β subunits. The nicked enzyme is enzymatically active and binds weakly to α subunits (Kaufmann *et al.*, 1991). Studies with conformational sensitive antibodies demonstrate that cleavage at Glu-296 prevents the conformational change that the native β subunit undergoes upon association with the α subunit (Linkens *et al.*, 1993).

The tryptophan synthase β subunit is an excellent model system for investigating protein folding and assembly. Studies by Michel Goldberg's group have exploited the proteolytic F1 and F2 fragments (Goldberg, 1985, 1991; Goldberg and Högberg-Raibaud, 1979; Högberg-Raibaud and Goldberg, 1977; Zakin *et al.*, 1980; Zetina and Goldberg, 1982) and monoclonal antibodies that recognize distinct epitopes on the F1 and F2 fragments (Blond and Goldberg, 1987; Blond-Elguindi and Goldberg, 1990; Friguet *et al.*, 1986, 1989; Goldberg *et al.*, 1990; Goldberg, 1991; Murry-Brelier and Goldberg, 1988). Refolding of the β subunit follows a series of folding events. More than half of the secondary structure is formed within 0.013 sec. An early transient folding intermediate ($t_{1/2}$ = 12 sec) is immunoreactive, binds a hydrophobic dye, and has the characteristics of a "molten globule" (Goldberg *et al.*, 1990). A subsequent, slower step results in an increase in the remaining far ultraviolet circular dichroism signal and acquisition of the near ultraviolet circular dichroism signal that reflects tight side-chain packing and disappearance of the molten globular form. The slowest steps may involve interdomain or intersubunit interactions in the β dimer.

A conformation-dependent monoclonal antibody has been used to investigate folding of nascent β chains on the ribosome (Federov *et al.*, 1992). The finding that this antibody binds to an N-terminal 11.5 kDa fragment either free or bound to the ribosome with an affinity close to that of the native β subunit suggests strongly that the polypeptide chain starts to fold during its biosynthesis and exhibits some local structural features of the native state.

3.3. Effects of Pyridoxal Phosphate on Conformation

Pyridoxal phosphate and certain pyridoxal phosphate analogues bind cooperatively to the *E. coli* apo β dimer and noncooperatively to the apo $\alpha_2\beta_2$ complex (Bartholmes *et al.*, 1976, 1980; Tschopp and Kirschner, 1980a, b). These observations suggest that the apo β subunit is a flexible molecule that can exist in several conformations and that one of these conformations is stabilized by binding pyridoxal phosphate. This hypothesis is supported by the occurrence of a slow isomerization reaction in the binding of pyridoxal phosphate analogs to the apo β subunit (Tschopp and Kirschner, 1980a). A difference in the conformation of the apo and holo forms of the β subunit is also supported by fluorescence studies (Lane, 1983) and by investigations with conformation-specific monoclonal antibodies (Djavadi-Ohaniance *et al.*, 1986). Furthermore, microcalorimetric studies demonstrate that binding of pyridoxal phosphate to the apo β subunit is accompanied by an unusually large negative change in heat capacity. This result indicates that the β subunit undergoes a major structural change or isomerization upon coenzyme binding that results in tightening of the protein structure (Wiesinger and Hinz, 1984b). Coenzyme binding to the apo $\alpha_2\beta_2$ complex is accompanied by a much smaller negative change in heat capacity. These results and others (Balk *et al.*, 1981) indicate that the α subunit is a positive effector for the apo β subunit and converts the apo β subunit from a low-affinity T state to a high-affinity R state. This model is supported by the finding that binding of α subunits to the apo β dimer also exhibits cooperativity (Bartholmes and Teuscher, 1979) (see also Section 4.1). Because the nicked β subunit binds pyridoxal phosphate noncooperatively, the nicked apo β subunit can be regarded as a stabilized form of the T state of the intact apo β subunit (Tschopp and Kirschner, 1980a). This conclusion is supported by studies with monoclonal antibodies (Djavadi-Ohaniance *et al.*, 1986). Calorimetric studies show that the association of the holo β subunit and the α subunit from *E. coli* is accompanied by a large negative enthalpy change, which is almost twice as large as that for complex formation with the apo β subunit (Wiesinger *et al.*, 1979). The results indicate energetic linkage between coenzyme binding and subunit association.

The luminescence properties of the single tryptophan 177 in the β subunit and of the pyridoxal phosphate coenzyme in the holoenzyme are useful probes of structural changes in the β subunit that result from pyridoxal phosphate binding and from association with the α subunit (Chaffotte and Goldberg, 1984; Lane, 1983). Recent studies have characterized some fluorescence and phosphorescence properties of the apo, holo, and reduced tryptophan synthase $\alpha_2\beta_2$ complex from *S. typhimurium* over a temperature range from 160 to 293 K (Strambini *et al.*, 1992a). The results show that tryptophan 177 is located in a compact and rigid environment. Because the lifetimes for phosphorescence decay are longer for the holo $\alpha_2\beta_2$ complex than for the apo $\alpha_2\beta_2$ complex, coenzyme binding results in a tightening of the N-domain in the β subunit. This result is consistent with results of calorimetric studies (Wiesinger and Hinz, 1984a) and of thermal inactivation studies (Chaffotte and Goldberg, 1983; Zetina and Goldberg, 1980). The nonexponential nature of the tryptophan phosphorescence decay provides direct evidence for structural heterogeneity in the bienzyme complex and supports the existence of two or more conformers that differ in the flexibility of the N-domain of the β subunit.

Kinetic studies of the irreversible thermal inactivation of the tryptophan synthase β subunit demonstrate that the β subunit is strongly stabilized by pyridoxal phosphate or by covalent attachment of the coenzyme by sodium borohydride reduction (Chaffotte and Goldberg, 1983; Zetina and Goldberg, 1980). Thermal inactivation studies of a hybrid β dimer containing one reduced protomer and one apo protomer show that the apo protomer is more stable within the hybrid dimer than within the apo dimer. Thus binding of one molecule of coenzyme modifies the stability of the entire dimeric enzyme (Zetina and Goldberg, 1980). Similar thermal inactivation experiments have been carried out to investigate the mechanism of 50% inactivation of the β subunit by β-chloro-L-alanine (Ahmed and Miles, 1992). The results show that the half-active enzyme exists as a mixture of homo- and heterodimers and provide evidence that two conformers of the β subunit exist in solution. Other aspects of the thermal unfolding of the β subunit are covered in Section 4.3.

Pyridoxal phosphate strongly influences the reversible high pressure-induced dissociation of the β subunit dimer to monomers (Seifert *et al.*, 1985; Silva *et al.*, 1986). The dissociation constants at atmospheric pressure and 25 °C are $\sim 1 \times 10^{-9}$ for apo β subunit, 3×10^{-10} for the holo β subunit, and 6×10^{-11} for the borohydride reduced β subunit (Silva *et al.*, 1986). Upon release of high pressure, the rates of dimer formation and the return of enzymatic activity, pyridoxal phosphate fluorescence and tryptophan fluorescence exhibit different time courses and hysteretic behavior. The results indicate that the separated monomers undergo "conformational drift," e.g., the progressive loss of free energy of association. The slow return of activity indicates that the isomerization of a "drifted" form must occur. Pyridoxal phosphate largely prevents the hysteretic behavior, indicating that cofactor binding stabilizes the protein in respect to the loss of subunit affinity upon dissociation (Silva *et al.*, 1986).

3.4. Active Site and Reaction Mechanism

Crystallographic studies reveal the location of pyridoxal phosphate and key residues close to it in the active center of the β subunit (Fig. 6) (Hyde *et al.*,

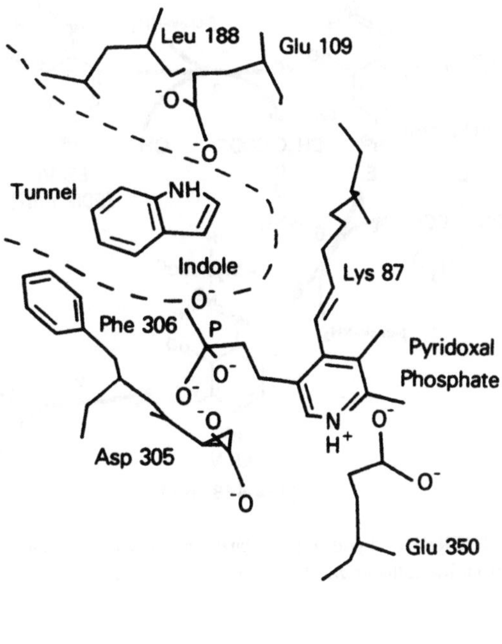

FIGURE 6. Active site of the β subunit based on X-ray crystallographic results showing the pyridoxal phosphate–Lys 87 Schiff base complex, the tunnel, the putative indole binding site, and some residues selected for site-directed mutagenesis. (From Miles *et al.*, 1991.)

1988). Pyridoxal phosphate forms a Schiff base with the ϵ-amino group of Lys-87. Although the substrate binding site has not yet been located by crystallographic studies using bound substrates or substrate analogues, with the wild-type $\alpha_2\beta_2$ complex, crystallographic data have been collected at high resolution (~2 Å) for complexes of L-serine (K. Parris, personal communication) and of L-tryptophan (C. C. Hyde, personal communication) with a mutant form of the $\alpha_2\beta_2$ complex in which Lys-87 of the β subunit was replaced by threonine (K87T) [see Lu *et al.*, (1993) and Section 3.5].

The mechanisms of pyridoxal phosphate-dependent reactions have been elucidated in model and enzymatic systems. The β subunit of tryptophan synthase belongs to a subclass of pyridoxal phosphate enzymes that catalyze β-elimination and β-replacement reactions (Miles, 1986). These reactions proceed though a series of pyridoxal phosphate intermediates shown in Fig. 7.

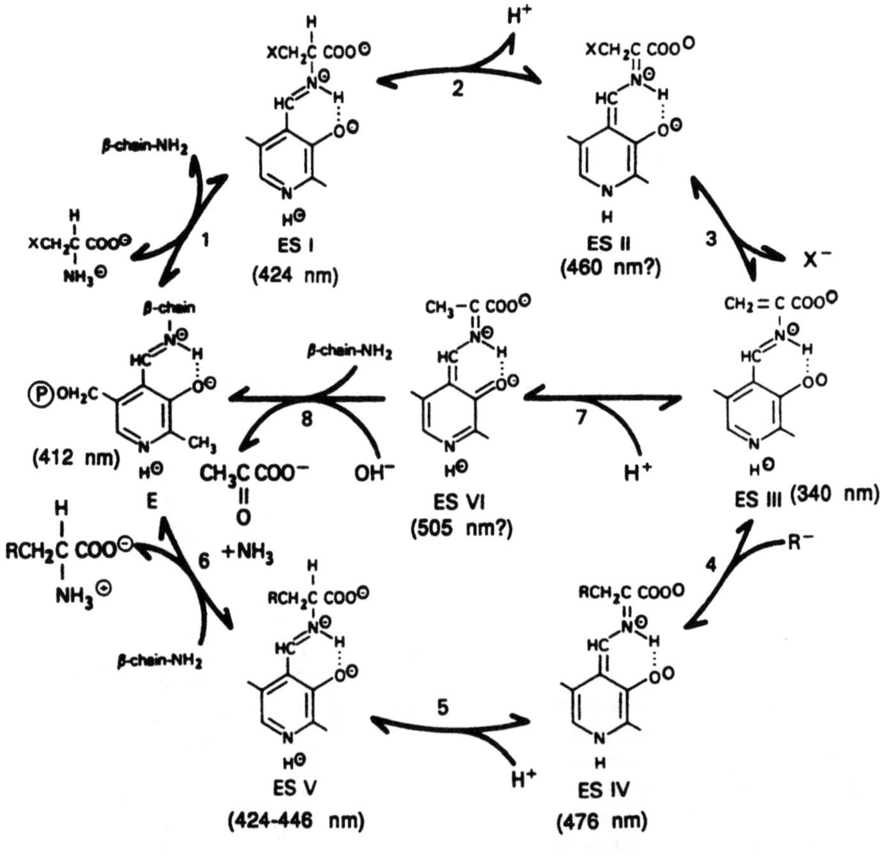

FIGURE 7. Pyridoxal phosphate intermediates in reactions catalyzed by the β subunit. (Adapted from Kayastha *et al.*, 1991, and Lu *et al.*, 1992).

Because each of these intermediates has characteristic spectroscopic properties, pyridoxal phosphate serves as a valuable spectrophotometric probe for detection of intermediates along the reaction pathway. Steady-state and rapid kinetic studies have detected many of these intermediates in pyridoxal phosphate-dependent reactions catalyzed by the β subunit and by the $\alpha_2\beta_2$ complex (Anderson *et al.*, 1991; Brzovic' *et al.*, 1992b, c; Drewe and Dunn, 1985, 1986; Dunn *et al.*, 1990, 1991; Faeder and Hammes, 1970, 1971; Goldberg and Baldwin, 1967; Goldberg *et al.*, 1968; Houben and Dunn, 1990; Houben *et al.*, 1989; Lane and Kirschner, 1981, 1983a, b, 1991; Miles *et al.*, 1968, 1986; Miles and McPhie, 1974; Mozzarelli *et al.*, 1991; Phillips *et al.*, 1984, 1985; Roy *et al.*, 1988a, b; York, 1972). Dunn and co-workers have used rapid-scanning stopped-flow UV-visible spectrophotometry to obtain rapid kinetic data over a wide range of wavelengths and to distinguish the formation and decay of several intermediates during the reaction time course. This method facilitates investigation of the effects of subunit association, allosteric ligands, and mutation on the rates of the individual steps and on the steady-state distribution of reaction intermediates.

The reaction of the wild-type $\alpha_2\beta_2$ complex (E in Fig. 7) with L-serine results in the rapid conversion of E to a transient intermediate (ES I), which has an absorbance maximum at 424 nm (Fig. 8) and an intense fluorescence emission at 520 nm. ES I decays to yield a complicated final spectrum with λ_{max} ~340 nm and with a broad envelope of absorbance extending out to approximately 525 nm. This spectrum represents the presence of a mixture of several intermediates; the peak at ~340 nm has been attributed to the Schiff base of aminoacrylate ES "I. The steady-state distribution of these L-serine intermediates is affect by pH, ' ligands of the α subunit, and by temperature (Mozzarelli *et al.*, 1989, 1991).

Addition of indole to the wild-type $\alpha_2\beta_2$ complex that has been previously ixed with L-serine ES III results in the rapid formation and disappearance of an termediate with λ_{max} at 476 nm. This intermediate is the quinonoid intermedi- e ES IV. Protonation of ES IV yields ES V, the Schiff base of L-tryptophan. ansimination with the ε-amino group of Lys-87 leads to release of the L-tryptophan product and regeneration of the internal aldimine E. Addition of the tryptophan product (or other substrate analogues, including L-histidine) to the β_2 complex E yields a complex equilibrium mixture of ES V and ES IV and her intermediates. The equilibrium distribution of ES V and ES IV formed om L-serine and indole, from L-tryptophan, or from other substrate analogues is ongly affected by ligands that bind to the α subunit (see Section 4.5). Addition strong nucleophiles (2-mercaptoethanol, indoline, aniline, hydroxylamine) to e aminoacrylate Schiff base intermediate ES III also leads to formation of the inonoid intermediates ES IV.

The reaction of indole with the Schiff base of aminoacrylate ES III requires tivation of indole because indole is a very weak nucleophile. A mechanism for tivation of indole in catalysis of the β reaction by tryptophan synthase is esented in Fig. 9. The formation of ES IVA is facilitated by a basic group, B_2,

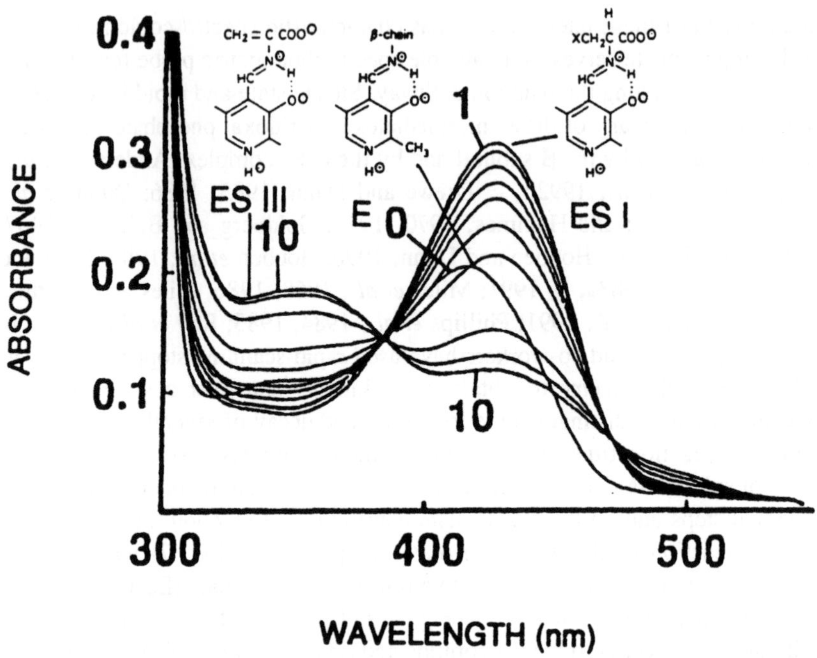

FIGURE 8. Rapid-sacanning stopped-flow spectra for the reaction of 10 μM wild-type $\alpha_2\beta_2$ complex from *S. typhimurium* with 40 mM L-serine at pH 7.8. The spectrum of the enzyme (E) (curve 0) is converted to that of the Schiff base intermediate ES I within 8.5 msec after mixing (curve 1). This intermediate decays to a complicated spectrum (curve 10) after 340 msec. The major peak at ~340 nm is ascribed to ES III. Adapted from Brzovic' *et al.* (1992c).

which abstracts the proton on N-1 of the heterocyclic ring with concomitant attack of indole on the Schiff base of aminoacrylate ES III. The tautomerization of ES IVA to yield ES IVB is carried out by abstraction of the proton at C_3 of indole by a basic group, B_1, concomitant with protonation of N-1 by the protonated B_2. Reprotonation of ES IVB by the protonated B_1 yields the Schiff base of L-tryptophan ES V. This mechanism is supported by studies with amino acid analogues, dihydro-L-alanine and oxindolyl-L-alanine (Miles *et al.*, 1986; Phillips *et al.*, 1984, 1985; Roy *et al.*, 1988b). These amino acids form quinonoid intermediates with pyridoxal phosphate that are analogues of the putative indolenine intermediate, ES IVA. The finding that dihydro-L-alanine and oxindolyl-L-alanine are strong competitive inhibitors of tryptophan synthase provides evidence for the intermediacy of ES IVA in the reaction of L-serine and indole.

Although the tryptophan synthase β subunit and $\alpha_2\beta_2$ complex catalyze pyridoxal phosphate-dependent reactions by the same basic mechanisms illustrated in Figs. 7 and 9, the two forms of the enzyme differ in reaction specificity,

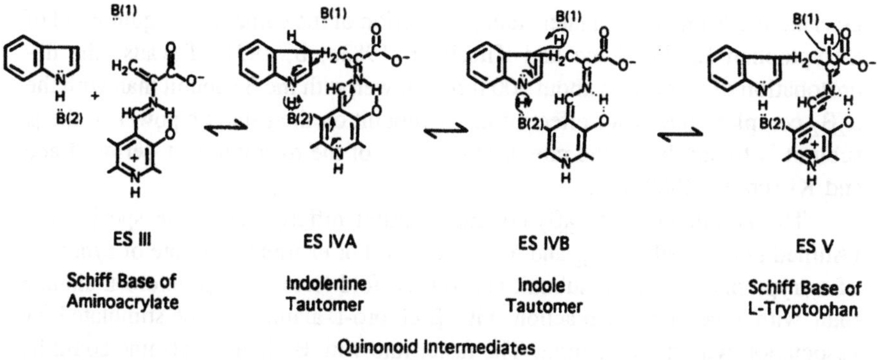

FIGURE 9. Mechanism for activation of indole in the β reaction (see also Fig. 7). The conversion of ES III to the indolenine tautomer of the quinonoid intermediate ES IVA is catalyzed by B_2, which abstracts the proton on N-1 of the heterocyclic ring concomitant with attack of indole on the aminoacrylate. The tautomerization of ES IVA to ES IVB is carried out by abstraction of the proton at C_3 of indole by a basic group, B_1, concomitant with protonation of N-1 by B_2H. Modified from Miles (1991a).

in substrate specificity, and in kinetic properties. The principal difference in reaction specificity is in the relative ability of the two forms of the enzyme to catalyze β-elimination and β-replacement reactions with L-serine. β-Elimination and β-replacement reactions involve hydration and substitution, respectively, of the common intermediate, the Schiff base of aminoacrylate (ES III in Fig. 7). Whereas the β subunit has a significant activity in the β-elimination reaction with L-serine, the $\alpha_2\beta_2$ complex has very low activity in this reaction and has been termed *β-replacement specific*. The decreased rate of the β-elimination reaction with L-serine in the presence of the α subunit could result from a decreased rate of any of the steps leading to (ES III) or from a very low rate of hydrolysis of ES III (Miles, 1986; Xu and Abeles, 1993). Recent studies conclude that the β-replacement specificity of the tryptophan synthase $\alpha_2\beta_2$ complex is due, at least in part, to slow hydrolysis of the aminoacrylate intermediate (Xu and Abeles, 1993). This conclusion is consistent with kinetic data that demonstrate that the $\alpha_2\beta_2$ complex forms the Schiff base of aminoacrylate ES III rapidly in the absence of indole (Drewe and Dunn, 1985) and that the rate of formation of ES III is more rapid than the rate of conversion of L-serine to pyruvate and ammonia (Ahmed *et al.*, 1986b).

The β subunit has a 30-fold lower k_{cat} than $\alpha_2\beta_2$ complex (Lane and Kirschner, 1983b). Analyses of primary isotope effects on the rates of reaction of the two forms of the enzyme with L-serine show that removal of the α-proton of L-serine is rate limiting with the β subunit but not with the $\alpha_2\beta_2$ complex (Lane and Kirschner, 1983b; Miles and McPhie, 1974). Association of the α subunit with the β subunit shifts the rate-limiting step from removal of the α-proton of

L-serine in ES I to a much later step, protonation of the carbanion or quinonoid of L-tryptophan ES IV (Lane and Kirschner, 1983a, b, 1991). Because the deprotonation of L-serine is about 500 times slower with the β subunit than with the $\alpha_2\beta_2$ complex, the major effect of the α subunit on the catalytic power of the β subunit is to accelerate the rate of abstraction of the α-proton of L-serine (Lane and Kirschner, 1983b).

The β subunit and $\alpha_2\beta_2$ complex exhibit different substrate specificities (Ahmed *et al.*, 1991; Yang and Miles, 1992). For example, the rate of synthesis of L-tryptophan by the β subunit is about 20-fold faster with β-chloro-L-alanine than with L-serine; the reaction with β-chloro-L-alanine is not stimulated by association with the α subunit. The faster rate with β-chloro-L-alanine could be due to the enhancement of the acidity of the α-proton by the chloride group. Alternatively, β-elimination of the hydroxyl group of L-serine may be the step catalyzed by the β subunit that is activated by the α subunit. β-Elimination of hydroxide requires protonation of the leaving group by a catalytic residue, whereas β-elimination of chloride does not (Borcsok and Abeles, 1982).

3.5. Protein Engineering

The crystal structure of the tryptophan synthase $\alpha_2\beta_2$ complex provides a useful framework for exploring structure–function relationships (Hyde *et al.*, 1988). This section describes investigations designed to probe the functional roles of residues located near pyridoxal phosphate in the active center of the β subunit (see Fig. 6).

The roles of Lys-87 in the active site of the β subunit have been evaluated by investigations of a mutant engineered by site-directed mutagenesis (Lu *et al.*, 1993; Miles *et al.*, 1989). The first role of Lys-87 in the wild-type enzyme is to form a Schiff base with pyridoxal phosphate (see Fig. 6 and E in Fig. 7). Substitution of Lys-87 by threonine yields a mutant form of the $\alpha_2\beta_2$ complex (K87T) that binds pyridoxal phosphate tightly as the free aldehyde. Thus Lys-87 is not essential for pyridoxal phosphate binding. A second role of Lys-87 is to undergo a transimination or transaldimination reaction with the incoming substrate or product (step 1 or 6 in Fig. 7). This reaction results in release of the unprotonated ε-amino group and formation of the substrate or product Schiff Base (ES I or ES V). The transimination reaction proceeds through a geminal diamine intermediate E(GD) in Fig. 10). In contrast, the K87T $\alpha_2\beta_2$ complex must form ES I directly from the amino acid and the aldehyde of the enzyme-bound pyridoxal phosphate (Fig. 10). the K87T $\alpha_2\beta_2$ complex forms Schiff base intermediates with L-serine (Fig. 11) and with other amino acids at extremely low rates. These results suggest that Lys-87 plays a critical but not essential role in facilitating transimination reactions. A third putative role for Lys-87 is to serve as a catalytic residue. The complete absence of catalytic activity of the

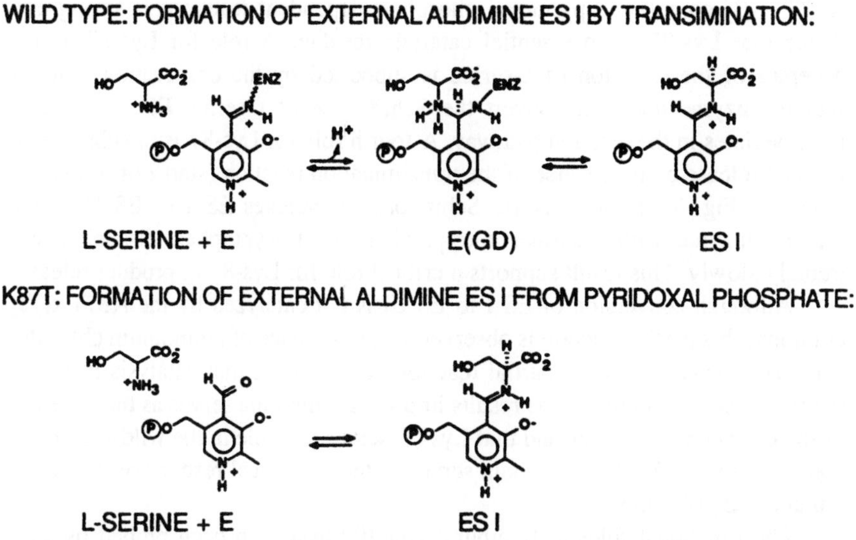

FIGURE 10. Formation of the external aldimine ES I by the wild-type and K87T $\alpha_2\beta_2$ complexes. The reaction of L-serine with the wild-type $\alpha_2\beta_2$ complex E proceeds by a transimination reaction through a geminal diamine intermediate E(GD). In the K87T $\alpha_2\beta_2$ complex, L-serine reacts directly with the free aldehyde of pyridoxal phosphate.

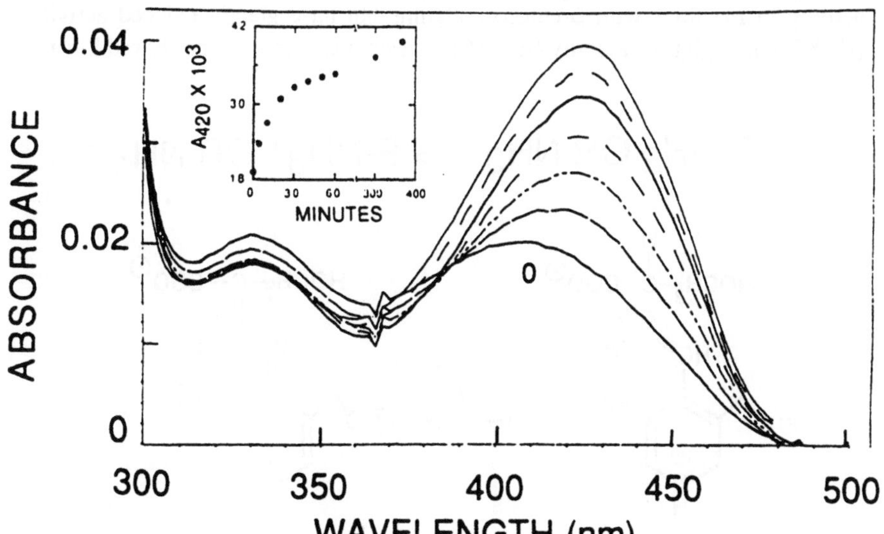

FIGURE 11. Reaction of the K87T $\alpha_2\beta_2$ complex with L-serine. Absorption spectra of the K87T $\alpha_2\beta_2$ complex (10 μM in 50 mM sodium bicine buffer, pH 7.8) at 0, 4, 10, 20, 60, 120, and 300 min after addition of 1 mM L-serine. The inset shows the time course of the absorbance at 420 nm.

K87T $\alpha_2\beta_2$ complex in pyridoxal phosphate-dependent reactions is strong evidence that Lys-87 is an essential catalytic residue. A role for Lys-87 as the acceptor of the α-proton of L-serine is supported by the observation that the mutant enzyme does not convert the Schiff base of L-serine ES I to further intermediates in the reaction pathway. A fourth role for Lys-87 is in substrate or product release by the reverse of the transimination reaction (step 2 or 6, respectively, in Fig. 7). Stoichiometric Schiff base complexes ES I or ES V of the mutant enzyme with L-serine, L-tryptophan, or D-tryptophan dissociate extremely slowly. This result supports a critical role for Lys-87 in product release.

Although conversion of ES I to ES III is not catalyzed by the K87T $\alpha_2\beta_2$ complex, this partial reaction is observed in the presence of ammonium chloride. The pH dependence of the partial reaction demonstrates that catalysis is due to the free base of ammonia. The results imply that ammonia serves as the acceptor of the α-proton of L-serine and that Lys-87 serves this role in the wild-type $\alpha_2\beta_2$ complex (Fig. 12). Thus ammonia serves in the "chemical rescue" of the K87T mutant $\alpha_2\beta_2$ complex.

The functional roles of β subunit Glu-109 have also been probed by site-directed mutagenesis (Ahmed *et al.*, 1991; Anderson *et al.*, 1991; Brzovic' *et al.*, 1992a; Kayastha and Miles, 1991). The γ-carboxyl of Glu-109 is located near the putative indole binding site (Fig. 6). Replacing Glu-109 by aspartate (E109D) or by alanine (E109A) greatly alters the substrate specificity and reaction specificity of the $\alpha_2\beta_2$ complex. The mutant enzymes exhibit catalytic activity in reactions with β-chloro-L-alanine but have greatly reduced activity (E109D) or negligible activity (E109A) in reactions with L-serine. Rapid kinetic

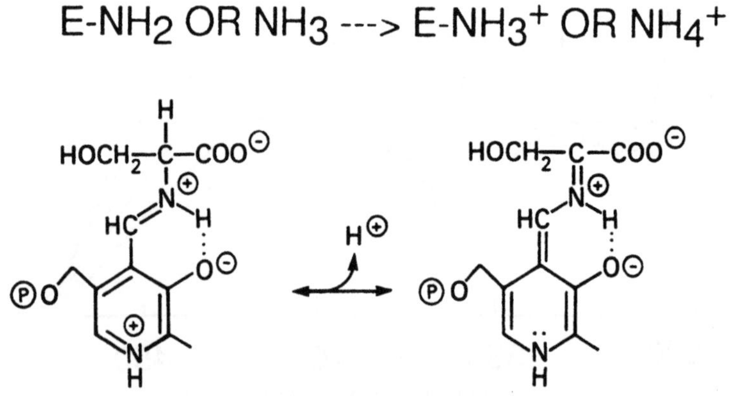

FIGURE 12. Chemical rescue of the K87T $\alpha_2\beta_2$ complex by ammonia. Results described in the text and Lu *et al.* (1993) provide evidence that the ε-amino group of Lys-87 serves as the acceptor of the α-proton of L-serine. In the K87T $\alpha_2\beta_2$ complex, ammonia may partially replace the deleted ε-amino group of Lys-87 by accepting the α-proton of L-serine.

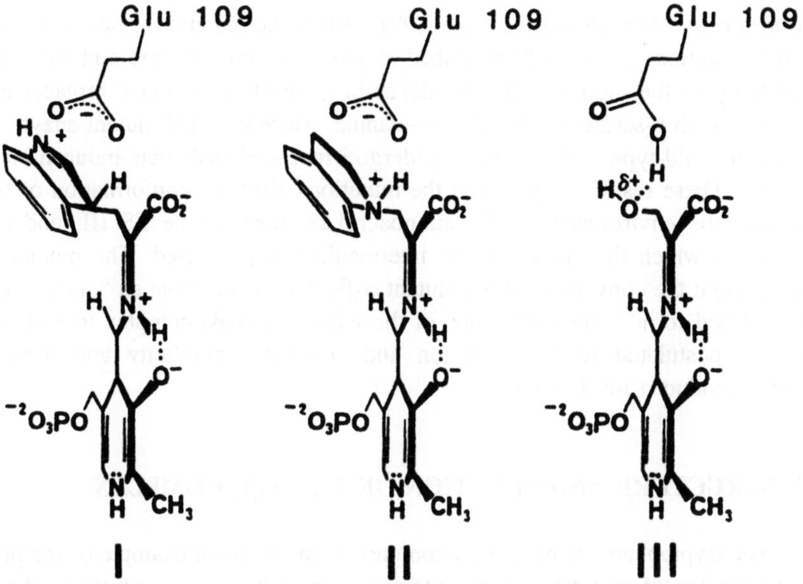

FIGURE 13. Possible catalytic roles of Glu-109 in the active site of the β subunit. (I) Glu-109 may play an important role in the activation of indole by promoting the formation of the iminium or indolenine tautomer of indole (see ES IVA in Fig. 9) or, as depicted here, in stabilization of the resulting positive charge on the indole ring. (II) Glu-109 may interact with the indoline quinonoid to stabilize the positive charge on the indolinium nitrogen atom. (III) Glu-109 may interact with the β-hydroxyl of L-serine, facilitating the β-elimination of this substituent. From Brzovic' *et al.* (1992a).

studies demonstrate that the E109D mutation does not seriously affect the rate of the pre–steady-state reaction with L-serine to form the aminoacrylate intermediates (ES III in Fig. 7) (Anderson *et al.*, 1991; Brzovic' *et al.*, 1992a). However, the E109D mutation drastically reduces the rate of reaction of indole with ES III to form the quinonoid intermediate ES IV and the produce L-tryptophan (Anderson *et al.*, 1991; Brzovic' *et al.*, 1992a). In contrast, substitution of Asp for Glu-109 increases the rate of reaction of indoline with ES III and stabilizes the quinonoid ES IV formed in this reaction (Brzovic' *et al.*, 1992a). Thus, the E109D mutation affects the covalent bond, forming addition reactions and the nucleophile specificity of the $\alpha_2\beta_2$ complex. Possible roles of Glu-109 in the β subunit are depicted in Fig. 13. Glu-109 may play a role in the activation of indole (I), may stabilize the indoline quinonoid (II), or may facilitate β-elimination of the hydroxyl group of L-serine (III).

Site-directed mutagenesis has been used to replace several other residues (Glu-350, Asp-305, Phe-306, and Leu-188) that are located in the active site of the β subunit close to pyridoxal phosphate or in the wall of the tunnel (Fig. 6)

(Ahmed *et al.*, 1991; Kayastha *et al.*, 1991). All of the mutations studied, as well as the mutations of Glu-109 described above, alter the reaction and substrate specificity of the mutant $\alpha_2\beta_2$ complexes in β-elimination and β-replacement reactions with L-serine and β-chloro-L-alanine. Moreover, the mutant enzymes, unlike the wild-type $\alpha_2\beta_2$ complex, undergo irreversible substrate-induced inactivation. These results suggest that the mutations alter the conformation of the enzyme, the environment of the aminoacrylate intermediate ES III, and the manner in which the aminoacrylate intermediate is processed. The mutations may prevent the conversion of the mutant $\alpha_2\beta_2$ complexes from an "open form" to a "closed form." The conversion of the wild-type $\alpha_2\beta_2$ complex to a closed form is postulated to alter reaction and substrate specificity and prevent substrate-induced inactivation.

4. STRUCTURE AND FUNCTION OF THE $\alpha_2\beta_2$ COMPLEX

The tryptophan synthase $\alpha_2\beta_2$ complex is an excellent example of the proposed benefits of enzyme–enzyme interaction in multienzyme complexes (Lane and Kirschner, 1991). These benefits include mutual activation and coordination of activities of the individual enzymes and metabolite channeling. To understand these features of tryptophan synthase, we must learn how formation of the $\alpha_2\beta_2$ complex alters the structure and function of the individual α and β subunits. Subunit association has striking effects on the reaction kinetics. The effects of subunit association are summarized here and described in more detail in the indicated sections. (1) The turnover numbers of the individual subunits in the α and β reactions are increased by 1 to 2 orders of magnitude. The rates of individual steps in the reactions catalyzed by the α and β subunits are also altered. These alterations in rates result from conformational changes induced by subunit association (Sections 4.1 and 4.3). (2) The reaction specificity of the β subunit is altered (Section 3.4). (3) Binding of a ligand to the active site of either subunit alters the reaction kinetics of the partner subunit. Since the active sites of the α and β subunits are about ~25 Å apart in the $\alpha_2\beta_2$ complex, the ligand-mediated interactions between the α and β sites are examples of allosteric interactions between heterologous sites (Sections 4.5 and 4.6). These reciprocal "site–site" interactions result from ligand-induced conformational changes that are transmitted from one subunit to the other. (4) The coordination of the α and β reactions results in the efficient catalysis of the overall αβ reaction (Section 4.7); indole is a channeled intermediate in the αβ reaction (Section 4.4).

4.1. Subunit Association

The tryptophan synthase $\alpha_2\beta_2$ complex from *S. typhimurium* has an extended αββα subunit arrangement (Hyde *et al.*, 1988) (see Section 1 and Fig. 1).

The contact region between the α and β subunits is about 1100 Å². Studies using ultracentrifugation and sucrose gradient centrifugation of tryptophan synthase from *E. coli* have characterized both the fully associated $\alpha_2\beta_2$ complex and a stable complex with an $\alpha\beta_2$ subunit composition (Adachi *et al.*, 1974; Creighton and Yanofsky, 1966; Goldberg *et al.*, 1966). The stability of the $\alpha\beta_2$ complex arises from negative cooperativity of binding of the α subunit to the β subunit dimer (Lane *et al.*, 1984). The holo $\alpha_2\beta_2$ complex contains two molecules of the cofactor, pyridoxal phosphate. Removal of pyridoxal phosphate from the holo $\alpha_2\beta_2$ complex yields the apo $\alpha_2\beta_2$ complex (Miles and Moriguchi, 1977). Pyridoxal phosphate, L-serine, and indole-3-glycerol phosphate increase the apparent association constants for the *E. coli* tryptophan synthase α and β subunits from 4×10^6/M to up to 3×10^9/M (Creighton and Yanofsky, 1966). An association constant of 10^{10}/M has been calculated from measurements of dissociation of the *E. coli* holo $\alpha_2\beta_2$ complex by iodide (Lane, 1983).

Kinetic studies demonstrate that subunit assembly involves slow synchronous conformational changes or isomerizations of both α and β protomers; these isomerization are mainly responsible for enhancing the rate of substrate turnover and for stabilizing the $\alpha_2\beta_2$ complex (Lane *et al.*, 1984). Ligands that bind to the α subunit (indole-3-propanol phosphate) or to the β subunit (L-serine) stabilize the complex by decreasing the rate of the reverse isomerization. Evidence that the β subunit undergoes a conformational change on association with the α subunit is also provided by studies using calorimetry (Wiesinger *et al.*, 1979), small angle X-ray scattering (Wilhelm *et al.*, 1982), fluorescence and hydrodynamic measurements (Lane, 1983; Lane and Kirschner, 1983c; Lane *et al.*, 1984), and neutron scattering (Ibel *et al.*, 1985). The results are consistent with conversion of the β subunit from an elongated into a more compact form upon association with the α subunit.

Immunochemical studies with monoclonal antibodies directed against the β subunit (Murry-Brelier and Goldberg, 1990) show that the β subunit dimer can bind only one Fab 93-6 while the $\alpha_2\beta_2$ complex can bind two such antibodies. The result implies that a major conformational rearrangement occurs upon subunit association. In contrast, the $\alpha_2\beta_2$ complex formed from β subunit nicked at Glu-296 can bind only one Fab 93-6 (Linkens *et al.*, 1993). This result implies that β subunit nicked at Glu-296 does not undergo conformational rearrangement upon association with the α subunit.

4.2. Effects of Mutation on Subunit Association

Several different mutations in the α and β subunits alter subunit association or intersubunit communication. A missense mutation in the β subunit, termed *trp*B8, results from the conversion of glycine 281 to arginine (G281R) (Zhao and Somerville, 1992, 1993). This mutation is located at a sharp turn in the trypsin-sensitive "hinge" region in the β subunit (see Section 3.1 and Fig. 5) in a position near the contact surface with the α subunit and near the indole tunnel. The G281R mutation reduces activity and substrate affinity and weakens association with the α subunit. The activity of the mutant β subunit is dramatically stimulated by a high concentration of ammonium ion or by a large excess of α subunit. Polyethylene glycol, a volume excluder, enhances association of the mutant β subunit with the α subunit. These studies are compatible with the proposal that the open versus closed conformational switch within the wild-type β subunit is crucial for catalytic efficiency and imply that the hinge region plays a critical role in the conformational switch. These conclusions are consistent with results of studies of proteolysis of the hinge described in Section 3.2.

Single amino acid replacement of α subunit proline residues 57 or 132 in the contact region with the β subunit also alters subunit association and mutual activation (Ogasahara *et al.*, 1992). Conversion of the *E. coli* α subunit proline 57 or proline 132 to alanine or glycine greatly reduces activities in the α and β reactions but has much smaller effects on activity in the overall αβ reaction. To investigate the possibility that the presence of the cosubstrate, indole-3-glycerol phosphate, contributes to the greater activity of the mutant complexes in the αβ reaction, the effects of indole-3-glycerol phosphate analogues (indole-3-propanol phosphate and α-glycerol 3-phosphate) were determined on the activity of the mutant $\alpha_2\beta_2$ complex in the β reaction (Table I). Whereas these α specific ligands inhibit the activity of the wild-type $\alpha_2\beta_2$ complex in the β reaction, they increase the activity of the $\alpha_2\beta_2$ complex containing P57A, P132A, or P132G α subunits. The activation by the analogues and apparent repair of the effects of mutation are explained by a decrease in the affinity of these mutant α subunits for the β subunits and by an increase in the affinity in the combined presence of ligands of both the α subunit and the β subunit. Isothermal calorimetric titrations show that both the affinity and the exothermic association enthalpy are greatly reduced in the mutant α subunit (P132G) although the stoichiometry of associa-

Table I
Effects of α Specific Ligands on the Rates of the Wild-Type and Mutant α Subunits in the β Reaction[a]

α Subunit	Units/mg −GP (1)	Units/mg +GP (2)	(2)/(1) × 100 (%)	Percent[b] +GP
Wild type	2890	530	18	[100]
P57A	1470	1380	94	260
P132A	491	550	110	100
P132G	347	472	136	90

[a] α Subunits from *E. coli* were assayed in the absence or presence of 80 mM DL-α-glycerol 3-phosphate (GP) and in the presence of an approximately fivefold molar excess of β chain from *E. coli*.
[b] Percent of the activity of the wild type α subunit in the presence of GP.

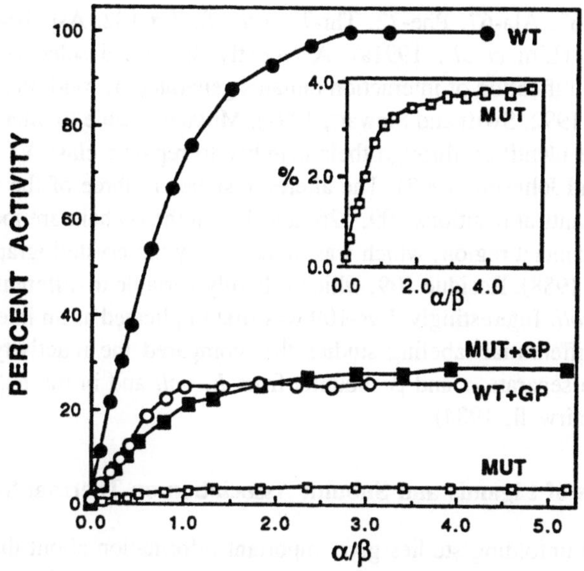

FIGURE 14. Effects of amino acid substitution of β subunit Lys-167 on interaction with the wild-type α subunit and on activity. The activities of the wild-type β subunit and of the K167T β subunit were measured in the β reaction in the presence of various amounts of the wild-type α subunit and in the presence or absence of 80 mM DL-α-glycerol 3-phosphate (GP). (From Yang *et al.*, 1993.)

tion is unchanged. Thus proline 132 plays a critical role in subunit interaction and in mutual activation.

An ion pair between Asp-56 in the α subunit and Lys-167 in the β subunit plays an important role in intersubunit communication. Replacement of β subunit Lys-167 by threonine results in a 25-fold decrease in the rate of synthesis of L-tryptophan from L-serine and indole but does not interfere with subunit association (Yang and Miles, 1993) (Fig. 14). Ligands that bind to the α subunit (indole-3-glycerol phosphate, indole-3-propanol phosphate, α-glycerol 3-phosphate, or potassium phosphate) largely restore the activity of the mutant $\alpha_2\beta_2$ complex in the β reaction (Fig. 14). The striking allosteric effects of α specific ligands may result from ligand-induced conformational changes in the α subunit that are transmitted to the β subunit and repair the mutational defect in the β subunit. A mutant form of the α subunit at Asp-56 (D56G) exhibits very low activity in the β reaction in the presence of excess β subunit and is also activated by addition of α-glycerol 3-phosphate (J. K. Hardman, personal communication).

Several mutant forms of the α subunit from *E. coli* engineered by use of a random single base pair saturation procedure (Milton *et al.*, 1986) appear to be involved in subunit interaction. These mutations are at Gly-51, Gly-56, Pro-53,

Asp-60, Pro-62, Ala-67, Phe-72, Thr-77, Pro-78, Tyr-102, Asp-104, Leu-105, and Asn-108 (Lim *et al.*, 1991a). A recently developed selection technique facilitates identification of interaction mutants generated by random mutagenesis (Swift *et al.*, 1992; Swift and Stewart, 1991). Mutants unable to interact with the β subunit are identified through their inability to repair a class VI β mutation (Crawford and Johnson, 1963). The altered residues in three of the interaction-deficient mutants at positions 109, 126, and 128 in the α chain are located in the intersubunit contact region, which was identified by the crystallographic studies (Hyde *et al.*, 1988). Residue 109, which is highly variable in different species, is lysine in *E. coli*. Interestingly, Lys-109 was first implicated as an interaction site residue by differential labeling studies that compared the reactivity of amino groups in the separate α and β subunits from *E. coli* and in the $\alpha_2\beta_2$ complex (Miles and Fairwell, 1984).

4.3. Effects of Ligands and Subunit Association on Thermal Stability

Thermal unfolding studies give important information about the effects of ligands and subunit association on the structure and conformational stability of tryptophan synthase. The thermal stability of the α subunit has been investigated by differential scanning calorimetry (DSC) (Lim *et al.*, 1992; Matthews *et al.*, 1980; Remeta *et al.*, 1992, 1993; Sugisaki *et al.*, 1990; Yutani *et al.*, 1982), circular dichroism (CD) (Kanzaki *et al.*, 1991; Ogasahara *et al.*, 1984; Yutani *et al.*, 1984b), and thermal inactivation studies (Lim *et al.*, 1992; Ruvinov and Miles, 1994) (Table I). The denaturation temperatures (T_m) of the α subunit depend on the bacterial source and are also altered by pH and single amino acid replacements. DSC studies show that the T_m values for the α subunit from *E. coli* are about 8 °C higher than the T_m values for the α subunit from *S. typhimurium* over a pH range of 7.5–9.5; the T_m for the protein from either source is about 4 °C higher at pH 8 than at pH 9 (Sugisaki *et al.*, 1990). However, these studies also demonstrate that the heat capacity change of denaturation is the same for α subunits from the *E. coli* and *S. typhimurium* enzymes and for an interspecies hybrid (ΔC_p = 0.15 cal•g^{-1}•K^{-1}). The denaturation calorimetric enthalpies (ΔH_{cal}) for the three proteins as a function of T_m lie on one line (Fig. 3) (Sugisaki *et al.*, 1990). This observation suggests that the net exposures of apolar and polar amino acid residues during thermal unfolding are similar. The T_m for the irreversible thermal inactivation of the α subunit from *S. typhimurium* (Table II: T_m = 50 °C at pH 7.8) (Ruvinov and Miles, 1994) is similar to T_m values obtained by DSC or CD measurements for the reversible thermal unfolding of the α subunit from *S. typhimurium* (Remeta *et al.*, in preparation; Sugisaki *et al.*, 1990; Yutani *et al.*, 1984b). Addition of an α subunit-specific ligand (α-glycerol 3-phosphate) results in a very small (~2 °C) stabilization of the α subunit (Table II).

Because thermal unfolding of the β subunit is irreversible, fewer studies

Table II
Thermal Inactivation of Tryptophan Synthase: Effects of Ligands and of Subunit Association[a]

Enzyme	β-Specific ligand	α-Specific ligand	α Subunit ~T_m, °C (1)	β Subunit ~T_m, °C (2)	(2) − (1) °C
α	None	None	50		
	None	GP	52		
Apo β	None	None		52	
Holo β	None	None		77	
	L-Ser	None		71	
	L-Trp	None		74	
	D-Trp	None		73	
		GP		75	
Holo $\alpha_2\beta_2$	None	None	54	78	24
	None	GP	57	73	16
	L-Ser	None	62	64	2
	L-Trp	None	56	72	16
	D-Trp	None	58	71	13
	L-Ser	GP	66	70	4
	L-Trp	GP	63	73	10
	D-Trp	GP	54	75	21

[a]Solutions of *S. typhimurium* tryptophan synthase α and apo or holo β subunits or of holo $\alpha_2\beta_2$ complex (1–2 mg/ml in 0.05 M sodium bicine buffer, pH 7.8, containing 1 mM EDTA, and 0.05 mM pyridoxal phosphate) were heated at each temperature for 5 min in the absence or presence of the indicated ligands: 80 mM DL-α-glycerol 3-phosphate (GP), 50 mM L-serine (L-Ser), and 1 mM D-tryptophan or L-tryptophan (D-Trp or L-Trp). Samples were chilled to 4 °C and centrifuged after 30 min. Aliquots of supernatant solutions were assayed in the β reaction in the presence of excess partner subunit. T_m values were estimated from plots of percent activity versus time (Ruvinov and Miles, 1994).

have been reported for the thermal stability of the β subunit than of the α subunit. Addition of the cofactor, pyridoxal phosphate, results in a large (~25 °C) stabilization of the *S. typhimurium* β subunit (see Table II) (Ruvinov and Miles, 1994). Pyridoxal phosphate has been found previously to stabilize the β subunit from *E. coli* from thermal inactivation (see Section 3.3). Addition of amino acids that form Schiff base intermediates with the holo β subunit partially destabilize the holo β subunit (Table II). The T_m values for thermal inactivation of the apo- and holo β subunit in Table II are similar to the T_m values of 57 and 80 °C, respectively, measured by DSC for unfolding these species (Remeta *et al.*, 1993, in preparation). Moreover, the binding of the cofactor pyridoxal phosphate increases the linkage (cooperativity) between unfolding domains in the β subunit (Remeta *et al.*, 1993, in preparation).

Thermal inactivation studies have been performed also on the tryptophan synthase $\alpha_2\beta_2$ complex from *S. typhimurium* (Ruvinov and Miles, 1994). Although the T_m values for thermal inactivation of the individual α and β subunits in the $\alpha_2\beta_2$ complex are similar to those of the isolated subunits, the α subunit is slightly stabilized in the complex (Table II). Ligands that promote subunit association (L-serine, L-tryptophan, or D-tryptophan alone or in combination with α-glycerol 3-phosphate) raise the inactivation temperature of the α subunit in the $\alpha_2\beta_2$ complex (Table II). In contrast, these ligands decrease the inactivation temperature of the β subunit alone in the $\alpha_2\beta_2$ complex (Table II). Thus ligands that strengthen subunit interactions lead to pronounced stabilization of the α subunit but not of the β subunit in the $\alpha_2\beta_2$ complex. Addition of 50 mM L-serine results in a larger decrease in the T_m of the β subunit in the $\alpha_2\beta_2$ complex (78 °C → 64 °C) than in the T_m of the β subunit alone (77 °C → 71 °C). Although the mechanism of the L-serine-induced destabilization of the β subunit is unclear, the altered stability of the β subunit in the $\alpha_2\beta_2$ complex provides evidence that the α and β subunits do not fully dissociate during thermal inactivation of the $\alpha_2\beta_2$ complex in the presence of L-serine.

The thermally induced unfolding of the tryptophan synthase $\alpha_2\beta_2$ complex has been studied also by DSC and UV spectroscopy (Remeta *et al.*, in preparation). DSC profiles of the $\alpha_2\beta_2$ complex in the presence of 0.2 mM pyridoxal phosphate and 10 mM K-PO_4 (pH 8.0) show two well-resolved endotherms centered at ~ 52 and ~ 81 °C. The low- and high-temperature endotherms approximate the unfolding transitions of isolated α chains and holo β subunit dimer, respectively. However, analysis of the DSC data for holo $\alpha_2\beta_2$ complex indicates that the unfolding pathway is more complicated than for the separated α and holo β subunits. The data are consistent with a model of sequential two-state transitions (without dissociation of the tetramer), where ΔH_{cal} values are expressed in kcal(mole of $\alpha_2\beta_2$ complex)$^{-1}$. The low-temperature endotherm includes relatively minor contributions from a conformational change involving perturbations of the Trp 177 residue in β chains ($T_m \cong 44$ °C; $\Delta H_{cal} \cong 50$ kcal/mole) together with major excess heat capacity changes from the sequential, partial unfolding of α chains ($T_m \cong 52$ and 54 °C; total $\Delta H_{cal} \cong 300$ kcal/mole). The high-temperature endotherm predominantly reflects the sequential, partial unfolding of β chains ($T_m \cong 78$ and 81 °C; total $\Delta H_{cal} \cong 400$ kcal/mole). The calorimetric enthalpy change at $T_m \sim 52$ °C for $\alpha_2\beta_2$ is more than twice that determined for the thermal unfolding of an isolated α chain, $\Delta H_{cal} \cong 100$ kcal mole of α)$^{-1}$ (Remeta *et al.*, in preparation; Sugisaki *et al.*, 1990) although the nondissociated α chains in the complex cannot be completely unfolded. The enthalpy change at $T_m \sim 81$ °C for $\alpha_2\beta_2$ complex also is larger than that observed for free holo β subunit dimer. Thus, unfolding domains in the $\alpha_2\beta_2$ complex can to some extent span α:β contacts. In summary, calorimetric and spectral measurements allow characterization of the unfolding–refolding pathway of the tryptophan synthase $\alpha_2\beta_2$ complex that permits correlation of thermodynamic and structural information.

4.4. Metabolite Channeling

Early studies with tryptophan synthase from *E. coli* (Crawford and Yanofsky, 1958), from *Neurospora crassa* (DeMoss, 1962; Matchett, 1974; Yanofsky and Rachmeler, 1958), and from *Saccharomyces cerevisiae* (Duntze and Manney, 1968; Manney, 1968, 1970; Manney *et al.*, 1969) showed that indole is not liberated as a free intermediate in the $\alpha\beta$ reaction (scheme I). The liberation of free indole in the synthesis of L-tryptophan might be detrimental to the organism since indole is uncharged and might escape the cell by passive diffusion across the cell membranes (Knowles, 1991; Manney, 1970). The results of the studies referred to above and other kinetic studies (Creighton, 1970) could not distinguish whether indole is transferred directly between adjacent active sites or diffuses through some sort of channel between adjoining active sites. Because the crystal structure of the $\alpha_2\beta_2$ complex shows that the active sites of the α and β subunits are not adjacent but are ~25 Å apart, indole must diffuse through a channel (Hyde *et al.*, 1988). Examination of the three-dimensional structure of the complex reveals that the active sites of the α and β subunits are connected by a tunnel (Fig. 1) that provides a putative pathway for the diffusion of the indole intermediate between the α and β sites. About one third of the tunnel is within the α subunit. Two thirds of the tunnel is within the β subunit and passes between the N-domain and the C-domain to the active site of the β subunit. The residues that line the wall of the tunnel are largely hydrophobic. The refined structures of tryptophan synthase reveal several polypeptide segments in both subunits that exhibit a high degree of disorder or flexibility and may be involved in transmitting allosteric information between the two sites and in altering the shape of the tunnel (C. C. Craig, personal communication; Brzovic' *et al.*, 1992c). Several of these disordered regions are in the surface of the tunnel. Two adjacent aromatic residues in the β subunit (Y279 and F280) are able to either fold along the wall of the tunnel or move into positions that might block the tunnel. These results suggest that the tunnel may undergo dynamic movement in response to ligands.

Recent rapid kinetic studies have addressed two questions about the channeling mechanism. (1) Is the indole intermediate in the $\alpha\beta$ reaction channeled intramolecularly? (2) Does the indole cosubstrate in the β reaction enter the α subunit and pass through the tunnel to the active site of the β subunit? The first question has been approached by studying the kinetics of the $\alpha\beta$ reaction under single-turnover conditions. Rates of individual steps in the reaction have been determined by monitoring the absorbance at 473 nm of the quininoid intermediates (ES IV in Fig. 7) (Lane and Kirschner, 1991), by determining the conversion of radiolabeled indole-3-glycerol phosphate to L-tryptophan by chemical quench-flow methods (Anderson *et al.*, 1991) and by following UV-visible absorption changes by rapid-scanning stopped-flow spectrophotometry (Brzovic' *et al.*, 1992b). Rapid mixing of the preformed aminoacrylate complex ES III with indole-3-glycerol phosphate leads to the appearance and subsequent disappearance of the quinonoid intermediate ES IV with absorbance at 473 nm (Lane and Kirschner, 1991). Because diffusion of indole into solution would result in a lag in formation of ES IV, the absence of any lag supports intramolecular channeling. Calculations of the rates of L-tryptophan formation under single-turnover conditions from indole-3-glycerol phosphate or from free indole in solution also support the conclusion that indole is a channeled intermediate in the αβ reaction (Anderson *et al.*, 1991; Lane and Kirschner, 1991).

Chemical quench-flow studies of the conversion of radiolabeled indole-3-glycerol phosphate to L-tryptophan under single-turnover conditions find no significant accumulation of a transient radiolabeled indole intermediate with the wild-type $\alpha_2\beta_2$ complex (Anderson *et al.*, 1991). Kinetic analysis shows that the rate of diffusion of indole through the tunnel and of the chemical reaction of indole with ES III are very rapid and are much faster than the rate-limiting protonation of L-tryptophan (Anderson *et al.*, 1991). This explanation has been tested by use of a mutant of tryptophan synthase (E109D) that has a 30-fold slower rate in the steady-state β reaction (see Section 3.5 and Fig. 6). Chemical quench-flow analysis of the conversion of indole-3-glycerol phosphate and L-serine to L-tryptophan by the mutant enzyme shows the formation and decay of a significant fraction of indole in a single turnover (Fig. 15). Analysis of the kinetic data shows that indole accumulates because the rate of condensation of indole with the aminoacrylate intermediate ES III is reduced 300-fold in the E109D mutant (Anderson *et al.*, 1991). Although this experiment provides direct evidence for formation of an indole intermediate, it does not prove that indole accumulates in the tunnel. Since the E109D $\alpha_2\beta_2$ complex binds indole more weakly than the wild-type enzyme, the indole formed may diffuse into the solution (Brzovic' *et al.*, 1992a).

The second question concerning the route of access of indole from solution to the active site of the β subunit in the β reaction has been examined by rapid-scanning stopped-flow UV-visible spectroscopy in combination with chemical probes (Brzovic' *et al.*, 1992b; Dunn *et al.*, 1987, 1990). These studies demonstrate that the rate of reaction of indole with the aminoacrylate intermediate ES III (step 4 in Fig. 7) at the β site is inhibited by an α subunit ligand, α-glycerol 3-phosphate. The results support the hypothesis that the preferred pathway of indole from solution into the β site is via the α site and the interconnecting tunnel. The decreased rate of reaction at the β site may result from a ligand-induced conformational change that obstructs direct access of indole into the tunnel or free passage of indole through the tunnel.

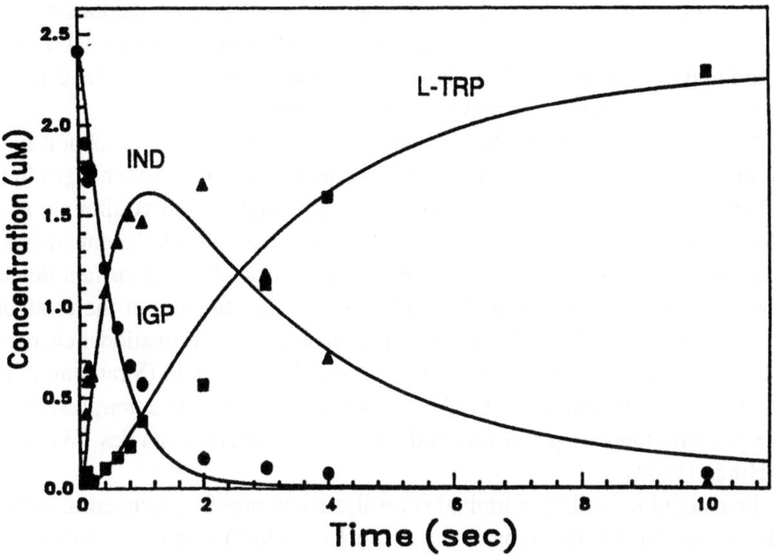

FIGURE 15. Effect of the E109D mutation on the kinetics of the $\alpha\beta$ reaction under single turnover conditions. The tryptophan synthase $\alpha_2\beta_2$ complex with a single amino acid replacement of Asp for Glu-109 in the β subunit (20 μM $\alpha\beta$) was mixed with 10 mM L-serine and 2.4 μM [^{14}C]indole-3-glycerol phosphate (IGP) in a rapid quench apparatus. The disappearance of indole-3-glycerol phosphate (IGP) and the formation of L-tryptophan (TRP) and Indole (IND) were analyzed by HPLC. Adapted from Anderson *et al.* (1991).

Further kinetic tests of the tunnel function have utilized mutant forms of the $\alpha_2\beta_2$ complex (E49F, D60Y, G51L, and R179L) with single amino acid replacements in the α subunit near the entrance to the tunnel or in loop 6 (Brzovic' *et al.*, 1992c, 1993). The most significant effect of these α subunit mutations on the β reaction is to greatly reduce the α-glycerol 3-phosphate-induced inhibition of the rate of quinonoid formation. The results lead to the hypothesis that the mutations inhibit the ligand-induced transition of the α subunit from an "open" to a "closed" conformation that serves to block the entrance to the tunnel. This conformational change may involve closure of loop 6 described in Section 2.1 and Fig. 3.

4.5. Information Transfer from the α Subunit to the β Subunit

Ligands that bind to the active site of the α subunit alter reaction kinetics at the active site of the β subunit ~25 Å distant. For example, α specific ligands (indole-3-propanol phosphate or α-glycerol 3-phosphate) increase the affinity of the β site for amino acids, alter the steady-state distribution of enzyme–substrate intermediates at the β site, and decrease the rate of the β reaction about threefold (Houben and Dunn, 1990; Kawasaki *et al.*, 1987; Kirschner *et al.*, 1991; Lane and Kirschner, 1981, 1983b). Recent studies of the αβ reaction demonstrate that indole-3-glycerol phosphate accelerates the rate of formation of ES III at the β site and that D-glyceraldehyde 3-phosphate increases the affinity of the β site for the product, L-tryptophan (Brzovic' *et al.*, 1992b).

Since several α specific ligands have similar effects on the β subunit in the $\alpha_2\beta_2$ complex, these ligands presumably produce conformational changes in the α subunit that are sensed by the β subunit, even though the active sites are about ~25 Å apart (Houben and Dunn, 1990; Kirschner *et al.*, 1991). Recent studies of the luminescence properties of the $\alpha_2\beta_2$ complex from *S. typhimurium* demonstrate that α-glycerol 3-phosphate binding affects the structure of the α subunit and, in the presence of pyridoxal phosphate, induces a conformational change in the β subunit that leads to a considerably more rigid structure (Strambini *et al.*, 1992b). In contrast, addition of indole, a substrate for the β reaction, results in a less rigid structure, as previously indicated by calorimetric studies (Wiesinger and Hinz, 1984a).

Investigations utilizing limited proteolysis and protein engineering provide evidence that the flexible loop in the α subunit (loop 6) (see Fig. 3 and Section 2.1) serves critical roles in substrate binding and in information transfer from the α subunit to the β subunit (Brzovic' *et al.*, 1992c, 1993; Kawasaki *et al.*, 1987; Miles, 1991c; Ruvinov and Miles, 1992; Ruvinov *et al.*, in preparation; Yang and Miles, 1992). Tryptic cleavage of the α subunit at Arg-188 in loop 6 (see sequence in Fig. 3) has minor effects on the rates of the $\alpha_2\beta_2$ complex in the α, β, and αβ reactions (Higgins *et al.*, 1979; Miles, 1991c; Miles and Higgins, 1978) but has major effects on the allosteric properties of the $\alpha_2\beta_2$ complex (Miles, 1991c; Ruvinov and Miles, 1992). Whereas α subunit ligands (indole-3-propanol phosphate and α-glycerol 3-phosphate) inhibit the activity of the native $\alpha_2\beta_2$ complex in the β reaction, the cleaved or "nicked" $\alpha_2\beta_2$ complex is desensitized to inhibition by this allosteric ligand. A role for loop 6 in binding α subunit ligands is supported by two results. (1) The presence of α-glycerol 3-phosphate decreases the rate of tryptic cleavage sixfold. (2) Tryptic cleavage increases the K_d for α-glycerol 3-phosphate. The results suggest that loop 6 moves from an "open" to a "closed" position upon substrate binding as illustrated in Fig. 3. This figure is based on crystallographic data for ligand-induced closure of the analogous loop in triosephosphate isomerase (Lolis and Petsko, 1990; Wierenga *et al.*, 1991). Loop closure would be expected to make the loop less flexible and less susceptible to proteolysis. Furthermore, loop closure may result in conformational changes that are communicated to the β subunit. If so, proteolysis of the loop would be expected to prevent loop closure, the concomitant conformational changes, and communication of these changes to the β subunit.

The effects of allosteric ligands and of the proteolysis of loop 6 on the conformation of tryptophan synthase have been probed by Nile Red (Ruvinov *et al.*, in preparation). Nile Red, which is a polarity-sensitive fluorescent probe of

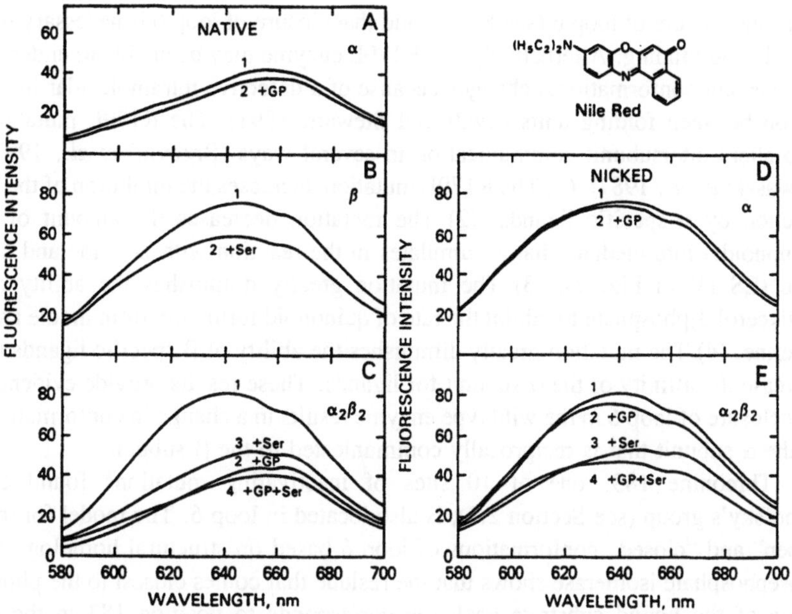

FIGURE 16. Effect of ligands and of proteolysis of the $\alpha_2\beta_2$ complex on interaction with Nile Red. (A–C) Native $\alpha_2\beta_2$ complex or trypsin treated or (D, E) "nicked" $\alpha_2\beta_2$ complex (5 μM protomer in 50 mM sodium bicine buffer, pH 7.8) were mixed with Nile Red (1 μM) in the absence or presence of 50 mM L-serine or 80 mM DL-α-glycerol 3-phosphate (GP) or both as indicated. After 15 min, emission spectra were taken with excitation at 550 nm. (From Ruvinov *et al.*, in preparation.)

hydrophobic regions in proteins (Sackett and Wolff, 1987), interacts with a hydrophobic β site in both the native and nicked $\alpha_2\beta_2$ complexes (Fig. 16). Addition of an allosteric ligand, α-glycerol 3-phosphate, dramatically decreases the intensity of the fluorescence emission of Nile Red at this β site in the native $\alpha_2\beta_2$ complex but not in the nicked $\alpha_2\beta_2$ complex (Fig. 16). Thus binding of a ligand to the active site of the α subunit alters the conformation of the β subunit in the $\alpha_2\beta_2$ complex. Furthermore, cleavage of the α subunit loop 6 blocks the transmission of this ligand-induced conformational change from the α subunit to the β subunit.

Substitution of leucine for arginine-179 (R179L) in α subunit loop 6 (see sequence in Fig. 3) does not significantly alter the rate of the $\alpha_2\beta_2$ complex in the β reaction, but does alter some of the allosteric properties of the enzyme (Brzovic' *et al.*, 1993; Kawasaki *et al.*, 1987). The primary effect of the R179L mutation is to reduce the binding affinity for α specific ligands (α-glycerol 3-phosphate, D-glyceraldehyde 3-phosphate, indole-3-glycerol phosphate, and indole-3-propanol phosphate). The results suggest that the R179L mutation prevents the closure of loop 6 (see Fig. 3) and that closure of loop 6 is necessary for tight ligand binding. Alternatively, the R179L enzyme may be unable to undergo the relevant conformational changes because of a defective intramolecular interaction between folding units (Swift and Stewart, 1991). The R179L mutation also alters intersubunit communication in several ways (Brzovic' *et al.*, 1993; Kawasaki *et al.*, 1987). (1) The R179L mutation decreases the inhibition of the β reaction by α-specific ligands. (2) The mutation decreases the amount of a quinonoidal intermediate that accumulates in the reaction with L-serine and indole (ES IV in Fig. 7). (3) The mutation greatly diminishes the ability of α-glycerol 3-phosphate to inhibit the rate of quinonoid formation from indole and L-serine. (4) The mutation greatly diminishes the ability of β specific ligands to increase the affinity of the α subunit for ligands. These results provide evidence that closure of loop 6 in the wild type enzyme results in a change in conformation in the α subunit that is reciprocally communicated to the β subunit.

Threonine 183, one of 10 sites of inactivating mutations found by Yanofsky's group (see Section 2.2), is also located in loop 6. The model for the "open" and "closed" conformations of loop 6 based on structural homology to triosephosphate isomerase shows that the residue that comes closest to the phosphate of the bound substrate analogue corresponds to position 183 in the α subunit sequence (Fig. 3). Changing threonine 183 to alanine (T183A) or deleting residues 185–187 (Δ185–187) greatly reduces the activities of the $\alpha_2\beta_2$ complex in the α reaction and the β reaction but does not alter subunit association (Fig. 17) (Yang and Miles, 1992). The results show the importance of the hydroxyl group at position 183 and support the model in Fig. 3 that shows interaction between threonine 183 and the phosphate of the active site ligand in the closed position of loop 6.

4.6. Information Transfer from the β Subunit to the α Subunit

Ligands that bind to the active site of the β subunit alter the kinetics of reactions at the active site of the α subunit ~25Å distant. For example, binding of L-serine at the β site decreases the K_m for indole-3-glycerol phosphate (Creighton, 1970; Kawasaki *et al.*, 1987; Lane and Kirschner, 1983b) and the K_d or K_I for indole-3-propanol phosphate at the α site (Kawasaki *et al.*, 1987; Lane and Kirschner, 1981, 1983b).

Because the turnover number of the $\alpha_2\beta_2$ complex in the αβ reaction is 20 times larger than the turnover number in the α reaction, L-serine may activate cleavage of indole-3-glycerol phosphate at the active site of the α subunit. However, it is not possible to test this hypothesis directly, because the nascent indole is immediately converted to L-tryptophan in the coupled αβ reaction. This obstacle has been circumvented by use of 6-nitroindole, which readily condenses with D-glyceraldehyde 3-phosphate at the α site but does not condense with

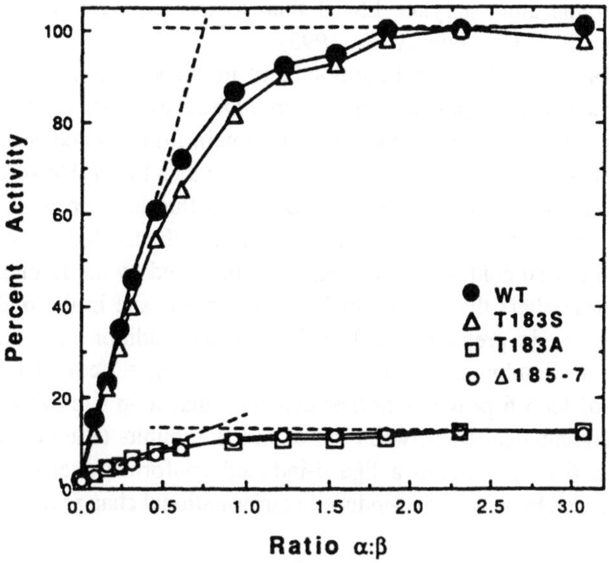

FIGURE 17. Effects of mutations in α subunit loop 6 on interaction with the wild-type β subunit and on activity. The activity of the wild-type β subunit in the β reaction was measured in the presence of various amounts of the wild-type or mutant α subunits. The extrapolated dashed lines indicate a stoichiometry of about 0.75 α chain per β chain in each case. (From Yang and Miles, 1992.)

L-serine at the β site (Kirschner *et al.*, 1991). Addition of L-serine results in a tenfold increase in the turnover number for 6-nitroindole-3-glycerol phosphate in the forward α reaction or for 6-nitroindole in the reverse α reaction. This result demonstrates directly that a derivative of L-serine at the β site activates catalysis at the α site (Kirschner *et al.*, 1991).

There are several types of evidence that the Schiff base formed between aminoacrylate and pyridoxal phosphate (ES III in Fig. 7) is the L-serine derivative that triggers the conformational change that activates the α subunit. (1) Amino acids that form covalent intermediates at the β site, but do not form ES III, do not appreciably stimulate the activity of the α subunit (Brzovic' *et al.*, 1991, 1992b). (2) L-Serine and amino acid analogues such as O-methyl-DL-serine, which form ES III, do stimulate the rate of the α reaction (Brzovic' *et al.*, 1991, 1992b). (3) The rate of formation of ES III from L-serine corresponds to the rate of activation of the α subunit under single turnover conditions (Anderson *et al.*, 1991). In the presence of L-serine the rate of cleavage of indole-3-glycerol phosphate increases 150-fold from 0.16 sec^{-1} to 24 sec^{-1}. (4) Formation of a stable ES III intermediate by a mutant form of the β subunit (K87T) (Lu *et al.*, 1993) results in activation of the α subunit (Banik *et al.*, in preparation). (5) D-Glyceraldehyde 3-phosphate binds 11 times morely tightly to the $\alpha_2\beta_2$ complex with bound aminoacrylate ES III than to the $\alpha_2\beta_2$ complex with bound L-tryptophan ES V (Brzovic' *et al.*, 1993).

There is also evidence that the β subunit in the $\alpha_2\beta_2$ complex undergoes a conformational change upon binding L-serine. Direct evidence is provided by the demonstration that a change in the intrinsic protein fluorescence occurs at a rate identical to that of ES III formation (Anderson *et al.*, 1991). However, fluorescence measurements may be complicated by the fluorescence of the enzyme–substrate intermediates and possible energy transfer effects. Indirect evidence for an L-serine-induced conformational change is provided by the finding that addition of L-serine alters the binding of Nile Red to a β site in the $\alpha_2\beta_2$ complex (Ruvinov *et al.*, in preparation). The finding that addition of L-serine in the presence of an α subunit ligand (α-glycerol 3-phosphate) completely prevents proteolysis of loop 6 provides further evidence that loop 6 plays a role in the reciprocal communication between the α and β subunits (Ruvinov and Miles, 1992). Loop 6 may undergo a ligand-induced conformational change that is further stabilized by an L-serine-induced conformational change in the β subunit.

4.7. Coordination of the Activities of the α and β Subunits

The sections above provide evidence that the tryptophan synthase α and β subunits undergo conformational changes upon subunit association and additional conformational changes upon ligand binding to the $\alpha_2\beta_2$ complex. The ligand-induced conformational changes are communicated reciprocally between the active sites of the α and β subunits in the $\alpha_2\beta_2$ complex over a distance of ~25Å (Sections 4.5 and 4.6). Figure 18 shows a proposed model depicting ligand- or intermediate-induced conformational changes in the $\alpha_2\beta_2$ complex that occur during the course of the $\alpha\beta$ reaction (Brzovic' *et al.*, 1992b). The model, which is based on results from several groups, explains how a series of conformational changes serve to activate the individual α and β sites, to increase the affinity of each site for its substrate, and to coordinate the activities of the two heterotrophic sites. This coordination of the activities prevents accumulation of the indole intermediate and escape of indole into solution.

Figure 18 shows that L-serine and indole-3-glycerol phosphate can bind to the $\alpha_2\beta_2$ complex (structure I) in random order, consistent with kinetic results (Creighton, 1970). Reaction of L-serine to form ES III at the β site induces a conformational change in the β subunit (structure II), which in turn triggers the conversion of the α subunit from the low activity open form (circle) to the high activity closed form (square) (structure IV), as described in Section 4.6 (Anderson *et al.*, 1991; Brzovic' *et al.*, 1992b; Dunn *et al.*, 1991; Kirschner *et al.*, 1991; Lane and Kirschner, 1991). Alternatively, binding of indole-3-glycerol phosphate results in conversion of the α subunit from an open form (circle) to a closed form (square), as described in Section 4.5. The proposed ligand-induced

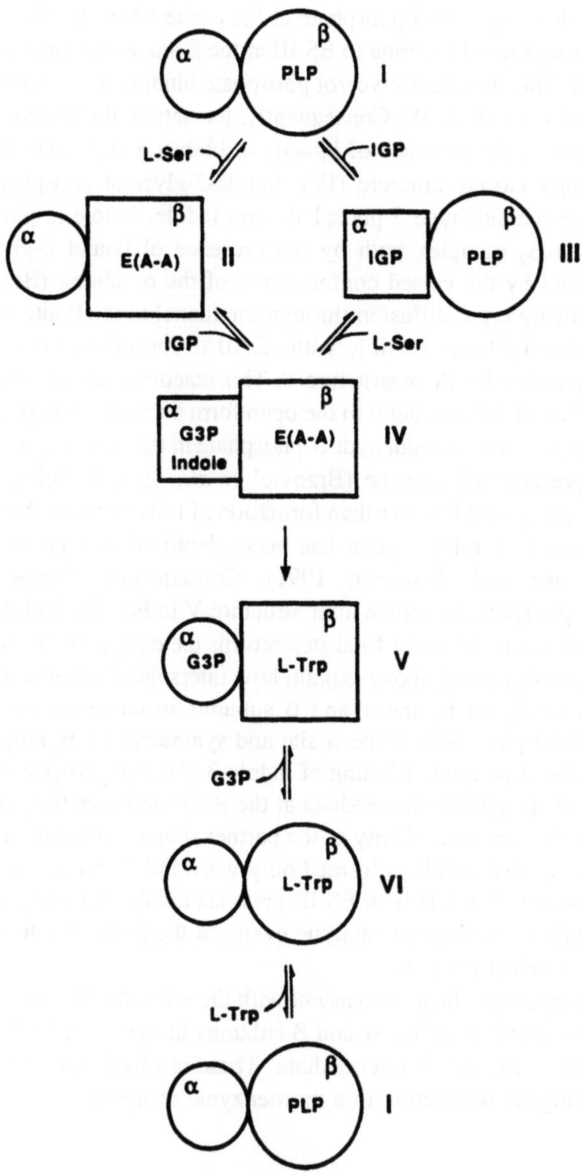

FIGURE 18. Model depicting ligand- or intermediate-induced conformational changes in the $\alpha_2\beta_2$ complex that occur during the course of the αβ reaction. Circles indicate open conformational states; squares indicate closed conformational states. (From Brzovic' *et al.*, 1992b. Reprinted with permission from American Chemical Society.)

closure of loop 6 during this conformational conversion is shown in Fig. 3. Binding of indole-3-glycerol phosphate at the α site (structure III) increases the rate of the conversion of L-serine to ES III at the β site about threefold (Brzovic' *et al.*, 1992b). Thus indole-3-glycerol phosphate binding triggers the conversion of structure III to structure IV. Consequently, formation of structure IV by either route is favored in the presence of ligands of both α and β subunits.

In the fully closed structure (IV), indole-3-glycerol phosphate is rapidly cleaved to D-glyceraldehyde 3-phosphate and indole. Indole is prevented from escaping the $\alpha_2\beta_2$ complex both by the presence of bound D-glyceraldehyde 3-phosphate and by the closed conformation of the α subunit (Brzovic' *et al.*, 1992b). Following rapid diffusion through the tunnel to the β site, indole reacts rapidly and essentially irreversibly with ES III to form the quinonoid or carbanion of L-tryptophan ES IV in structure V. This reaction must be accompanied by an isomerization of the α subunit to the open form because the $\alpha_2\beta_2$ complex has a lower affinity for D-glyceraldehyde 3-phosphate in the presence of L-tryptophan than in the presence of L-serine (Brzovic' *et al.*, 1992b). Release of D-glyceraldehyde 3-phosphate is faster than formation of L-tryptophan; the rate-limiting step in formation of L-tryptophan has been identified as reprotonation of the quinonoid (Lane and Kirschner, 1991). Consequently, release of D-glyceraldehyde 3-phosphate is shown after structure V in Fig. 18. Release of L-tryptophan after structure VI in the final step returns the $\alpha_2\beta_2$ complex to structure I.

The results described above explain how intersubunit communication keeps the reactions catalyzed by the α and β subunits in phase so that cleavage of indole-3-glycerol phosphate at the α site and synthesis of L-tryptophan at the β site occur at the same rates. Binding of indole-3-glycerol phosphate to the α site or formation of the ES III intermediate at the β site activates the partner subunit and increases the substrate affinity of the partner subunit. Activation of the α site by ES III insures that indole is formed only when indole can react at the β site. The rapid reaction of indole with ES III prevents indole accumulation and diffusion into solution. Subsequent catalytic events at the β site deactivate the α site and facilitate product release.

In summary, the data are consistent with the view that allosteric interactions coordinate the activities of the α and β subunits in the overall αβ reaction and prevent escape of the indole intermediate. These results demonstrate the benefit of enzyme–enzyme interaction in a multienzyme complex.

ACKNOWLEDGMENTS AND DEDICATION. I thank members of my group, including S. A. Ahmed, S. B. Ruvinov, X.-J. Yang, Z. Lu, and U. Banik, for their recent contributions to the work and P. S. Brzovic, M. F. Dunn, C. C. Hyde, D. R. Davies, K. Parris, K. Ogasahara, K. Yutani, P. McPhie, K. Johnson, K. Anderson, D. L. Sackett, D. P. Remeta, and A. Ginsburg for their work in several collaborations cited. I dedicate this chapter to Dr. Esmond E. Snell, my Ph. D. advisor, on the occasion of his 80th birthday in 1994.

5. REFERENCES

Adachi, O., Kohn, L. D., and Miles, E. W., 1974, Crystalline $\alpha_2\beta_2$ complexes of tryptophan synthetase of *Escherichia coli.* A comparison between the native complex and the reconstituted complex, *J. Biol. Chem.* **249:**7756–7763.

Ahmed, S. A., and Miles, E. W., 1992, Evidence for two conformers of the β subunit of tryptophan synthase in solution, *J. Biol. Chem.* **267:**23309–23317.

Ahmed, S. A., Fairwell, T., Dunn, S., Kirschner, K., and Miles, E. W., 1986a, Identification of three sites of proteolytic cleavage in the hinge region between the two domains of the β_2 subunit of tryptophan synthase of *Escherichia coli* or *Salmonella typhimurium, Biochemistry* **25:**3118–3124.

Ahmed, S. A., Martin, B., and Miles, E. W., 1986b, β-Elimination of indole from L-tryptophan catalyzed by bacterial tryptophan synthase: A comparison between reactions catalyzed by tryptophanase and tryptophan synthase, *Biochemistry* **25:**4233–4240.

Ahmed, S. A., Ruvinov, S. B., Kayastha, A. M., and Miles, E. W., 1991, Mechanism of mutual activation of the tryptophan synthase α and β subunits. Analysis of the reaction specificity and substrate-induced inactivation of active site and tunnel mutants of the β subunit, *J. Biol. Chem.* **266:**21540–21557.

Allen, M. K., and Yanofsky, C., 1963, A biochemical and genetic study of reversion with the A-gene A-protein system of *Escherichia coli* tryptophan synthetase, *Genetics* **48:**1065–1083.

Anderson, K. S., Miles, E. W., and Johnson, K. A., 1991, Serine modulates substrate channeling in tryptophan synthase: A novel intersubunit triggering mechanism, *J. Biol. Chem.* **266:**8020–8033.

Balk, H., Frank, A., Bartholmes, P., and Jaenicke, R., 1981, Equilibrium and kinetic measurements of the binding of pyridoxal 5′-phosphate to hybrid tryptophan synthase from *Escherichia coli, Eur. J. Biochem.* **121:**105–112.

Banik, U., Zhu, D.-M., Chock, P. B., and Miles, E. W. (in preparation), Allosteric activation of the tryptophan synthase $\alpha_2\beta_2$ complex: Kinetic studies with a mutant enzyme (β K87T) provide evidence for triggering by a serine intermediate.

Bartholmes, P., and Teuscher, B., 1979, Cooperative binding of α subunit to the apoβ_2 subunit of tryptophan synthase from *Escherichia coli, Eur. J. Biochem.* **95:**323–326.

Bartholmes, P., Kirschner, K., and Gschwind, H.-P., 1976, Cooperative and noncooperative binding of pyridoxal 5′-phosphate to tryptophan synthase from *Escherichia coli, Biochemistry* **15:**4712–4717.

Bartholmes, P., Balk, H., and Kirschner, K., 1980, Mechanism of reconstitution of the apoβ subunit and the $\alpha_2$apoβ_2 complex complex of tryptophan synthase with pyridoxal 5′-phosphate: Kinetic studies, *Biochemistry* **19:**4527–4533.

Beasty, A. M., and Matthews, C. R., 1985, Characterization of an early intermediate in the folding of the α subunit of tryptophan synthase by hydrogen exchange measurement, *Biochemistry* **24:**3547–3553.

Beasty, A. M., Hurle, M. R., Manz, J. T., Stackhouse, T., Onuffer, J. J., and Matthews, C. R., 1986, Effects of the phenylalanine-22 → leucine, glutamic acid-49 → methionine, glycine-234 → aspartic acid, and glycine-234 → lysine mutations on the folding and stability of the α subunit of tryptophan synthase from *Escherichia coli, Biochemistry* **25:**2965–2974.

Blond, S., and Goldberg, M., 1987, Partly native epitopes are already present on early intermediates in the folding of tryptophan synthase, *Proc. Natl. Acad. Sci. USA* **84:**1147–1151.

Blond-Elguindi, S., and Goldberg, M. E., 1990, Kinetic characterization of early immunoreactive intermediates during the refolding of guanidine-unfolded *Escherichia coli* tryptophan synthase β_2 subunits, *Biochemistry* **29:**2409–2417.

Borcsok, E., and Abeles, R. H., 1982, Mechanism of action of cystathionine synthase, *Arch. Biochem. Biophys.* **213:**695–707.

Brzovic', P., Miles, E. W., and Dunn, M. F., 1991, Interactions between heterologous subunits coordinate catalytic activity in the tryptophan synthase bienzyme complex, in *Enzymes Dependent on Pyridoxal Phosphate and Other Carbonyl Compounds as Cofactors* (T. Fukui, H. Kagamiyama, K. Soda, and H. Wada, eds.), pp. 277–279, Pergamon Press, Oxford.

Brzovic', P. S., Kayastha, A. M., Miles, E. W., and Dunn, M. F., 1992a, Substitution of glutamic acid 109 by aspartic acid alters the substrate specificity and catalytic activity of the β-subunit in the tryptophan synthase bienzyme complex from *Salmonella typhimurium*, *Biochemistry* **31:**1180–1190.

Brzovic', P. S., Ngo, K., and Dunn, M. F., 1992b, Allosteric interactions coordinate catalytic activity between successive metabolic enzymes in the tryptophan synthase bienzyme complex, *Biochemistry* **31:**3831–3839.

Brzovic', P. S., Sawa, Y., Hyde, C. C., Miles, E. W., and Dunn, M. F., 1992c, Evidence that mutations in a loop region of the α-subunit inhibit the transition from an open to a closed conformation in the tryptophan synthase bienzyme complex, *J. Biol. Chem.* **267:**13028–13038.

Brzovic', P. S., Hyde, C. C., Miles, E. W., and Dunn, M. F., 1993, Characterization of the functional role of a flexible loop in the α subunit of tryptophan synthase from *Salmonella typhimurium* by rapid-scanning stopped-flow spectroscopy and site-directed mutagenesis, *Biochemistry* **32:**10404–10413.

Chaffotte, A. F., and Goldberg, M. E., 1983, Immunochemical evidence for conformational flexibility and its modulation by specific ligands in the β_2 subunit of *Escherichia coli* tryptophan synthase, *Biochemistry* **22:**2708–2714.

Chaffotte, A. F., and Goldberg, M. E., 1984, Fluorescence-quenching studies on a conformational transition within a domain of the β_2 subunit of *Escherichia coli* tryptophan synthase, *Eur. J. Biochem.* **139:**47–50.

Chen, X., Rambo, R., and Matthews, C. R., 1992, Amino acid replacements can selectively affect the interaction energy of autonomous folding units in the α subunit of tryptophan synthase, *Biochemistry* **31:**2219–2223.

Chrunyk, B. A., and Matthews, C. R., 1990, Role of diffusion in the folding of the α subunit of tryptophan synthase from *Escherichia coli*, *Biochemistry* **29:**2149–2154.

Crawford, I. P., 1989, Evolution of a biosynthetic pathway: The tryptophan paradigm, *Annu. Rev. Microbiol.* **43:**567–600.

Crawford, I. P., and Johnson, L. M., 1963, Mutants of *Escherichia coli* defective in the B protein of tryptophan synthetase. I. Selection and classification, *Genetics* **48:**725–736.

Crawford, I. P., and Yanofsky, C., 1958, On the separation of the tryptophan synthetase of *Escherichia coli* into two protein components, *Proc. Natl. Acad. Sci. USA* **44:**1161–1170.

Creighton, T. E., 1970, A steady-state kinetic investigation of the reaction mechanism of the tryptophan synthetase of *Escherichia coli*, *Eur. J. Biochem.* **13:**1–10.

Creighton, T. E., and Yanofsky, C., 1966, Association of the α and β_2 subunits of the tryptophan synthetase of *Escherichia coli*, *J. Biol. Chem.* **241:**980–990.

Davenport, R. C., Bash, P. A., Seaton, B. A., Karplus, M., Petsko, G. A., and Ringe, D., 1991, Structure of the triosephosphate isomerase–phosphoglycolohydroxamate complex: An analogue of the intermediate on the reaction pathway, *Biochemistry* **30:**5821–5826.

DeMoss, J. A., 1962, Studies on the mechanism of the tryptophan synthetase reaction, *Biochim. Biophys. Acta* **62**:279–293.

Djavadi-Ohaniance, L., Friguet, B., and Goldberg, M. E., 1986, Conformational effects of ligand binding on the β_2 subunit of *Escherichia coli* tryptophan synthase analyzed with monoclonal antibodies, *Biochemistry* **25**:2502–2508.

Drewe, W. J., and Dunn, M. F., 1985, Detection and identification of intermediates in the reaction of L-serine with *Escherichia coli* tryptophan synthase via rapid-scanning ultraviolet-visible spectroscopy, *Biochemistry* **24**:3977–3987.

Drewe, W. J., and Dunn, M. F., 1986, Characterization of the reaction of L-serine and indole with *Escherichia coli* tryptophan synthase via rapid-scanning ultraviolet-visible spectroscopy, *Biochemistry* **25**:2494–2501.

Dunn, M. F., Roy, M., Robustell, B., and Aguilar, V., 1987, Chemical probes of substrate channeling in the *E. coli* tryptophan synthase $\alpha_2\beta_2$ bienzyme complex, in *Biochemistry of Vitamin B6, Proceedings of the 7th International Congress of Chemical and Biological Aspects of Vitamin B6 Catalysis* (T. Korpela and P. Christen, eds.), pp. 171–181, Birkhauser Verlag AG, Basel.

Dunn, M. F., Aguilar, V., Brzovic', P. S., Drewe, W. F. J., Houben, K. F., Leja, C. A., and Roy, M., 1990, The tryptophan synthase bienzyme complex transfers indole between the α and β-sites via a 25–30 Å long tunnel, *Biochemistry* **29**:8598–8607.

Dunn, M. F., Brzovic', P. S., Leja, C., Houben, K., Roy, M., Aguilar, A., and Drewe, W. F., Jr., 1991, Allosteric interactions coordinate covalent steps in catalysis and modulate indole transfer between the α and β-sites of the tryptophan synthase bienzyme complex, in *Enzymes Dependent on Pyridoxal Phosphate and Other Carbonyl Compounds as Cofactors* (T. Fukui, H. Kagamiyama, K. Soda, and H. Wada, eds.), pp. 257–264, Pergamon Press, Oxford.

Duntze, W., and Manney, T. R., 1968, Two mechanisms of allelic complementation among tryptophan synthetase mutants of *Saccharomyces cerevisiae*, *J. Bacteriol.* **96**:2085–2093.

Eder, J., and Kirschner, K., 1992, Stable substructures of eightfold βα-barrel proteins: Fragment complementation of phosphoribosylanthranilate isomerase, *Biochemistry* **31**:3617–3625.

Faeder,, E. J., and Hammes, G. G., 1970, Kinetic studies of tryptophan synthetase. Interaction of substrates with the B subunit, *Biochemistry* **9**:4043–4039.

Faeder, E. J., and Hammes, G. G., 1971, Kinetic studies of tryptophan synthetase. Interaction of L-serine, indole, and tryptophan with the native enzyme, *Biochemistry* **10**:1041–1045.

Farber, G. K., and Petsko, G. A., 1990, The evolution of α/β barrel enzymes, *Trends Bioch. Sci.* **15**:228–234.

Federov, A. N., Friguet, B., Djavadi-Ohaniance, L., Alakhov, Y. B., and Goldberg, M. E., 1992, Folding on the ribosome of *Escherichia coli* tryptophan synthase β subunit nascent chains probed with a conformation-dependent monoclonal antibody, *J. Mol. Biol.* **228**:351–358.

Friguet, B., Djavadi-Ohaniance, L., and Goldberg, M. E., 1986, Conformational changes induced by domain assembly within the β_2 subunit of *Escherichia coli* tryptophan synthase analysed with monoclonal antibodies, *Eur. J. Biochem.* **160**:593–597.

Friguet, B., Djavadi-Ohaniance, L., and Goldberg, M. E., 1989, Polypeptide–antibody binding mechanism: Conformational adaptation investigated by equilibrium and kinetic analysis, *Res. Immunol.* **140**:355–376.

Goldberg, M. E., 1985, The second translation of the genetic message: Protein folding and assembly, *TIBS* **10**:388–391.

Goldberg, M. E., 1991, Investigating protein conformation, dynamics and folding with monoclonal antibodies, *TIBS* **16**:358–362.

Goldberg, M. E., and Baldwin, R. L., 1967, Interactions between the subunits of the tryptophan synthetase of *Escherichia coli*. Optical properties of an intermediate bound to the $\alpha_2\beta_2$ complex, *Biochemistry* **6**:2113–2119.

Goldberg, M. E., and Högberg-Raibaud, A., 1979, Conformational and ligand binding properties of the isolated domains from the β_2 subunit of *Escherichia coli* tryptophan synthetase investigated by the reactivity of their cysteines, *J. Biol. Chem.* **254:**7752–7757.

Goldberg, M. E., Creighton, T. E., Baldwin, R. L., and Yanofsky, C., 1966, Subunit structure of the tryptophan synthetase of *Escherichia coli, J. Mol. Biol.* **21:**71–82.

Goldberg, M. E., York, S., and Stryer, L., 1968, Fluorescence studies of substrate and subunit interactions of the β_2 protein of *Escherichia coli* tryptophan synthetase, *Biochemistry* **7:**3662–3667.

Goldberg, M. E., Semisotnov, G. V., Friguet, B., Kuwajima, K., Ptitsyn, O. B., and Sugai, S., 1990, An early immunoreactive folding intermediate of the tryptophan synthase β_2 subunit is a "molten globule", *FEBS* **263:**51–56.

Helinski, D. R., and Yanofsky, C., 1963, A genetic and biochemical analysis of second site reversion, *J. Biol. Chem.* **238:**1043–1048.

Higgins, W., Fairwell, T., and Miles, E. W., 1979, An active proteolytic derivative of the α subunit of tryptophan synthase. Identification of the site of cleavage and characterization of the fragments, *Biochemistry* **18:**4827–4835.

Högberg-Raibaud, A., and Goldberg, M. E., 1977, Isolation and characterization of independently folding regions of the β chain of *Escherichia coli* tryptophan synthetase, *Biochemistry* **16:**4014–4020.

Houben, K. F., and Dunn, M. F., 1990, Allosteric effects acting over a distance of 20–25 Å in the *Escherichia coli* tryptophan synthase bienzyme complex increase ligand affinity and cause redistribution of covalent intermediates, *Biochemistry* **29:**2421–2429.

Houben, K. F., Kadima, W., Roy, M., and Dunn, M. F., 1989, L-Serine analogues form Schiff base and quinonoidal intermediates with *Escherichia coli* tryptophan synthase, *Biochemistry* **28:**4140–4147.

Hurle, M. R., Tweedy, N. B., and Matthews, C. R., 1986, Synergism in folding of a double mutant of the α subunit of tryptophan synthase, *Biochemistry* **25:**6356–6360.

Hyde, C. C., Ahmed, S. A., Padlan, E. A., Miles, E. W., and Davies, D. R., 1988, Three-dimensional structure of the tryptophan synthase $\alpha_2\beta_2$ multienzyme complex from *Salmonella typhimurium, J. Biol. Chem.* **263:**17857–17871.

Ibel, K., May, R. P., Kirschner, K., Lane, A. N., Szadkowski, H., Dauvergne, M. T., and Zulauf, M., 1985, The domain structure of tryptophan synthase. A neutron scattering study, *Eur. J. Biochem.* **151:**505–514.

Kanzaki, H., McPhie, P., and Miles, E. W., 1991, Effect of single amino acid substitutions at positions 49 and 60 on the thermal unfolding of the tryptophan synthase α subunit from *Salmonella typhimurium, Arch. Biochem. Biophys.* **284:**174–180.

Kaufmann, M., Schwartz, T., Jaenicke, R., Schnackerz, K. D., Meyer, H. E., and Bartholmes, P., 1991, Limited proteolysis of the β_2-dimer of tryptophan synthase yields an enzymatically active derivative that binds α subunits, *Biochemistry* **30:**4173–4179.

Kawasaki, H., Bauerle, R., Zon, G., Ahmed, S. A., and Miles, E. W., 1987, Site-specific mutagenesis of the α subunit of tryptophan synthase from *Salmonella typhimurium.* Changing arginine 179 to leucine alters the reciprocal transmission of substrate-induced conformational changes between the α and β subunits, *J. Biol. Chem.* **262:**10678–10683.

Kayastha, A. M., and Miles, E. W., 1991, Role of active site glutamic acid 109 in the active site of the β subunit of tryptophan synthase from *Salmonella typhimurium* in controlling the reaction and substrate specificity of the $\alpha_2\beta_2$ complex, in *Enzymes Dependent on Pyridoxal Phosphate and Other Carbonyl Compounds as Cofactors* (T. Fukui, H. Kagamiyama, K. Soda, and H. Wada, eds.), pp. 265–267, Pergamon Press, Oxford.

Kayastha, A. M., Sawa, Y., Nagata, S., and Miles, E. W., 1991, Site-directed mutagenesis of the β subunit of tryptophan synthase from *Salmonella typhimurium.* Role of active site glutamic acid 350, *J. Biol. Chem.* **266:**7618–7625.

Kirschner, K., Lane, A. N., and Strasser, A. W. M., 1991, Reciprocal communication between the lyase and synthase active sites of the tryptophan synthase bienzyme complex, *Biochemistry* **30:**472–478.

Knowles, J. R., 1991, Calmer waters in the channel, *J. Theor. Biol.* **152:**53–56.

Lane, A. N., 1983, The accessibility of the active site and conformation states of the β_2 subunit of tryptophan synthase studied by fluorescence quenching, *Eur. J. Biochem.* **133:**531–538.

Lane, A. N., and Kirschner, K., 1981, The mechanism of tryptophan binding to tryptophan synthase from *Escherichia coli, Eur. J. Biochem.* **120:**379–387.

Lane, A. N., and Kirschner, K., 1983a, The catalytic mechanism of tryptophan synthase from *Escherichia coli.* Kinetics of the reaction of indole with the enzyme—L-serine complexes, *Eur. J. Biochem.* **129:**571–582.

Lane, A. N., and Kirschner, K., 1983b, The mechanism of binding of L-serine to tryptophan synthase from *Escherichia coli, Eur. J. Biochem.* **129:**561–570.

Lane, A. N., and Kirschner, K., 1983c, The quaternary structure of tryptophan synthase from *Escherichia coli.* Fluorescence and hydrodynamic studies, *Eur. J. Biochem.* **129:**675–684.

Lane, A. N., and Kirschner, K., 1991, Mechanism of the physiological reaction catalyzed by tryptophan synthase from *Escherichia coli, Biochemistry* **30:**479–484.

Lane, A. N., Paul, C. H., and Kirschner, K., 1984, The mechanism of self-assembly of the multienzyme complex tryptophan synthase from *Escherichia coli, EMBO J.* **3:**279–287.

Lim, W. K., Sarkar, S. K., & Hardman, J. K., 1991a, Enzymatic properties of mutant *Escherichia coli* tryptophan synthase α subunits, *J. Biol. Chem.* **266:**20205–20212.

Lim, W. K., Shin, H. J., Milton, D. L., and Hardman, J. K., 1991b, Relative activities and stabilities of mutant *Escherichia coli* tryptophan synthase α subunits, *J. Bacteriol.* **173:**1886–1893.

Lim, W. K., Brouillette, C., and Hardman, J. K., 1992, Thermal stabilities of mutant *Escherichia coli* tryptophan synthase α subunit, *Arch. Biochem. Biophys.* **292:**34–41.

Linkens, H.-J., Djavadi-Ohaniance, L., and Goldberg, M. E., 1993, Nicking of the tryptophan synthase β_2-subunit at Glu-296 prevents the conformational change undergone on binding the α-subunit, *FEBS Lett.* **320:**224–228.

Lolis, E., and Petsko, G., 1990, Crystallographic analysis of the complex between triosephosphate isomerase and 2-phosphoglycolate at 2.5 Å resolution: Implications for catalysis, *Biochemistry* **29:**6619–6625.

Lolis, E., Alber, T., Davenport, R. C., Rose, D., Hartman, F. C., and Petsko, G., 1990, Structure of yeast triosephosphate isomerase at 1.9 Å resolution, *Biochemistry* **29:**6609–6618.

Lu, Z., Nagata, S., McPhie, P., and Miles, E. W., 1993, Lysine-87 in the β subunit of tryptophan synthase that forms an internal aldimine with pyridoxal phosphate serves critical roles in transimination, catalysis, and product release, *J. Biol. Chem.* **268:**8727–8734.

Manney, T. R., 1968, Regulation of factors that influence the in vitro stability of tryptophan synthetase from yeast, *J. Bacteriol.* **96:**403–408.

Manney, T. R., 1970, Physiological advantage of the mechanism of the tryptophan synthetase reaction, *J. Bacteriol.* **102:**483–488.

Manney, T. R., Duntze, W., Janosko, N., and Salazar, J., 1969, Genetic and biochemical studies of partially active tryptophan synthetase mutants of *Saccharomyces cerevisiae, J. Bacteriol.* **99:**590–596.

Matchett, W. H., 1974, Indole channeling by tryptophan synthase of *Neurospora, J. Biol. Chem.* **249:**4041–4049.

Matthews, C. R., and Crisanti, M. M., 1981, Urea-induced unfolding of the α subunit of tryptophan synthase: evidence for a multistate process, *Biochemistry* **20:**784–792.

Matthews, C. R., Crisanti, M. M., Gepner, G. L., Velicelebi, G., and Sturtevant, J. M., 1980, Effect of single amino acid substitutions on the thermal stability of the α subunit of tryptophan synthase, *Biochemistry* **19:**1290–1293.

Matthews, C. R., Crisanti, M. M., Manz, J. T., and Gepner, G. L., 1983, Effect of a single amino acid substitution on the folding of the α subunit of tryptophan synthase, *Biochemistry* **22:**1445–1452.

Miles, E. W., 1979, Tryptophan synthase: Structure, function, and subunit interaction, *Adv. Enzymol.* **49:**127–186.

Miles, E. W., 1986, Pyridoxal phosphate enzymes catalyzing β-elimination and β-replacement reactions, in *Pyridoxal Phosphate and Derivatives* (D. Dolphin, R. Poulson, and O. Avramovic, eds.), pp. 253–310, John Wiley, New York.

Miles, E. W., 1991a, Structural basis for catalysis by tryptophan synthase, *Adv. Enzymol. Relat. Areas Mol. Biol.* **64:**93–172.

Miles, E. W., 1991b, The tryptophan synthase $\alpha_2\beta_2$ complex multienzyme complex: Relationship of the amino acid sequence and folding domains to the three-dimensional structure, in *Conformations and Forces in Protein Folding* (B. T. Nall and K. A. Dill, eds.), pp. 115-124, AAAS, Waldorf, MD.

Miles, E. W., 1991c, The tryptophan synthase $\alpha_2\beta_2$ complex: Cleavage of a flexible loop in the α subunit alters allosteric properties, *J. Biol. Chem.* **266:**10715–10718.

Miles, E. W., and Fairwell, T., 1984, The tryptophan synthase $\alpha_2\beta_2$ complex: A comparison of the reactivity of amino groups in the α and β_2 subunits in the complex by differential labeling studies, *Arch. Biochem. Biophys.* **230:**430–439.

Miles, E. W., and Higgins, W., 1978, An active $\alpha'_2\beta_2$ derivative of tryptophan synthase formed by limited proteolysis, *J. Biol. Chem.* **253:**6266–6269.

Miles, E. W., and McPhie, P., 1974, Evidence for a rate-determining proton abstraction in the serine deaminase reaction of the β_2 subunit of tryptophan synthase, *Biochemistry* **249:**2852–2857.

Miles, E. W., and Moriguchi, M., 1977, Tryptophan synthase of *Escherichia coli.* Removal of pyridoxal 5′-phosphate and separation of the α and β_2 subunits, *J. Biol. Chem.* **252:**6594–6599.

Miles, E. W., Hatanaka, M., and Crawford, I. P., 1968, A new thiol-dependent transamination reaction catalyzed by the B protein of *Escherichia coli* tryptophan synthetase, *Biochemistry* **7:**2742–2753.

Miles, E. W., Yutani, K., and Ogasahara, K., 1982, Guanidine hydrochloride induced unfolding of the α subunit of tryptophan synthase and of the two α proteolytic fragments: Evidence for stepwise unfolding of the two α domains, *Biochemistry* **21:**2586–2592.

Miles, E. W., Phillips, R. S., Yeh, H. J., and Cohen, L. A., 1986, Isomerization of (3S)-2,3-dihydro-5-fluoro-L-tryptophan and of 5-fluoro-L-tryptophan catalyzed by tryptophan synthase: Studies using fluorine-19 nuclear magnetic resonance and difference spectroscopy, *Biochemistry* **25:**4240–4249.

Miles, E. W., Kawasaki, H., Ahmed, S. A., and Bauerle, R., 1987, Bacterial tryptophan synthase: A multienzyme complex which contains pyridoxal phosphate, in *Biochemistry of Vitamin B6, Proceedings of the 7th International Congress of Chemical and Biological Aspects of Vitamin B6 Catalysis* (T. Korpela and P. Christen, eds.), pp. 73–81, Birkhauser Verlag AG, Basel.

Miles, E. W., McPhie, P., and Yutani, K., 1988, Evidence that glutamic acid 49 of tryptophan synthase α subunit is a catalytic residue. Inactive mutant proteins substituted at position 49 bind ligands and transmit ligand-dependent to the β subunit, *J. Biol. Chem.* **263:**8611–8614.

Miles, E. W., Kawasaki, H., Ahmed, S. A., Morita, H., Morita, H., and Nagata, S., 1989, The β subunit of tryptophan synthase. Clarification of the roles of histidine 86, lysine 87, arginine 148, cysteine 170, and cysteine 230, *J. Biol. Chem.* **264:**6280–6287.

Miles, E. W., Ahmed, S. A., and Kayastha, A. M., 1991, Exploring the tryptophan synthase $\alpha_2\beta_2$ multienzyme complex by X-ray crystallography and site-directed mutagenesis, in *Enzymes Dependent on Pyridoxal Phosphate and Other Carbonyl Compounds as Cofactors; Proceedings of the 8th International Symposium on Vitamin B6 and Carbonyl Catalysis* (T. Fukui, H. Kagamiyama, K. Soda, and H. Wada, eds.), pp. 249–256, Pergamon Press, Oxford.

Miles, E. W., Ahmed, S. A., Hyde, C. C., Kayastha, A. M., Yang, X.-J., Ruvinov, S. B., and Lu,

Z., 1994, Tryptophan synthase, in *Molecular Aspects of Enzyme Catalysis* (K. Soda, T. Fukui, and K. Soda, eds.), pp. 127–146, Kodansha, Ltd., Tokyo, Japan.

Milton, D. L., Napier, M. L., Myers, R. M., and Hardman, J. K., 1986, In vitro mutagenesis and overexpression of the *Escherichia coli trpA* gene and the partial characterization of the resultant tryptophan synthase mutant α-subunits, *J. Biol. Chem.* **261**:16604–16615.

Mozzarelli, A., Peracchi, A., Rossi, G. L., Ahmed, S. A., and Miles, E. W., 1989, Microspectrophotometric studies on single crystals of the tryptophan synthase $\alpha_2\beta_2$ complex demonstrate formation of enzyme–substrate intermediates, *J. Biol. Chem.* **264**:15774–15780.

Mozzarelli, A., Peracchi, A., Bettati, S., and Rossi,, G. L., 1991, Allosteric regulation of tryptophan synthase: A pKa change at the β-active site induced by α-subunit ligands, in *Enzymes Dependent on Pyridoxal Phosphate and Other Carbonyl Compounds as Cofactors* (T. Fukui, H. Kagamiyama, K. Soda, and H. Wada, eds.), pp. 273–275, Pergamon Press, Oxford.

Murry-Brelier, A., and Goldberg, M. E., 1988, Kinetics of appearance of an early immunoreactive species during the refolding of acid-denatured *Escherichia coli* tryptophan synthase β_2 subunit, *Biochemistry* **27**:7633–7640.

Murry-Brelier, A., and Goldberg, M. E., 1990, Mechanism of inactivation of the β_2 subunit of *Escherichia coli* tryptophan synthase by monoclonal antibodies, *J. Biol. Chem.* **265**:7987–7993.

Nagata, S., Hyde, C. C., and Miles, E. W., 1989, The α subunit of tryptophan synthase. Evidence that aspartic acid 60 is a catalytic residue and that the double alteration of residues 175 and 211 in a second-site revertant restores the proper geometry of the substrate binding site, *J. Biol. Chem.* **264**:6288–6296.

Ogasahara, K., Yutani, K., Suzuki, M., and Sugino, Y., 1984, Thermal denaturation of tryptophan synthase α subunit. Comparison of the values of thermodynamic parameters of unfolding obtained from van't Hoff analysis of CD measurement with those from calorimetry, *Int. J. Pept. Protein Res.* **24**:147–54.

Ogasahara, K., Hiraga, K., Ito, W., Miles, E. W., and Yutani, K., 1992, Origin of the mutual activation of the α and β subunits in the $\alpha_2\beta_2$ complex of tryptophan synthase. Effect of alanine or glycine substitutions at proline residues in the α subunit, *J. Biol. Chem.* **267**:5222–5228.

Phillips, R. S., Miles, E. W., and Cohen, L. A., 1984, Interactions of tryptophan synthase, tryptophanase, and pyridoxal phosphate with oxindolyl-L-alanine and 2,3-dihydro-L-tryptophan: Support for an indolenine intermediate in tryptophan metabolism, *Biochemistry* **23**:6228–6234.

Phillips, R. S., Miles, E. W., and Cohen, L. A., 1985, Differential inhibition of tryptophan synthase and of tryptophanase by the two diastereoisomers of 2,3-dihydro-L-tryptophan. Implications for the stereochemistry of the reaction intermediates, *J. Biol. Chem.* **260**:14665–14670.

Remeta, D. P., Miles, E. W., and Ginsburg, A., 1992, Thermally induced unfolding of the tryptophan synthase multienzyme complex, *Biophys. J.* **61**:A213.

Remeta, D. P., Miles, E. W., and Ginsburg, A., 1993, Differential scanning calorimetric studies of the tryptophan synthase multienzyme $\alpha_2\beta_2$ complex, *Biophys. J.* **64**:A175.

Remeta, D. P., Miles, E. W., and Ginsburg, A., in preparation, Differential scanning calorimetric studies of the tryptophan synthase $\alpha_2\beta_2$ complex and the isolated α and β subunits.

Ross, C. M., and Winkler, M. E., 1988, Structure of the *Caulobacter crescentrus* trpFBA operon, *J. Bacteriol.* **170**:757–768.

Roy, M., Keblawi, S., and Dunn, M. F., 1988a, Stereoelectronic control of bond formation in *Escherichia coli* tryptophan synthase: Substrate specificity and enzymatic synthesis of the novel amino acid dihydroisotryptophan, *Biochemistry* **27**:6698–6704.

Roy, M., Miles, E. W., Phillips, R. S., and Dunn, M. F., 1988b, Detection and identification of transient intermediates in the reactions of tryptophan synthase with oxindolyl-L-alanine and 2,3-dihydro-L-tryptophan. Evidence for a tetrahedral (gem-diamine) intermediate. *Biochemistry* **27**:8661–8669.

Ruvinov, S. B., and Miles, E. W., 1992, Subunit communication in the tryptophan synthase $\alpha_2\beta_2$ complex. Effects of β subunit ligands on proteolytic cleavage of a flexible loop in the α subunit, *FEBS Lett.* **299:** 197–200.

Ruvinov, S. B., and Miles, E. W., 1994, Thermal inactivation of tryptophan synthase: Stabilization by protein–protein interaction and protein–ligand interaction, *J. Biol. Chem.* **269:**11703–11706.

Ruvinov, S. B., Yang, X.-J., Parris, K. D., Banik, U., Ahmed, S. A., Miles, E. W., and Sackett, D. L., in preparation, The tryptophan synthase $\alpha_2\beta_2$ complex: Alteration in the tunnel probed by the interaction of Nile Red.

Sackett, D. L., and Wolff, J., 1987, Nile red as a polarity-sensitive fluorescent probe of hydrophobic protein surfaces, *Anal. Biochem.* **167:**228–234.

Seifert, T., Bartholmes, P., and Jaenicke, R., 1985, Influence of cofactor pyridoxal 5′-phosphate on reversible high-pressure denaturation of isolated β_2 dimer of tryptophan synthase bienzyme complex from *Escherichia coli, Biochemistry* **24:**339–345.

Shirvanee, L., Horn, V., and Yanofsky, C., 1990, *Escherichia coli* mutant *trp*34 has an Asp → Asn change at active site residue 60 of the tryptophan synthase α chain, *J. Biol. Chem.* **265:**6624–6625.

Silva, J. L., Miles, E. W., and Weber, G., 1986, Pressure dissociation and conformational drift of the β dimer of tryptophan synthase, *Biochemistry* **25:**5780–5786.

Strambini, G. B., Cioni, P., Peracchi, A., and Mozzarelli, A., 1992a, Characterization of tryptophan and coenzyme luminescence in tryptophan synthase from *Salmonella typhimurium, Biochemistry* **31:**7527–7534.

Strambini, G. B., Cioni, P., Peracchi, A., and Mozzarelli, A., 1992b, Conformational changes and subunit communication in tryptophan synthase: Effect of substrates and substrate analogs, *Biochemistry* **31:**7535–7542.

Sugisaki, Y., Ogasahara, K., Miles, E. W., and Yutani, K., 1990, Scanning calorimetric study of tryptophan synthase α-subunits from *Escherichia coli, Salmonella typhimurium,* and an interspecies hybrid, *Thermochim. Acta* **163:**117–122.

Swift, S., and Stewart, G. S., 1991, The molecular biology of tryptophan synthase: A model for protein–protein interaction, *Biotechnol. Genet. Eng. Rev.* **9:**229–294.

Swift, S., Kuhn, J., and Stewart, G. S., 1992, Selection and analysis of noninteractive mutants in the *Escherichia coli* tryptophan synthase α subunit, *Mol. Gen. Genet.* **233:**129–135.

Tschopp, J., and Kirschner, K., 1980a, Kinetics of cooperative ligand binding to the apo β_2 subunit of tryptophan synthase and its modulation by the α subunit, *Biochemistry* **19:**4521–4527.

Tschopp, J., and Kirschner, K., 1980b, Subunit interactions of tryptophan synthase from *Escherichia coli* as revealed by binding studies with pyridoxal phosphate analogues, *Biochemistry* **19:**4514–4521.

Tweedy, N. B., Hurle, M. R., Chrunyk, B. A., and Matthews, C. R., 1990, Multiple replacements at position 211 in the α subunit of tryptophan synthase as a probe of the folding unit association reaction, *Biochemistry* **29:**1539–1545.

Weischet, W. O., and Kirschner, K., 1976a, The mechanism of the synthesis of indoleglycerol phosphate catalyzed by tryptophan synthase from *Escherichia coli.* Steady-state kinetic studies, *Eur. J. Biochem.* **65:**365–373.

Weischet, W. O., and Kirschner, K., 1976b, Steady-state kinetic studies of the synthesis of indoleglycerol phosphate catalyzed by the α subunit of tryptophan synthase from *Escherichia coli.* Comparison with the $\alpha_2\beta_2$ complex, *Eur. J. Biochem.* **65:**375–385.

Wierenga, R. K., Noble, M. E. M., Postma, J. P. M., Groendijk, H., Kalk, K. H., Hol, W. G. J., and Opperdoes, F. R., 1991, The crystal structure of the "open" and the "closed" conformation of the flexible loop of trypanosomal triosephosphate isomerase, *Proteins* **10:**33–49.

Wiesinger, H., and Hinz, H. J., 1984a, Energetic adaptation of ligand binding to subunit structure of tryptophan synthase from *Escherichia coli*, *Biochemistry* **23:**4928–4934.

Wiesinger, H., and Hinz, H. J., 1984b, Linkage of subunit interactions, structural changes, and energetics of coenzyme binding in tryptophan synthase, *Biochemistry* **23:**4921–4928.

Wiesinger, H., Bartholmes, P., and Hinz, H. J., 1979, Subunit interaction in tryptophan synthase of *Escherichia coli:* Calorimetric studies on association of α and β_2 subunits, *Biochemistry* **18:**1979–1984.

Wilhelm, P., Pilz, I., Lane, A. N., and Kirschner, K., 1982, Small-angle X-ray scattering studies of tryptophan synthase from *Escherichia coli* and its α and β_2 subunits, *Eur. J. Biochem.* **129:**51–56.

Wilmanns, M., Hyde, C. C., Davies, D. R., Kirschner, K., and Jansonius, J. N., 1991, Structural conservation in parallel βα-barrel enzymes that catalyse three sequential reactions in the pathway of tryptophan biosynthesis, *Biochemistry* **30:**9161–9169.

Wilmanns, M., Priestle, J. P., Niermann, T., and Jansonius, J. N., 1992, Three-dimensional structure of the bifunctional enzyme phosphoribosylanthranilate isomerase: Indoleglycerolphosphate synthase from *Escherichia coli* refined at 2.0 Å resolution, *J. Mol. Biol.* **233:**477–507.

Xu, Y., and Abeles, R. H., 1993, Inhibition of tryptophan synthase by (1-fluorovinyl)glycine, *Biochemistry* **32:**806–811.

Yang, X.-J., and Miles, E. W., 1992, Threonine 183 and adjacent flexible loop residues in the α subunit of tryptophan synthase have critical roles in modulating the enzymatic activities of the β subunit in the $\alpha_2\beta_2$ complex, *J. Biol. Chem.* **267:**7520–7528.

Yang, X.-J., and Miles, E. W., 1993, A novel intersubunit repair mechanism in the tryptophan synthase $\alpha_2\beta_2$ complex: Critical role of β subunit lysine 167 in intersubunit communication, *J. Biol. Chem.* **268:**22269–22272.

Yanofsky, C., 1971, Tryptophan biosynthesis in *Escherichia coli.* Genetic determination of the proteins involved, *JAMA* **218:**1026–1038.

Yanofsky, C., 1987, Tryptophan synthetase: Its charmed history, *Bioessays* **6:**133–137.

Yanofsky, C., 1989, Tryptophan synthetase of *E. coli:* A multifunctional, multicomponent enzyme; a commentary by Charles Yanofsky, *Biochim. Biophys. Acta* **1000:**133–137.

Yanofsky, C., and Crawford, I. P., 1972, Tryptophan synthetase, in *The Enzymes* (P. D. Boyer, eds.), pp. 1–31, Academic Press, New York.

Yanofsky, C., and Horn, V., 1972, Tryptophan synthetase α chain positions affected by mutations near the ends of the genetic map of *trp*A of *Escherichia coli*, *J. Biol. Chem.* **247:**4494–4498.

Yanofsky, C., and Rachmeler, M., 1958, The exclusion of free indole as an intermediate in the biosynthesis of tryptophan in *Neurospora crassa*, *Biochim. Biophys. Acta* **28:**641–642.

Yanofsky, C., Helinski, D. R., and Maling, B. D., 1961, The effects of mutation on the composition and properties of the A protein of *Escherichia coli* tryptophan synthetase, *Cold Spring Harbor Symp. Quant. Biol.* **26:**11–24.

Yanofsky, C., Yee, M. C., and Horn, V., 1993, Partial revertants of tryptophan synthetase α chain active site mutant Asp60 → Asn, *J. Biol. Chem.* **268:**8213–8220.

York, S. S., 1972, Kinetic spectroscopic studies of substrate and subunit interactions of tryptophan synthetase, *Biochemistry* **11:**2733–2740.

Yutani, K., Ogasahara, K., Suzuki, M., and Sugino, Y., 1979, Comparison of denaturation by guanidine hydrochloride of the wild type tryptophan synthase α-subunit of *Escherichia coli* and two mutant proteins (Glu 49 → Met or Gln), *J. Biochem (Tokyo)* **85:**915–921.

Yutani, K., Ogasahara, K., and Sugino, Y., 1980, pH dependence of stability of the wild-type

Calorimetric study of tryptophan synthase α-subunit and two mutant proteins, *Int. J. Pept. Protein Res.* **20:**331–336.

Yutani, K., Ogasahara, K., Aoki, K., Kakuno, T., and Sugino, Y., 1984a, Effect of amino acid residues on conformational stability in eight mutant proteins variously substituted at a unique position of the tryptophan synthase α-subunit, *J. Biol. Chem.* **259:**14076–14081.

Yutani, K., Sato, T., Ogasahara, K., and Miles, E. W., 1984b, Comparison of denaturation of tryptophan synthase α-subunits from *Escherichia coli, Salmonella typhimurium,* and an interspecies hybrid, *Arch. Biochem. Biophys.* **229:**448–454.

Yutani, K., Ogasahara, K., Tsujita, T., Kanemoto, K., Matsumoto, M., Tanaka, S., Miyashita, T., Matsushiro, A., Sugino, Y., and Miles, E. W., 1987a, Tryptophan synthase α subunit glutamic acid 49 is essential for activity. Studies with 19 mutants at position 49, *J. Biol. Chem.* **262:**13429–13433.

Yutani, K., Ogasahara, K., Tsujita, T., and Sugino, Y., 1987b, Dependence of conformational stability on hydrophobicity of the amino acid residue in a series of variant proteins substituted at a unique position of tryptophan synthase α subunit, *Proc. Natl. Acad. Sci. USA* **84:**4441–4444.

Zakin, M. M., Boulot, G., and Goldberg, M. E., 1980, Immunochemical study of the β chain of *Escherichia coli* tryptophan synthetase and its proteolytic fragments, *Eur. J. Immunol.* **10:**16–21.

Zetina, C. R., and Goldberg, M. E., 1980, A comparative study of the thermal inactivation of the isolated and associated domains within the β_2 subunit of *Escherichia coli* tryptophan synthetase. Evidence for strong interdomain interactions, *J. Biol. Chem.* **255:**4381–4385.

Zetina, C. R., and Goldberg, M. E., 1982, Kinetics of renaturation and self-assembly of intermediates on the pathway of folding of the β_2-subunit of *Escherichia coli* tryptophan-synthetase, *J. Mol. Biol.* **157:**133–148.

Zhao, G.-P., and Somerville, R. L., 1992, Genetic and biochemical characterization of the *trp*B8 mutation of *Escherichia coli* tryptophan synthase: An amino acid switch at the sharp turn of the trypsin sensitive "hinge" region diminishes substrate binding and alters solubility, *J. Biol. Chem.* **267:**526–541.

Zhao, G.-P., and Somerville, R. L., 1993, A single amino acid switch within the "hinge" region of the tryptophan synthase β subunit of *Escherichia coli* that leads to diminished association with α subunit and arrested conversion of ES III to product, *J. Biol. Chem.* **268:**14921–14931.

Chapter 9

Structure and Function in the Tubulin Dimer and the Role of the Acidic Carboxyl Terminus

Dan L. Sackett

1. INTRODUCTION

Microtubules are intracellular polymeric structures found in nearly all eukaryotic cells. Two other major intracellular fiber systems—the intermediate filaments and the actin-based microfilaments—combine with microtubules (MTs) to form the cytoskeleton, the dynamic architectural coordinate system of the cell. MTs are steady-state dynamic polymers, constantly exchanging with a pool of the subunit protein tubulin. In addition to tubulin, the intracellular polymer should be considered to contain the many other proteins that bind to MTs, move along the surface, and change their properties. Recent reviews and monographs on MTs and tubulin have emphasized different aspects, such as MT structure (Amos and Amos, 1991; Linck, 1989; Wade and Chretien, 1993), tubulin structure (Luduena *et al.*, 1992; Ponstingl *et al.*, 1984; Serrano and Avila, 1990), sequence and isotype analysis (Burns, 1991a; Burns and Surridge, 1990; Little and Seehaus, 1988; Luduena, 1993), tubulin/MT cell biology (Byard and Lange,

Dan L. Sackett Laboratory of Biochemical Pharmacology, National Institute of Diabetes and Digestive and Kidney Diseases, National Institutes of Health, Bethesda, Maryland 20892.

Subcellular Biochemistry, Volume 24. Proteins: Structure, Function, and Engineering, edited by B. B. Biswas and Siddhartha Roy. Plenum Press, New York, 1995.

1991; Huang, 1990), membrane tubulin (Stephens, 1986), and plant MT and tubulin (Fosket and Morejohn, 1992). Reviews of other topics will be cited in appropriate sections. This chapter covers material published through October 1993.

1.1. Tubulin Is the Subunit Protein of Microtubules

The term *microtubule* was coined by Slautterback (1963) and the structures described have subsequently been reported in many cells, largely due to the advent of glutaraldehyde fixation (Dustin, 1984; Porter, 1966). Microtubules are characteristic of all eukaryotic cells, although exceptions are known, such as mature mammalian erythrocytes. A typical MT is a long, unbranched, hollow cylinder, 25-nm outer diameter, 14-nm inner diameter, and variable in length. The wall consists of 13 "protofilaments," each composed of ~4-nm diameter protein subunits. These subunits are of two chemical types—α- and β-tubulin—and are arranged along the protofilament in alternating fashion (αβαβαβ . . .). Microtubules are found in a wide variety of arrangements including single MTs, arrays such as the interphase and mitotic cytoskeleton, the long, parallel arrays of the neuronal cytoskeleton (Hirokawa, 1991), the ordered architecture of cilia, flagella, and centrioles, and the dazzling diversity of geometric arrays found in many protoctists (reviewed by Grain, 1986). Many variations on the prototypical MT structure are known (reviewed by Unger *et al.*, 1990). MTs are important to a number of cellular functions, including cellular motility, intracellular movements such as mitosis, and axonal transport, secretion, and establishment of cellular anisotropy and polarity (reviewed by Dustin, 1984). The alkaloid colchicine has long been recognized as cytotoxic due to its interference with the mitotic spindle, even before the MT was recognized ultrastructurally (reviewed by Eigsti and Dustin, 1955; see also Dustin, 1984). Tracing colchicine binding through steps of protein purification with radioactive colchicine led to the isolation of the protein tubulin, the subunit protein of MT (Borisy and Taylor, 1967; Wilson and Friedkin, 1967). Tubulin is composed of two quite similar but nonidentical subunits known as α and β tubulin. Tubulin can be made to polymerize reversibly *in vitro:* Mg, GTP, warming, and accessory proteins (MAPs) promote assembly and Ca, GDP, and cooling inhibit it (reviewed by Burns, 1991b; Correia and Williams, 1983; Engelborghs, 1990).

1.2. Tubulin Occurs in Cells in Multiple Structural Forms

Tubulin occurs in cells in more than one form. Polymerized tubulin occurs in MT arrays which differ remarkably in architecture. The interphase arrays of MT in mammalian cells are composed of long MTs which usually originate at a MT organizing center near the nucleus (DeBrabander *et al.*, 1985; Huang, 1990).

The mitotic array of MT in the same cells has a different geometry, the classic mitotic spindle. In other cells, the MT arrays may have quite different forms, including ordered solid arrays of many MTs, as well as ordered planar arrays of MT tightly associated with cell plasma membranes (Gilbert *et al.*, 1991; Grain, 1986; Murray, 1984). MTs are steady-state structures and are in equilibrium with a critical concentration of unpolymerized tubulin. Unpolymerized tubulin occurs as a soluble, 6S protein. It may be found in cells at various concentrations, including concentrations sufficient to polymerize, suggesting active control of polymerization (Bibring and Baxandall, 1977; Harris *et al.*, 1989; Raff and Kaumeyer, 1973; Suprenant and Rebhun, 1983). Unpolymerized (dimeric as opposed to polymeric) tubulin may have biological roles beyond supporting polymer stability, since it is known to interact with a number of non-MT proteins, such as the G-proteins (Wang and Rasenick, 1991). In addition to these two forms, tubulin has been found in a membrane-bound form in a number of cells (Bhattacharyya and Wolff, 1975; Steiner, 1983; Stephens, 1981; reviewed by Stephens, 1986). In some cells, tubulin may be a major membrane protein component of certain membranes (Dentler, 1980; Matesic *et al.*, 1992; Stephens, 1977, 1991). Membrane tubulin has been reported to differ from cytoplasmic tubulin in isotype distribution and posttranslational modifications (Beltramo *et al.*, 1992; Matten *et al.*, 1990; Stephens, 1991; see also Section 1.5). Membrane tubulin has thus far only been reported in animals and a few protoctists, and its biological role is unclear.

1.3. Tubulin Forms Fall into Three Families: α, β, and γ

Soluble cytoplasmic tubulin, polymerized microtubule tubulin, and membrane tubulin are all composed of α and β tubulins. A third, recently identified form of tubulin is known as γ-tubulin. This tubulin is found associated with microtubule organizing centers such as centrioles and has been found in cells from mammals, fungi, and maize (Horio *et al.*, 1991; Joshi *et al.*, 1992; Oakley and Oakley, 1989; Oakley *et al.*, 1990; Stearns *et al.*, 1991). Compared to α- and β-tubulin, γ-tubulin is a relatively low-abundance protein. The structural studies reviewed here have all been done with α- and β-tubulin. However, the structure of γ-tubulin may be similar to that of α- and β-tubulin. Predicted structural features calculated based on the sequences of an α-, β-, and γ-tubulin are shown in Fig. 1. Comparison of tubulin sequences from a spectrum of organisms demonstrate the evolutionary conservation of α- and β-tubulin (Little and Seehaus, 1988; MacRae and Langdon, 1989; Siflow *et al.*, 1987) and α-, β-, and γ-tubulin (Burns, 1991a), and suggest that all three have similar tertiary structures (Burns, 1991a). In general, the α and β forms are heterogeneous. Heterogeneity is due to expression of multiple genes for α and β, as well as to posttranslational modifications of both subunits.

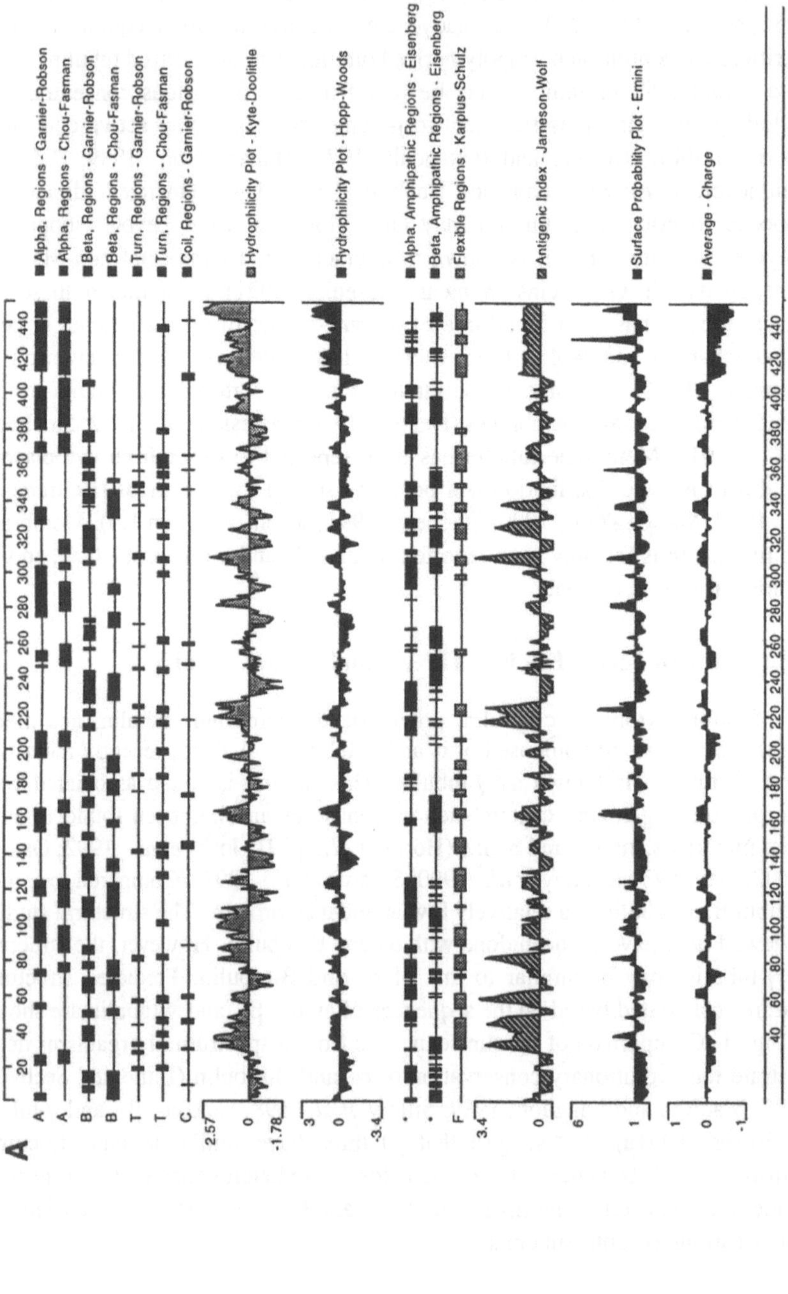
A
Alpha, Regions - Garnier-Robson
Alpha, Regions - Chou-Fasman
Beta, Regions - Garnier-Robson
Beta, Regions - Chou-Fasman
Turn, Regions - Garnier-Robson
Turn, Regions - Chou-Fasman
Coil, Regions - Garnier-Robson
Hydrophilicity Plot - Kyte-Doolittle
Hydrophilicity Plot - Hopp-Woods
Alpha, Amphipathic Regions - Eisenberg
Beta, Amphipathic Regions - Eisenberg
Flexible Regions - Karplus-Schulz
Antigenic Index - Jameson-Wolf
Surface Probability Plot - Emini
Average - Charge

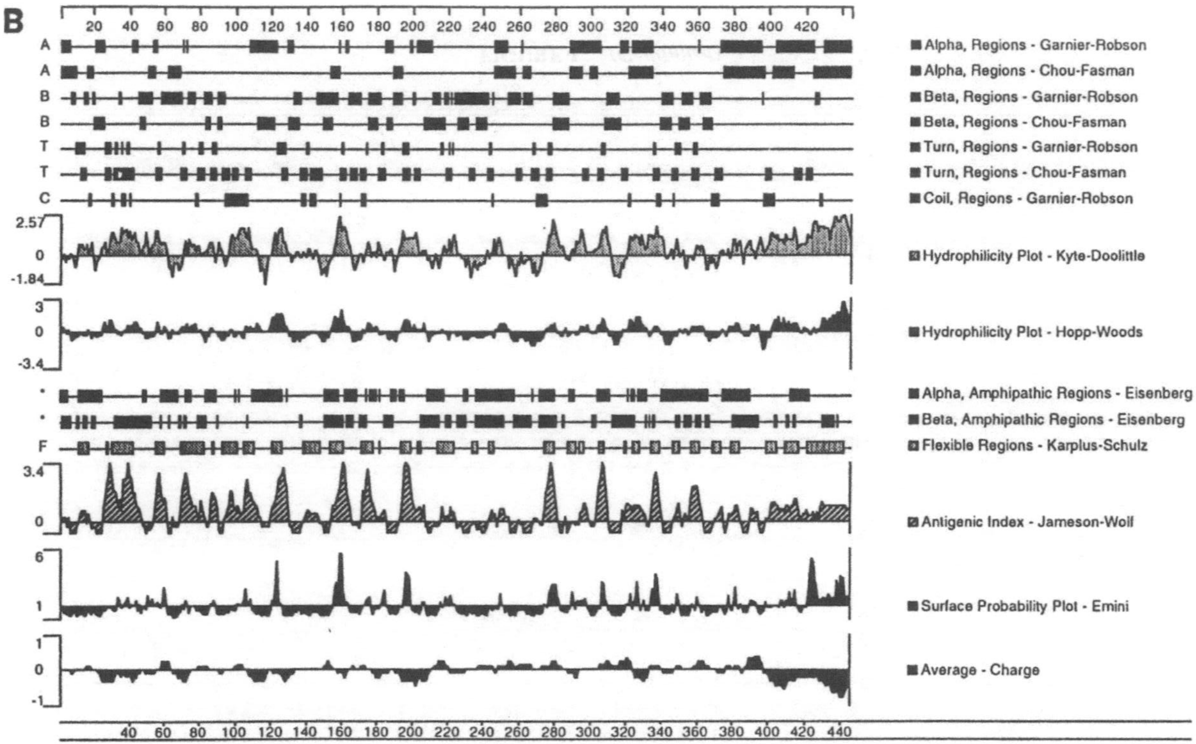

FIGURE 1. Structural features of α-, β-, and γ-tubulin. Plots of structural features calculated from sequence are presented for (A) α-tubulin; (B) β-tubulin; and (C) γ-tubulin. Sequence position is indicated by the numbers at the top and bottom. The sequences used are rat alpha (Lemischka *et al.*, 1981); rat beta (II) (Ginzburg *et al.*, 1985); and human gamma (Zheng *et al.*, 1991). Calculations were performed using the algorithms of Protean software (DNAstar, Madison, WI). Note that for each tubulin type, the plot labeled "alpha regions" means alpha helix; the plot labeled "beta regions" means beta strand, and so on.

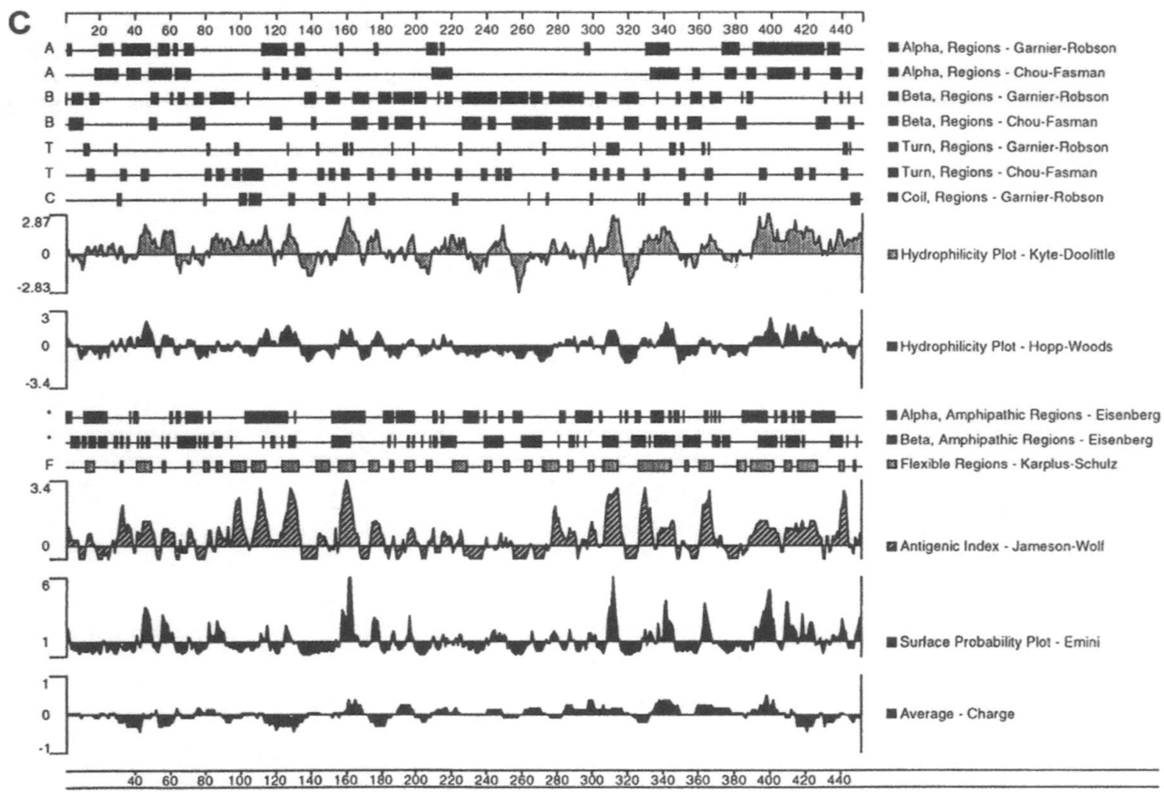

FIGURE 1. *(Continued)*

1.4. Multiple Genes Are Known for α and β Tubulin: Isotype Classes

Most eukaryotes have multiple genes for α or β or both. This is true of animals, plants, fungi, and protoctists, though animals and plants tend to have larger numbers of genes than do fungi and protoctists (reviewed by Cleveland and Sullivan, 1985; Lewis and Cowan, 1990; Luduena, 1993; Siflow *et al.*, 1987). Examples are known in which only a single gene exists for α and for β, or in which multiple genes produce proteins of identical sequence (Gaertig *et al.*, 1993; Siflow, 1991). Vertebrate genomes typically contain a highly conserved set of ~ 6 α- and ~ 6 β-tubulin genes (as well as numerous pseudogenes). These differ from each other primarily in the sequence of the final ~15 carboxyl-terminal amino acid residues. The sequence of this region in a given gene allows that gene to be classified in one of several recognized isotypic classes (Sullivan, 1988). This is shown in Table I for β sequences. The fraction of brain tubulin and avian erythrocyte tubulin that is made up of each isotype is also given. The major β isotype in brain is β II (the rat β sequence in Fig. 1 is a type II isotype). Substitutions at other positions in the sequence correlates with the isotype-defining C-terminus and these sequences may interact in concert during polymerization (Burns and Surridge, 1990; Luduena, 1993). Particular isotypes may

Table I
Beta Tubulin Isotype Characteristic C-Terminal Sequences

Type	Expression	Animal[a]	Sequence	Percent in brain[b]	Percent in RBC[c]
I	Major constitutive	Chicken cβ7	EEEEDFGEEAEEEA	3	10
	Many tissues	Rat rβ3			
II	Major brain	Chicken cβ1/cβ2	DEQGEFEEEEGEDEA	58	0
	Many tissues	Rat rβ1			
III	Minor brain	Chicken cβ4	EEEGEMYEDDEEESEQGAK	25	0
	Brain specific				
IVA	Major brain	Rat rβ2	EEGEFEEEAEEEVA	13	
	Brain specific				
IVB	Major testis	Chick cβ3	EEEGEFEEEAEEEVA		10
	Other tissues				
V	Minor constitutive	Chicken cβ5	NDGEEAFEDDEEEINE	0	0
	Not brain	(no mammalian)			
VI	Hemopoietic cells	Chicken cβ6	DVEEYEEAEASPEKET	0	80

[a]Only chicken and rat sequences are shown. Information is from Sullivan (1988). Sequences start at position 430 of type I.
[b]Bovine brain (Banerjee *et al.*, 1988).
[c]Avian erythrocytes (Cleveland *et al.*, 1990).

show quite similar tissue- or cell-type-specific distributions in different species (reviewed by Sullivan, 1988). The defining sequence for a given isotype class may be conserved across many species, for example, even between birds and mammals. The biological significance of this isotype diversity is unsettled (Joshi and Cleveland, 1990; Luduena, 1993). Distinct tubulin isotypes with different assembly properties nonetheless coassemble efficiently into the same polymer (Baker *et al.*, 1990; Lopata and Cleveland, 1987). Divergent isotypes of α and β are functionally interchangeable in some organisms (May, 1989; Schatz *et al.*, 1986). Furthermore, some organisms do not have multiple isoforms. Ciliates, for example, have a wide variety of tubulin-containing structures performing functions as varied as those in multicellular animals, and yet may have only one α- and one β-tubulin (Gaertig *et al.*, 1993; Siflow, 1991).

The carboxyl-terminal region of tubulin accounts for a significant amount of the variation found among the more than 150 known tubulin sequences (as of this writing there were 61 α, 91 β, and 6 γ sequences in the PIR database (Protein Identification Resource, National Biomedical Research Foundation, Washington, DC). Even including this region, the sequences of α- and β-tubulin are highly conserved evolutionarily (Burns, 1991a; Gaertig *et al.*, 1993). This conservation in sequence is reflected in conservation of function, demonstrated, for example, by the ability of *Physarum* myxamoebal tubulin to incorporate into interphase and mitotic MT of mammalian cells (Prescott *et al.*, 1989). Combinations within a single protein can also be functional: a β-tubulin constructed of the amino part of chicken βII and the carboxyl 25% of yeast β sequences appeared to fold properly and to incorporate into mouse MTs (Bond *et al.*, 1986). In addition to sequence conservation, tubulin gene structure shows remarkable conservation across many species, though differences are seen between kingdoms (Fosket and Morejohn, 1992; Siflow *et al.*, 1987).

1.5. Posttranslational Modifications Increase Tubulin Heterogeneity

Tubulin is subject to several posttranslational modification reactions (reviewed by Greer and Rosenbaum, 1989; Purich and Scaife, 1985). The carboxyl terminal tyrosine on α-tubulin is removed by a carboxypeptidase and may be replaced by an RNA-independent tubulin-tyrosine ligase (reviewed by Barra *et al.*, 1988; Thompson, 1982). Consequently, intracellular tubulin is a mixture of molecules with and without tyrosine at the α-carboxyl terminus. α-Tubulin can also be acetylated on Lys40 (LeDizet and Piperno, 1987; L'Hernault and Rosenbaum, 1985). The proportion of neural membrane tubulin that is acetylated is much higher than that for soluble tubulin, while the proportion tyrosinated was similar in both tubulin fractions (Beltramo *et al.*, 1992). Phosphorylation of tubulin has been repeatedly described, on both α and β subunits (Eipper, 1974; Ikeda and Steiner, 1979; Kadowaki *et al.*, 1985; Wandosell *et al.*, 1987). The α subunit is apparently not phosphorylated *in vivo*, and in brain tubulin only the βIII isotype is phosphorylated on Ser444 (Alexander *et al.*, 1991; Luduena *et al.*, 1988). γ-Glutamylation is an unusual and recently described modification. As many as six glutamic acids may be added to Glu445 of α- and Glu435 of β-tubulin (Alexander *et al.*, 1991; Edde *et al.*, 1990; Redeker *et al.*, 1991; Rudiger *et al.*, 1992). Following the initial linkage to the γ-carboxyl, subsequent glutamic acid residues (at least the next two) are added through the α-carboxyl of the preceding unit (Redeker *et al.*, 1991). This variable modification will clearly result in increased heterogeneity, as well as increasing the already considerable concentration of acidic residues in the extreme carboxyl termini of both α and β. Though close to the carboxyl-terminus, this modification does not prevent tyrosination/detyrosination at the terminus (Edde *et al.*, 1992). Tubulin has also been reported to be modified by ADP-ribosylation of α and β by eukaryotic ADP-ribosyltransferase (Raffaelli *et al.*, 1992; Scaife *et al.*, 1992).

The consequences of these modifications can be hard to detect. Acetylation and tyrosination are reported to be associated with MT stability (Bulinski *et al.*, 1988; Kreis, 1987; Piperno *et al.*, 1987; Schulze *et al.*, 1987; Webster and Borisy, 1989), though the cause-and-effect relationships are unsettled (Webster *et al.*, 1990). Phosphorylation has been reported to increase tubulin affinity for membranes (Hargreaves *et al.*, 1986) and to decrease incorporation into MTs (Wandosell *et al.*, 1987). In general, however, modified and unmodified tubulin behave quite similarly, at least *in vitro*. These modifications do not all occur in all cells and may occur in developmentally significant patterns. Tubulin heterogeneity has been found to change (increase) during development in animals (Denoulet *et al.*, 1988; George *et al.*, 1981; reviewed by Cambray-Deakin, 1991), and plants (Hussey *et al.*, 1988). However, the significance of these modifications to developmental processes is not clear.

1.6. Tubulin Contains Bound Guanine Nucleotides That Regulate Polymerization

Polymerization of tubulin is dependent on the presence of GTP and is accompanied by the hydrolysis of GTP to GDP (Carlier, 1989; Hamel, 1990; Weisenberg *et al.*, 1976). It currently seems likely that hydrolysis of GTP is coupled to the addition of a dimer to the growing polymer. It has long been established that the tubulin dimer contains two molecules of guanine nucleotide. One is GTP, which is nonexchangeable with exogenous nucleotide, bound to α, and cannot be released without denaturing the protein. The other molecule is bound to β, is exchangeable with exogenous nucleotide, and may be GTP or GDP. This site has been probed by photoaffinity labeling with azido analogues of GTP as well as directly with GTP itself. Several sequence regions have been identified as contributing to this site by photoaffinity labeling: β7-15 (GTP bound to Cys12), (Shivanna *et al.*, 1993), β63-77 (Chavan *et al.*, 1990; Linse and Mandelkow, 1988), and β155-174 (Hesse *et al.*, 1987). GTP-binding properties of altered β-tubulins produced by site-directed mutagenesis indicated that β-Asp224 may interact with the Mg^{2+}, which binds with GTP (Farr and Sternlicht, 1992). Conversion of Glu69, Asp177, and Asp197 to the corresponding amides increased affinity for GTP in the absence of Mg^{2+} by about 20-fold. The binding properties of the mutant tubulins indicates that the residues that comprise the GTP site cannot be predicted by analogy with other GTP binding proteins (Farr and Sternlicht, 1992; Sternlicht *et al.*, 1987).

If this site contains GTP, the dimer is polymerization competent, and it is this GTP that is hydrolyzed consequent to polymerization. Since nonhydrolyzable analogues of GTP support polymerization, hydrolysis of GTP is not necessary for polymerization. Rather, GTP maintains a polymerization competent conformation, and polymers assembled with non- or poorly hydrolyzable GTP analogues are stable (Hyman *et al.*, 1992). Hydrolysis to GTP is necessary for rapid depolymerization, and therefore is at the heart of the "dynamic instability" that MTs exhibit (Bayley and Martin, 1991; Mitchison and Kirschner, 1984; O'Brien *et al.*, 1987; Stewart *et al.*, 1990).

1.7. Tubulin Binds Several Classes of Pharmacologic Agents

A bewildering number of small molecules are known that bind to tubulin. These include both natural products and synthesized agents, and bind to a number of different sites (reviewed by Hamel, 1990; Iwasaki, 1993). These agents, which are active in micromolar concentrations, may be broadly classified as polymerization promoting and polymerization poisoning agents. The former class currently contains one example, taxol, which promotes polymerization of tubulin and inhibits depolymerization. The location of the taxol binding site is unknown, but underivitized taxol directly photolabels β-tubulin preferentially (Rao *et al.*, 1992). Taxol is currently under intense study because of promising antitumor activity (reviewed by Rowinsky and Donehower, 1991).

Many of the depolymerizing agents bind to one of two sites, the colchicine site and the vinblastine site. Colchicine site agents include colchicine (reviewed by Hastie, 1991), podophyllotoxin, steganacin, combretastatin (latter three reviewed by Sackett, 1993), benzimidazole carbamates (reviewed by Lacey, 1988), and others. Direct photoaffinity labeling of tubulin with colchicine labels β preferentially, indicating that the primary binding site is here (Wolff *et al.*, 1991). Some labeling of α is seen and is enhanced by certain pretreatments and by extended irradiation and by the presence of a spacer group between colchicine and the photoactivatable moiety (Floyd *et al.*, 1989). This, plus the ability of colchicine to photosensitize covalent dimerization of tubulin (Wolff *et al.*, 1992), indicates that the binding site may be near the α/β interface (Floyd *et al.*, 1989;

Lacey, 1988; Lin *et al.*, 1989; Wolff *et al.*, 1991). Peptides that are photoaffinity labeled in the binding site on β have been isolated and localized to residues 1-36 and 214-241 (Uppuluri *et al.*, 1993). Binding of colchicine to tubulin induces conformational changes in tubulin, recently identified as a local unfolding of a helical region near β (390) (Sackett and Varma, 1993).

Vinblastine site agents include vinblastine and vincristine, maytansine, rhizoxin, and others (reviewed by Himes, 1991; Hamel, 1992). Direct photoaffinity labeling with vinblastine labels primarily the α subunit (threefold over β) (Wolff *et al.*, 1991). Labeling with a photoactive analogue of vinblastine labels α 1.5-fold greater than β (Safa *et al.*, 1987). Nonetheless, the inhibition of GTP exchange by vinblastine site agents has led to the suggestion that this binding site is also primarily on the β subunit (Hamel, 1992). In order to accommodate the photoaffinity labeling results, however, the binding site must be very close to α, again possibly at the α/β interface.

The potency of agents of both classes varies widely between species and particularly between kingdoms. Synthetic herbicides that poison plant but not animal tubulin assembly are known (reviewed by Morejohn and Fosket, 1991). In addition to these agents, tubulin is sensitive to sulfhydryl-active compounds such as mercurials (Graff *et al.*, 1993; Sager *et al.*, 1983; Vogel *et al.*, 1985), the benzophenanthridine alkaloid sanguinarine (Wolff and Knipling, 1993), and other sulfhydryl-directed agents (Bai *et al.*, 1989; Potchoo *et al.*, 1986), which poison polymerization.

1.8. Many Proteins Bind to Tubulin

Microtubules are not composed only of tubulin. A large number of other proteins interact with the surface of the MT, stabilizing the polymer, covalently modifying it, moving along it, or linking it to other parts of the cytoskeleton. In addition, many proteins likely organize themselves on the large surfaces presented by MTs in many cells (Zimmerman and Minton, 1993). The microtubule associated proteins (MAPs) bind tightly enough to copurify with tubulin through cycles of temperature-induced assembly and disassembly and generally enhance polymer formation (reviewed by Burgoyne, 1991; Chapin and Bulinski, 1992; Lee, 1993). ATPase translocator or motor proteins such as kinesin and dynein are responsible for intracellular transport along MTs (reviewed by Skoufias and Scholey, 1993; Walker and Sheetz, 1993). Other proteins that bind to MTs include various kinases, including c-mos (Zhou *et al.*, 1991) and v-mos (Bai *et al.*, 1992), other kinases (Crute and VanBuskirk, 1992), G-proteins (Roychowdhury *et al.*, 1993), glycolytic enzymes (Somers *et al.*, 1990; Walsh *et al.*, 1989), and others (Kambadur *et al.*, 1990). Many, possibly most, of these proteins interact with the C-terminal region of α- and/or β-tubulin. MAPs have been shown to interact here by several lines of evidence. Proteolysis with subtilisin indicates that MAPs bind to the C-termini (Paschal *et al.*, 1989; Rodionov *et al.*, 1990; Sackett *et al.*, 1985; Serrano *et al.*, 1984a), and proteolysis with carboxypeptidase indicates that the binding site is not in the final 6-8 residues (Vera *et al.*, 1989). A C-terminal binding site has also been indicated by displacement of MAPs by antibodies to C-terminal sequences (Vera *et al.*, 1988) and by drug binding (Mazumdar *et al.*, 1992; Wallin *et al.*, 1985). Interactions with peptides of defined sequence indicate the importance of residues β (434-440) for MAP1, MAP2, and tau binding (Cross *et al.*, 1991; Littauer *et al.*, 1986). Peptides with nearby sequences, such as β(422-434) (Maccioni *et al.*, 1988; Cross *et al.*, 1991) show lower binding by MAP1 and 2 but still significant binding to tau (Littauer *et al.*, 1986). Peptides from the α-tubulin C-terminus were less potent, though a tau binding site at the N-terminus of α was found (Littauer *et al.*, 1986). ATPase motor proteins have also been shown to bind to the C-terminal region by antibody binding (Goldsmith *et al.*, 1991) and by the effects of proteolysis and by mutual inhibition of binding with MAPs, which indicate that MAPs such as MAP2 do share a binding site with ATPase motor proteins (kinesin, dynein) (Lopez and Sheetz, 1993; Paschal *et al.*, 1989) or do not (Rodionov *et al.*, 1990). One kinesin molecule can bind per tubulin heterodimer (Harrison *et al.*, 1993). Proteolysis studies also indicate that G-proteins interact with tubulin C-terminal regions, probably in a site different from that for MAPs (Wang and Rasenick, 1991).

2. STRUCTURE OF THE TUBULIN DIMER

2.1. Size of the Dimer

The size of the tubulin dimer is known from three types of data: sequence, physical measurements, and imaging techniques. The coding sequence of α- and β-tubulin genes yields proteins of very nearly 50,000 Da each (50,135 for rat α and 49,937 for rat β sequences used in Fig. 1) (Ginzburg *et al.*, 1985; Lemischka *et al.*, 1981). A number of calculated structural features are shown in Fig. 1 for each sequence. There is little variation in the size of the protein coded by the more than 150 known tubulin sequences (yeast β is an exception; see Section 2.5.3). This is perhaps not surprising given the high degree of sequence homology characteristic of tubulin. The carboxyl terminal region of α and β has significant excess negative charge due to the density of acidic residues; γ-tubulin lacks this feature (Fig. 2).

Tubulin exists in solution as a dimer and α- and β-tubulin are always isolated in near-stoichiometric amounts. Hence tubulin is very likely an αβ heterodimer. Bifunctional cross-linkers can covalently couple α and β, as expected for an αβ heterodimer (Boekelheide, 1987; Sioussat and Boekelheide,

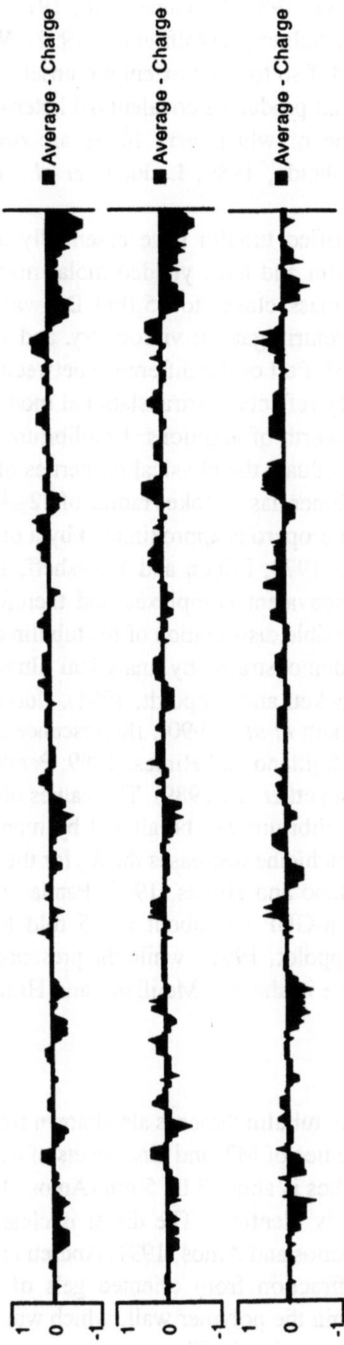

FIGURE 2. Comparison of charge distribution of β-, α-, and γ-tubulin. Plots of average charge versus sequence position are shown for β, α, and γ (top to bottom), aligned to normalized length. The plots were produced with Protean software (DNAstar, Madison, WI) using a nine-residue moving window average.

1989; Kirchner and Mandelkow, 1985; Luduena *et al.*, 1977; Serrano and Avila, 1985) as can photoactivated colchicine (Wolff *et al.*, 1992). Whether αα and ββ homodimers form *in vivo* and if so to what extent are unsettled questions. Many of the cross-linking studies that produce a covalent αβ heterodimer also produce heterogeneous products some of which very likely are αα and ββ covalent dimers (Sioussat and Boekelheide, 1989; Luduena *et al.*, 1977; Wolff *et al.*, 1992).

Physical studies on purified tubulin have essentially all been performed using mammalian brain tubulin and have yielded molar masses larger than the sequence mass. A subunit mass closer to 55,000 Da was obtained by light scattering, equilibrium ultracentrifugation, viscometry, and size exclusion chromatography (Lee *et al.*, 1973). Part of the difference between the sequence mass and the measured mass surely reflects posttranslational modifications, but there is no evidence for 5000 Da worth of additions. Equilibrium and hydrodynamic methods have been used to evaluate the physical properties of the tubulin dimer. These data indicate that the dimer has a Stokes radius of 42–44 nm, a sedimentation coefficient of 5.8 S, and properties approximated by a prolate ellipsoid with an axial ratio of 4–6 (Bryan, 1974; Frigon and Timasheff, 1975).

Tubulin dimers are noncovalent complexes and therefore dissociate with increasing dilution. The reversible dissociation of the tubulin αβ dimer into the α and β monomers has been demonstrated by analytical ultracentrifugation (Detrich and Williams, 1978; Sackett and Lippoldt, 1991), fluorescence of noncovalently bound Nile Red (Sackett *et al.*, 1990), fluorescence anisotropy of covalently bound fluorophores (Mejillano and Himes, 1989; Panda *et al.*, 1992), and by kinetics of proteolysis (Sackett *et al.*, 1989). The values obtained for K_d range from 0.2 to 1 μM. This equilibrium can be altered by manipulation of conditions. Notably, addition of colchicine decreases the K_d for the dimer 2.5 to 4-fold (Detrich *et al.*, 1982; Mejillano and Himes, 1989; Panda *et al.*, 1992; Sackett and Lippoldt, 1991). Tubulin-GDP has about a 3.5 fold lower K_d than does tubulin-GTP (Sackett and Lippoldt, 1991), while the presence of excess calcium results in about a twofold rise in the K_d (Mejillano and Himes, 1989).

2.2. Shape of the Dimer

The size and shape of the tubulin dimer is also known from electron microscopy and from diffraction studies of MT and zinc sheets of tubulin. The unit cell for the dimer from these studies is about 8 by 5 nm (Amos, 1982). The shape of the α and β subunits is nearly identical. The dimer is clearly asymmetric and each subunit is multilobed (Amos and Amos, 1991; Andreu *et al.*, 1992; Beese *et al.*, 1987a). X-ray fiber diffraction from oriented gels of MTs yielded nonuniform electron density within the polymer wall, which was consistent with the existence of three domains per subunit. The arrangement suggested significant intradimer and intrasubunit flexibility (Beese *et al.*, 1987a). The presence of soluble additives such as 1 M sucrose causes conformational changes in the structure of tubulin in MTs, possibly related to this flexibility (Beese *et al.*, 1987b).

The detailed three-dimensional structure of the tubulin dimer is not known since efforts to produce crystals of tubulin suitable for X-ray diffraction studies have thus far not succeeded. Proteolysis of the dimer and polymer has provided insights into the general folding architecture of tubulin. Antibody binding studies have identified sequences exposed on the surface of the dimer and polymer. These data, combined with that from electron microscopy and diffraction, inform the current understanding of dimer structure.

2.3. Proteolysis of the Dimer

Proteolysis has also been used to probe the structure of the tubulin dimer. Native tubulin is relatively resistant to proteolysis, as are most native globular proteins. Nonetheless, native tubulin is cleaved by proteases, but only in a few positions (Brown and Erickson, 1983). The fragments produced are quite similar for proteases of differing sequence specificity, suggesting that the cleavage occurs due to particularities of tubulin folding rather than to the sequence specificity of the enzyme. In particular, such cleavage is taken to reveal a region of the protein that is especially exposed, perhaps a loop or single strand of polypeptide connecting more globular compact domains or a terminal sequence that extends into the aqueous surroundings. Native tubulin appears to contain three such protease-sensitive regions: one very near the C-terminus and two internal regions that divide the protein approximately into thirds. These are regions I, II, and III on Fig. 3. Details of cleavage at regions II and III are better known than for region I. Sodium dodecyl sulfate (SDS) gel separations of digestions of tubulin by several proteases are shown in Fig. 4, coupled with native (nondenaturing) agarose gels of the same digests.

Not all known cleavages occur in these regions. Thermolysin cleaves α-tubulin in the dimer or polymer (Brown and Erickson, 1983). The cleavage occurs on the amino side of Ile42 (Paschal *et al.*, 1989). Other evidence also indicates that this site is accessible on the polymer. Acetylation of polymerized tubulin occurs in α-tubulin at Lys40 (LeDizet and Piperno, 1987). When tubulins from different species are compared, insertions are frequently found in this region (Burns, 1991a). An α gene construct containing an insertion of 17 residues at residue α41 preserves normal *in vivo* MT function (Schatz *et al.*, 1987), arguing that this site is on the surface of the dimer and the polymer. A region adjacent to the corresponding site in β-tubulin appears to be inaccessible in the polymer since an antibody to β28-38 recognizes the dimer but not tubulin polymers (Chene *et al.*, 1992).

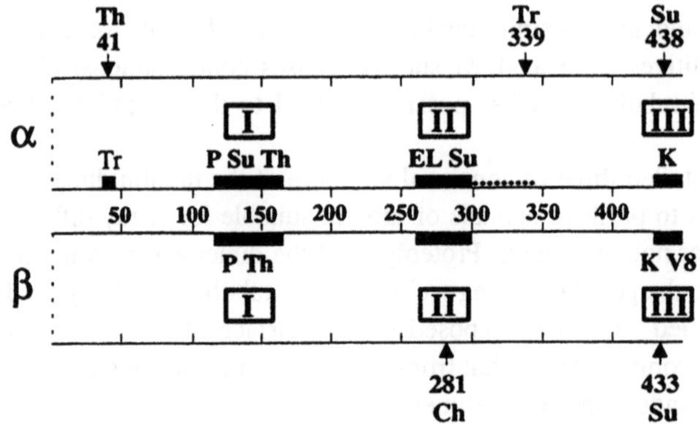

FIGURE 3. Principal proteolytic cleavage sites on α- and β-tubulin. The sequences of α- and β-tubulin are depicted by the pairs of labeled lines and the residue numbers are given in the middle. The principal cut sites are given for a number of proteases. The outer lines give the positions of cleavages for which the exact cleavage position is known. The inner lines give the approximate positions of cleavage for protease products for which the exact site is unknown. Regions I, II, and III are indicated (see text). Region II of α is shown extended somewhat C-terminal by a dashed line to include the trypsin cleavage. Proteases are: Ch, chymotrypsin; EL, endoprotease lys; K, protease K; P, pepsin; Su, subtilisin; Th, thermolysin; Tr, trypsin; V8, V8 protease.

2.3.1. Proteolysis in Region II

Cleavage in region II will be discussed first, because more is known about cleavage here than in region I. Both α and β have such protease-sensitive sites approximately two thirds of the way from the amino to the carboxyl ends of the proteins (region II). These sites are usually defined by the action of trypsin and chymotrypsin. Trypsin cuts α-tubulin after Arg339, producing amino and carboxyl terminal fragments αN and αC, with sequence mass of 37.6 kDa and 12.5 kDa, respectively. Chymotrypsin cuts β-tubulin after Tyr281, producing fragments βN and βC, with sequence mass of 31 kDa and 18.9 kDa, respectively (Mandelkow *et al.*, 1985; Sackett and Wolff, 1986). Sequence masses refer to rat α and β sequences (Ginzburg *et al.*, 1985; Lemischka *et al.*, 1981), but would not be very different for other tubulins. These cleavage sites are not accessible in polymerized tubulin (Brown and Erickson, 1983; Sackett and Wolff, 1986). Following cleavage, the fragments remain tightly associated (Sackett and Wolff, 1986) (see Fig. 4B), and the cleaved protein can still polymerize (Brown and Erickson, 1983), although the polymers include incomplete MTs with hooked and S-shaped cross section (Mandelkow *et al.*, 1985; Serrano *et al.*, 1988).

Other proteases also cleave tubulin in region II (see Fig. 3). The exact sites of these cleavages are not known, but cleavages occur in the region 280 ± 20 (de la Vina *et al.*, 1988). Secondary structure calculations indicate that a surface loop

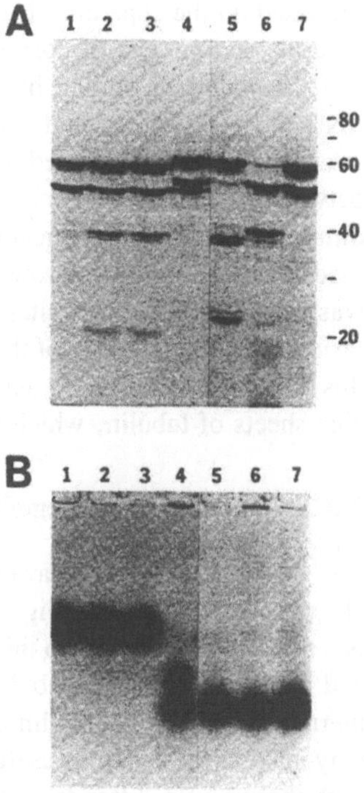

FIGURE 4. Gel electrophoretic analysis of proteolytic cleavages of native tubulin by several proteases. Tubulin was cleaved by several proteases at 30 °C for 1 hr and the reactions stopped by addition of specific inhibitors. Aliquots were separated by electrophoresis under denaturing or native conditions. (A) Denaturing electrophoresis. Samples are separated in SDS polyacrylamide gels following boiling with SDS and mercaptoethanol. Proteolytic enzymes used were: (1) protease K, (2) subtilisin carlsberg, (3) subtilisin BPN, (4) V8 protease, (5) chymotrypsin, (6) trypsin, and (7) no enzyme. Approximate molecular weights are shown on the right. (B) Native electrophoresis. Samples are separated in 1% agarose gels made in 0.1 M Mes assembly buffer. Samples were cooled on ice and the gels run at 4 °C. The origin of electrophoresis is at the top of the gel and migration is toward the anode. The loss of charge upon cleavage with subtilisin or protease K is clear. From Sackett and Wolff (1986).

around position 280 seems likely in β, corresponding to a peak in the hydrophilicity, antigenic index, and surface probability plots (see Fig. 1). This also corresponds to the position of cleavage of chymotrypsin. In α, the peak around 280 is still present, although a more prominent peak occurs around position 310. However, trypsin cleavage occurs at position 339, and this fact prescribes caution in such analysis.

Of the four fragments produced by trypsin and chymotrypsin (αN, αC, βN, and βC), all behave on SDS gels as expected based on mass, with the exception of αC, which yields an anomalously high M_r of about 17,000 (Sackett and Wolff, 1986; de la Vina *et al.*, 1988). The separation of intact α and β on SDS gels is anomalous (since they have essentially the same mass) and depends on pH, ionic strength, and detergent content of the gel (Best *et al.*, 1981; Bryan, 1974; Sackett *et al.*, 1985). Separation is due to anomalous behavior of α, which runs with an apparent M_r of 55,000–58,000, probably due to altered detergent binding (Bryan, 1974). Since both α and αC exhibit a similar "excess M_r" of ca. 5000, it is likely that the site of altered detergent binding responsible for this is contained in the 12.5-kDa fragment of αC (Sackett and Wolff, 1986). It should be noted that due to this behavior of α-tubulin, determining the site of cleavage in α-tubulin based on the apparent M_r of αC-terminal fragments on SDS gels can lead to incorrect assignment of position (Sackett and Wolff, 1986).

Cross-linking studies have revealed that intradimer contacts occur between αN and βC (Kirchner and Mandelkow, 1985; Serrano and Avila, 1985), while interdimer contacts (formed along the axis of the protofilaments in MTs) occur between αC and βN (Kirchner and Mandelkow, 1985). Proteolysis of the free dimer versus MTs demonstrates that the trypsin and chymotrypsin sites are inaccessible in the MT (Brown and Erickson, 1983; Sackett and Wolff, 1986). It was suggested that these sites are incorporated in the axis of the protofilaments (as opposed to the lumen of the MT) and thereby protected from proteolysis, and this was confirmed by the demonstration of similar protection from cleavage in zinc sheets of tubulin, which have no lumen (Sackett and Wolff, 1986).

2.3.2. Proteolysis in Region I

A second internal cleavage region has been reported for both α and β, and cleavage here (region I, Fig. 3) divides the proteins into approximately amino one third and carboxyl two thirds (de la Vina *et al.*, 1988). Papain, thermolysin, and subtilisin cleave α-tubulin in this region and elsewhere, but papain and thermolysin cleave β-tubulin only in this region (de la Vina *et al.*, 1988). This may make these enzymes useful probes of β-tubulin conformation. These results with thermolysin differ from those of others, who found that thermolysin only cleaved α (Brown and Erickson, 1983; Paschal *et al.*, 1989). The explanation for these differences is not clear. The details of cleavage in this region are much less clear than for region II. It is not clear if these cleavage sites are accessible on the polymer nor is it known if the fragments remain associated following cleavage or if tubulin cleaved in region I can polymerize. It seems likely that region I is not accessible for cleavage in the polymer. Extended proteolysis of MTs with subtilisin produces few of the fragments indicating internal cleavage (Rodionov *et al.*, 1990; Sackett and Wolff, 1986), although they are present and are produced at low yield from polymers produced in the presence of 4 M glycerol or taxol (Lobert and Correia, 1992). These fragments are produced upon extended digestion of the dimer and can be clearly seen in Figs. 4 and 5. In addition, some antibodies to this region do not bind to the polymer (Arevalo *et al.*, 1990) (see below).

Other aspects of cleavage in this region also remain to be defined. For example, the exact site of cleavage is not known for any enzyme that cleaves here in α or β. Residues 140 ± 25 define region I based on SDS gel fragment weights (de la Vina *et al.*, 1988). From the secondary structure calculations, a few possible restrictions on likely cleavage sites in this region may be suggested. In α tubulin, a prominent peak around residue 160 occurs in the hydrophilicity, antigenic index, and surface probability plots (this is particularly clear in the surface probability plot), has net positive charge, and may represent a surface loop candidate for proteolytic cleavage. A possible alternative is provided by a less prominent peak around position 125. A peak occurs in β around position 160, which also may contain the cleavage sites, but here the second peak around 125 is much more prominent. Antibodies to the peptides α(155-168) and β(153-165) both react with dimer, indicating surface exposure of this region. Since the α(155-168) epitope is not accessible on MTs, it has been suggested that this epitope is near a subunit–subunit interface (Arevalo *et al.*, 1990; Burns, 1991a). Additionally, this region (155-174) in β can be photolabeled with GTP and an antibody to this sequence prevents GTP incorporation and tubulin polymerization, results that also indicate a role for this region in interdimer contacts (Hesse *et al.*, 1987). In discussing the probable location of region I cleavages, it is important to remember the caution prescribed in the discussion of region II cleavage site assignment. Precise determination of position for cleavages in region I would be very helpful.

2.3.3. Proteolysis in Region III: Carboxyl-Terminal Cleavage

The third region of proteolytic cleavage in tubulin is at the C-terminus (region III, Fig. 3). The most commonly used protease for cleavage in this region is subtilisin, although other endopeptidases also cleave in this region (see Fig. 3). The exopeptidase, carboxypeptidase Y, has also been used to remove 6-8 residues from both subunits (Vera *et al.*, 1989; Mejillano and Himes, 1991). The C-terminil of α and β tubulin are very rich in acidic residues, particularly Glu, and thus have a significant negative charge (see Fig. 2). Subtilisin cleaves a small, acidic peptide from the end of both α- and β-tubulin (Sackett *et al.*, 1985; Serrano *et al.*, 1984a,b). The products of cleavage do not remain associated, with the result that the net charge on the resulting tubulin dimer is reduced by about 40% at pH 7 (Sackett *et al.*, 1985). This result can be clearly seen in Fig. 4B and Fig. 5. Coupling native gel electrophoresis (to evaluate charge) with SDS gel electrophoresis (to evaluate mass) provides a simple and informative method to follow the reaction and to compare different digests, especially digestions at sites in the C-termini. The loss of the charged peptide after digestion suggests that the C-terminal ends of the proteins do not have extensive noncovalent contacts with the rest of the protein and are connected to the body of the protein only by the polypeptide backbone (Sackett *et al.*, 1985; Sackett and Wolff, 1986). Unlike digestion of regions II and (probably) I, subtilisin cleavage of region III occurs whether tubulin is dimeric or polymerized, which indicates that the C-termini of α and β extend from the surface of the MT as well as the body of the isolated dimer (Lobert and Correia, 1992; Sackett and Wolff, 1986). Antibody studies also indicate surface exposure of the C-terminus in dimer and polymer since antibodies with several different epitopes on the α and β C-termini react with

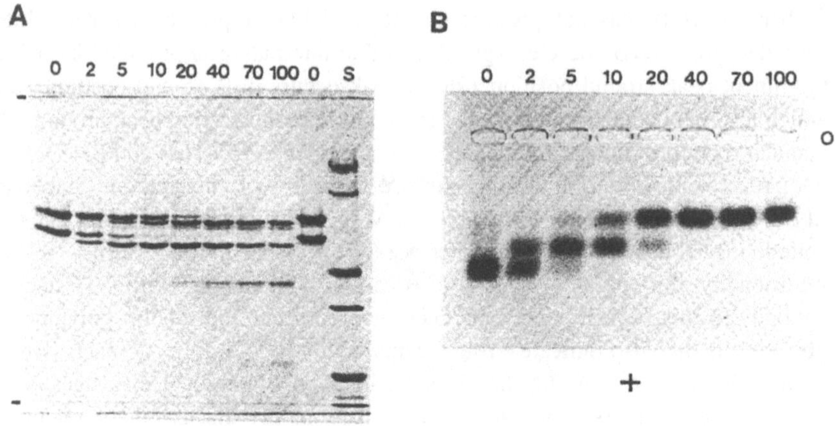

FIGURE 5. Time course of digestion of tubulin by subtilisin. Tubulin was digested by subtilisin at 30 °C for the time indicated above each lane in minutes. Samples were then electrophoresed in (A) denaturing or (B) native conditions as discussed in the legend to Fig. 6. The origin of electrophoresis is indicated in B by the "O" to the right. From Bhattacharyya *et al.* (1985).

unfixed MTs (Breitling and Little, 1986; Draber *et al.*, 1989; Vera *et al.*, 1988; Wehland *et al.*, 1984). Subtilisin digests β-tubulin much faster if it is dissociated, monomeric β than if it is present in the dimer, and this can be used to estimate dimer dissociation (Sackett *et al.*, 1989).

Upon extended digestion or under some buffer conditions, subtilisin also cleaves the unpolymerized dimer internally (de la Vina *et al.*, 1988; Sackett and Wolff, 1986; Serrano *et al.*, 1984b). The primary use of subtilisin has been for C-terminal cleavage, however. Cleavage is faster for β than α, resulting in first the production of dimers with intact α and cleaved β ends, termed $\alpha\beta_S$, and then cleavage of α, yielding the double cleaved $\alpha_S\beta_S$, or tubulin S (Bhattacharyya *et al.*, 1985; Sackett *et al.*, 1985). Native gel electrophoresis is at least as useful as SDS gel electrophoresis in following this process (Fig. 5). The fully cleaved $\alpha_S\beta_S$ is termed *tubulin S* (Bhattacharyya *et al.*, 1985), after the convention for subtilisin-cleaved RNase, termed *RNase S* (Richards and Vithayathil, 1959). The properties of the three types of dimers ($\alpha\beta$, $\alpha\beta_S$, and $\alpha_S\beta_S$) are different and are summarized in Table II.

The site of subtilisin cleavage has been the subject of some disagreement. Based on SDS gels and charge loss, it was estimated that the cleavage site was within 7–15 residues of the carboxyl end of α and β (Sackett *et al.*, 1985), which would produce a peptide of ≤ ca 1.5 kDa. Alternatively, it has been suggested (Maccioni *et al.*, 1986; Serrano *et al.*, 1984a,b) that subtilisin cleavage releases 4-kDa peptides, and this number has been widely quoted in the literature. Much evidence now exists to indicate that the primary cleavage from both α and β

Table II
Effect of C-terminus on Dimer Properties

	$\alpha\beta$	$\alpha\beta_s$	$\alpha_s\beta_s$	Reference[a]
Dimer				
Normalized net charge	1	0.8	0.6	1, 2
Dimer-monomer K_d, μM	0.2–1.0	n.d.	0.07	3–9
Colchicine binding				
pH maximum	6.8	6.8	Independent	13
K_d, μM (pH 6.8)	0.14	0.15	0.53	13
k_{-1}, sec^{-1} (pH 6.8)	0.019	0.02	0.07	13
Polymer				
Major form	Microtubule	Rings	Sheets, hooks, double rings, double micro-tubules	1, 2, 10–12
GTP required	+	−	+	1
Critical concentration, g/L	2	0.4	0.04	1
Cold (0°) sensitive	+	−	+	
Calcium (1mM) sensitive	+	−	−	1
Salt (240 mM NaCl) sensitive	+	−	−	1
Podophyllotoxin (50 μM) sensitive	+	−	+	1

[a] References for Table II: 1. Bhattacharyya *et al.* (1985); 2. Sackett *et al.* (1985); 3. Detrich and Williams (1978); 4. Detrich *et al.* (1982); 5. Panda *et al.* (1992); 6. Mejillano and Himes (1989); 7. Sackett *et al.* (1989); 8. Sackett *et al.* (1990); 9. Sackett and Lippoldt (1991); 10. Serrano *et al.* (1988); 11. White *et al.* (1987); 12. Mandelkow *et al.* (1985); 13. Mukhopadhyay *et al.* (1990).

yields a peptide that is much smaller than this, and is closer to 1.4 kDa. If cleavage removed 4 kDa from the C-terminus, cleavage would be near residue 415 in α and 410 in β. This should remove the epitope for antibodies DM1A and DM1B at α(426-450; probably 426-430) and β(416-430), respectively (Breitling and Little, 1986; Ponstingl *et al.*, 1990), but tubulin S ($\alpha_S\beta_S$) in fact reacts with both antibodies (Paschal *et al.*, 1989; D. L. Sackett, unpublished data). Tubulin S also reacts with antibodies raised against peptides with sequence α(415-430) and β(412-431) (de la Vina *et al.*, 1988) and with a monoclonal antibody reactive against a peptide β(410-430) (Lee *et al.*, 1990; Lobert and Correia, 1992). Based on size (SDS gels) and charge (native agarose and denaturing isoelectric focusing gels), the subtilisin fragments with these properties (consistent with cleavage in the carboxyl terminal 15-20 residues) represent the clear majority of the cleavage products (de la Vina *et al.*, 1988; Lee *et al.*, 1990; Lobert and Correia, 1992; Sackett and Wolff, 1986; Sackett *et al.*, 1985). If there had been significant production of a product with more than twice the mass loss and more than twice the charge loss, as would occur for cleavage around residue 410, the band would be clearly resolved from the observed bands by these techniques (see

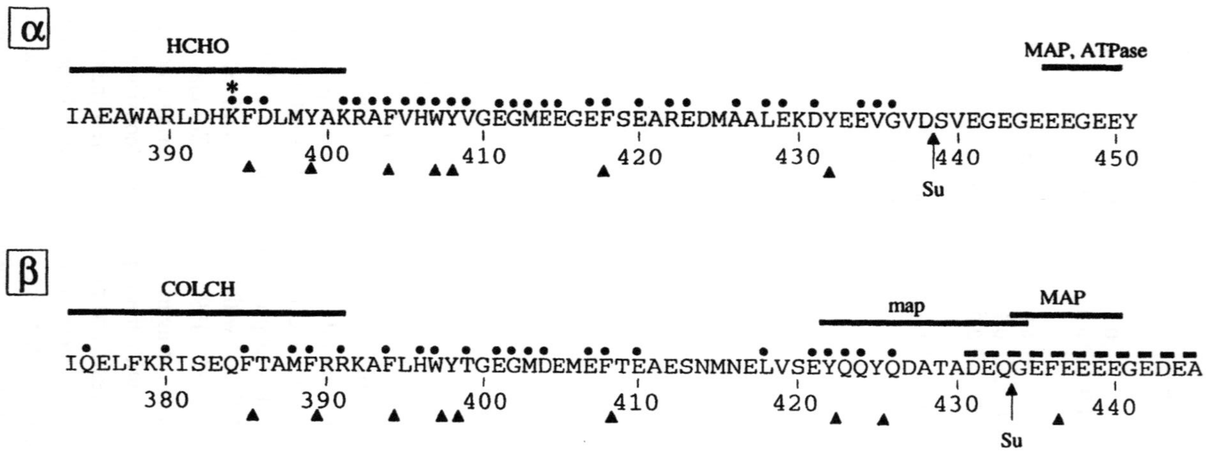

FIGURE 6. Carboxyl terminal sequence features. Carboxyl terminal sequences of α and β tubulin are shown, beginning with residue 384 in α and 374 in β. The sequences are those in Fig. 1. The residues in α that have a dot over them are identical in all known α sequences; likewise, residues in β that have a dot over them are identical in all known β tubulins (Burns, 1991a). The dashed line above the sequence indicates the variable, isotype-specific region. The triangles under the sequence indicate the positions of aromatic residues in α and β. The arrow labeled Su indicates the major site of subtilisin cleavage (Redeker *et al.*, 1992). The line above the α sequence labeled "MAP, ATPase" indicates a possible binding site for MAP2 and dynein (Paschal *et al.*, 1989). The line above the β sequence labeled "MAP" identifies the major MAP binding site, the adjacent line labeled "map" covers sequences with lower affinity for MAP2 and MAP1 (Cross *et al.*, 1991; Littauer *et al.*, 1986). The line above the α sequence labeled "HCHO" indicates the helical region that contains basic residues involved in polymerization, including the lysine that reacts with formaldehyde (labeled with a *) (Szasz *et al.*, 1986). The line above the β sequences indicates the helical region that is partially unfolded at the carboxyl end by binding of colchicine (Sackett and Varma, 1993).

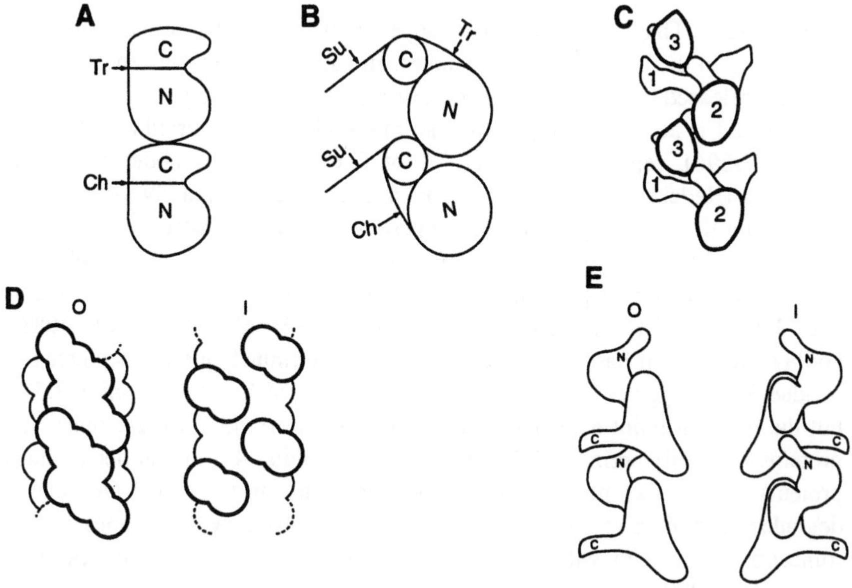

FIGURE 7. Models of the tubulin dimer. A number of models have been constructed that emphasize different features of the tubulin dimer. All have been drawn so that the dimer is about the same size. (A) This model shows α at the top, β below. Both have a larger N and smaller C domain, based on cleavage in region II (see Fig. 3). The intradimer contact between βC and αN is shown, as is the site of the trypsin (Tr) cleavage of α and chymotrypsin (Ch) cleavage of β that define the N and C domains (modified from Kirchner and Mandelkow, 1985). (B) This model shows the same features as A, but the trypsin and chymotrypsin sites are shown on an exposed loop connecting the N and C domains. In addition, the trypsin and chymotrypsin sites are shown in positions such that addition of another dimer above the α and below the β (as in a protofilament) would protect these sites from digestion, as is observed. The C-terminal, charged "tail" is shown on both α and β, with cleavage by subtilisin (Su) indicated (region III cleavage, see Fig. 3). The "tail" is extended (from the dimer and from the polymer), and is shown with no noncovalent contacts with the rest of the protein, unlike the N and C domains of each subunit, which remain together after cleavage by trypsin or chymotrypsin (redrawn from Sackett and Wolff, 1986; Sackett *et al.*, 1989). (C) This model emphasizes slightly different features than does A or B, and is based on a combination of X-ray diffraction data and proteolysis. The X-ray diffraction model (Beese *et al.*, 1987a) contains three structural domains per subunit and does not distinguish between the subunits. This is combined in this model with the three regions (1, 2, 3 on the model) of both subunits produced by cleavage in both regions I and II. Thus, domain I is roughly the first (amino) third, 2 is the second third, and 3 is the carboxyl third of each subunit. Other mappings of the proteolytically defined domains onto the X-ray model are possible (redrawn from de la Vina *et al.*, 1988). (D) X-ray solution scattering indicated the features shown in this model. The symmetry between the two subunits is clear. Dark lines indicate nearer features. The asymmetry of each subunit and of the overall dimer is also clear. The difference between the solution and lumen face of the subunit in the microtubule is shown. O = outside, I = inside (redrawn from Andreu *et al.*, 1992). (E) X-ray fiber diffraction and electron microscopic data are combined in this model. In this model, α is the lower subunit and the αN-βC intradimer contact is clear. A GTP binding site is proposed in the constriction in the N domain of both subunits, so the exchangeable site is the upper one. This model employs essentially a two-domain structure like A and B, with the N domain composed of two subdomains separated by the nucleotide binding site. As in D, the subunits present distinct faces to the outside (O) and inside (I) of the microtubule (redrawn from Amos and Amos, 1991).

Figs. 4 and 5). That such bands do not appear indicates that the primary subtilisin cleavage is in the terminal 15-20 residues. In addition, direct sequencing of the peptides released by subtilisin showed that the primary cleavage site in α is after Asp438 and is at Gln433 in β, although minor cleavage occurred at α(His406) and β(His396), representing about 2% of β cleavages and about 15% of α cleavages (Redeker *et al.*, 1992). Tubulin S polymerizes readily (see Section 2.5.1), and thus the polymerization properties of tubulin constructs truncated at different positions may provide further evidence regarding the extent of subtilisin deletion. Constructs of β-tubulin have been described that are truncated at residues 396, 411, 431, and 430 (Fackenthal *et al.*, 1993; Matsuzaki *et al.*, 1988), and a CHO mutant has been described with a C-terminal deletion beginning at residue 330 (Boggs *et al.*, 1988). These are discussed further in Section 2.5.3, but the relevant point here is that truncations at 430 or 431 produce functional proteins that polymerize, while truncations to residue 411 or further produce proteins that are not polymerization competent, are unstable, and are quickly degraded. Tubulin S resembles the 430-431 truncations more than it does the 411 truncation, consistent with the data discussed above, and a primary cleavage after residue 430.

Subtilisin does not have high sequence specificity for cleavage, and it prob-

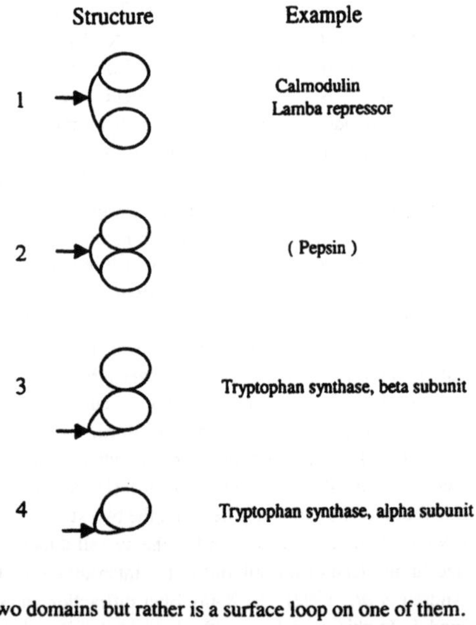

FIGURE 8. Domain models and proteolysis. Protease sensitivity does not necessarily imply a linker between two folding units. Different geometries of folded domains, loops, and linker regions are shown. In (1) two independent folded regions are connected by an exposed linker which is protease sensitive (arrow). Following cleavage, the two domains separate. Examples are calmodulin and lambda repressor. In structure (2), the two folded domains have extensive contacts and would remain together even after the linker is cleaved. Pepsin provides an imperfect example. Pepsin has two quite similar domains connected by a linker. These two regions may be separately synthesized and mixed. They will then combine to form an active enzyme, even though the linker is not connected. In (3) the two folded domains are tightly bound, and a single protease sensitive site is found, but it is not a linker connecting the two domains but rather is a surface loop on one of them. This is found in the case of tryptophan synthase beta subunit. In (4), there is only a single structural domain, but it contains a protease-sensitive surface loop. Following proteolysis, structures 2, 3, and 4 will all yield two bands on an SDS gel and a single band on a native gel or a single peak on a native size exclusion chromatographic column. Structure 1 is the only one that would be expected to yield two fragments in the native and denatured state.

ably can cleave α- and β-tubulin at more than one place in region III. Evidence indicates that there may be more than one cleavage site near residues 430-435, and that one site is accessible in the polymer that is not accessible in the dimer (Lobert and Correia, 1992). Furthermore, sequencing of the released peptides demonstrated that there is minor cleavage at α(406) and β(396) in addition to the major cleavage at α(438) and β(433) (Redeker *et al.*, 1992). It is possible that differences in cleavage conditions, tubulin source, purification protocols, storage conditions, and so forth can enhance cleavage at the minor site, and this may be the explanation for the reports of subtilisin cleavage at this or other sites more internal than the major sites after residue 430 (Maccioni *et al.*, 1986; Serrano *et al.*, 1984a,b).

2.4. Conformation of the C-Termini

The carboxyl terminus of a protein logically begins at the first residue after the middle and contains everything between there and the actual carboxyl terminal amino acid residue. The somewhat elastic meaning of the term *carboxyl terminus* has probably led to some confusion. Usually what is meant by the C-terminus, or C-terminal region, of tubulin is the terminal 5–10% of the protein. Most of the properties that have been assigned to the C-terminus are due to the final 5% of the protein, probably less than 20 residues.

Figure 6 presents the C-terminal 60-70 residues of the rat α and β sequences from Fig. 1. The overall structure is very similar for both of these sequences (note, for example, the similar distribution of aromatic amino acids marked by the triangles below the sequences). Both sequences may usefully, if somewhat arbitrarily, be divided into three regions. These are (1) a helical region with surface exposed basic residues required for polymerization, (2) a possibly helical region that is highly conserved evolutionarily, and (3) the highly charged terminal region, which contains the isotype specific sequences and MAP binding sites. The entire region shown in Fig. 6 has high calculated helical potential (see Fig. 1, alpha helical regions plot), but the likelihood of its actually being helical in the neutral aqueous solutions in which MTs occur decreases as the C-terminus is approached, due to the increasing density of acidic residues and therefore negative charges (see Fig. 1, average charge plot).

The first region in α is probably helical and contains surface-exposed basic charges involved in polymerization (these can be seen in the average charge plot in Figs. 1 and 2 as small bumps of positive charge around residue 400 in α and 390 in β). A lysine residue in this region (394) is highly reactive, and reductive methylation of this residue renders the dimer polymerization incompetent. The corresponding region in β is also probably helical and also is amphipathic, with a number of basic residues exposed on the surface. The folding of the carboxyl end of this helix is disrupted by binding of colchicine, forming an unfolded loop that may prevent polymerization (Sackett and Varma, 1993).

The second region in both subunit sequences is highly conserved evolutionarily. The dots over the sequences in Fig. 6 indicate residues that are identical in all known α or β sequences. This region also has high calculated helical potential, except for a small region after 410 in α and 400 in β. This region also has an increasing density of acidic residues, with a small minimum around 430 in α and 420 in β. The structure of this region is unclear, but studies with peptides from this region suggest that a helical structure is probably not achieved (see below).

The final region is the highly charged "tail" of the protein. The sequences in this region are highly variable evolutionarily, but are always very acidic. The isotype sequences are here, as are the MAP binding sites. The major cleavage site for subtilisin is also in this region. These features are shown in Fig. 6.

As discussed, calculations indicate that the C-termini of both subunits are likely to be folded, with a high content of alpha helix (see Fig. 1). Based on these calculations, schemes have been proposed invoking multiple alpha helical folds and the involvement of these helixes in interactions with other proteins (Lupus *et al.*, 1991; Ponstingl *et al.*, 1979; Serrano and Avila, 1990; Serrano *et al.*, 1986). Despite this, most experimental data indicate that these regions are extended and unstructured. A number of antibodies recognize this region in the native or denatured state, suggesting that epitopes are available in the native state. Some of these antibodies were raised against peptides, but still recognize the intact protein, suggesting that the epitope on the protein is not more folded than in the peptide (Laver *et al.*, 1990). [^{13}C]-NMR spectral features indicate that the C-termini are conformationally flexible, suggesting a lack of order (Ringel and Sternlicht, 1984). This region is also of course subject to proteolysis, by both endo- and exoproteases (see Section 2.3.2), hardly a property expected of a highly folded structure. Following cleavage, the peptides do not remain associated with the protein (easily seen by the loss of charge, Figs. 4 and 5), indicating that there are not extensive interactions between this region and the rest of the protein. Weak interaction between subtilisin-cleaved peptides and the resulting tubulin S has been observed (Ortiz *et al.*, 1993), but in this study the peptides could be separated from tubulin S by simple gel filtration, indicating that the interactions were not strong or extensive.

The folding of the C-terminal region has been addressed directly by determining the conformational properties of peptides corresponding to residues in this region. Circular dichroic (CD) measurements of the peptides released from tubulin by subtilisin revealed an unordered structure in aqueous solution between pH 4.6 and 6.8. Addition of 50% trifluoroethanol induced α-helical spectral features (Maccioni *et al.*, 1986). Addition of trifluoroethanol to tubulin also induces a significant increase in helical content, from 23 to 66% as trifluoroethanol is increased from 0 to 85% (Bayley *et al.*, 1983); the structures induced by this cosolvent are obviously quite different from those in the native protein. Peptides corresponding to defined sequences in this region predicted to be helical have also been examined. Peptides corresponding to residues α(430-441) (Otter and Kotovych, 1988) and β(422-434) (Otter *et al.*, 1991) were synthesized and their conformation evaluated by CD and NMR techniques. Both peptides are essentially completely unstructured and extended in water but adopt conformations that are at least partially helical though still largely extended in 80–90% methanol. Much larger peptides corresponding to the entire C-terminal region gave similar results: peptides corresponding to β400-436 and β400-445 of the type II isotype of porcine brain tubulin were analyzed in aqueous solution at neutral pH by CD spectroscopy and one- and two-dimensional NMR (Reed *et al.*, 1992). Both techniques indicated that the peptides were quite flexible and lacked regular secondary structure. These results were unaffected by addition of a large molar excess of Ca^{2+} of Mg^{2+}, but α-helical structure was observed in the presence of trifluoroethanol or on lowering the pH to 3.0. These data indicate that the C-termini of tubulin are charged and extended in neutral aqueous solution. This is also congruent with physical measurements indicating that tubulin S is more compact and less asymmetric than tubulin (Kanazawa and Timasheff, 1989; Maccioni *et al.*, 1986). In a study comparing tubulin S with tubulin, the axial ratio of the protein was reduced from 3.9 to 2.8 following cleavage, indicating that the peptide removed was extended into the solution from the body of the protein (Kanazawa and Timasheff, 1989; Sackett and Wolff, 1986).

The results indicate that at neutral pH in aqueous solution, significant charge repulsion between the many ionized acidic residues in the C-terminal sequence prevents formation of structure. The C-termini are probably charged and extended away from the protein. It has been earlier argued that charge repulsion between the C-termini of separate dimers represents a barrier to polymerization and that it is reduction in this charge that allows tubulin S and chemically charge-neutralized tubulin to assemble at much lower critical concentrations than intact tubulin (Mejillano and Himes, 1991; Sackett *et al.*, 1985, 1986). Under conditions that neutralize the charged acidic residues, such as low pH or possibly the presence of appropriately arranged basic charges on another polymer (e.g., MAPs), this region could assume helical conformation in a complex (Ponstingl *et al.*, 1979; Reed *et al.*, 1992). Thus this region could adopt two quite different conformations depending on conditions: charged, extended, and unordered, or uncharged (or charge neutralized), more compact, and structured.

2.5. Covalent Modification of Tubulin C-Termini and Functional Changes

The properties and functions of the C-termini have been evaluated by examining the results of chemical modification of this region and by genetic alteration of the sequence, as well as by removing the C-terminal region proteolytically. The results of these three types of alterations are considered in turn.

2.5.1. Proteolysis of C-Termini

Proteolysis of C-termini, usually with subtilisin, alters some properties of tubulin aside from the charge. If affects tubulin conformation, interaction with small molecules, formation of *intradimer* and *interdimer* contacts (i.e., polymerization), and interaction with other proteins. These will be considered in turn, and many are summarized in Table II.

The net charge on the protein is reduced about 40% by subtilisin cleavage, as estimated by native agarose electrophoresis (Sackett *et al.*, 1985) (see Figs. 4 and 5). Slightly more charge loss occurs on cleavage of α than on cleavage of β (see Fig. 5). This is accompanied by a reduction of 50 to 60% in the number of isotypes found on isoelectric focusing (Lobert and Correia, 1992). This loss of charge probably explains the lowered solubility of tubulin S relative to tubulin in solutions containing a variety of cosolvents (Lobert and Correia, 1991).

The conformation of tubulin S has been reported to differ from that of intact tubulin, although some reports on this subject differ. The negative far UV CD is decreased in tubulin S (Kanazawa and Timasheff, 1989; Maccioni *et al.*, 1986; Mukhopadhyay *et al.*, 1990), indicating a reduction in α helical content, but the extent of this may be only 10 to 15% (Kanazawa and Timasheff, 1989). The absorbance spectra of tubulin and tubulin S are coincident within experimental error (Kanazawa and Timasheff, 1989). Reports vary regarding a change in tryptophan fluorescence properties. Relative to intact tubulin, it has been reported that tubulin S tryptophan emission is blue-shifted 3 nm, from 332 to 329 nm (Maccioni *et al.*, 1986), red-shifted 3 nm, from 332 to 335 nm (Kanazawa and Timasheff, 1989), or unaffected, at 338 nm (Mukhopadhyay *et al.*, 1990). In the latter study, it was reported that addition of the fluorescence quencher acrylamide resulted in a larger shift (3.5 nm) with tubulin S than with tubulin (2 nm), and that a smaller fraction of tubulin S tryptophan can be quenched relative to tubulin, though it is quenched more effectively (1.6 times higher Stern-Volmer quenching constant) (Mukhopadhyay *et al.*, 1990). It is more stable, as judged by time-dependent loss of CD (Maccioni *et al.*, 1986) and colchicine binding (Mukhopadhyay *et al.*, 1990). Tubulin S is more compact than tubulin (Kanazawa and Timasheff, 1989; Maccioni *et al.*, 1986), with an axial ratio (prolate ellipsoid) of 2.8 versus 3.9 for tubulin (Kanazawa and Timasheff, 1989). This result is likely due to the extension of the C-termini away from the body of the protein in tubulin; their loss results in a lowered frictional coefficient ratio (Kanazawa and Timasheff, 1989; Sackett and Wolff, 1986; see also discussion in Section 3.2).

Tubulin interaction with colchicine and calcium is altered by C-terminal cleavage. Most notably, the pH dependence of colchicine binding is nearly eliminated by subtilisin cleavage, and the time-dependent decay of binding capacity is very much reduced (Mukhopadhyay *et al.*, 1990). These changes appear to be linked to cleavage of the α C-terminus, since $\alpha\beta_S$ has characteristics like those of intact tubulin. In addition, the affinity for colchicine is lowered, due to an increase in the dissociation rate constant of the complex (Mukhopadhyay *et al.*, 1990). The decrease of interaction with calcium upon C-terminal cleavage indicates that the calcium binding site is located here. This effect has been observed by the ability of tubulin S to polymerize in the presence of up to 1 mM calcium (Bhattacharyya *et al.*, 1985; Serrano *et al.*, 1986). Calcium binding studies indicate the loss of calcium sites following cleavage (Serrano *et al.*, 1986). Digestion with carboxypeptidase Y, sufficient to remove about 7 residues per subunit, also removed the inhibitory effect of 0.1 mM calcium on polymerization, suggesting that calcium binding occurs at the extreme C-termini (Vera *et al.*, 1989). It is surprising that addition of a large molar excess of calcium did not alter the conformation of peptides corresponding to this region [β(400-445)] (Reed *et al.*, 1992). Perhaps these data taken together indicate that the C-terminal sequences necessary for calcium binding and removed by protease digestion only provide part of the binding site, another part being provided by sequences that are not removed by subtilisin.

Tubulin *intradimer* contacts form during association of monomers to form the heterodimer (see Section 2.1). Subtilisin cleavage results in the heterodimer $\alpha_S\beta_S$ or tubulin S. The dissociation constant for the tubulin S heterodimer was measured by fluorescence anisotropy to be about tenfold lower (K_d about 0.07 μM) than that obtained for the intact heterodimer (Panda *et al.*, 1992). The increased dimer stability may result from the reduction of charge repulsion between the charged C-termini of the intact α and β subunits.

Tubulin *interdimer* contacts form during polymerization, and C-terminal cleavage has striking effects on polymerization. The critical concentration is significantly lowered (Sackett *et al.*, 1985; Serrano *et al.*, 1984b), and a small mole fraction of tubulin S added to a solution of tubulin markedly lowers the critical concentration of the mixture (Bhattacharyya *et al.*, 1985). The enhancement of polymerization is likely due to reduction in charge repulsion and occurs in steps: the critical concentration decreases in the order $\alpha\beta > \alpha\beta_S > \alpha_S\beta_S$ (see Table II) (Bhattacharyya *et al.*, 1985). Tubulin S assembles into polymorphic structures including open MTs, sheets, ribbons, and hooks, and some MTs (Bhattacharyya *et al.*, 1985; Mandelkow *et al.*, 1985; Sackett *et al.*, 1985; Serrano *et al.*, 1988; White *et al.*, 1987). The lattice parameters are the same as for MTs (Bhattacharyya *et al.*, 1985), so lateral interactions required for MT closure must be disturbed (Serrano *et al.*, 1984b). In the presence of GDP or elevated magnesium, ring formation is enhanced (Peyrot *et al.*, 1990; White *et al.*, 1987). Rings are the predominant polymer formed by $\alpha\beta_S$, regardless of nucleotide (Bhattacharyya *et al.*, 1985). Tubulin S polymerization is sensitive to cold, GDP, and podophyllotoxin (Sackett *et al.*, 1985), but polymerization of $\alpha\beta_S$ is not (Bhattacharyya *et al.*, 1985) (see Table II). Tubulin S assembly is insensitive to added calcium, MAPs, or high ionic strength (Bhattacharyya *et al.*, 1985; Serrano *et al.*, 1984b, 1986; White *et al.*, 1987), indicating that the calcium and MAP binding sites have been removed (Serrano *et al.*, 1984b, 1986; Maccioni *et al.*, 1986). Removal of the terminal 6-8 residues with carboxypeptidase Y also removes sensitivity to inhibition of polymerization by calcium, though stimulation by MAPs remains, suggesting that the calcium binding occurs at the extreme termini, while MAP binding is more internal (Vera *et al.*, 1989). Interestingly, tubulin S can still form Zn^{2+}-induced sheets, indicating that the binding sites for zinc ions are not removed by subtilisin cleavage, in contrast to the Ca^{2+} sites (White *et al.*, 1987).

The influence of the C-termini on polymerization may be more nuanced that just simple charge repulsion. Tubulin purified by β isotype reveals assembly properties that differ by isotype, and each isotype polymerizes better than the original mixture of isotypes (Banerjee *et al.*, 1990, 1992; Luduena, 1993).

2.5.2. Chemical Modification of Charge

The properties of the C-termini have been altered by chemical amidation of carboxyls exposed on polymerized tubulin (Mejillano and Himes, 1991). Reaction modified 8–12 carboxyls per subunit and occurred mainly in the C-termini, though not primarily at the extreme end since carboxypeptidase Y digestion of the terminal 6-8 residues released only about 15% of the amidated residues. CD indicated that the overall structure of the protein was not affected. It polymerized with a sevenfold lowered critical concentration (0.3 mg/ml) when 22 groups per dimer were modified; this is not as low as the critical concentration for tubulin S (see Table II). Polymers appeared to be normal MTs and were cold sensitive. Polymer formation was not inhibited by high ionic strength or by 1 mM calcium; binding studies indicated loss of the high affinity calcium site. MAP binding was 70% reduced by modification of 24 residues per dimer. Polymers of the modified tubulin were more stable to depolymerization (Mejillano *et al.*, 1992). Dynamic instability was suppressed. The off rate constant for the modified polymer was decreased threefold and the steady-state GTP hydrolysis rate constant was decreased 12-fold. MAP binding to the acidic residues of the C-termini may neutralize exposed charges and stabilize MTs similarly (Mejillano *et al.*, 1992).

2.5.3. Directed Mutation and Carboxyl-Terminal Function

Much of the extensive genetic work done on tubulin structure and expression has focused on understanding the significance of tubulin isotypes and their tissue and developmental specificity (see Section 1.4). The properties of the C-terminal region of tubulin have also been examined. A number of different β-tubulin mutations have been constructed, with truncations at different positions in the C-terminal region. *Saccharomyces cerevisiae* contains a single β-tubulin gene that is unusual in that it encodes 12 amino acids beyond the more common C-terminus. Strains expressing only a β-tubulin lacking these additional residues grow normally and show normal mitotic and meiotic division and mating, but are hypersensitive to the antimitotic agent benomyl (Katz and Solomon, 1988). Strains have also been constructed that express β-tubulin truncated after residues 430 and 411 (Matsuzaki *et al.*, 1988). The 430-truncated β-tubulin is similar to subtilisin cleaved β, since the major subtilisin site on β is at position 432 (see Section 2.3.3). This protein was expressed normally, copolymerized with brain tubulin, and supported normal growth at 30 but not at 37 °C, at which temperature progress of mitosis was retarded. At permissive temperatures the truncated β resulted in hypersensitivity to antimitotic drugs (thiabendazole and methylbenzimidazolecarbamate) and low spore viability, but the cells were not cold sensitive. Thus, the 430-truncated tubulin supported normal MT function required for vegetative growth, but lacked properties required for suboptimal growth conditions. Truncation after residue 411 apparently resulted in an unstable protein, not polymerization competent, since cells expressing this construct were not viable at any temperature (Matsuzaki *et al.*, 1988).

A similar study has been reported in *Drosophila*. A *Drosophila* testis-specific isotype of β-tubulin, β2, is known to be required for male fertility (Hoyle and Raff, 1990). Constructs of the β2 gene were prepared containing variable truncation at the C-terminus (15, 50, or 171 residues) and their functional properties examined (Fackenthal *et al.*, 1993). A large C-terminal truncation (50 residues) was equivalent to truncation in the middle of the protein (171 residues): no detectable protein accumulated. This suggests that β-tubulin with 50 or more C-terminal residues removed is not stable and is degraded. If only 15 residues are removed, the protein does accumulate, although it is less stable than intact β. It is assembly-competent and is incorporated into morphologically normal MTs, but MT-containing superstructures, notably sperm tail axonemes, do not assembly properly (Fackenthal *et al.*, 1993; see also Curry and Rosenbaum, 1993).

Mutant lines of Chinese Hamster ovary (CHO) cells have been described that express forms of β-tubulin that are assembly-defective due to C-terminal deletions (Boggs *et al.*, 1988). One mutant β is truncated at about residue 330, does not polymerize, is unstable, and is degraded. Another mutant β containing an internal deletion of residues 250-350 is even more unstable (Boggs *et al.*, 1988).

A chimeric chicken–yeast–β-tubulin has been constructed in which the C-terminal 25% is from yeast (starting at residue 344). This chimeric β was expressed in mouse 3T3 cells and readily incorporated into endogenous MTs (Bond *et al.*, 1986). A number of alterations have been produced in the C-terminus of this protein and their properties examined (Fridovich-Keil *et al.*, 1987). If the proportion of yeast sequence was increased by moving the chicken–yeast junction closer to the amino end of the protein, the resulting proteins became less stable and assembled much less well. An insertion of four amino acids after residue 438 produced a stable protein that assembled into MTs. An insertion at residue 409 or before resulted in proteins that could not assemble even with taxol. A deletion near the extreme C-terminus, which moved the extra 12 terminal residues of yeast β-tubulin into the position usually occupied by the isotype-specific residues of chicken β had no effect: the protein was stable and was incorporated into all MTs (Fridovich-Keil *et al.*, 1987).

Taken together, these data suggest several conclusions. Removal of as few as 34 residues from the C-terminus (i.e., truncation after residue 411) produces a β-tubulin that is unstable and unable to polymerize. Tubulin lacking fewer residues, for example, the C-terminal 15 residues, can polymerize into morphologically normal MTs that may be unable, however, to perform specialized functions that require larger structures of which MTs are a part. Finally, the terminal 15 residues of β-tubulin are important in regulating sensitivity to antimitotic drugs. The latter *in vivo* finding differs slightly from *in vitro* observations, indicating that the C-terminus of α, not β, regulates this. Removal of the C-terminus of β with subtilisin (i.e., $\alpha\beta_S$) had little effect on interaction with colchicine, while further removal of the α C-terminus (i.e., $\alpha_S\beta_S$ or tubulin S) did alter interaction with colchicine (see Section 2.5.1) (Mukhopadhyay *et al.*, 1990). On the other hand, different brain β-tubulin isotype preparations have quite different colchicine binding properties, despite a presumably common array of α subunits (Banerjee and Luduena, 1992).

3. MODELS OF THE TUBULIN DIMER

3.1. General Models of Dimer

A number of models have been proposed for the tubulin dimer (see Fig. 7, page 277). As might be expected, all feature significant symmetry between the α and β subunit structure. Each model emphasizes different features of the dimer, reflecting the data on which each is based. A feature shared by all of the models is that the structure of α is very similar to the structure of β. Another point illustrated by most of these models is the existence of multiple points of flexibility within the dimer, where one region of the dimer or subunit can easily move relative to another. Models based on X-ray diffraction and electron microscopy illustrate the difference between the face the dimer presents to the solution and that which it presents to the lumen of the MT (Andreu *et al.*, 1992; Beese *et al.*, 1987a; Amos and Amos, 1991). These are shown in C, D, and E. One combination of the three-domain X-ray model of Beese *et al.* (1987a) with the three regions of tubulin defined by proteolysis in regions I and II (that is, the amino third, the middle third, and the carboxyl third) is shown in C; other mappings of the proteolysis results onto the X-ray model are possible (de la Vina *et al.*, 1988). Model E indicates two main domains per subunit, but the N-domain is separated into two subdomains by a constriction that is proposed to contain the nucleotide binding site (Amos and Amos, 1991). The C-domain also has substructure, containing a projection that is slightly larger than the projection from the N-domain involved in subunit-subunit contact. The relationship between this C-domain projection and the C-terminal "tail" in model B is not clear. Proteolysis results indicate that the C-termini are a separate structural element from the rest of the protein, regardless of whether the rest is considered to be composed of two or three domains. There is no N-terminal equivalent in terms of cleavage and release, so the C-domain projection in E and the C-termini in B may be different. Since data indicate that the C-termini shown in B are extended, flexible, and unstructured (see Section 2.4), these regions would not be expected to have a strong signature by X-ray diffraction. Whether the rest of the subunit structure is described by a three-domain model or a two-domain model with substructure is somewhat a matter of linguistic convenience, since either description can incorporate the known data.

Studies of proteolytic susceptibility do not provide unambiguous data regarding the architecture of the protein under study. A region of the protein that is sensitive to proteolysis surely is likely to be both exposed on the surface of the protein and unstructured enough to provide a good substrate for binding by a protease. However, this does not require that such an extended loop necessarily separates two structural domains. The term *domain* in current usage can denote a number of things of varying specificity. Thus domain may refer to an independently folding unit of structure, an element of structure recognized by some symmetry element, a region of a molecule defined by its involvement in binding of another molecule, and so forth.

A number of possible relationships between structural domains and regions of proteolysis are illustrated in Fig. 8 (see page 278). Certainly examples are known in which proteolysis clearly separates two structural units of the whole molecule. An example is the bacteriophage λ repressor protein, which is cleaved by papain into two fragments that remain folded and are easily separated by gel filtration (Pabo *et al.*, 1979). Here the region connecting the two domains is exposed and the domains are free of each other. If the two domains interact noncovalently, the results of proteolysis are less clear. An imperfect example is provided by pepsin, which is composed of two nearly identical domains linked by a loop (as in Fig. 8, No. 2). The domains can be synthesized separately and then will combine tightly to form active enzyme (Lin *et al.*, 1992). In this case, even though the link between the two domains is broken, noncovalent interactions between the domains keep them together under nondenaturing conditions. A protein in which the results of proteolysis are more problematic is the β-subunit of bacterial tryptophan synthase (see Chapter 8, this volume). This protein is composed of two nearly superimposable structural elements, as shown by the crystal structure (Hyde *et al.*, 1988). An exposed loop can be cleaved by trypsin, but the two fragments so produced remain associated in the absence of denaturants. The fragments resulting from cleavage will independently fold, following separation, and can reassemble into the (cleaved) native structure. Nonetheless, the loop region is not a "hinge" region between the two structural domains clearly seen in the crystal structure. Rather, the loop occurs in the middle of the carboxyl element, separating the protein 70:30 instead of the 50:50 corresponding to the structural units (like Fig. 8, # 3). Actin provides a further example. Actin is composed of two domains that are nearly equal in size and are connected by two crossovers (Kabsch *et al.*, 1990), and therefore overall resembles model 2 in Fig. 8, except that there are two strands between the domains. A number of proteolytic sites have been characterized on actin (Kabsch *et al.*, 1990; Mornet and Ue, 1984). All of these sites are exposed on the surface of the protein in loops, exposed strands, and near the C-terminus. None of them occur in the crossover strands (the "hinge" between the domains) nor do they occur in regions connecting the two subdomains in each of the main domains (Kabsch *et al.*, 1990). A final cautionary example is provided by the α-subunit of tryptophan synthase. This protein is composed of a single structural unit, based on the crystal structure (but folding studies indicate more than one folding element; see Chapter 8, this volume). An extended loop is protease sensitive, but the protein remains intact in the absence of denaturants, and therefore corresponds to model 4 in Fig. 8.

Thus, a protease-sensitive region may indicate two independent structural elements, easily separated; two interdigitated structural elements, not easily separated (in which the loop may or may not define the junction between these elements); or a single element with a loop on the surface of the molecule. Exposed, protease sensitive loop regions are correlated with regions of segmental mobility in proteins (Fontana *et al.*, 1986), and thus may provide structural clues to the mode of action of the protein (see Chapter 8, this volume, for discussion of this with regard to the two loop regions of tryptophan synthase). Domain models constructed on the basis of proteolytic cleavages should be thought of as heuristic tools (and therefore useful), not as structural outlines.

3.2. The Structure of the C-Termini

The carboxyl termini of tubulin are (1) accessible, (2) on the surface of the dimer and polymer, (3) not involved in extensive noncovalent interactions with the rest of the protein, and (4) probably charged and extended at neutral pH. Point 1 follows from the fact that antibodies and other proteins (MAPS, proteases) can interact. Point 2 follows from the fact that subtilisin can cleave the C-termini of dimer and polymer and from the fact that C-terminal epitopes are recognized on MT. Point 3 follows from the fact that the peptides produced by cleavage here are easily released from the protein. This does not mean that there are no interactions between C-termini and the rest of the protein; indeed, these have been demonstrated between the cleaved peptides and tubulin S (Ortiz *et al.*, 1993) and may be important in polymerization. Point 4 follows from the data on the conformation of C-terminal peptides and from the decrease in asymmetry found after cleavage.

It is useful to consider the structural consequences of these properties. For the moment, let us assume that the properties described for the C-terminal region at neutral pH (surface exposed, highly charged, extended into the aqueous solvent, and fairly separate from the rest of the protein except for the peptide backbone) apply only to the terminal 15 residues. The linear extension of this 15-residue segment from the rest of the protein could be quite considerable (see Fig. 9). If fully extended in the trans configuration, this peptide could extend nearly 5.5 nm from the surface of the protein. This is more than the diameter of the subunit from which it extends (about 4 nm) (see Fig. 9A). This "tail" could interact with other regions of tubulin, interact with other proteins, or extend freely into the surrounding solvent. As mentioned, there is evidence for weak interaction between subtilisin-cleaved peptides and the rest of the protein (Ortiz *et al.*, 1993), and there are clusters of basic residues that are known to be important to polymerization. A cluster of basic residues has been described in α tubulin, and reductive methylation here results in loss of polymerizability (Szasz *et al.*, 1986). Another group of basic residues occurs in β tubulin, and disruption of the organization of these residues may underlie colchicine's ability to poison polymerization (Sackett and Varma, 1993). It is possible that the C-terminal acidic residues may interact with one or both of these regions, or with basic residues in the non-C-terminal isotype-associated regions (Burns and Surridge, 1990; Luduena, 1993). Whatever the site of interaction, the reach of this "tail" would easily allow interaction with a significant fraction of the surface of the subunit of origin, as well as easily reaching the other subunit of the same dimer, or subunits in dimers adjacent laterally or longitudinally (see Fig. 9A).

Instead of interacting with itself or other tubulin dimers, the C-termini could extend out from the surface of the MT into the surrounding solvent. This is shown in Fig. 9B. In the absence of interactions with basic residues (either on tubulin as suggested above or on other proteins such as MAPs), it seems quite likely that charge repulsion between the many charged C-termini would force some similar arrangement to minimize approach of like charges. As illustrated in the Fig. 9, it is significant that the volume that would be defined by these

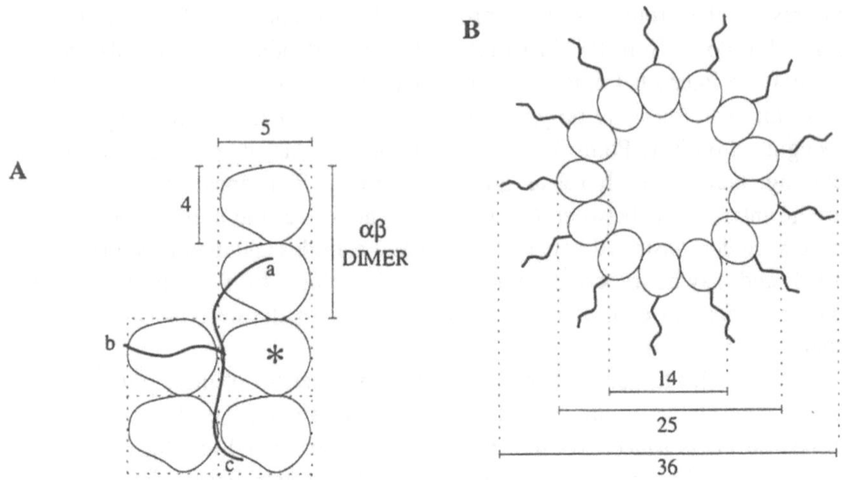

FIGURE 9. Possible extent of C-terminal region of tubulin. The extreme carboxyl terminal regions of α and β tubulin are cleaved by subtilisin and released from the rest of the protein. This region of about 15 residues is highly charged and probably extended. If these 15 residues are extended to a maximum extent, the terminal residue will be as much as 5.5 nm from the main part of the subunit. This figure shows some consequences of this in the context of the microtubule polymer. (A) The C-terminus could reach neighboring dimers. Parts of two protofilaments are shown and the dimensions of the αβ dimer are indicated in nanometers. The C-terminal region is shown extended in three possible configurations from the subunit marked by the asterisk. In all cases the contour length is approximately 5.5 nm. In (a) it is shown reaching a subunit of the neighboring dimer in the protofilament (a longitudinal association); in (b) it is shown reaching a subunit of the neighboring dimer in an adjacent protofilament (a lateral association); in (c) it is shown reaching the other subunit of its own dimer (a self-association). (B) The microtubule is shown in cross section. The lumen is about 14 nm in diameter and the outer surface of the main body of the subunits is 25 nm in diameter. If fully extended in all directions, the 15 residues of the carboxy termini would extend another 5.5 nm out in all directions, yielding a diameter of 36 nm. The volume of this C-termini space is larger than the volume of the rest of the microtubule combined.

residues is larger than the volume of the rest of the MT from which they extend. In addition to interacting with tubulin, the extended regions interact with other protein, such as MAPs (see Section 1.8). MAP interaction with this region could result in structuring of the complex (see Section 2.4), with associated shortening of the tail toward the body of the microtubule.

It is important to remember that this discussion is based on two assumptions: (1) the region extended from the MT is about 15 residues, and (2) it is fully extended. If the number of residues in assumption 1 is low, then the extension distance will increase. If the region is not fully extended (for example by interaction with basic groups on the dimer body), it will decrease. Dynamic variation of both of these variables seems likely. In any case, the data reviewed indicate that the C-terminal regions have properties represented in the figure. Varying assumptions 1 or 2 will change the extent but not the existence of these properties.

These results have biological implications. One is that the volume occupied by the MT in the cytoplasm is increased significantly over that calculated from the body of the MT alone. The fraction of solution that is occupied by macromolecules, especially by a network of macromolecular fibers, profoundly influences the chemical activity of other proteins in the surrounding solution. This effect can cause an increase in association constants for macromolecular association of several powers of ten, with the result that associations that would be considered "nonspecific" in dilute solution become significant in these crowded and confined solutions (Zimmerman and Minton, 1993).

Another result of biological consequence is not unrelated to the first; that is, the binding surface available on a given MT is very much increased by extending the binding region out into the solution. Many more proteins, even quite large proteins like MAPs and ATPase motor proteins, can bind to a MT if they do not all have to abut the surface of the MT body. This principle is, of course, why airport terminal buildings are not circular. Perhaps nature discovered and deployed this docking strategy in microtubules long ago.

4. REFERENCES

Alexander, J. E., Hunt, D. F., Lee, M. K., Shabanowitz, J., Michel, H., Berlin, S. C., MacDonald, T. L., Sundberg, R. J., Rebhun, L. I., and Frankfurter, A., 1991, Characterization of post-translational modification in neuron-soecific class III β tubulin by mass spectrometry, *Proc. Natl. Acad. Sci. USA* **88**:4685–4689.

Amos, L. A., 1982, Tubulin and associated proteins, in *Electron Microscopy of Proteins*, vol. 3 (Harris, J. R., ed.), pp. 207–250, Academic Press, New York.

Amos, L. A., and Amos, W. B., 1991, *Molecules of the Cytoskeleton*, MacMillan, London.

Andreu, J. M., Bordas, J., Diaz, J. F., Garcia de Ancos, J., Gil, R., Medrano, F. J., Nogales, E., Pantos, E., and Towns-Andrews, E., 1992, Low resolution structure of microtubules in solution: Synchrotron X-ray scattering and electron microscopy of taxol-induced microtubules assembled from purified tubulin in comparison with glycerol and MAP-induced microtubules, *J. Mol. Biol.* **226**:169–184.

Arevalo, M. A., Nieto, J. M., Andreu, D., and Andreu, J. M., 1990, Tubulin assembly probed with antibodies to synthetic peptides, *J. Mol. Biol.* **214**:105–120.

Bai, R., Duanmu, C., and Hamel, E., 1989, Mechanism of action of the antimitotic drug 2,4-dichlorobenzyl thiocyanate: alkaylation of sulfhydryl group(s) of beta-tubulin, *Biochim. Biophys. Acta* **994**:12–20.

Bai, W., Singh, B., Yang, Y., Ramagli, L. S., Nash, M., Herzog, N. K., and Arlinghaus, R. B., 1992, The physical interaction between p37(env-mos) and tubulin structures, *Oncogene* **7**:493–500.

Baker, H. N., Rothwell, S. W., Grasser, W. A., Wallis, K. T., and Murphy, D. B., 1990, Copolymerization of two distinct tubulin isotypes during microtubule assembly in vitro, *J. Cell Biol.* **110**:97–104.

Banerjee, A., and Luduena, R. F., 1992, Kinetics of colchicine binding to purified β-tubulin isotypes from bovine brain, *J. Biol. Chem.* **267**:13335–13339.

Banerjee, A., Roach, M. C., Wall, K. A., Lopata, M. A., Cleveland, D. W., and Luduena, R. F., 1988, A monoclonal antibody against the type II isotype of β-tubulin. Preparation of isotypically altered tubulin, *J. Biol. Chem.* **263**:3029–3034.

Banerjee, A., Roach, M. C., Trcka, P., and Luduena, R. F., 1990, Increased microtubule assembly in bovine brain tubulin lacking the type III isotype of β-tubulin, *J. Biol. Chem.* **265**:1794–1799.

Banerjee, A., Roach, M. C., Trcka, P., and Luduena, R. F., 1992, Preparation of a monoclonal antibody specific for the class IV isotype of β-tubulin. Purification and assembly of αβII, αβIII, and αβIV tubulin dimers from bovine brain, *J. Biol. Chem.* **267**:5625–5630.

Barra, H. S., Arce, C. A., and Argarana, C. E., 1988, Posttranslational tyrosination/detyrosination of tubulin, *Mol. Neurobiol.* **2**:133–153.

Bayley, P. M., and Martin, S. R., 1991, Microtubule dynamic instability: Some possible physical mechanisms and their implications, *Biochem. Soc. Trans.* **19**:1023–1028.

Bayley, P., Clark, D. C., and Martin, S. R., 1983, Conformation properties of microtubule protein: Their relation to the self-assembly process in vitro, *Biopolymers* **22**:87–91.

Beese, L., Stubbs, G., and Cohen, C., 1987a, Microtubule structure at 18 Å resolution, *J. Mol. Biol.* **194**:257–264.

Beese, L., Stubbs, G., Thomas, J., and Cohen, C., 1987b, Structure of microtubules with reduced hydration: Comparison of results from X-ray diffraction and electron microscopy, *J. Mol. Biol.* **196**:575–580.

Beltramo, D. M., Alonso, A. C., and Barra, H. S., 1992, Tyrosinated, detyrosinated and acetylated tubulin isotypes in rat brain membranes. Their proportions in comparison with those in cytosol, *Mol. Cell. Biochem.* **112**:173–180.

Best, D., Warr, P. J., and Gull, K., 1981, Influence of the composition of commercial sodium dodecyl sulfate preparations on the separation of alpha and beta tubulin during polyacrylamide gel electrophoresis, *Anal. Biochem.* **114**:281–284.

Bhattacharyya, B., and Wolff, J., 1975, Membrane-bound tubulin in brain and thyroid tissue, *J. Biol. Chem.* **250**:7639–7646.

Bhattacharyya, B., Sackett, D. L., and Wolff, J., 1985, Tubulin, hybrid dimers and tubulin S: Stepwise charge reduction and polymerization, *J. Biol. Chem.* **260**:10208–10216.

Bibring, T., and Baxandall, J., 1977, Tubulin synthesis in sea urchin embryos: Almost all tubulin of the first cleavage mitotic apparatus derives from the unfertilized egg, *Dev. Biol.* **55**:191–195.

Boekelheide, K., 1987, 2,5-hexanedione alters microtubule assembly. II Enhanced polymerization of crosslinked tubulin, *Toxicol. Appl. Pharmacol.* **88**:383–396.

Boggs, B. A., Minotti, A. M., Loeb, L. M., Cook, R., and Cabral, F., 1988, Mutations affecting assembly of beta-tubulin localize to a region near the carboxyl terminus, *J. Biol. Chem.* **263**:14566–14573.

Bond, J. F., Fredovich-Keil, J. L., Pillus, L., Mulligan, R. C., and Solomon, F., 1986, A chicken–yeast chimeric beta-tubulin protein is incorporated into mouse microtubules in vivo, *Cell* **44**:461–468.

Borisy, G. G., and Taylor, E. W., 1967, The mechanism of action of colchicine. Binding of colchicine-^{3}H to cellular protein, *J. Cell Biol.* **34**:525–533.

Breitling, F., and Little, M., 1986, Carboxy-terminal regions on the surface of tubulin and microtubules: Epitope locations of YOL/34, DM1A and DM1B, *J. Mol. Biol.* **189**:367–370.

Brown, H. R., and Erickson, H. P., 1983, Assembly of proteolytically cleaved tubulin, *Arch. Biochem. Biophys.* **220**:46–51.

Bryan, J., 1974, Biochemical properties of microtubules, *Fed. Proc.* **33**:152–157.

Bulinski, J. C., Richards, J. E., and Piperno, G., 1988, Posttranslational modification of α-tubulin: Detyrosination and acetylation differentiate populations of interphase microtubules in cultured cells, *J. Cell Biol.* **106**:1213–1220.

Burgoyne, R. D., 1991, High molecular weight microtubule-associated proteins of brain, in *The Neuronal Cytoskeleton* (R. D. Burgoyne, ed.), pp. 75–91, Wiley-Liss, New York.

Burns, R. G., 1991a, α-, β-, and γ-tubulins: Sequence comparisons and structural constraints, *Cell Motil. Cytoskel.* **20:**181–189.

Burns, R. G., 1991b, Properties and assembly of neuronal microtubules in vitro, in *The Neuronal Cytoskeleton* (R. D. Burgoyne, ed.), pp. 93–119, Wiley-Liss, New York.

Burns, R. G., and Surridge, C., 1990, Analysis of β-tubulin sequences reveals highly conserved, coordinated amino acid substitutions. Evidence that these "hot spots" are directly involved in the conformational change required for dynamic instability, *FEBS Lett.* **271:**1–8.

Byard, E. H., and Lange, B. M., 1991, Tubulin and microtubules, *Essays Biochem.* **26:**13–25.

Cambray-Deakin, M. A., 1991, Cytoskeleton of the growing axon, in *The Neuronal Cytoskeleton* (R. D. Burgoyne, ed.), pp. 233–255, Wiley-Liss, New York.

Carlier, M.-F., 1989, Role of nucleotide hydrolysis in the dynamics of actin filaments and microtubules, *Int. Rev. Cytol.* **115:**139–170.

Chapin, S. J., and Bulinski, J. C., 1992, Microtubule stabilization by assembly-promoting microtubule-associated proteins: A repeat performance, *Cell Motil. Cytoskel.* **23:**236–243.

Chavan, A. J., Kim, H., Haley, B. E., and Watt, D. S., 1990, A photoreactive phosphonamide derivative of GTP for the identification of the GTP-binding domain of tubulin, *Bioconjugat. Chem.* **1:**337–344.

Chene, P., Mazarguil, H., and Wright, M., 1992, Microtubule assembly protects the region 28-38 of the β-tubulin subunit, *Cell Motil. Cytoskel.* **22:**25–37.

Cleveland, D. W., and Sullivan, K. F., 1985, Molecular biology and genetics of tubulin, *Annu. Rev. Biochem.* **54:**331–365.

Cleveland, D. W., Joshi, H. C., and Murphy, D. B., 1990, Tubulin site interpretation (letter), *Nature* **344:**389.

Correia, J. J., and Williams, R. C., 1983, Mechanisms of assembly and disassembly of microtubules, *Annu. Rev. Biophys. Bioeng.* **12:**211–235.

Cross, D., Dominguez, J., Maccioni, R. B., and Avila, J., 1991, MAP-1 and MAP-2 binding sites at the C-terminus of β-tubulin. Studies with synthetic tubulin peptides, *Biochemistry* **30:**4362–4366.

Crute, B. E., and VanBuskirk, R. G., 1992, A casein kinase-like kinase phosphorylates beta-tubulin and may be a microtubule-associated protein, *J. Neurochem.* **59:**2017–2023.

Curry, A. M., and Rosenbaum, J. L., 1993, Flagellar radial spoke: A model molecular genetic system for studying organelle assembly, *Cell Motil. Cytoskel.* **24:**224–232.

DeBrabander, M., DeMey, J., Geuns, G., Nuydens, R., Aerts, F., Willebrords, R., and Moeremans, M., 1985, Microtubules in cell organization and motility, in *Hormones and Cell Regulation* (J. E. Dumont, B. Hamprecht, and J. Nunez, eds.), pp. 85–104, Elsevier/North Holland, New York.

de la Vina, S., Andreu, D., Medrano, F. J., Nieto, J. M., and Andreu, J. M., 1988, Tubulin structure probed with antibodies to synthetic peptides. Mapping of three major types of limited proteolysis fragments, *Biochemistry* **27:**5352–5365.

Denoulet, P., Edde, B., Pinto-Henrique, D., Koulakoff, A., Berwald-Netter, Y., and Gros, F., 1988, The increase of tubulin heterogeneity during differentiation of neurons in primary culture is controlled mainly at the post-translational level, in *Structure and Functions of the Cytoskeleton* (B. A. F. Rousset, ed.), pp. 231–237, John Libbey Eurotext Ltd, London.

Dentler, W. L., 1980, Microtubule-membrane interactions in cilia. I. Isolation and characterization of ciliary membranes from *Tetrahymena pyriformis*, *J. Cell Biol.* **84:**364–380.

Detrich, H. W., III, and Williams, R. C., Jr., 1978, Reversible dissociation of the αβ dimer of tubulin from bovine brain, *Biochemistry* **17:**3900–3907.

Detrich, H. W., III, Williams, R. C., Jr., and Wilson, L., 1982, Effect of colchicine binding on the reversible dissociation of the tubulin dimer, *Biochemistry* **21:**2392–2400.

Draber, P., Draberova, E., Linhartova, I., and Viklicky, V., 1989, Differences in the exposure of C- and N-terminal tubulin domains in cytoplasmic microtubules detected with domain-specific monoclonal antibodies, *J. Cell Sci.* **92**:519–528.

Dustin, P., 1984, *Microtubules*, Springer-Verlag, New York.

Edde, B., Rossier, J., LeCaer, J. P., Desbruyeres, E., Gros, F., and Denoulet, P., 1990, Posttranslational glutamylation of alpha-tubulin, *Science* **247**:83–85.

Edde, B., Rossier, J., LeCaer, J. P., Prome, J. C., Desbruyeres, E., Gros, F., and Denoulet, P., 1992, Polyglutamylated alpha-tubulin can enter the tyrosination/detyrosination cycle, *Biochemistry* **31**:403–410.

Eigsti, O. J., and Dustin, P., 1955, *Colchicine—In Agriculture, Medicine, Biology, and Chemistry*, Iowa State College Press, Ames.

Eipper, B. A., 1974, Properties of rat brain tubulin, *J. Biol. Chem.* **249**:1407–1416.

Engelborghs, Y., 1990, Dynamic aspects of microtubule assembly, in *Microtubule Proteins* (J. Avila, ed.), pp. 1–36, CRC Press, Boca Raton, Florida.

Fackenthal, J. D., Turner, F. R., and Raff, E. C., 1993, Tissue-specific microtubule functions in *Drosophila* spermatogenesis require the β2-tubulin isotype-specific carboxy terminus, *Dev. Biol.* **158**:213–227.

Farr, G. W., and Sternlicht, H., 1992, Site-directed mutagenesis of the GTP-binding domain of β-tubulin, *J. Mol. Biol.* **227**:307–321.

Floyd, L. F., Barnes, L. D., and Williams, R. F., 1989, Photoaffinity labeling of tubulin with (2-nitro-4-azidophenyl) deacetylcolchicine: Direct evidence for two colchicine binding sites, *Biochemistry* **28**:8515–8525.

Fontana, A., Fassina, G., Vita, C., Dalzoppo, D., Zamai, M., and Zambonin, M., 1986, Correlation between sites of limited proteolysis and segmental mobility in thermolysin, *Biochemistry* **25**:1847–1851.

Fosket, D. E., and Morejohn, L. C., 1992, Structural and functional organization of tubulin, *Annu. Rev. Plant Physiol. Plant Mol. Biol.* **43**:201–240.

Fridovich-Keil, J. L., Bond, J. F., and Solomon, F., 1987, Domains of β-tubulin essential for conserved function in vivo, *Mol. Cell. Biol.* **7**:3792–3798.

Frigon, R. P., and Timasheff, S. N., 1975, Magnesium-induced self-association of calf brain tubulin. I. Stoichiometry, *Biochemistry* **14**:4559–4566.

Gaertig, J., Thatcher, T. H., McGrath, K. E., Callahan, R. C., and Gorovsky, M. A., 1993, Perspectives on tubulin isotype and function and evolution based on the observation that *Tetrahymena thermophila* microtubules contain a single α- and β-tubulin, *Cell Motil. Cytoskel.* **25**:243–253.

George, H. J., Misra, L., Field, D. J., and Lee, J. C., 1981, Polymorphism of brain tubulin, *Biochemistry* **20**:2402–2409.

Gilbert, T., LeBivic, A., Quaroni, A., and Rodriguez-Boulan, E., 1991, Microtubular organization and its involvement in the biogenetic pathways of plasma membrane proteins in Caco-2 intestinal epithelial cells, *J. Cell Biol.* **113**:275–288.

Ginzburg, I., Teichman, A., Dodmont, H. J., Behar, L., and Littauer, U. Z., 1985, Regulation of three beta-tubulin mRNAs during rat brain development, *EMBO J.* **4**:3667–3673.

Goldsmith, M., Connolly, J. A., Kumar, N., Wu, J., Yarbrough, L. R., and van der Kooy, D., 1991, Conserved β-tubulin binding domain for the microtubule-associated motors underlying sperm motility and fast axonal transport, *Cell Motil. Cytoskel.* **20**:249–262.

Graff, R. D., Philbert, M. A., Lowndes, H. E., and Reuhl, K. R., 1993, The effect of glutathione depletion on methyl mercury-induced microtubule disassembly in cultured embryonal carcinoma cells, *Toxicol. Appl. Pharmacol.* **120**:20–28.

Grain, J., 1986, The cytoskeleton in protists: Nature, structure, and functions, *Int. Rev. Cytol.* **104**:153–249.

Greer, K., and Rosenbaum, J. L., 1989, Post-translational modifications of tubulin, in *Cell Move-*

ment, vol. 2: *Kinesin, dynein, and microtubule dynamics* (F. D. Warner and J. R., McIntosh, eds.), pp. 47–66. Alan R. Liss, New York.

Hamel, E., 1990, Interactions of tubulin with small ligands, in *Microtubule Proteins* (J. Avila, ed.), pp. 89–192, CRC Press, Boca Raton, Florida.

Hamel, E., 1992, Natural products which interact with tubulin in the vinca domain: Maytansine, rhizoxin, phomopsin A, dolastatins 10 and 15 and halichondrin B, *Pharmacol. Ther.* **55**:31–51.

Hargreaves, A. J., Wandosell, F., and Avila, J., 1986, Phosphorylation of tubulin enhances its interaction with membranes, *Nature* **323**:827–828.

Harris, P. J., Clason, E. L., and Prier, K. R., 1989, Tubulin polymerization in unfertilized sea-urchin eggs induced by elevated temperature, *J. Cell Sci.* **93**:9–17.

Harrison, B. C., Marchese-Ragona, S. P., Gilbert, S. P., Cheng, N., Steven, A. C., and Johnson, K. A., 1993, Decoration of the microtubule surface by one kinesin head per tubulin heterodimer, *Nature* **362**:73–75.

Hastie, S. B., 1991, Interactions of colchicine with tubulin, *Pharmacol. Ther.* **51**:377–401.

Hesse, J., Thierauf, M., and Ponstingl, H., 1987, Tubulin sequence region β155-174 is involved in binding exchangable guanosine triphosphate, *J. Biol. Chem.* **262**:15472–15475.

Hiller, G., and Weber, K., 1978, Radioimmunoassay for tubulin: A quantitative comparison of the tubulin content of different established tissue culture cells and tissues, *Cell* **14**:795–804.

Himes, R. H., 1991, Interactions of the catharanthus (Vinca) alkaloids with tubulin and microtubules, *Pharmacol. Ther.* **51**:257–267.

Hirokawa, N., 1991, Molecular architecture and dynamics of the neuronal cytoskeleton, in *The Neuronal Cytoskeleton* (R. D. Burgoyne, ed.), pp. 5–74, Wiley-Liss, New York.

Horio, T., Uzawa, S., Jung, M. K., Oakley, B. R., Tanaka, K., and Yanagida, M., 1991, The fission yeast γ-tubulin is essential for mitosis and is localized at microtubule organizing centers, *J. Cell Sci.* **99**:693–700.

Hoyle, H. D., and Raff, E. C., 1990, *Drosophila* beta tubulin isoforms are not functionally equivalent, *J. Cell Biol.* **111**:1009–1026.

Huang, B., 1990, Genetics and biochemistry of centrosomes and spindle poles, *Curr. Opinion Cell Biol.* **2**:28–32.

Hussey, P. J., Lloyd, C. W., and Gull, K., 1988, Differential and developmental expression of β-tubulins in a higher plant, *J. Biol. Chem.* **263**:5474–5479.

Hyde, C. C., Ahmed, S. A., Padlan, E. A., Miles, E. W., and Davies, D. R., 1988, Three-dimensional structure of the tryptophan synthase α2β2 multienzyme complex from *Salmonella typhimurium*, *J. Biol. Chem.* **263**:17857–17871.

Hyman, A. A., Salser, S., Drechsel, D. N., Unwin, N., and Mitchison, T. J., 1992, Role of GTP hydrolysis in microtubule dynamics: Information from a slowly hydrolyzable analogue, GMPCPP, *Mol. Biol. Cell* **3**:1155–1167.

Ikeda, Y., and Steiner, M., 1979, Phosphorylation and protein kinase activity of platelet tubulin, *J. Biol. Chem.* **254**:66–74.

Iwasaki, S., 1993, Antimitotic agents: Chemistry and recognition of tubulin molecule, *Med. Res. Rev.* **13**:183–198.

Joshi, H. C., and Cleveland, D. W., 1990, Diversity among tubulin subunits: Toward what functional end? *Cell Motil. Cytoskel.* **16**:159–163.

Joshi, H. C., Palacios, M. J., McNamara, L., and Cleveland, D. W., 1992, γ-Tubulin is a centrosomal protein required for cell cycle-dependent microtubule nucleation, *Nature* **356**:80–83.

Kabsch, W., Mannherz, H. G., Suck, D., Pai, E. F., and Holmes, K. C., 1990, Atomic structure of the actin:DNase I complex, *Nature* **347**:37–44.

Kadowaki, T., Fujita-Yamaguchi, Y., Nishida, E., Takaku, F., Akiyama, T., Kathuria, S., Akanuma, Y., and Kasuga, M., 1985, Phosphorylation of tubulin and microtubule-associated proteins by the purified insulin receptor kinase, *J. Biol. Chem.* **260**:4016–4020.

Kambadur, R., Lewis, M., Chang, S., and Flavin, M., 1990, Characterization of putative cytoskeletal proteins from a trypanosomatid and their comparative binding to microtubules and soluble tubulin, *J. Biol. Chem.* **265**:20959–20965.

Kanazawa, K., and Timasheff, S. N., 1989, Preparation and characterization of des-C-terminal tubulin, *J. Protein Chem.* **8**:131–147.

Katz, W. S., and Solomon, F., 1988, Diversity among beta-tubulins: A carboxy-terminal domain of yeast beta-tubulin is not essential in vivo, *Mol. Cell Biol.* **8**:2730–2736.

Kirchner, K., and Mandelkow, E.-M., 1985, Tubulin domains responsible for assembly of dimers and protofilaments, *EMBO J.* **4**:2397–2402.

Kreis, T. E., 1987, Microtubules containing detyrosinated tubulin are less dynamic, *EMBO J.* **9**:2597–2606.

Lacey, E., 1988, The role of the cytoskeletal protein, tubulin, in the mode of action and mechanism of drug resistance to benzimidazoles, *Int. J. Parasitol.* **18**:885–936.

Laver, W. G., Air, G. M., Webster, R. G., and Smith-Gill, S. J., 1990, Epitopes on protein antigens: Misconceptions and realities, *Cell* **61**:553–556.

LeDizet, M., and Piperno, G., 1987, Identification of an acetylation site of *Chlamydomonas* α-tubulin, *Proc. Natl. Acad. Sci. USA* **84**:5720–5724.

Lee, G., 1993, Non-motor microtubule-associated proteins, *Curr. Opinion Cell Biol.* **5**:88–94.

Lee, J. C., Frigon, R. P., and Timasheff, S. N., 1973, The chemical characterization of calf brain microtubule protein subunits, *J. Biol. Chem.* **248**:7253–7262.

Lee, M. K., Tuttle, J. B., Rebhun, L. I., Cleveland, D. W., and Frankfurter, A., 1990, The expression and posttranslational modification of a neuron-specific β-tubulin isotype during chick embryogenesis, *Cell Motil. Cytoskel.* **17**:118–132.

Lemischka, I. R., Farmer, S.,, Racaniello, V. R., and Sharp, P. A., 1981, Nucleotide sequence and evolution of a mammalian alpha-tubulin messenger RNA, *J. Mol. Biol.* **151**:101–120.

Lewis, S. A., and Cowan, N. J., 1990, Tubulin genes: Structure, expression, and regulation, in *Microtubule Proteins* (J. Avila, ed.), pp. 37–66, CRC Press, Boca Raton, Florida.

L'Hernault, S. W., and Rosenbaum, J. L., 1985, *Chlamydomonas* α-tubulin is posttranslationally modified by acetylation on the eta-amino group of a lysine, *Biochemistry* **24**:473–478.

Lin, C. M., Ho, H. H., Petit, G. R., and Hamel, E., 1989, Antimitotic natural products combretastatin A-4 and combretastatin A-2: Studies on the mechanism of their inhibition of the binding of colchicine to tubulin, *Biochemistry* **28**:6984–6991.

Lin, X.-L., Lin, Y.-Z., Koelsch, G., Gustchina, A., Wlodawer, A., and Tang, J., 1992, Enzymatic activities of two-chain pepsinogen, two-chain pepsin, and the amino-terminal lobe of pepsinogen, *J. Biol. Chem.* **267**:17257–17263.

Linck, R. W., 1989, Microtubule structure and function, in *Cell Movement*, vol. 2, *Kinesin, Dynein, and Microtubule Dynamics* (F. D. Warner and J. R. McIntosh, eds.), pp. 13–21, Alan R. Liss, New York.

Linse, K., and Mandelkow, E.-M., 1988, The GTP-binding peptide of β-tubulin: Localization by direct photoaffinity labeling and comparison with nucleotide-binding proteins, *J. Biol. Chem.* **263**:15205–15210.

Littauer, U. Z., Biveon, D., Thierauf, M., Ginzburg, I., and Ponstingl, H., 1986, Common and distinct tubulin binding sites for microtubule-associated proteins, *Proc. Natl. Acad. Sci. USA* **83**:7162–7166.

Little, M., and Seehaus, T., 1988, Comparative analysis of tubulin sequences, *Comp. Biochem. Physiol.* **90B**:655–670.

Lobert, S., and Correia, J. J., 1991, Studies of crystallization conditions for native and subtilisin-cleaved pig brain tubulin, *Arch. Biochem. Biophys.* **290**:93–102.

Lobert, S., and Correia, J. J., 1992, Subtilisin cleavage of tubulin heterodimers and polymers, *Arch. Biochem. Biophys.* **296**:152–160.

Lopata, M. A., and Cleveland, D. W., 1987, In vivo microtubules are copolymers of available β-tubulin isotypes: Localization of each of six vertebrate β-tubulin isotypes using polyclonal antibodies elicited by synthetic peptide antigens, *J. Cell Biol.* **105:**1707–1720.

Lopez, L. A., and Sheetz, M. P., 1993, Steric inhibition of cytoplasmic dynein and kinesin motility by MAP2, *Cell Motil. Cytoskel.* **24:**1–16.

Luduena, R. F., 1993, Are tubulin isotypes functionally significant, *Mol. Biol. Cell* **4:**445–457.

Luduena, R. F., Shooter, E. M., and Wilson, L., 1977, Structure of the tubulin dimer, *J. Biol. Chem.* **252:**7006–7014.

Luduena, R. F., Zimmerman, H.-P., and Little, M., 1988, Identification of the phosphorylated β-isotype in differentiated neuroblastoma cells, *FEBS Lett.* **230:**142–146.

Luduena, R. F., Banerjee, A., and Khan, I. A., 1992, Tubulin structure and biochemistry, *Curr. Opinion Cell Biol.* **4:**53–57.

Lupas, A., VanDyke, M., and Stock, J., 1991, Predicting coiled coils from protein sequences, *Science* **252:**1162–1164.

Maccioni, R. B., Serrano, L., Avila, J., and Cann, J. R., 1986, Characterization and structural aspects of the enhanced assembly of tubulin after removal of its carboxyl-terminal domain, *Eur. J. Biochem.* **156:**375–381.

Maccioni, R. B., Rivas, C. I., and Vera, J. C., 1988, Differential interaction of synthetic peptides from the carboxyl-terminal regulatory domain of tubulin with microtubule-associated proteins, *EMBO J.* **7:**1957–1963.

MacRae, T. H., and Langdon, C. M., 1989, Tubulin synthesis, structure, and function: What are the relationships? *Biochem. Cell. Biol.* **67:**770–790.

Mandelkow, E.-M., Herrmann, M., and Ruhl, U., 1985, Tubulin domains probed by limited proteolysis and subunit-specific antibodies, *J. Mol. Biol.* **185:**311–327.

Matesic, D. F., Philp, N. J., Murray, J. M., and Liebman, P. A., 1992, Tubulin in bovine rod outer segments, *J. Cell Sci.* **103:**157–166.

Matsuzaki, F., Matsumoto, S., and Yahara, I., 1988, Truncation of the carboxy-terminal domain of yeast β-tubulin causes temperature-sensitive growth and hypersensitivity to antimitotic drugs, *J. Cell Biol.* **107:**1427–1435.

Matten, W. T., Aubry, M., West, J., and Maness, P. F., 1990, Tubulin is phosphorylated at tyrosine by pp60c-src in nerve growth cone membranes, *J. Cell Biol.* **111:**1959–1970.

May, G. S., 1989, The highly divergent β-tubulins of *Aspergillus nidulans* are functionally interchangeable, *J. Cell Biol.* **109:**2267–2274.

Mazumdar, M., Parrack, P. K., Mukhopadhyay, K., and Bhattacharyya, B., 1992, Bis-ANS as a specific inhibitor for microtubule-associated protein induced assembly of tubulin, *Biochemistry* **31:**6470–6474.

Mejillano, M. R., and Himes, R. H., 1989, Tubulin dimer dissociation detected by fluorescence anisotropy, *Biochemistry* **28:**6518–6524.

Mejillano, M. R., and Himes, R. H., 1991, Assembly properties of tubulin after carboxyl group modification, *J. Biol. Chem.* **266:**657–664.

Mejillano, M. R., Tolo, E. T., Williams, R. C. Jr., and Himes, R. H., 1992, The conversion of tubulin carboxyl groups to amides has a stabilizing effect in microtubules, *Biochemistry* **31:**3478–3483.

Mitchison, T., and Kirschner, M. W., 1984, Dynamic instability of microtubule growth, *Nature* **312:**237–242.

Morejohn, L. C., and Fosket, D. E., 1991, The biochemistry of compounds with anti-microtubule activity in plant cells, *Pharmacol. Ther.* **51:**217–230.

Mornet, D., and Ue, K., 1984, Proteolysis and structure of skeletal muscle actin, *Proc. Natl. Acad. Sci. USA* **81:**3680–3684.

Mukhopadhyay, K., Parrack, P. K., and Bhattacharyya, B., 1990, The carboxy terminus of the α subunit of tubulin regulates its interaction with colchicine, *Biochemistry* **29:**6845–6850.

Murray, J. M., 1984, Disassembly and reconstitution of a membrane-microtubule complex, *J. Cell Biol.* **98:**1481–1487.

Oakley, B. R., Oakley, C. E., Yoon, Y., and Jung, M. K., 1990, γ-Tubulin is a component of the spindle pole body that is essential for microtubule function in *Aspergillus nidulans*, *Cell* **61:**1289–1301.

Oakley, C. E., and Oakley, B. R., 1989, Identification of γ-tubulin, a new member of the tubulin superfamily encoded by *mip*A gene of *Aspergillus nidulans*, *Nature* **338:**662–664.

O'Brien, E. T., Voter, W. A., and Erickson, H. P., 1987, GTP hydrolysis during microtubule assembly, *Biochemistry* **26:**4148–4156.

Ortiz, M., Lagos, R., and Monasterio, O., 1993, Interaction between the C-terminal peptides of tubulin and tubulin S detected with the fluorescent probe 4′,6-diamidino-2-phenylindole, *Arch. Biochem. Biophys.* **303:**159–164.

Otter, A., and Kotovych, G., 1988, The solution conformation of the synthetic tubulin fragment Ac-tubulin-α(430-441)-amide based on two-dimensional ROESY experiments, *Can J. Chem.* **66:**1814–1820.

Otter, A., Scott, P. G., Maccioni, R. B., and Kotovych, G., 1991, The solution conformation of tubulin β(422–434)-NH2 and its Nac-DATADEQG-NH2 fragment based on NMR, *Biopolymers* **31:**449–458.

Pabo, C. O., Sauer, R. T., Sturtevant, J. M., and Ptashne, M., 1979, The λ repressor contains two domains, *Proc. Natl. Acad. Sci. USA* **76:**1608–1612.

Panda, D., Roy, S., and Bhattacharyya, B., 1992, Reversible dimer dissociation of tubulin S and tubulin detected by fluorescence anisotropy, *Biochemistry* **31:**9709–9716.

Paschal, B. M., Obar, R. A., and Valee, R. B., 1989, Interaction of brain cytoplasmic dynein and MAP2 with a common sequence at the C terminus of tubulin, *Nature* **342:**569–572.

Peyrot, V., Briand, C., and Andreu, J. M., 1990, C-terminal cleavage of tubulin by subtilisin enhances ring formation, *Arch. Biochem. Biophys.* **279:**328–337.

Piperno, G., LeDizet, M., and Chang, X. J., 1987, Microtubules containing acetylated α-tubulin in mammalian cells in culture, *J. Cell Biol.* **104:**289–302.

Ponstingl, H., Little, M., Krauhs, E., and Kempf, T., 1979, Carboxy-terminal amino acid sequence of α-tubulin from porcine brain, *Nature* **282:**423–424.

Ponstingl, H., Little, M., and Krauhs, E., 1984, Tubulins, *Peptide Protein Rev.* **2:**1–81.

Ponstingl, H., Little, M., Kempf, T., Krauhs, E., and Ade, W., 1990, Mapping of ligand binding sites, epitopes, and posttranslational modifications on protein sequences—Tubulin as an example, in *Laboratory Methodology in Biochemistry: Amino Acid Analysis and Protein Sequencing* (C. Fini, A. Floridi, U. N. Finelli, and S. Wittman-Liebold, eds.), pp. 151–171, CRC Press, Boca Raton, Florida.

Porter, K., 1966, Cytoplasmic microtubules and their function, in *CIBA Foundation Symposium: Principles of Biomolecular Organization* (G. E. W. Wolstenholme and M. O'Connor, eds.), pp. 308–356, Little, Brown, Boston.

Potchoo, Y., Braguer, D., Peyrot, V., Chauvet-Monges, A. M., Sari, J. C., and Crevat, A., 1986, In vitro inhibition of microtubule assembly by disulfiram, *Int. J. Clin. Pharmacol.* **24:**499–504.

Prescott, A. R., Foster, K. E., Warn, R. M., and Gull, K., 1989, Incorporation of tubulin from an evolutionarily diverse source, *Physarum polycephalum*, into the microtubules of a mammalian cell, *J. Cell Sci.* **92:**595–605.

Purich, D. L., and Scaife, R. M., 1985, Microtubule cytoskeleton proteins as targets for covalent interconverting enzymes, *Curr. Top. Cell. Regul.* **27:**107–116.

Raff, R. A., and Kaumeyer, J. F., 1973, Soluble microtubule proteins of the sea urchin embryo: Partial characterization of the proteins and behaviour of the pool in early development, *Dev. Biol.* **32:**309–320.

Raffaelli, N., Scaife, R. M., and Purich, D. L., 1992, ADP ribosylation of chicken red cell tubulin and inhibition of microtubule self-assembly in vitro by the NAD$^+$-dependent avian ADP ribosyl transferase, *Biochem. Biophys. Res. Commun.* **184**:414–418.

Rao, S., Horwitz, S. B., and Ringel, I., 1992, Direct photoaffinity labeling of tubulin with taxol, *J. Natl. Cancer Inst.* **84**:785–788.

Redeker, V., LeCaer, J. P., Rossier, J., and Prome, J. C., 1991, Structure of the polyglutamyl side chain posttranslationally added to α-tubulin, *J. Biol. Chem.* **266**:23461–23466.

Redeker, V., Melki, R., Prome, D., LeCaer, J. P., and Rossier, J., 1992, Structure of tubulin C-terminal domain obtained by subtilisin treatment. The major alpha and beta tubulin isotypes from pig brain are glutamylated, *FEBS Lett.* **313**:185–192.

Reed, J., Hull, W. E., Ponstingl, H., and Himes, R. H., 1992, Conformational properties of the β(400-436) and β(400-445) C-terminal peptides of porcine brain tubulin, *Biochemistry* **31**:11888–11895.

Richards, F. M., and Vithayathil, P. J., 1959, The preparation of subtilisin-modified ribonuclease and the separation of the peptide and protein components, *J. Biol. Chem.* **234**:1459–1465.

Ringel, I., and Sternlicht, H., 1984, Carbon-13 nuclear magnetic resonance study of microtubule protein: Evidence for a second colchicine site involved in the inhibition of microtubule assembly, *Biochemistry* **23**:5644–5653.

Rodionov, V. I., Gyoeva, F. K., Kashina, A. S., Kuznetsov, S. A., and Gelfand, V. I., 1990, Microtubule-associated proteins and microtubule-based translocatores have different binding sites on tubulin molecule, *J. Biol. Chem.* **265**:5702–5707.

Rowinsky, E. K., and Donehower, R. C., 1991, The clinical pharmacology and use of antimicrotubule agents in cancer chemotherapeutics, *Pharmacol. Ther.* **52**:35–84.

Roychowdhury, S., Wang, N., and Rasenick, M. M., 1993, G protein binding and G protein activation by nucleotide transfer involve distinct domains on tubulin: Regulation of signal transduction by cytoskeletal elements, *Biochemistry* **32**:4955–4961.

Rudiger, M., Plessman, U., Kloppel, K.-D., Wehland, J., and Weber, K., 1992, Class II tubulin, the major brain β tubulin isotype is polyglutamylated on glutamic acid residue 435, *FEBS Lett.* **308**:101–105.

Sackett, D. L., 1993, Podophyllotoxin, steganacin, and combretastatin; natural products which interact with tubulin at the colchicine site, *Pharmacol. Ther.* **59**:163–228.

Sackett, D. L., and Lippoldt, R. E., 1991, Thermodynamics of the reversible monomer–dimer association of tubulin, *Biochemistry* **30**:3511–3517.

Sackett, D. L., and Varma, J. K., 1993, Molecular mechanism of colchicine action: Induced unfolding of beta tubulin, *Biochemistry* **32**:13560–13565.

Sackett, D. L., and Wolff, J., 1986, Proteolysis of tubulin and the substructure of the tubulin dimer, *J. Biol. Chem.* **261**:9070–9076.

Sackett, D. L., and Wolff, J., 1987, Nile red as a polarity sensitive fluorescent probe of hydrophobic surfaces, *Anal. Biochem.* **167**:228–234.

Sackett, D. L., Bhattacharyya, B., and Wolff, J., 1985, Tubulin carboxyl terminii determine polymerization efficiency, *J. Biol. Chem.* **260**:43–45.

Sackett, D. L., Bhattacharyya, B., and Wolff, J., 1986, Promotion of tubulin assembly by carboxyterminal charge reduction, *Ann. NY Acad. Sci.* **466**:460–466.

Sackett, D. L., Zimmerman, D. A., and Wolff, J., 1989, Tubulin dimer dissociation and proteolytic accessibility, *Biochemistry* **28**:2662–2667.

Sackett, D. L., Knutson, J. R., and Wolff, J., 1990, Hydrophobic surfaces of tubulin studied by steady state and time-resolved fluorescence of Nile red, *J. Biol. Chem.* **25**:14889–14906.

Safa, A. R., Hamel, E., and Felsted, R. L., 1987, Photoaffinity labeling of tubulin subunits with a photoactive analogue of vinblastine, *Biochemistry* **26**:97–102.

Sager, P. R., Doherty, R. A., and Olmsted, J. B., 1983, Interaction of methylmercury with microtubules in cultured cells and in vitro, *Exp. Cell Res.* **146**:127–137.

Scaife, R. M., Wilson, L., and Purich, D. L., 1992, Microtubule protein ADP-ribosylation in vitro leads to assembly inhibition and rapid depolymerization, *Biochemistry* **31**:310–316.

Schatz, P. J., Solomon, F., and Botstein, D., 1986, Genetically essential and nonessential α-tubulin genes specify functionally interchangeable proteins, *Mol. Cell Biol.* **6**:3722–3733.

Schatz, P. J., Georges, G. E., Solomon, F., and Botstein, D., 1987, Insertions of up to 17 amino acids into a region of α-tubulin do not disrupt function in vivo, *Mol. Cell Biol.* **7**:3799–3805.

Schulze, E., Asai, D. J., Bulinski, J. C., and Kirschner, M., 1987, Posttranslational modification and microtubule stability, *J. Cell Biol.* **105**:2167–2177.

Serrano, L., and Avila, J., 1985, The interaction between subunits in the tubulin dimer, *Biochem. J.* **230**:551–556.

Serrano, L., and Avila, J., 1990, Structure and function of tubulin regions, in *Microtubule Proteins* (J. Avila, ed.), pp. 67–88, CRC Press, Boca Raton, Florida.

Serrano, L., Avila, J., and Maccioni, R. B., 1984a, Controlled proteolysis of tubulin by subtilisin: Location of the site for MAP2 interaction, *Biochemistry* **23**:4675–4681.

Serrano, L., de la Torre, J., Maccioni, R. B., and Avila, J., 1984b, Involvement of the carboxyl-terminal domain of tubulin in the regulation of its assembly, *Proc. Natl. Acad. Sci. USA* **81**:5989–5993.

Serrano, L., Valencia, A., Caballero, R., and Avila, J., 1986, Localization of the high affinity calcium-binding site on tubulin molecule, *J. Biol. Chem.* **261**:7076–7081.

Serrano, L., Wandosell, F., de la Torre, J., and Avila, J., 1988, Effect of specific proteolytic cleavages on tubulin polymer formation, *Biochem. J.* **252**:683–691.

Shivanna, B. D., Mejillano, M. R., Williams, T. D., and Himes, R. H., 1993, Exchangeable GTP binding site of β-tubulin: Identification of cysteine 12 as the major site of cross-linking by direct photoaffinity labeling, *J. Biol. Chem.* **268**:127–132.

Siflow, C. D., 1991, Why do tubulin gene families lack diversity in flagellate/ciliate protists? *Protoplasma* **164**:9–11.

Siflow, C. D., Oppenheimer, D. G., Kopczak, S. D., Ploense, S. E., Ludwig, S. R., Haas, N., and Snustad, D. P., 1987, Plant tubulin genes: structure and differential expression during development, *Dev. Genet.* **8**:435–460.

Sioussat, T. M., and Boekelheide, K., 1989, Selection of a nucleation-promoting element following chemical modification of tubulin, *Biochemistry* **28**:4435–4443.

Skoufias, D. A., and Scholey, J. M., 1993, Cytoplasmic microtubule-based motor proteins, *Curr. Opin. Cell Biol.* **5**:95–104.

Slautterback, D. B., 1963, Cytoplasmic microtubules. I. Hydra, *J. Cell Biol.* **18**:367–388.

Somers, M., Engelborghs, Y., and Baert, J., 1990, Analysis of the binding of glyceraldehyde-3-phosphate dehydrogenase to microtubules, the mechanism of bundle formation and the linkage effect, *Eur. J. Biochem.* **193**:437–444.

Steiner, M., 1983, Membrane-bound tubulin in human platelets, *Biochim. Biophys. Acta* **729**:17–22.

Stearns, T., Evans, L., and Kirschner, M., 1991, Gamma-tubulin is a highly conserved component of the centrosome, *Cell* **65**:825–836.

Stephens, R. E., 1981, Chemical differences distinguish ciliary membrane and axonemal tubulins, *Biochemistry* **20**:4716–4723.

Stephens, R. E., 1986, Membrane tubulin, *Biol. Cell* **57**:95–110.

Stephens, R. E., 1991, Tubulin in sea urchin embryonic cilia: characterization of the membrane-periaxonemal matrix, *J. Cell Sci.* **100**:521–531.

Sternlicht, H., Yaffe, M. B., and Farr, G. W., 1987, A model of the nucleotide-binding site in tubulin, *FEBS Lett.* **214**:226–235.

Stewart, R. J., Farrell, K. W., and Wilson, L., 1990, Role of GTP hydrolysis in microtubule polymerization: Evidence for a coupled hydrolysis mechanism, *Biochemistry* **29**:6489–6498.

Sullivan, K. F., 1988, Structure and utilization of tubulin isotypes, *Annu. Rev. Cell. Biol.* **4**:687–716.

Suprenant, K. A., and Rebhun, L. I., 1983, Assembly of unfertilized sea urchin egg tubulin at physiological temperatures, *J. Biol. Chem.* **258**:4518–4525.

Szasz, J., Yaffe, M. B.,, Elzinga, M., Blank, G. S., and Sternlicht, H., 1986, Microtubule assembly is dependent on a cluster of basic residues in α-tubulin, *Biochemistry* **25**:4572–4582.

Thompson, W. C., 1982, The cyclic tyrosination/detyrosination of alpha tubulin, *Methods Cell Biol.* **24**:235–255.

Unger, E., Bohm, K. J., and Vater, W., 1990, Structural diversity and dynamics of microtubules and polymorphic tubulin assemblies, *Electron Microscop. Rev.* **3**:355–395.

Uppuluri, S.,, Knipling, L., Sackett, D., and Wolff, J., 1993, Location of the colchicine binding site of tubulin, *Proc. Natl. Acad. Sci. USA* **90**:11598–11602.

Vera, J., Rivas, C. I., and Maccioni, R. B., 1988, Antibodies to synthetic peptides from the tubulin regulatory domain interact with tubulin and microtubules, *Proc. Natl. Acad. Sci. USA* **85**:6763–6767.

Vera, J. C., Rivas, C. I., and Maccioni, R. B., 1989, Biochemical dissection of the role of the one-kilodalton carboxyl-terminal moiety of tubulin in its assembly into microtubules, *Biochemistry* **28**:333–339.

Vogel, D. G., Margolis, R. L., and Mottet, N. K., 1985, The effects of methyl mercury binding to microtubules, *Toxicol. Appl. Pharmacol.* **80**:473–486.

Wade, R. H., and Chretien, D., 1993, Cryoelectron microscopy of microtubules, *J. Struct. Biol.* **110**:1–27.

Walker, R. A., and Sheetz, M. P., 1993, Cytoplasmic microtubule-associated motors, *Annu. Rev. Biochem.* **62**:429–451.

Wallin, M., Deinum, J., and Friden, B., 1985, Interaction of estramustine phosphate with microtubule-associated proteins, *FEBS Lett.* **179**:289–293.

Walsh, J. L., Keith, T. J., and Knull, H. R., 1989, Glycolytic enzyme interactions with tubulin and microtubules, *Biochim. Biophys. Acta* **999**:64–70.

Wandosell, F., Serrano, L., and Avila, J., 1987, Phosphorylation of tubulin carboxy terminal tyrosine prevents its incorporation into microtubules, *J. Biol. Chem.* **262**:8268–8273.

Wang, N., and Rasenick, M. M., 1991, Tubulin-G protein interactions involve microtubule polymerization domains, *Biochemistry* **30**:10957–10965.

Webster, D. R., and Borisy, G. G., 1989, Microtubules are acetylated in domains that turn over slowly, *J. Cell Sci.* **92**:57–65.

Webster, D. R., Wehland, J., Weber, K., and Borisy, G. G., 1990, Detyrosination of alpha tubulin does not stabilize microtubules in vivo, *J. Cell Biol.* **111**:113–132.

Wehland, J., Schroder, H. C., and Weber, K., 1984, Amino acid sequence requirements in the epitope recognized by the α-tubulin specific rat monoclonal antibody YL 1/2, *EMBO J.* **3**:1295–1300.

Weisenberg, R. C., Deery, W. J., and Dickinson, P. J., 1976, Tubulin–nucleotide interactions during the polymerization and depolymerization of microtubule, *Biochemistry* **15**:4248–4254.

White, E. A., Burton, P. R., and Himes, R. H., 1987, Polymorphic assembly of subtilisin-cleaved tubulin, *Cell Motil. Cytoskel.* **7**:31-38.

Wilson, L., and Friedkin, M., 1967, The biochemical events of mitosis. II. The *in vivo* and *in vitro* binding of colchicine in grasshopper embryos and its possible relation to inhibition of mitosis, *Biochemistry* **6**:3126–3135.

Wolff, J., and Knipling, L., 1993, The antimicrotubule properties of benzophenanthridine alkaloids, *Biochemistry* **32**:13334–13339.

Wolff, J., Knipling, L., Cahnmann, H. J., and Palumbo, G., 1991, Direct photoaffinity labeling of tubulin with colchicine, *Proc. Natl. Acad. Sci. USA* **88**:2820–2824.

Wolff, J., Hwang, J., Sackett, D. L., and Knipling, L., 1992, Colchicine photosensitizes tubulin dimerization, *Biochemistry* **31**:3935–3940.

Zheng, Y., Jung, M. K., and Oakley, B. R., 1991, Gamma-tubulin is present in *Drosophila melanogaster* and Homo sapiens and is associated with the centrosome, *Cell* **65**:817–823.

Zhou, R. P., Oskarsson, M., Paules, R. S., Schulz, N., Cleveland, D., and Vande Woude, G. F., 1991, Ability of the c-mos product to associate with and phosphorylate tubulin, *Science* **251**:671–675.

Zimmerman, S. B., and Minton, A., 1993, Macromolecular crowding: biochemical, biophysical, and physiological consequences, *Annu. Rev. Biophys. Biomol. Struct.* **22**:27–65.

Chapter 10

Role of Allosteric Changes in Cyclic AMP Receptor Protein Function

Sankar Adhya, Sangryeol Ryu, and Susan Garges

1. INTRODUCTION

The idea that conformational changes may be controlled by a ligand binding was first pointed out in explaining the homotropic cooperative binding of O_2 to hemoglobin which has four O_2 binding sites (Wyman, 1948; Wyman and Allen, 1951). The term *reversible allosteric transition,* which is an expression of such ligand-induced conformational changes, was introduced later (Monod and Jacob, 1961; Monod *et al.*, 1963). An ideal system to study allostery is the interaction of cyclic AMP (cAMP) and its receptor protein (CRP). cAMP and CRP are two "master elements" of a vast global regulatory network in *Escherichia coli* (reviewed in Kolb *et al.*, 1993). The level of cAMP dramatically changes in response to the nature of energy sources for cell growth. The CRP level also varies in the cell depending on the energy sources. When present in effective concentrations, cAMP binds to CRP, and the resulting complex regulates transcription of many genes. The most studied role of the cAMP–CRP complex is its role as an activator of transcription initiation. The protein is a homodimer of a 209 amino acid subunit. Two molecules of cAMP can bind to CRP, one to each subunit.

Sankar Adhya, Sangryeol Ryu, and Susan Garges Laboratory of Molecular Biology, National Cancer Institute, National Institutes of Health, Bethesda, Maryland 20892.

Subcellular Biochemistry, Volume 24. Proteins: Structure, Function, and Engineering, edited by B. B. Biswas and Siddhartha Roy. Plenum Press, New York, 1995.

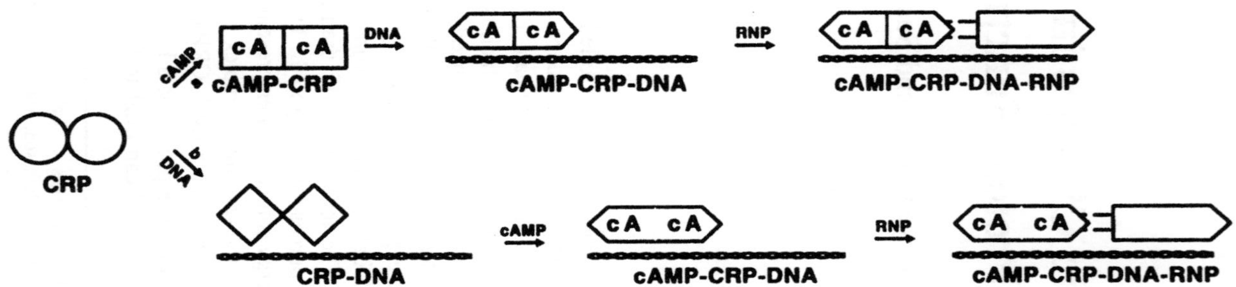

FIGURE 1. The pathway of CRP in transcription activation. The potential or demonstrated allosteric changes in CRP are shown by change in shape. As discussed in the text, pathway b is more likely than pathway a. The DNA binding shown reflects specific sites.

cAMP binding to CRP allosterically changes the structure of the protein, making it proficient first to bind to specific sequences on DNA located near the promoter and then to contact the RNA polymerase at the promoter of the gene. Without cAMP, CRP binds to DNA nonspecifically at a much lower efficiency. cAMP–CRP binding induces structural changes in the DNA, and DNA reverberates back conformational alterations in the protein. The resulting cAMP–CRP–DNA complex activates transcription initiation by influencing both the promoter and RNA polymerase. The pathway to transcription activation by cAMP and CRP is diagrammed in Fig. 1. In this chapter, we will describe the steps of the pathway and emphasize the allosteric changes induced in CRP by cAMP.

2. THREE-DIMENSIONAL STRUCTURE AND FUNCTION OF CRP

The understanding of the allosteric changes in CRP that accompany cAMP binding and their role in DNA binding and transcription activation must take into account the three-dimensional structure of the free CRP, of the cAMP–CRP complex as well as of the specific cAMP–CRP–DNA ternary complex. Whereas the structures of $(cAMP)_2$–CRP and $(cAMP)_2$–CRP–DNA have been determined by X-ray diffraction of the corresponding crystals to reasonable resolution (Weber and Steitz, 1987; Schultz *et al.*, 1991), one has not been able to crystallize free CRP. It has been suggested that the transcriptionally active complex is $(cAMP)_1$–CRP and not $(cAMP)_2$–CRP (Heyduk and Lee, 1989). The structure of $(cAMP)_1$–CRP, which appears to be significantly different from that of $(cAMP)_2$–CRP by physiochemical studies, is also not known. Figure 2 shows that structure of $(cAMP)_2$–CRP, which is totally consistent with characteristics of the protein previously determined by biochemical means, which are summarized in Section 5.2. As indicated in Fig. 2, each subunit of the CRP dimer possesses two domains: the larger N-terminal domain has the dimerization and cAMP binding features. The dimerization occurs mainly by interaction between the two long C α-helixes and cAMP binds through its ribose and phosphate in the pocket of a β-roll structure in the N-terminal domain. The smaller C-terminal domain contains the helix-turn-helix DNA binding motif and the main surface loops for RNA polymerase contact. The two domains are connected by a hinge region. Since the alteration of the property of the DNA and RNA polymerase binding to the C-terminal domain is induced by cAMP binding to the N-terminal domain, the conformational changes by definition are allosteric in nature. In bringing about the structural changes, cAMP binding also affects, as discussed later, the interaction between the two subunits and the orientation of the two domains within each subunit. There is disagreement about the stereoisomer (*syn* or *anti*) of the two cAMP molecules present in cAMP–CRP. Whereas the X-ray diffraction data indicate a *syn* conformation for the two bound cyclic nucleotides,

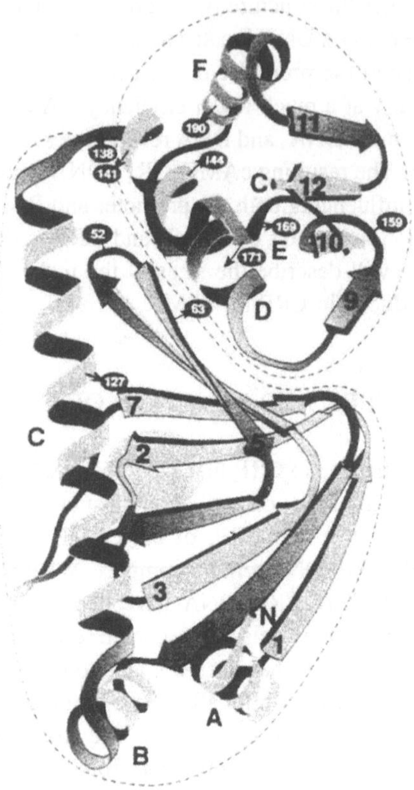

FIGURE 2. Structure of CRP monomer. The six α-helices are lettered and the twelve β-sheets are numbered. The hinge (residues 35-138) connecting the two domains (shown by dotted lines) is between C and D α-helices. E and F constitute the helix-turn-helix DNA binding motif. cAMP binding is within the roll structure of the β-sheets 1-8. The C α-helix interacts with its counterpart from the other subunit for dimerization. The amino acids identified to participate in cAMP binding, allostery, and RNA polymerase contacts are marked by their position numbers.

nuclear magnetic resonance (NMR) experiments suggest an *anti* form for the bound cAMP. The two subunits are slightly asymmetrical, the hinge in one subunit being more "open" and the one in the other subunit "closed."

The structure of the $(cAMP)_2$–CRP complexed with a 30 base pairs (bp) DNA containing the consensus binding sequence has confirmed the amino acid–nucleotide contacts determined genetically and biochemically and the DNA bending discussed in Sections 3 and 4.2. It should be mentioned that the structure of the $(cAMP)_2$–CRP in the DNA complex is symmetrical, both subunits being in "closed" form of the hinge region.

3. CRP–DNA INTERACTIONS

The intracellular concentration of CRP is 1–2 μM and that of nonspecific binding sites is about 7 mM. Since the nonspecific binding of free CRP to DNA occurs with a K_D of 10^{-6} M, CRP would almost exclusively be present bound to DNA in the absence of cAMP. Thus, when cAMP is made available, it mostly binds to DNA-bound CRP and rarely to free CRP. The nonspecific binding is solely ionic with the phosphate backbone making contact with basic amino acids involving six ion pairs per CRP dimer (Hudson *et al.*, 1990; Takahashi *et al.*, 1983). The binding of preformed cAMP–CRP complex with specific DNA sequence is three orders of magnitude higher, with a K_D in the range of 10^{-9} M. The target for specific binding is a dyad symmetry motif of the consensus sequence 5′TGTGAN$_X$TCACA3′. The length (x) of the spacer sequence between the two inverted repeat recognition boxes is usually six and occasionally eight (Shumilov, 1987; Barber and Zhurkin, 1990). It has been suggested that translocation of a CRP dimer from nonspecific to a specific site is accompanied by the loss of a cAMP molecule from CRP dimer (Fried and Crothers, 1984). This is consistent with the observation that binding of CRP to specific DNA sequences decreases at higher cAMP concentration, suggesting the active specific complex may be $(cAMP)_1$–CRP rather than $(cAMP)_2$–CRP (Heyduk and Lee, 1989, 1990). Whereas the nonspecific binding of the cAMP–CRP complex to DNA backbone involves four to five ion pairs per dimer, the increase in the affinity to the specific site is partly because of two additional ion pairs (eight instead of six) (Takahashi *et al.*, 1983). But the higher affinity is largely attributed to nonelectrostatic contributions. The extra interactions are through amino acid side chains of the two helix-turn-helix motifs in CRP (one per subunit) contacting the recognition box (5′TGTGA3′) base edges in the two successive major groups located on the same face of the DNA (Schultz *et al.*, 1991). The main contacts are between the side chains of Arg180, Glu181, and Arg185 in the recognition helixes of the two helix-turn-helix motif in the CRP dimer and the base edges of the consensus boxes as shown in Fig. 3. These contacts were mostly predicted from genetic analysis of the CRP mutants (Ebright *et al.*, 1987;

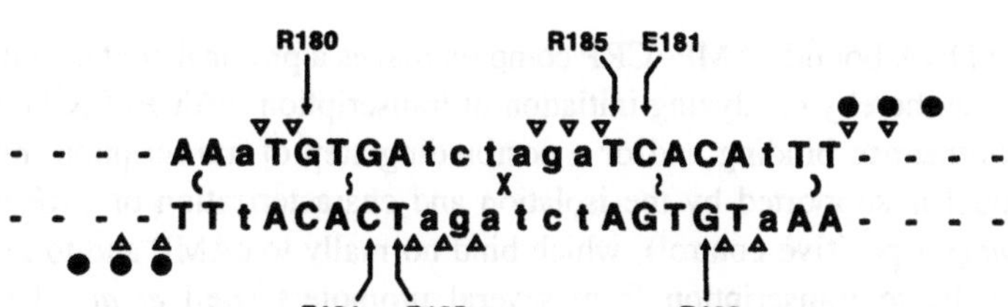

FIGURE 3. Chemical contacts between CRP and its specific DNA binding site. The DNA sequence shown is a consensus dyad symmetry. The data are from Schultz *et al.* (1991), Gunasekera *et al.* (1992) and Ebright *et al.* (1990). The open triangles positioned between bases represent phosphates that interferes with CRP binding after ethylation. The closed circles positioned between bases represent phosphates that increase the binding affinity in the Liu-Johnson *et al.* (1986) assay. The "X" represents the center of DNA dyad symmetry. The "S" and "C" symbols represent bend points.

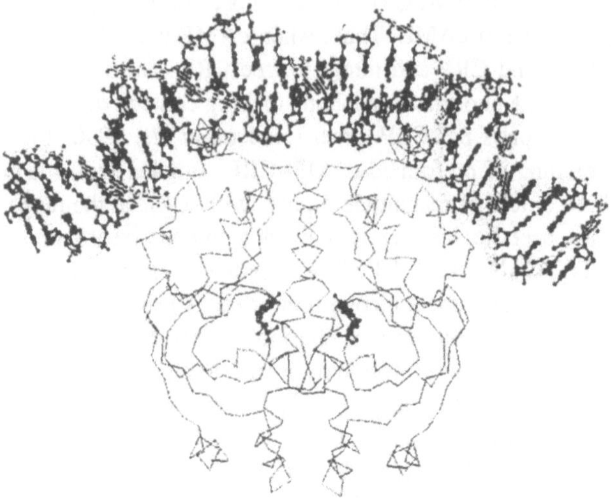

FIGURE 4. The structure of $(cAMP)_2$–CRP–DNA complex, adapted from Schultz *et al.* (1991). The α-carbon backbone of CRP is shown.

Jansen *et al.*, 1987; Zhang *et al.*, 1990; Gunasekera *et al.*, 1992). The structure of the entire $(cAMP)_2$–CRP–DNA complex is shown in Fig. 4.

4. TRANSCRIPTION ACTIVATION

Two models have been proposed to explain the mode by which cAMP–CRP complex activates transcription from a promoter.

4.1. Contact with RNA Polymerase

The DNA-bound cAMP–CRP complex makes a physical contact with RNA polymerase thereby catalyzing initiation of transcription. cAMP–CRP may help RNA polymerase binding and/or a postbinding step of transcription initiation. This model is supported by the isolation and characterization of CRP mutants called *pc* (for positive control), which bind normally to cAMP and to DNA but do not activate transcription from several promoters (Bell *et al.*, 1990; Eschenlauer and Reznikoff, 1991; Zhou *et al.*, 1993b). It has been suggested that these *pc* mutations define the RNA polymerase-contact sites in CRP. The available results indicate that there are different contact points in CRP for different kinds of promoters. For promoters (type I) in which the CRP binding site is centered at −61.5 position from the start site of transcription, e.g., the *lac* promoter, the contact site is a loop structure on the surface of CRP covering the amino acids at position 157-162. Mutations in this loop fail to activate transcription from type I promoters.

Recent biochemical and genetic studies have also suggested a role for the C-terminus region of the α-subunit of RNA polymerase as a target for CRP action (Igarashi and Ishihama, 1991; Ishihama, 1993). Reconstituted RNA polymerase carrying mutation or deletion of this region is capable of transcribing from several promoters but not from the cAMP–CRP-dependent type I promoters, suggesting that this region may contact the 157-162 loop in CRP. These mutant RNA polymerases are still capable of stimulating transcription from cAMP–CRP-dependent type II promoters, where the CRP binding site is at −41.5. Thus, in these cases CRP contacts a different segment of α or entirely another subunit of RNA polymerase (Ishihama, 1993; Ebright, 1993).

A CRP–RNA polymerase direct contact is also suggested by observations both *in vivo* and *in vitro* that cAMP–CRP and RNA polymerase bind cooperatively and must be on the same face of the DNA for action (Ren *et al.*, 1988; Gaston *et al.*, 1990; Ushida and Aiba, 1990; S. Ryu *et al.*, 1994). If the angular orientation of the CRP binding site is altered with respect to the promoter by, say, 5 bp insertion, CRP fails to act. The CRP binding site must be located at an integral number of B-DNA helical turns away from the start site of transcription for an effect. Interestingly, it has also been shown that the subunit of the DNA-bound CRP dimer that is proximal to the RNA polymerase makes the productive contact with the polymerase (Zhou *et al.*, 1993a). By the use of a heterodimeric CRP molecule in which one subunit is wild type and the other a *pc* mutant, it has been shown that only when the wild-type subunit in the DNA-bound heterodimer is oriented toward the promoter, transcription is activated in the type I *lac* promoter but not when the wild-type subunit in the heterodimer is facing away from the promoter.

4.2. Change the DNA Structure

An alternative but not mutually exclusive mode of cAMP–CRP action is through changing the DNA structure. It has been shown that cAMP–CRP binding to a specific sequence on DNA bends the polynucleotide sequence around the protein (Fried and Crothers, 1983; Kolb *et al.*, 1983; Wu and Crothers, 1984; Schultz *et al.*, 1991). Such bent DNA was initially detected by reduced electrophoretic mobility of the DNA–protein complex compared to a control unbent complex by gel electrophoresis. cAMP–CRP binding sites present in different promoters show different amounts of bending ranging from 80 to 100° (Kim *et al.*, 1989). The total bend estimated from gel electrophoresis is the sum of two major kinks, each in the center of the major groove recognition box, i.e., between the second T and G in the 5′TGTGA sequence as revealed by X-ray structural determination (Schultz *et al.*, 1991). Whether such a bend plays any role in transcription activation is not known. However, a role of cAMP–CRP-induced DNA allostery in transcription activation in the *lac* promoter has been established by the observation that interruption of either strand of the spacer DNA segment between the cAMP–CRP binding site and the RNA polymerase binding site in the *lac* promoter by, for example, four bases abolishes transcription activity (Ryu *et al.*, 1994). This result shows that cAMP–CRP may send a signal to the promoter or to the bound RNA polymerase through the spacer DNA. The signal transmission requires an intact double-stranded DNA. This mode of CRP action is consistent with the demonstration that the DNA site to which cAMP–CRP binds must be close and *cis* to the promoter for effective transcription activation (Ryu *et al.*, 1994). CRP is unable to activate transcription from the *lac* promoter when the CRP site is either moved more than four helical turns away from the transcription start site or placed in *trans* on a DNA template that is catenated with the promoter DNA. The nature of this DNA allostery and how it influences transcription is unknown.

5. CONFORMATION CHANGES IN CRP

5.1. DNA-Induced Allostery

The following evidence suggests that DNA binding may affect the structure of CRP:

1. Free CRP binds to nonspecific DNA sequences in a cooperative fashion, which is a result of protein–protein interactions between adjacent DNA-bound CRP molecules (Saxe and Revzin, 1979; Takahashi *et al.*, 1979). In the absence of DNA, CRP dimers do not show polymerization. Thus, DNA must induce dimer–dimer interactions in free CRP.
2. The binding of cAMP to CRP shows negative cooperativity at low ionic strengths. With increasing ionic strengths, the binding becomes noncooperative and finally cooperative. However, in the presence of DNA at low ionic strengths, cAMP binding is enhanced showing positive cooperativity (Takahashi *et al.*, 1980). These results suggest a conformational change in CRP by DNA for favorable cAMP binding.
3. DNA binding increased the Trp fluorescence which is unaffected by cAMP, suggesting a DNA-induced conformational change around a Trp residue (Hudson *et al.*, 1990).
4. The crystal structure of the cAMP–CRP shows an asymmetry between the two subunits of CRP. The cAMP–CRP–DNA cocrystal shows CRP to be

symmetrical, again showing DNA induced structural change (Weber and Steitz, 1987; Schultz *et al.*, 1991).

5. Circular dichroism spectral analysis has shown that specific DNA binding causes a change in the cAMP component of the CD spectra of the cAMP–CRP–DNA complex (Ivanov *et al.*, 1994). This suggests that a signal is sent to the cAMP binding pocket of the N-terminal domain of CRP by DNA binding to the C-terminal domain. However, it is not clear whether there is any physiological significance to the conformational changes in CRP (or cAMP) induced by DNA.

5.2. cAMP-Induced Allostery

That cAMP binding brings about structural alterations in CRP has been suggested by both biochemical and biophysical experiments:

1. The sensitivity to several proteases—trypsin, subtilisin, chymotrypsin, and V-8 protease—has been a diagnostic of the cAMP-induced conformational changes in CRP. Whereas free CRP is resistant to the attacks under limited conditions of proteolysis, cAMP–CRP complex is readily digested (Eilen *et al.*, 1978; Krakow and Pastan, 1973; Aiba and Krakow, 1981).

2. Dithio-*bis*-2-nitrobenzoic acid readily forms an intersubunit disulfide bond between two cysteine residues (at position 178 in each monomer) in the cAMP–CRP complex, but not in the free protein dimer (Eilen and Krakow, 1977).

3. Neutron-scattering experiments have shown that the radius of gyration of CRP decreases in the presence of cAMP, indicating that cAMP binding brings about an overall compaction of the protein dimer (Kumar *et al.*, 1980). This is supported by fluorescence polarization experiments that show a decrease in the rotational correlation time of CRP induced by cAMP binding.

4. The fluorescent probe ANS, when bound to CRP, exhibits an increase in the fluorescence intensity with a blue shift in the emission maxima. The presence of cAMP decreases the fluorescence signal (Saxe and Revzin, 1979).

5. There are analogues of cAMP, which bind to CRP but either do not show any alteration in CRP properties or show some alteration in CRP by one or more biochemical or biophysical properties, but do not promote high affinity DNA binding or transcription activation (Ebright *et al.*, 1985). These results are consistent with several conformational states induced by cAMP.

Quantitative studies of structural changes and DNA binding of CRP at different concentrations of cAMP show biphasic dependence of CRP conformation and activity on the ligand concentration. The results are summarized in Table

Table I
Structural Changes in CRP in the Presence of Low and High Concentrations of cAMP

	Properties of CRP in the presence of cAMP		
	None	μM cAMP	mM cAMP
Proteolysis	No	High rate	Low rate
ANS-fluorescence intensity	Basal	High	Low
Tryptophan fluorescence	Basal	Low	High
Specific DNA binding	None	High	Low
Transcription in S30 extracts	Basal	High	Very low
[^{2}H]-NMR spectra	—	No change	Change
Stoke's radius	Normal	Normal	Decrease

I. Compared to the behavior in the presence of micromolar concentrations of cAMP, the rates of proteolysis and chemical modification of CRP and the quenching of the fluorescence intensity of ANS–CRP complex decrease and the tryptophan fluorescence intensity increase at millimolar concentrations of cAMP (Heyduk and Lee, 1989, 1990). The affinity of CRP for specific DNA sites as well as the ability to transcribe the *lac* operon in purified system and to synthesize β-galactosidase in crude S-30 system also decreases at higher cAMP concentrations (Heyduk and Lee, 1989, 1990; Emmer *et al.*, 1970; Ryu *et al.*, 1994). Although such results are subject to other interpretations, they suggest the equilibria CRP $\longleftrightarrow$ $(cAMP)_1$–CRP $\longleftrightarrow$ $(cAMP)_2$–CRP in which the conformation of the protein is different from one another in the three states like the one proposed by Koshland *et al.*, 1966) for sequential allosteric changes. It is clear the structure of CRP bound to one molecule of cAMP is different from that of CRP with two cAMP molecules occupying the protein. The latter situation corresponds to the currently available crystallographic data. Clearly, X-ray data of CRP and $(cAMP)_1$–CRP are needed before one could know the nature of the difference.

Raman spectral analysis of free CRP in solution and of $(cAMP)_2$–CRP crystals have large differences, suggesting structural changes in CRP induced by binding of two molecules of cAMP, although the difference in the conformational-sensitive Raman bands of free CRP and of $(cAMP)_1$–CRP in solution is small (DeGrazia *et al.*, 1990; Tan *et al.*, 1991). Similarly, the binding of one molecule of cAMP does not change the [^{2}H]-NMR spectra of CRP, but when two molecules are bound, a change is observed (Gronenborn and Clore, 1982). $(cAMP)_1$–CRP has the same Stoke's radius as does CRP as measured by analytical gel chromatography, but $(cAMP)_2$–CRP shows a significant contraction (Heyduk and Lee, 1992). It is likely that the complex that binds to specific DNA sequence and is transcriptionally active is $(cAMP)_1$–CRP, which predominates at micromolecular concentrations of cAMP, whereas $(cAMP)_2$–CRP complex formed at millimolar cAMP concentrations is inactive for DNA binding and transcription activation. A definitive experiment to prove this concept is to use heterodimers of CRP with one mutant subunit unable to bind cAMP and the other wild type and to show that such a dimer is active at all cAMP concentrations.

The main purpose of allosteric changes introduced into CRP by cAMP has been surmised to be the following: In free CRP the interacting solvent-exposed amino acid side chains of the recognition helix of the helix-turn-helix DNA binding motif are inwardly located on the surface and are unable to reach the interacting distance of the base edges in the major groove of the DNA recognition box. cAMP binding helps the F α-helix protrude out of the surface. The new position of the F α-helix side chains allows CRP to make base-specific contacts.

6. SITES OF ALLOSTERIC CHANGES IN CRP INDUCED BY cAMP

Even when the structure of free CRP and of the active cAMP-liganded CRP are determined, thereby making all of the allosteric changes that occur following ligand binding known, it would be difficult to predict the aspects of the structural alterations that are essential for the pathway to transcription activation. These must be determined by genetic analysis. Three classes of CRP mutants have been isolated and characterized for such purpose: (1) *crp** mutants that can activate transcription in the absence of cAMP and presumably conform CRP to an active state without the ligand binding (Garges and Adhya, 1985; Aiba *et al.*, 1985); (2) *allo*$^-$ mutants that bind cAMP normally but are defective at one or more steps of the allosteric changes (Ryu *et al.*, 1993); and (3) intragenic suppressor, i.e., second site mutations that restore wild-type phenotype by neutralizing the effect of *crp** or *allo*$^-$ mutations (Garges and Adhya, 1988). These studies combined with the X-ray structure of the $(cAMP)_2$–CRP and $(cAMP)_2$–CRP–DNA complexes have led to the following specific predictions about the ligand-induced intramolecular allostery needed for transcription activation: (1) proper alignment between the two subunits; (2) an adjustment between the positions of the two domains within a subunit; and (3) a reorientation of the hinge connecting the two domains that help the DNA binding F α-helix to protrude. These conclusions were based on genetic identification of the sites (amino acids) that are involved in one way or another in generating structures competent in transcription activation, at least partially. Some of these sites are shown by their position numbers in the polypeptide chain in Fig. 2. The role of some of these amino acids in the proposed structural changes and the purpose of the changes are explained below.

6.1. F α-Helix Protrusion

One premise of the cAMP-induced conformational change in CRP is that the new structure will be able to make the proper side-chain contacts within the DNA major groove. Thus, it is expected that the mutations that cause cAMP independence change CRP conformation to fulfill minimally this goal: the protrusion of F α-helix normally achieved by cAMP.

A number of the *crp** mutants that can turn on transcription *in vivo* in the absence of cAMP are clustered on the D α-helix close to the hinge. One of these is Ala144Thr (Garges and Adhya, 1985; Harman *et al.*, 1986). The role of the amino acid at position 144 in causing cAMP independence has been tested by changing the Ala residue at this location by site-directed mutagenesis. All of the 16 substitutions tested at position 144 exhibited a cAMP-independent phenotype (*crp**) (Garges and Adhya, 1985; Kim *et al.*, 1992). Since all of the substituted amino acids have bulkier side chains than the wild-type amino acid Ala, the following model has been suggested to explain cAMP independence. The side chains of position 144 in the D α-helix face the F α-helix and those of position 190 in the F α-helix face directly toward the 144 side chain and away from the solvent-exposed side chains of the F α-helix in the $(cAMP)_2$–CRP structure. The van der Waal's radius of the bulkier side chain of the substituted amino acids at position 144 overlaps with that of the side chain of X at position 190 and very likely causes a repulsion, thus ejecting the F α-helix away from the D α-helix. This protrusion allows the solvent-exposed side chains of the F α-helix to enter the DNA major groove.

6.2. Hinge Reorientation

Another set of *crp** mutations on the D α-helix have provided an important clue to a critical structural rearrangement induced by cAMP. These mutations confer protease sensitivity to the hinge region of CRP in the absence of cAMP similar to the cleavage of wild-type CRP conferred by binding of cAMP (Tan *et al.*, 1991; Aiba *et al.*, 1985; Harman *et al.*, 1986; Ren *et al.*, 1988). The *crp** mutation Gly141Ser is at the hinge end of the D α-helix and oriented inward facing the hinge region. The side chain of 141Ser faces and is within interacting distance of the side chains of amino acids at positions 137 and 138 in the hinge region. From the location of this mutation and its ability to confer protease sensitivity to the hinge region, it has been proposed that the *crp** Gly141Ser mutation mimics cAMP-induced conformational change by reorienting the hinge, which involves an interaction between the amino acids at positions 141 and 137 and/or 138 (Garges and Adhya, 1985; Kim *et al.*, 1992). The model of such interaction(s) and hinge reorientation has been tested by amino acid replacements at position 141 as well as at positions 137 (Leu in wild type) or 138 (Asp in wild type) (Kim *et al.*, 1992). Use of combinations of amino acids at positions 141 and 137 showed that the Crp^+ or Crp* phenotype caused by the amino acid at 141 persisted regardless of the nature of the amino acid at position 137. These results suggest that there is no essential interaction between amino acids at 141 and 137. On the other hand, a duplicate study of amino acid combinations between positions 141 and 138 gave different results. First, single amino acid substitutions at position 141 alone or 138 alone gave all three phenotypes: cAMP-dependent wild-type Crp^+, cAMP-independent Crp*, or allosterically defective Crp^-. For there is at least one amino acid at position 138 that retained the original Crp^+, Crp*, or Crp^- phenotype due to a single change at position 141. Similarly, there is at least one amino acid at position 141 that did not change the Crp^+, Crp*, or Crp^- phenotype caused by an amino acid change at 138. Many combinations at 138 and 141 showed that a given amino acid at one position can suppress, i.e., neutralize, the mutational effect of the amino acid at the other position. The nature of the phenotype of various amino acid combinations at 141-138, described in Table II, clearly show that the amino acids at these two positions interact. This follows the principle of Murgola and Yanofsky (1974), who implicated interaction between two amino acids in the α-subunit of tryptophan synthetase from *E. coli* by second-site mutation studies. Their interpretation has been verified by structural studies (Hyde *et al.*, 1988; Nagata *et al.*, 1989). As explained below, the interactions between amino acids at 138 and 141 reorient the hinge. The mutational studies have provided an insight into the nature of the interactions that reorient the hinge to make CRP cAMP independent: (1) When both sites are occupied by polar amino acids, e.g., 141Arg-138Ser combination, CRP almost always mimics what is normally achieved by the presence of cAMP, i.e., has a Crp* phenotype. In other words, if there is a polar interaction between residues at 138 and 141, it shapes the hinge into a conformation, similar to the one induced by cAMP. (2) Occupancy of one of the two sites by a polar amino acid and the other by a hydrophobic residue, e.g., 141Val-138Ser seems to cause a wild type, i.e., cAMP-dependent Crp^+ phenotype. This is true regardless of the relative positions of the polar and hydrophobic amino acids. Finally, (3) if two hydrophobic amino acids are inserted at 141 and 138, e.g., 141Ile-138Leu combination, CRP becomes totally or partially Crp^-, i.e., is inactive even in the presence of cAMP. In those cases tested in this group, cAMP binds but does not produce the required structural changes. For example, they remain protease resistant in the presence of cAMP (Ryu *et al.*, 1993).

From these studies it has been proposed that polar interactions at 141-138 help the hinge reorientation that moves the D α-helix close to the C α-helix across the hinge (Kim *et al.*, 1992; Ryu *et al.*, 1993). This also happens when there is a polar–hydrophobic combination in the presence of cAMP. In either case, the movement allows access by proteases. In this model, the D α-helix is repositioned or swiveled during the hinge bending, changing the side toward the

Table II
Interactions between Amino Acids 141 and 138[a]

Amino acid at 141–138	Type of amino acid 141–138	Phenotype
G - D		
A - D		
V - D		
L - D		
I - D		
Y - D		CRP+
G - S		
V - S		
I - S		
G - Q		
V - Q	Hydrophobic–polar	
S - L	Polar–hydrophobic	
R - L		
R - Y		
* V - L	Hydrophobic–hydrophobic	
* G - L		
* S - Q	Polar–Polar	
* R - Q		
S - D		
R - D		
K - D		
Q - D		
S - S	Polar–Polar	
R - S		Crp*
G - L		
I - L		
G - Y	Hydrophobic–hydrophobic	Crp−
V - Y		
I - Y		
I - Q	Hydrophobic–polar	
* S - Y	Polar–hydrophobic	

[a]The single letter amino acid code is used in this table. Data from Kim *et al.* (1992). The asterisks denote deviation from the arbitrary classification of the nature of the amino acid pairs into Crp$^+$, Crp*, and Crp$^-$ groups.

F α-helix. The side of the D α-helix on which the 144 side chain is located pushes against the F α-helix through amino acid Leu at position 190, setting the F α-helix away from the body of the protein where it can bind into a major groove of the DNA. A mutation at position 144 can produce a Crp* phenotype and a protein that becomes sensitive to proteases in the absence of cAMP (Harman *et al.*, 1986) when the D–F helix interaction is set by such a mutation, rather than by cAMP, then the hinge reorientation and a C–D helix interaction can be accomplished even in the absence of cAMP. An alteration in the hinge near the bend is not likely to compensate completely for absence of cAMP perhaps because the hinge angle is unstable. A change further from the hinge angle could allow for a stable bend without cAMP.

Structural change by hinge bending has been well documented for bacteriophage T4 lysozyme (Matthews, 1987) and *E. coli* Trp repressor (Luisi and Sigler, 1990). Like CRP, Trp repressor is an allosteric protein that binds to specific DNA sequence after liganding with L-tryptophan. By contrasting the three-dimensional structure of the Trp repressor and its liganded complex, it has been shown that L-tryptophan binding results in a change in the orientation of the DNA-binding helix-turn-helix motif by affecting the hinge region connecting these helixes. Consistently, when Ala in the hinge was replaced by Val with a bulkier side chain, the mutant repressor did not require L-tryptophan for specific DNA binding. Thus, hinge reorientation by ligand binding or by a proper amino acid substitution in CRP is similar to the one that has been shown in Trp repressor.

6.3. Subunit–Subunit Alignment

A second-site alteration at position 127 (Thr127Ala) in the C α-helix in the *crp** Gly141Ser mutant allows CRP to regain the requirements of cAMP for activity and thus defines an additional region that participates in CRP allostery (Garges and Adhya, 1988). It is believed that the hydroxyl group of Thr127 forms a bond with the six-amino group of cAMP. The six-amino group also forms a H-bond to Ser128 in the other subunit. Thr127 appears to be involved in subunit–subunit alignment (Weber and Steitz, 1987). Thus, Thr127 very likely disrupts the alignment in the *crp** mutant, but it can be corrected by cAMP binding.

6.4. Domain–Domain Adjustment

It has been suggested that amino acids at 169 and/or 171 are involved in domain–domain alignment in the cAMP-independent *crp** Ala144Thr mutant (Garges and Adhya, 1988). This mutant is found to be suppressed, i.e., becomes cAMP dependent by Arg169Cys and Glu171Gly changes. Arg169 establishes an electrostatic contact with phosphate backbone of DNA and Glu171 forms a H-bond with Tyr63 in β-sheet in the cAMP binding pocket. In addition to the interactions between the two C α-helix backbones, the H-bond is a major interaction point between the small and large domains in each subunit. Thus, in the triple mutant, the protein lost its ability to function in the absence of cAMP because the two domains cannot align properly. cAMP at least partially restores the domain–domain adjustment through other contacts, e.g., the 2 β-sheet and the 7 β-sheet, which are located in the N-terminal domain but oriented toward the C-terminal domain (Fig. 5).

7. CONCLUSION

Genetic, biochemical, and crystallographic studies available so far suggest the following allosteric changes in CRP. The binding of each cAMP molecule aligns the subunits of a CRP dimer by contacting the large domains of both subunits at the middle of the long C α-helix. These contacts transmit a signal to the hinge end of the D α-helix and alters the relative orientation of the large and small domains by changing the hinge angle. These motions position the F α-helix for favorable sequence interaction with base edges in the major groove of the recognition box major groove. Some of the cAMP-independent CRP mutant partially mimics this conformational change by introducing a bulkier side chain on the D α-helix, which pushes the F α-helix the correct way because of a steric clash between residues of 144 in D and at 191 in the F α-helixes. The three-dimensional structure determination of unliganded wild-type CRP, allosterically defective mutant CRP liganded with cAMP, and of cAMP-independent mutant CRP without cAMP will test these models and facilitate our understanding of these processes.

8. REFERENCES

Aiba, H., and Krakow, J. S., 1981, Isolation and characterization of the amino and carboxyl proximal fragments of the adenosine cyclic 3′,5′-phosphate receptor protein of *Escherichia coli*, *Biochemistry* **20**:4774–4780.

Aiba, H., Nakumura, T., Mitani, H., and Mori, H., 1985, Mutations that alter the allosteric nature of cAMP receptor protein of *Escherichia coli*, *EMBO J.* **4**:3329–3332.

Barber, A. M., and Zhurkin, V. B., 1990, CAP binding sites reveal pyrimidine–purine pattern characteristic of DNA bending. *J. Biomol. Struct. Dyn.* **8**:213–232.

Bell, A., Gaston, K., Williams, R., Chapman, K., and Kolb, A., 1990, Mutations that alter the ability of the *Escherichia coli* cyclic AMP receptor protein to active transcription, *Nucleic Acids Res.* **18**:7243–7250.

DeGrazia, H., Harman, J. G., Tan, G. S., and Wartell, R., 1990, Investigation of the cAMP receptor protein secondary structure by Raman spectroscopy, *Biochemistry* **29**:3557–3562.

Ebright, R. H., 1993, Transcription activation at Class I CAP-dependent promoters, *Mol. Microbiol.* **8**:792–802.

Ebright, R. H., LeGrice, S. F., Miller, J. P., and Krakow, J. S., 1985, Analysis of cyclic AMP analogues that elicit the biochemically defined conformational change in catabolite gene activator protein (CAP) but do not stimulate binding to DNA, *J. Mol. Biol.* **182**:91–107.

Ebright, R. H., Kolb, A., Buc, H., Kunkel, T. A., Krakow, J. S., and Beckwith, J., 1987, Role of glutamic acid-181 in DNA-sequence recognition by the catabolite gene activator protein (CAP) of *Escherichia coli:* Altered DNA-sequence-recognition properties of [Val[181]]CAP and [Leu[181]]CAP, *Proc. Natl. Acad. Sci. USA* **84:**6083–6087.

Ebright, R. H., Gunasekera, A., Zhang, X., Kunkel, T. A., and Krakow, J. S., 1990, Lysine 188 of the catabolite gene activator protein (CAP) plays a role in specificity at base pair 7 of the DNA half site, *Nucleic Acids Res.* **18:**1457–1464.

Eilen, E., and Krakow, J. S., 1977, Cyclic AMP-mediated intersubunit disulfide crosslinking of the cyclic AMP receptor protein of *Escherichia coli, J. Mol. Biochem.* **114:**47–60.

Eilen, E., Pampeno, C., and Krakow, J. S., 1978, Production and properties of the α core derived from the cyclic adenosine monophosphate receptor protein of *Escherichia coli, Biochemistry* **17:**2469–2473.

Emmer, M., deCrombrugghe, B., Pastan, I., and Perlman, R., 1970, Cyclic AMP receptor protein of *E. coli:* Its role in the synthesis of inducible enzymes, *Proc. Natl. Acad. Sci. USA* **66:**80–487.

Eschenlauer, A. C., and Reznikoff, W. S., 1991, *Escherichia coli* catabolite gene activator protein mutants defective in positive control of *lac* operon transcription, *J. Bacteriol.* **173:**5024–5029.

Fried, M. G., and Crothers, D. M., 1983, CAP and RNA polymerase interactions with the lac promoter: Binding stoichiometry and long range effects, *Nucleic Acids Res.* **11:**141–158.

Fried, M. G., and Crothers, D. M., 1984, Equilibrium studies of the cyclic AMP receptor protein–DNA interaction, *J. Mol. Biol.* **172:**241–262.

Garges, S., and Adhya, S., 1985, Sites of allosteric shift in the structure of the cyclic AMP receptor protein, *Cell* **41:**745–751.

Garges, S., and Adhya, S., 1988, Cyclic AMP-induced conformational change of cyclic AMP receptor protein (CRP): Intragenic suppressors of cyclic AMP-independent CRP mutations, *J. Bacteriol.* **170:**1417–1422.

Gaston, K., Bell, A., Kolb, A., Buc, H., and Busby, S., 1990, Stringent spacing requirements for transcription activation by CRP, *Cell* **62:**733–743.

Gronenborn, A., and Clore, G. M., 1982, Proton nuclear magnetic resonance studies on cyclic nucleotide binding to the *Escherichia coli* adenosine cyclic 3′,5′-phosphate receptor protein, *Biochemistry* **21:**4040–4048.

Gunasekera, A., Ebright, Y. W., and Ebright, R. H., 1992, DNA sequence determinants for binding of the *Escherichia coli* catabolite gene activator protein, *J. Biol. Chem.* **267:**14713–14720.

Harman, J. G., McKenney, K., and Peterkofsky, A., 1986, Structure–function analysis of three cAMP-independent forms of the cAMP receptor protein, *J. Biol. Chem.* **261:**16332–16339.

Heyduk, T., and Lee, J. C., 1989, *Escherichia coli* cAMP receptor protein: Evidence for three protein conformational states with different promoter binding affinities, *Biochemistry* **28:**6914–6924.

Heyduk, T., and Lee, J. C., 1990, Application of fluorescence energy transfer and polarization to monitor *Escherichia coli* cAMP receptor protein and *lac* promoter interaction, *Proc. Natl. Acad. Sci. USA* **87:**1744–1748.

Heyduk, T., and Lee, J. C., 1992, Global conformational changes in allosteric proteins, *J. Biol. Chem.* **267:**3200–3204.

Heyduk, T., Lee, J. C., Ebright, Y. W., Blatter, E. E., Zhou, Y., and Ebright, R. H., 1993, CAP interacts with polymerase in solution in the absence of promoter DNA, *Nature* **364:**548–549.

Hudson, J. M., Crowe, L. G., and Fried, M. G., 1990, A new DNA binding mode for CAP, *J. Biol. Chem.* **265:**3219–3224.

Hyde, C., Ahmed, S., Padlan, E., Miles, E. W., and Davies, D., 1988, Three-dimensional structure of the tryptophan synthase $\alpha_2\beta_2$ multienzyme complex from *Salmonella typhimurium, J. Biol. Chem.* **263:**17857–17871.

Igarashi, K., and Ishihama, A., 1991, Bipartite functional map of the *E. coli* RNA polymerase α subunit: Involvement of the C-terminal region in transcriptional activity by cAMP-CRP, *Cell* **65:**1015–1022.

Ishihama, A., 1993, Protein–protein communication within the transcription apparatus, *J. Bacteriol.* **175**:2483–2489.

Ivanov, V. I., Minchenkova, L. E., Chernov, B. K., McPhie, P., Ryu, S., Garges, S., Barber, A. M., Zhurkin, V. B., and Adhya, S., 1993, CRP-DNA complexes: Inducing the A-like form in the binding sites with an extended central spacer. *J. Mol. Biol.* (submitted).

Jansen, C., Gronenborn, A., and Clore, G. M., 1987, The binding of cyclic AMP receptor protein to synthetic DNA sites containing permutations in the consensus sequence TGTGA, *Biochem. J.* **246**:227–232.

Kim, J., Zwieb, C., Wu, C., and Adhya, S., 1989, Binding of DNA by gene regulatory proteins: Construction and use of a DNA bending vector, *Gene* **85**:15–23.

Kim, J., Adhya, S., and Garges, S., 1992, Allosteric changes in the cAMP receptor protein of *Escherichia coli:* Hinge reorientation, *Proc. Natl. Acad. Sci. USA* **89**:9700–9704.

Kolb, A., Spassky, A., Chapon, C., Blazy, B., and Buc, H., 1983, On the different binding affinities of CRP at the *lac, gal* and *malT* promoter regions, *Nucleic Acids Res.* **11**:7833–7852.

Kolb, A., Busby, S., Buc, H., Garges, S., and Adhya, S., 1993, Transcriptional regulation by cAMP and its receptor protein, *Annu. Rev. Biochem.* **62**:749–795.

Koshland Jr., D. E., Nemethy, G., and Filmer, D., 1966, Comparison of experimental binding data and theoretical models in proteins containing subunits, *Biochemistry* **5**:365–385.

Krakow, J. S., and Pastan, I., 1973, Cyclic adenosine monophosphate receptor: Loss of cAMP-dependent DNA binding activity after proteolysis in the presence of cyclic adenosine monophosphate, *Proc. Natl. Acad. Sci. USA* **70**:2529–2533.

Kumar, S., Murthy, N., and Krakow, J. S., 1980, Ligand-induced change in the radius of gyration of cAMP receptor protein from *Escherichia coli, FEBS Lett.* **109**:121–124.

Liu-Johnson, H. N., Gartenberg, M., and Crothers, D., 1986, The DNA binding domain and bending of *E. coli* CAP protein, *Cell* **47**:995–1005.

Luisi, B., and Sigler, P., 1990, The stereochemistry and biochemistry of the *trp* repressor-operator complex, *Biochem. Biophys. Acta* **1048**:113-126.

Matthews, B. W., 1987, Genetic and structural analysis of the protein stability problem, *Biochemistry* **26**:6885–6888.

Monod, J., and Jacob, F., 1961, *Cold Spring Harbor Symp. Quant. Biol.* **26**:389–401.

Monod, J., Changeux, J. P., and Jacob, F., 1963, Allosteric proteins and cellular control systems, *J. Mol. Biol.* **6**:306–329.

Murgola, E., and Yanofsky, C., 1974, Selection of new amino acids at position 211 of the tryptophan synthetase α chain of *Escherichia coli, J. Mol. Biol.* **86**:775–784.

Nagata, S., Hyde, C., and Miles, E., 1989, The alpha subunit of tryptophan synthase. Evidence that aspartic acid 60 is a catalytic residue and that the double alteration of residues 175 and 211 in a second-site revertant restores the proper geometry of the substrate binding site, *J. Biol. Chem.* **264**:6288–6296.

Ren, Y. L., Garges, S., Adhya, S., and Krakow, J. S., 1988, Cooperative binding of heterologeric proteins: Evidence for contact between the cyclic AMP receptor protein and RNA polymerase, *Proc. Natl. Acad. Sci. USA* **85**:4138–4142.

Ryu, S., Kim, J., Adhya, S., and Garges, S., 1993, Pivotal role of amino acid at position 138 in the allosteric hinge reorientation of cAMP receptor protein, *Proc. Natl. Acad. Sci. USA* **90**:75–79.

Ryu, S., Garges, S., and Adhya, S., 1994, An arcane role of DNA in transcription activation, *Proc. Natl. Acad. Sci. USA* **91**:8582–8586.

Saxe, S., and Revzin, A., 1979, Cooperative binding to DNA of catabolite activator protein of *Escherichia coli, Biochemistry* **18**:255–263.

Schultz, S. C., Shields, G. C., and Steitz, T. A., 1991, Crystal structure of a CAP–DNA complex: The DNA is bent by 90 degrees, *Science* **253**:1001–1007.

Shumilov, V. Y., 1987, Minor promoters of bacteriophage ϕX174 are controlled by CRP-cAMP, lexA, glnG, and several other global regulatory systems of the host cell. *Mol. Biol.* (Engl. transl.) **21**:168–187.

Takahashi, M., Blazy, B., and Baudras, A., 1979, Non-specific interactions of CRP from *E. coli* with native and denatured DNAs: Control of binding by cAMP and cGMP and by cation concentration, *Nucleic Acids Res.* **7**:1699–1712.

Takahashi, M., Blazy, B., and Baudras, A., 1980, An equilibrium study of the cooperative binding of adenosine cyclic 3′,5′-monophosphate and guanosine cyclic 3′,5′-monophosphate to the adenosine cyclic 3′,5′-monophosphate receptor protein from Escherichia coli. *Biochemistry* **19**:5124–5130.

Takahashi, M., Blazy, B., Baudras, S., and Hillen, W., 1983, On the origin of selectivity in recognition by cyclic adenosine 3′,5′-monophosphate receptor protein of its specific binding site of the lactose promoter, *J. Mol. Biol.* **167**:895–899.

Tan, G. S., Kelly, P., Kim, J., and Wartell, R. M., 1991, Comparison of cAMP receptor protein (CRP) and a cAMP-independent form of CRP by Raman spectroscopy and DNA binding, *Biochemistry* **30**:5076–5080.

Ushida, C., and Aiba, H., 1990, Helical phase dependent action of CRP: Effect of the distance between the CRP site and the −35 region on promoter activity, *Nucleic Acids Res.* **18**:6325–6330.

Weber, I. T., and Steitz, T. A., 1987, Structure of a complex of catabolite gene activator protein and cyclic AMP refined at 2.5 Å resolution, *J. Mol. Biol.* **198**:311–326.

Wu, H. M., and Crothers, D. M., 1984, The locus of sequence-directed and protein-induced DNA bending, *Nature* **308**:509–513.

Wyman, J., 1948, *Adv. Protein Chem.* **4**:407–531.

Wyman, J., and Allen, D. W., 1951, *J. Polymer Sci.* **7**:499–518.

Zhang, X. P., and Ebright, R. H., 1990, Identification of a contact between arginine-180 of the catabolite gene activator protein (CAP) and base pair 5 of the DNA site in the CAP-DNA complex, *Proc. Natl. Acad. Sci. USA* **87**:4717–4721.

Zhou, Y., Busby, S., and Ebright, R. H., 1993a, Identification of the functional subunit of a dimeric transcription activator protein by use of oriented heterodimers, *Cell* **73**:375–379.

Zhou, Y., Zhang, X., and Ebright, R. H., 1993b, Identification of the activating region of catabolite gene activator protein (CAP): Isolation and characterization of mutants of CAP specifically defective in transcription activation, *Proc. Natl. Acad. Sci. USA* **90**:6081–6085.

Chapter 11

A Family of RNA-Binding Enzymes
The Aminoacyl–tRNA Synthetases

Yves Mechulam, Thierry Meinnel, and Sylvain Blanquet

Since their early characterization, aminoacyl–tRNA synthetases (aaRS)* have been the object of successive extensive reviews (see, for instance, Berg, 1961; Novelli, 1967; Lotfield, 1972; Söll and Schimmel, 1974; Schimmel and Söll, 1979; Schimmel, 1987; Meinnel *et al.*, 1995). The aim of this new chapter, dealing with the general features of this family of enzymes, is to integrate many recent results, in particular the structural studies describing the binding between aaRS and tRNAs.

1. FUNCTIONS OF AMINOACYL–tRNA SYNTHETASES IN THE LIVING CELLS

1.1. Early Characterization of Aminoacyl–tRNA Synthetases

The discovery that aaRS were responsible for the aminoacylation of tRNAs, in the mRNA-directed protein synthesis, came out in the late 1950s. Firstly,

*A given aminoacyl–tRNA synthetase is abbreviated by its cognate amino acid (three-letter code) followed by RS; for instance, AlaRS stands for alanyl–tRNA synthetase.

Yves Mechulam, Thierry Meinnel, and Sylvain Blanquet Laboratoire de Biochimie, Unité de Recherche Associée au CNRS n°240, Ecole Polytechnique, F-91128 Palaiseau Cedex, France

Subcellular Biochemistry, Volume 24. Proteins: Structure, Function, and Engineering, edited by B. B. Biswas and Siddhartha Roy. Plenum Press, New York, 1995.

enzyme activities able to activate the amino acid with ATP to give an aminoacyl adenylate were evidenced (Hoagland, 1955), and then shown to be each specific of one amino acid (see, for instance, Berg, 1956; Cole *et al.*, 1957). The enzyme-bound aminoacyl adenylate was found to react with hydroxylamine, allowing thereby the regeneration of the enzyme. Later, these amino acid-specific enzymes were shown to support the attachment of the amino acid to an RNA molecule. Consequently, such RNAs appeared to be each specific for a given amino acid (Berg and Ofengand, 1958; Schweet *et al.*, 1958). The carboxyl groups of amino acids are bound to the acceptor RNA through an ester linkage with the 2′- or 3′-hydroxyl group of the 3′ terminal adenosyl ribose moiety of the RNA (Zachau *et al.*, 1958; Hecht *et al.*, 1959; Preiss *et al.*, 1959). Finally, the amino acid moiety, bound to its cognate RNA molecule or transfer RNA (tRNA), was demonstrated to be subsequently transferred into nascent proteins (Hoagland *et al.*, 1958; Allen and Schweet, 1960).

The first purification of amino acid-activating enzymes was reported in the early 1960s (Bergmann *et al.*, 1961). The designation of this class of enzymes as aaRS was proposed by Berg *et al.* (1961). The two-step reaction catalyzed by aaRS, summarized in Eqs. (1) and (2) (where aa stands for a natural amino acid), is now generally admitted, although a one-step concerted mechanism was also envisaged in a few cases (Lotfield, 1972):

$$\text{aaRS} + \text{ATP:Mg}^{2+} + \text{aa} \rightarrow \text{aaRS:aa} \sim \text{AMP} + \text{PPi:Mg}^{2+} \quad (1)$$

$$\text{aaRS:aa} \sim \text{AMP} + \text{tRNA} \rightarrow \text{aaRS} + \text{aa-tRNA} + \text{AMP} \quad (2)$$

1.2. Aminoacyl-tRNA Synthetases and the Synthesis of Proteins

The aaRS are responsible for a large part of the fidelity of the translation process since these enzymes dictate which amino acid will correspond to a given tRNA (Chapeville *et al.*, 1962). The direct correspondence between the tRNA and the trinucleotide codon on the mRNA is solely insured by the complementarity of the codon and a tRNA triplet: the anticodon (Crick, 1958; Holley *et al.*, 1965; Goodman *et al.*, 1968; Dube *et al.*, 1969) (see Fig. 1). The critical role in the synthesis of protein of the aminoacylation reaction catalyzed by aaRS is well illustrated by the experiment of Chapeville *et al.* (1962), which established that after a misaminoacylation reaction has occurred, the mRNA–ribosome–aminoacyl–tRNA complex is unable to correct for the entry of a wrong amino acid. Such an absence of proofreading for the amino acid at the level of the ribosome imposes the aaRS to be highly specific enzymes, both at the level of the activation of the amino acid and at that of the recognition of the tRNA.

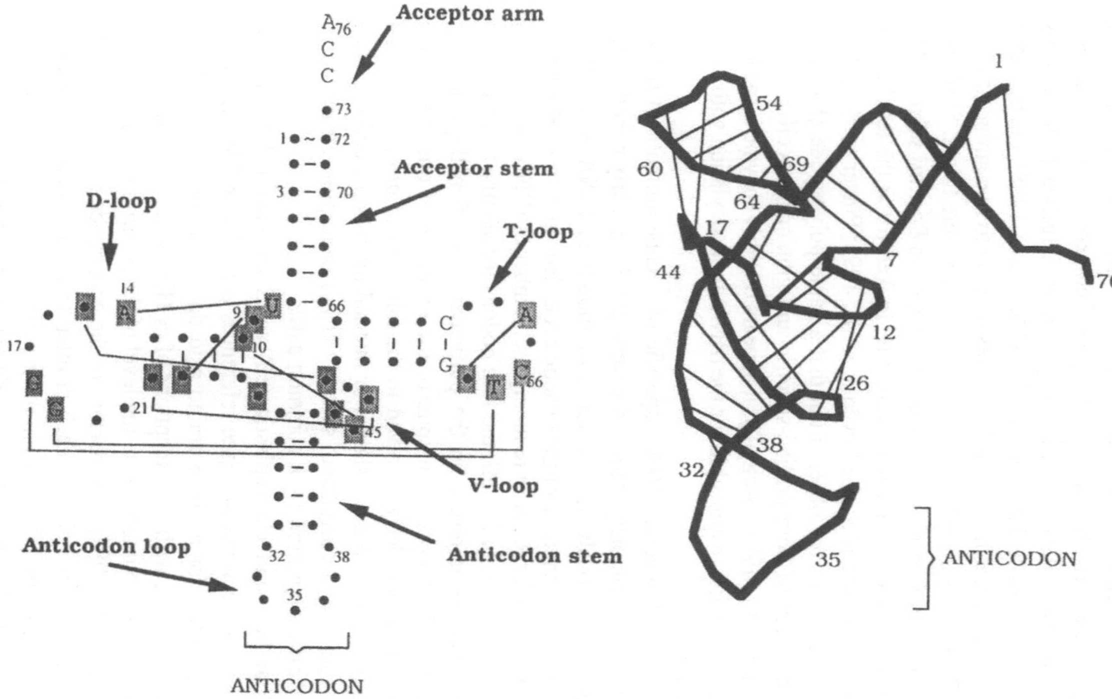

FIGURE 1. Secondary and tertiary structure of tRNAs. (A) The cloverleaf secondary structure of tRNAs with the constant stem and loop regions is shown. Nucleotides, numbered from the 5′- to the 3′-end (according to Sprinzl *et al.*, 1987), are figured with dots. The conserved bases among all tRNAs are indicated. A dash indicates a secondary interaction. A tertiary interaction is schematized with a line between two shaded nucleotides. (B) The tertiary structure of tRNAs is figured. This structure is acquired thanks to the combination of secondary and tertiary interactions. The phosphate backbone of the tRNA molecule is boldfaced, while secondary and tertiary noncovalent interactions are figured with a normal line.

To each of the 20 amino acids, there corresponds one aaRS. However, a GlnRS activity could not be evidenced in mitochondria, *Bacillus subtilis* and cyanobacteria. Similarly, in *Escherichia coli,* selenocysteine, an additional natural amino acid, lacks the aaRS capable of activating it. It is known now that the tRNAs specific for selenocysteine or glutamine are firstly aminoacylated by SerRS and GluRS, respectively, and that the esterified serine or glutamate are further modified to give selenocysteine and glutamine attached to the tRNAs (Lapointe *et al.*, 1986; Leinfelder *et al.*, 1988; Schön *et al.*, 1988). On the other hand, there may occur more than one aaRS corresponding to one amino acid. In *E. coli,* two LysRS species encoded by two distinct genes were found. Similarly, two ThrRS and two TyrRS species have been recognized in *Bacillus subtilis*. The reasons for such functional redundancies might be looked for at the level of the genetic expression of these aaRS (see Lévêque *et al.*, 1991; Pützer *et al.*, 1992). A last exception deals with the observation that a same polypeptide chain sustains the aminoacylation of both proline-specific and glutamate-specific tRNAs in mammals (Cerini *et al.*, 1991). Actually, it appears that this enzyme is made of two separate domains, each responsible for the tRNA aminoacylation with either proline or glutamate.

The possibility that several tRNAs, specific of a same amino acid (tRNA isoacceptors), might be aminoacylated by the same aaRS was early hypothesized (Berg *et al.*, 1961). Obviously, the existence of subsets of tRNA isoacceptors appear dictated by the degeneracy of the genetic code. However, one tRNA may read more than one codon, therefore reducing the number of tRNA isoacceptors (Crick, 1966). For instance, in *E. coli,* the four codons for alanine correspond to only two tRNAs. On the contrary, in the particular case of methionine, two different methionine tRNAs occur, in spite of the existence of one methionine codon: one isoacceptor is strictly involved in the initiation of protein synthesis and the other participates in the translation of internal methionine codons. Nevertheless, in all cases, the occurrence of several tRNA isoacceptors clearly raises the problem of their recognition by a common aaRS.

Mitochondrial (*mt*) and chloroplastic (*ch*) aaRS are encoded by nuclear most aaRS genes have been isolated (see references compiled in Meinnel *et al.*, 1995), *mt* aaRS specific for a given amino acid do not strongly resemble the corresponding *cy* enzyme, with the exceptions of *cy* HisRS and *cy* ValRS, which only differ from their *mt* counterparts at the level of the N-terminus (Natsoulis *et al.*, 1986; Jordana *et al.*, 1987). The reason for the occurrence of distinct aaRS, specific for the same amino acid but each devoted to a given cellular compartment, is likely to arise from differences in the genetic codes in the organelles and in the cytoplasm and also from the requirement to aminoacylate organelle-specific tRNAs. Indeed, the structures of organellar tRNAs are in some cases markedly different from those of *cy* tRNAs (Okimoto and Wolstenholme, 1990). Such differences reflect the particular mechanism of protein synthesis in organelles, as compared to cytoplasm. As a consequence of the above considerations, the identity rules governing the recognition of a tRNA by its cognate synthetase may therefore vary following the considered compartment. In agreement with this, the absence of cross-reactivities between *cy, mt,* and *ch* aaRS and their cognate tRNAs has been often reported (e.g., Jacobson, 1971).

1.3. The Synthesis of Adenylated *bis* (5′-Nucleosidyl)Polyphosphates

Ap_4A, a component of the adenylated *bis* (5′-nucleosidyl)polyphosphates family (or Ap_4N and Ap_3N, with N = A, C, G, U), was early shown to be produced *in vitro* by *E. coli* LysRS at the expense of ATP (Zamecnik *et al.*, 1966). Numerous aaRS have since been demonstrated to also direct the synthesis of such nucleotide molecules *in vitro* (for a review, see Plateau and Blanquet, 1992). The following two-step scheme accounts for the production of Ap_4N and Ap_3N, from ATP and NTP or from ATP and NDP:

$$\text{aaRS} + \text{ATP:Mg}^{2+} + \text{aa} \rightarrow \text{aaRS:aa}{\sim}\text{AMP} + \text{PPi:Mg}^{2+} \quad (1)$$

$$\text{aaRS:aa}{\sim}\text{AMP} + \text{NTP} \rightarrow \text{aaRS} + \text{Np}_4\text{A} + \text{aa} \quad (2)$$

or

$$\text{aaRS:aa}{\sim}\text{AMP} + \text{NDP} \rightarrow \text{aaRS} + \text{Np}_3\text{A} + \text{aa} \quad (2)$$

The involvement of aaRS in Ap_4N and Ap_3N synthesis *in vivo* is now firmly demonstrated (Brevet *et al.*, 1989). On overexpression of an aaRS, the cellular concentration of Ap_4N increases. In parallel, the cellular concentration of non-adenylylated dinucleotides, which the aaRS cannot produce, remains unmodified. However, the physiological relevance for the occurrence of these 5′, 5′-dinucleotides is still unknown. Since the intracellular concentrations of Ap_4N and Ap_3N increase in response to various stimuli including heat shock, it was hypothesized that these compounds could act as alarmones (Bochner *et al.*, 1984). This attractive idea was weakened, however, by the observation that *E. coli* cells overproducing Ap_4A hydrolase still expressed the heat-shock response of the wild-type strain (Plateau *et al.*, 1987).

1.4. Other Functions of Aminoacyl–tRNA Synthetases

Many recent results have indicated that aaRS might possess additional functions beyond their primary role in the synthesis of proteins. This includes (1) a possible role of *E. coli* LysRS in RNaseH-independent $ColE_1$ replication (Nakamura *et al.*, 1990), (2) the controversial activity of mammalian TrpRS as release factor specific for UGA codons (Lee *et al.*, 1990; Bange *et al.*, 1992; Frolova *et al.*, 1993), and (3) the function of *Saccharomyces* spp. LeuRS and

Neurospora crassa or *Podospora anserina* in the splicing of group I introns (Akins and Lambowitz, 1987; Herbert *et al.*, 1988; Kämper *et al.*, 1992).

At the present time, the molecular bases for these functions are still unexplained, except in the case of the splicing of group I introns. Indeed, the splicing reaction is abolished by deletions in *N. crassa* TyrRS that disrupt the N- and C-terminal regions; the same deletions prevent the aminoacylation of $tRNA^{Tyr}$ (Kittle *et al.*, 1991; Guo and Lambowitz, 1992). Contrarily, the catalytic center of the synthetase is likely not to be involved in the splicing process, as probed by point substitutions in *N. crassa* TyrRS or yeast LeuRS affecting the amino acid activation reaction, not the splicing activity (Kittle *et al.*, 1991; Li *et al.*, 1992). It appears therefore that the aaRS binds a part of the intron core, which would fold into a tRNA-like structure (see references in Section 3.3.2). The consequence could be that the intron adopts a three-dimensional structure favoring the splicing reaction. By assisting the splicing reaction, which occurs spontaneously *in vitro*, an aaRS would act therefore as a catalyst.

2. FROM THE THREE-DIMENSIONAL STRUCTURES TO THE CATALYTIC MECHANISMS

2.1. Two Classes of aaRS as Revealed by Three-Dimensional Structures

It was early noticed that the native homodimeric MetRS from *E. coli* (2 × 676 residues) could be converted into a fully active monomeric form upon mild trypsinolysis (Cassio and Waller, 1971). This enzymatic form, lacking the 125 ± 5 C-terminal residues, showed an increased thermal stability compared to its native counterpart. The truncated enzyme readily crystallized in a form suitable for high-resolution X-ray analysis (Waller *et al.*, 1971). A few years later, Reid *et al.* (1973) carried out the successful crystallization of TyrRS from the thermophilic *Bacillus stearothermophilus*. The first three-dimensional data obtained with MetRS (Risler *et al.*, 1981) demonstrated the presence of a structural element, termed the "Rossman fold," already characterized as the nucleotide-binding domain in the dehydrogenases (Rossman *et al.*, 1974). This structural element is composed of a parallel β-sheet with helixes connecting the strands (Fig. 2A). Interestingly, the three-dimensional structure of the TyrRS also showed a Rossman fold (Bhat *et al.*, 1982). This provided the first evidence for a structural resemblance between two aaRS. In addition, comparison of the amino acid sequences of MetRS and TyrRS revealed a conserved region composed of seven residues, comprising a HXGH motif, where X can be I or L (Barker and Winter, 1982). Finally, the Rossman folds of TyrRS and MetRS were shown to remarkably superimpose, with the HXGH motifs occupying the same positions

(Blow *et al.*, 1983). The structural homology (Fig. 2A and Fig. 3) was confirmed after the refinements of the structure of MetRS complexed with ATP (Brunie *et al.*, 1987, 1990) and of that of TyrRS complexed with tyrosyl adenylate (Brick *et al.*, 1989). This also confirmed that the Rossman fold indeed corresponded to the active site for the catalysis of aminoacyl adenylate formation.

Identification of lysine residues at the active sites of *E. coli* MetRS and TyrRS was achieved by affinity labeling techniques, using either tRNA derivatives rendered reactive by oxidation of their 3′-terminal adenosine (Fayat *et al.*, 1979; Hountondji *et al.*, 1979) or an analogue of ATP carrying a pyridoxal group on the γ-phosphoryl (Ap_3PL) (Tagaya and Fukui, 1986). These labeling experiments mainly targeted Lys335 of MetRS (Hountondji *et al.*, 1985, 1990) and Lys229, 234, and 237 of TyrRS (Hountondji *et al.*, 1986b), corresponding to Lys225, 230, and 233 in the crystallized *B. stearothermophilus* TyrRS. It was noticed that the sequences around the labeled lysines in the MetRS on one hand and in the TyrRS on the other displayed similarities (Hountondji *et al.*, 1986b). This led these authors to observe that this conserved motif (KMSKS in the *E. coli* MetRS sequence) was present in the sequences of many aaRS (Hountondji *et al.*, 1986a). The observations that (1) the second lysine of the KMSKS sequence motif in the *E. coli* ValRS was also the target of affinity labeling using Ap_3PL (Hountondji *et al.*, 1990) and that (2) this motif was present at identical positions of the Rossman folds of MetRS and TyrRS (Brunie *et al.*, 1990) reinforce the functional significance of its conservation.

The third solved three-dimensional structure of an aaRS was that of the monomeric *E. coli* GlnRS complexed with ATP and $tRNA^{Gln}$ (Rould *et al.*, 1989; Rould and Steitz, 1992). Again, the active site (in which the ATP bound) comprised a Rossman fold, superimposable to those of MetRS or TyrRS (Perona *et al.*, 1991) (Figs. 2A and 3), with a perfect match of a HXGH sequence that the GlnRS also contains. Moreover, GlnRS comprises a motif resembling KMSKS, located in the same region of the Rossman fold as the corresponding residues of MetRS and TyrRS.

In fact, the HXGH sequence, later termed as the *HIGH signature sequence*, was also observed in the *E. coli* IleRS (Webster *et al.*, 1984; Starzyk *et al.*, 1987; Burbaum *et al.*, 1990). Among the *E. coli* aaRS, whose sequences are now all known, the ten specific for Arg, Cys, Gln, Glu, Ile, Leu, Met, Trp, Tyr, and Val do contain the HIGH signature sequence. Notably, the corresponding homologous aaRS from other organisms display this motif (Burbaum *et al.*, 1990). Finally, it should be underlined that the aaRS that contain the HIGH motif are exactly those that have the KMSKS one (Eriani *et al.*, 1990).

In this context, it was striking that the three-dimensional structure of the homodimeric *E. coli* SerRS (Cusack *et al.*, 1990) did not show a Rossman fold. Instead, the active site of SerRS is built around a seven-stranded antiparallel

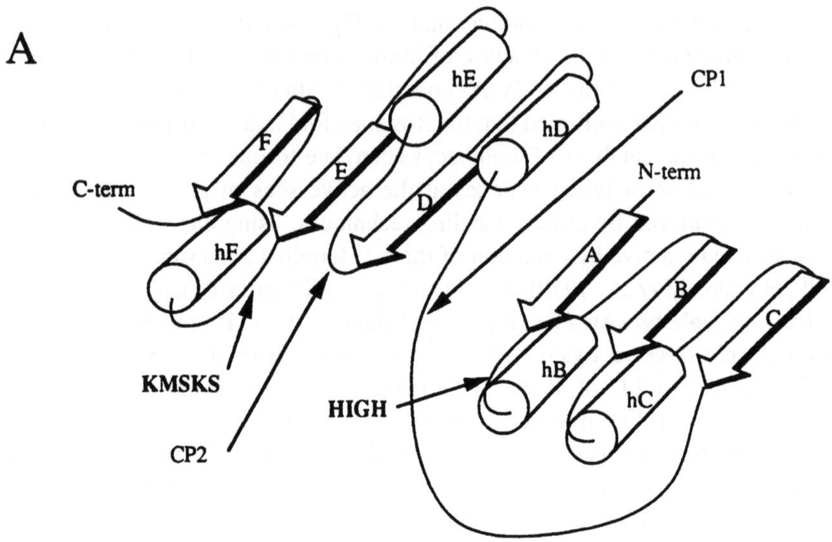

	MetRS	TyrRS	GlnRS
ßA	ßA (6-12)	ßB (32-37)	ß-1 (27-35)
hB	HA (24-40)	H3 (49-60)	αB (42-57)
ßB	ßB (44-52)	ßC (64-69)	ß-2 (59-68)
hC	HB (73-79)	H5 (91-105)	αC (77-83)
ßC	ßC (92-98)	ßD (119-122)	ß-3 (95-99)
hD	HC (200-210)	H10 (164-184)	αD (212-223)
ßD	ßD (224-232)	ßE (187-192)	ß-9 (226-230)
hE	HD (303-313)	H11 (196-210)	αH (233-247)
ßE	ßE (318-328)	ßF (216-220)	ß10 (254-264)
hF	HE (341-348)		αK (302-312)
ßF	ßF (354-359)		ß11 (317-323)

FIGURE 2. Schematic drawing of the active sites of class 1 and class 2 aminoacyl–tRNA synthetases. The structures of the conserved secondary elements of the active sites of (A) class 1 and (B) class 2 aaRS are schematized. β-Strands are figured with arrows and α-helix with cylinders. The direction of the peptidic backbone is also shown. (A) The Rossman fold of class 1 aminoacyl–tRNA synthetases is schematized at the top of the figure. The locations of the two signature sequences of class 1 aaRS are indicated (HIGH and KMSKS), as well as those of the two connective peptides (CP1 and CP2), whose length depends on each aaRS (see text and Burbaum *et al.*, 1990). Each secondary structure element is labeled with a simple capital letter from A to F (β-strands) or with an "h" followed by a capital letter from A to F (α-helixes). The table at the bottom of the figure indicates the limits of the above secondary structures within the amino acid sequence of the three class 1 aaRS, whose three-dimensional structure are known. The table uses the designations of these elements as they were originally introduced (Brick *et al.*, 1989; Rould *et al.*, 1989; Brunie *et al.*, 1990). (B) Schematic representation of the most conserved elements (according to Cusack *et al.*, 1991) of the antiparallel β-sheet of the active site of the two class 2 aaRS of known three-dimensional structures (SerRS and AspRS). Each secondary structure element is labeled with an A followed by a number

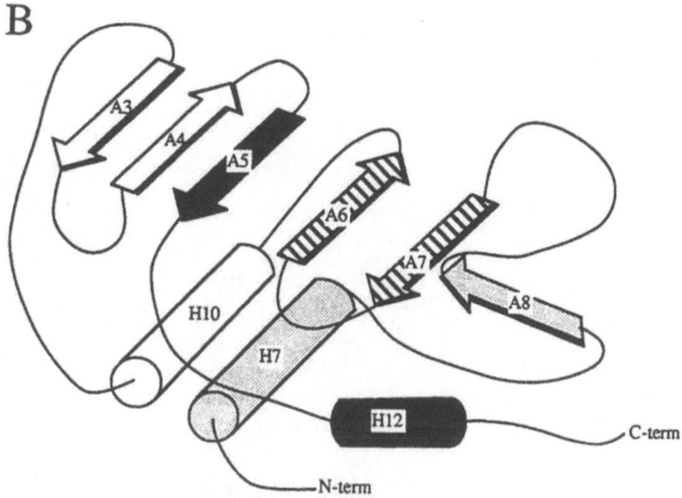

AspRS	SerRS
H1 (244-264)	H7 (167-186)
S1 (268-271)	A8 (191-194)
S4 (316-324)	A7 (258-267)
S5 (338-348)	A6 (286-296)
H3 (352-372)	H10 (301-319)
S8 (469-474)	A3 (339-347)
S9 (477-485)	A4 (352-363)
S10 (522-528)	A5 (386-392)
H9 (529-537)	H12 (395-406)

FIGURE 2. ***(Continued)***

from 3 to 8 (β-barrels) or with an H followed by a number (α-helixes); this numbering corresponds to that in SerRS (Cusack *et al.*, 1990). The positions of motifs 1, 2, and 3 are indicated as follows: light grey, motif 1; hatches, motif 2; dark grey, motif 3. The location of each of the above elements in the amino acid sequence of AspRS (Ruff *et al.*, 1991) is indicated at the bottom of the panel.

β-sheet (Fig. 2B). The same active site topology was found in the yeast AspRS, as deduced from the three-dimensional structure of this homodimeric enzyme complexed with $tRNA^{Asp}$ (Ruff *et al.*, 1991) (Fig. 2B).

In fact, the two types of active site organization found up to now in aaRS (Rossman fold for MetRS, TyrRS, and GlnRS, and antiparallel β-sheet for SerRS and AspRS) (Fig. 2) may coincide with two subsets of aaRS as deduced from amino acid sequence analysis (Eriani *et al.*, 1990). AspRS and SerRS do not contain the HIGH and KMSKS sequences, but instead share in common three motifs, termed 1, 2, and 3. Interestingly, the ten aaRS species that do not possess HIGH and KMSKS related sequences are characterized by at least motif 3, and

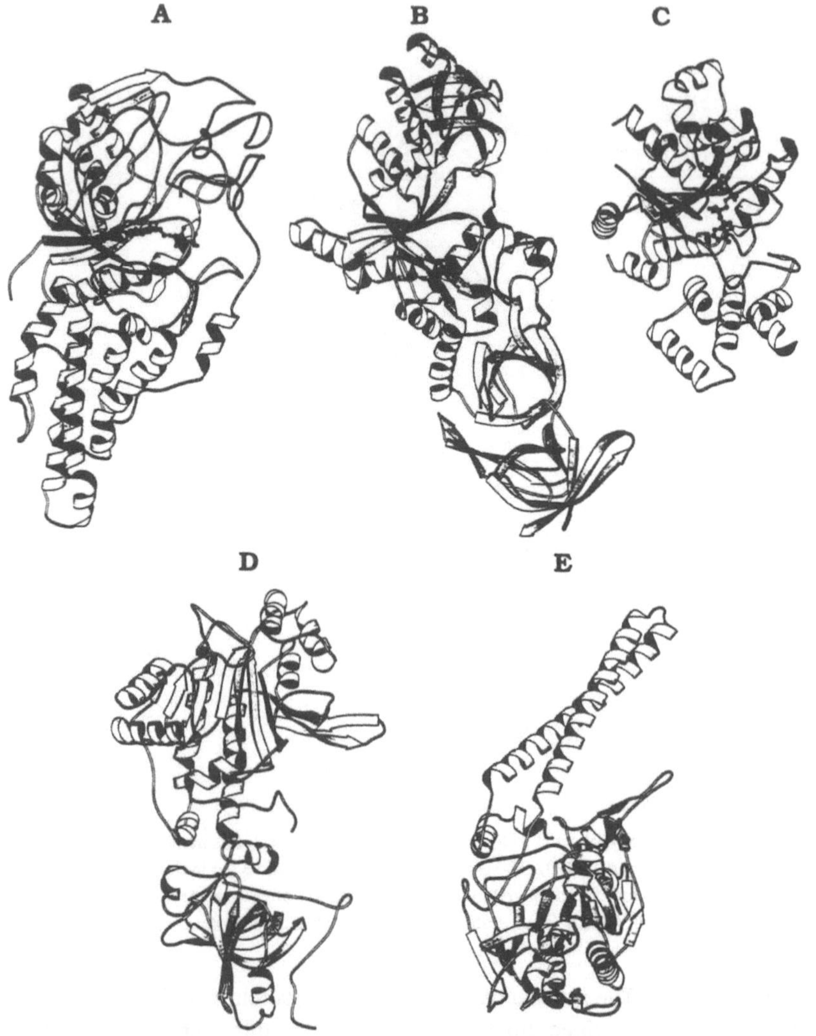

FIGURE 3. Peptide backbone folding of five aminoacyl–tRNA synthetases. Shown are the three-dimensional structure of (A) *E. coli* MetRS (Brunie *et al.*, 1990), (B) *E. coli* GlnRS (Rould *et al.*, 1989, 1991), (C) the active site domain of *B. stearothermophilus* TyrRS (Brick *et al.*, 1989), (D) yeast AspRS (Ruff *et al.*, 1991; Cavarelli *et al.*, 1993), (E) and *E. coli* SerRS (Cusack *et al.*, 1990). The pictures were drawn using the Molscript program (Kraulis, 1991). The peptidic backbones only are shown. In the case of dimeric enzymes only one subunit is represented. The Rossman folds of class 1 aaRS (A–C) have been given the same orientation. The ATP molecule in MetRS and the tyrosyl adenylate in TyrRS are also shown using the ball-and-stick representation.

often by motifs 1, 2, and 3. Each motif was shown to contain a conserved core, with at least one invariant residue, defined as gΦXXΦXXPΦΦ* (motif 1), (F or Y or H)RX(E or D)(4 to 12 residues)(R or H)XXXFXXX(D or E) (motif 2), and λXΦgΦgΦgΦeRΦΦΦΦΦ (motif 3) (Eriani *et al.*, 1990). This led to the proposal that aaRS could be partitioned into two classes, according to the presence of either the HIGH and KMSKS signature sequences (class I), or the three above motifs (class II) (Eriani *et al.*, 1990). It was also noted that class I aaRS systematically corresponded to enzymes aminoacylating the 2′OH group of the terminal adenosine of their cognate tRNA, whereas class II enzymes would aminoacylate the 3′OH group, thereby suggesting a common ancestor for the enzymes belonging to each class (Eriani *et al.*, 1990).

Multiple sequence alignments were performed in order to further classify the aaRS and to gain information on the possible evolutionary relationships between them (Tzagoloff *et al.*, 1990; Cusack *et al.*, 1991; Nagel and Doolittle, 1991). AsnRS, AspRS, and LysRS, and to a lesser extent GlyRS and PheRS (these two latter aaRS are $\alpha_2\beta_2$ tetramers), appear related to each other (class IIb according to Cusack *et al.*, 1991). Another class II subclass includes ProRS, SerRS, and ThrRS (Class IIa according to Cusack *et al.*, 1991). HisRS was proposed to be related to either class IIa enzymes (Cusack *et al.*, 1991) or to the homotetrameric AlaRS (Nagel and Doolittle, 1991). An important conclusion from these alignments is that class II aaRS are likely to possess an active site formed around an antiparallel β-sheet, whereas those belonging to class I should possess a Rossman fold (Burbaum and Schimmel, 1991; Cusack *et al.*, 1991; Hou *et al.*, 1991). Moreover, the presence of two connective polypeptides of variable lengths might account for the large differences of sizes of the active site domains of class I aaRS (Hou *et al.*, 1991). The first connective polypeptide, termed CP1, links the two halves of the Rossman fold, whereas the second one, termed CP2, is located between the fourth β-strand and the fourth α-helix (Fig. 2A).

2.2. Functional Domains in aaRS

The five aaRS of known three-dimensional structure contain a separate domain in addition to that containing the active center. Such an additional domain is composed of the C-terminal residues in the cases of the three class I aaRS (Met, Tyr, Gln), whereas it is composed of N-terminal residues in the two class II ones (Ser, Asp).

In the case of GlnRS, the anticodon loop of $tRNA^{Gln}$ is bound to the C-terminal domain formed by two β-barrels (Rould *et al.*, 1991). In the case of

*The following symbols are used: Φ stands for a hydrophobic amino acid (I,L,M,F,W,V), λ stands for a small amino acid (A,C,G,P,S,T), and X stands for any amino acid. Lower case letters are used for residues that are not fully invariant.

TyrRS, the C-terminal domain is disordered in the crystal. However, its function in tRNA recognition was demonstrated by the construction of a truncated enzyme lacking this domain (residues 320 to 419), which remained fully able to activate tyrosine while it failed to aminoacylate $tRNA^{Tyr}$ (Waye *et al.*, 1983). This deletion mutant was crystallized and the three-dimensional structure of its complex with tyrosine was solved (Brick and Blow, 1987). In trypsin-modified MetRS, the additional C-terminal domain is essentially helical and composed of residues 360 to 525. The 526 to 551 remaining C-terminal residues form an arm folding back toward the active center (Brunie *et al.*, 1987, 1990). From biochemical (Leon and Schulman, 1987), mutagenesis (Ghosh *et al.*, 1990; Meinnel *et al.*, 1991b), and model building (Perona *et al.*, 1991) experiments, the helical C-terminal domain was shown to serve in the binding of the anticodon of $tRNA^{Met}$. However, attempts to obtain a truncated MetRS devoid of tRNA aminoacylation capacity but still able to activate methionine were unsuccessful. Indeed, upon progressive deletion of residues belonging to the C-terminal arm, the enzyme became thermolabile and then inactive (Mellot *et al.*, 1989). It is therefore likely that this arm (526–551) is essential for the folding of the active center domain (Brunie *et al.*, 1987).

The 150 N-terminal residues of AspRS form a β-barrel domain (Ruff *et al.*, 1991) involved in the binding of the anticodon loop of $tRNA^{Asp}$ (Cavarelli *et al.*, 1993). SerRS, the other class II synthetase whose structure was solved, features a 60-Å arm exposed to the solvent. This arm consists in an antiparallel coiled coil of two α-helixes and is composed of the 100 N-terminal residues. This domain is involved in tRNA recognition (Price *et al.*, 1993).

Many aaRS behave as multimers. In some cases, multimerization could be shown to involve a specialized domain. This is clearly the case for the three studied prokaryotic MetRS (*E. coli*, *B. stearothermophilus*, and *T. thermophilus*), which can be converted into fully active monomers upon the proteolytic removal of a hundred C-terminal residues (Cassio and Waller, 1971; Kalogerakos *et al.*, 1980; Kohda *et al.*, 1987). In turn, TyrRS dimerizes through contacts between residues of the active site domain (Jones *et al.*, 1985). This is also the case for SerRS and AspRS, in which dimerization involves an α-helix comprising residues belonging to motif 1. Actually, the seven class II aaRS possessing a clear-cut motif 1 are exactly those of the α_2 type (Cusack *et al.*, 1991). This correspondence reinforces the idea of a common structural organization of this subset of aaRS.

An organization into functional domains composed of distinct polypeptidic regions might be a general feature of aaRS. Indeed, a distribution of functions along the polypeptidic chain has early been demonstrated in the case of *E. coli* AlaRS, whose three-dimensional structure is, however, still unknown. This enzyme behaves as a homotetramer (Putney *et al.*, 1981b), each polypeptidic chain being composed of 875 residues (Putney *et al.*, 1981a). The organization of this enzyme was studied by introducing deletions at the 3′-end of the corresponding gene, resulting in the production of AlaRS truncated at its C-terminus (Jasin *et al.*, 1983). The results showed that as few as 461 amino acids out of 875 were still enough to sustain enzymatic activity *in vivo*, although the *in vitro* catalytic activity of the truncated enzyme was reduced by more than 1000-fold compared to that of the wild-type AlaRS (Regan *et al.*, 1987). Notably, the truncated enzyme was monomeric, thereby demonstrating that the C-terminal part of the enzyme is essential for multimerization, but dispensable for keeping a small part of the tRNA aminoacylation activity. This situation contrasts with that of MetRS, where a C-terminal deletion monomerizes the enzyme, while keeping intact the tRNA aminoacylation activity. Further deletion of the gene for AlaRS showed that a polypeptide reduced to the 385 N-terminal residues completely lost the ability to bind and aminoacylate $tRNA^{Ala}$, but remained able to activate alanine (Jasin *et al.*, 1983; Regan *et al.*, 1987). From this, it could be deduced that the region between residues 385 and 461 contained residues essential for tRNA recognition but dispensable for alanine activation. The AlaRS polypeptide can therefore be viewed as the sequential arrangement of three regions, responsible for amino acid activation, tRNA binding, and subunit tetramerization, respectively.

2.3. The Catalytic Centers of aaRS

2.3.1. Mechanism of the Amino Acid Activation Reaction

Although three aaRS (ArgRS, GlnRS, and GluRS) require the presence of their cognate tRNA to sustain the [^{32}P]-PPi-ATP isotopic exchange, it is admitted that the tRNA aminoacylation reaction proceeds through the synthesis of an aminoacyl adenylate intermediate tightly bound to the enzyme. The mixed anhydride bond of the aminoacyl adenylate is highly susceptible to nucleophilic attack. Therefore, the transfer of the amino acid onto tRNA can schematically be viewed to result from the correct positioning of the acceptor arm of the nucleic acid on the enzyme and the placing of the 2′ or 3′ hydroxyl of the terminal ribose in the vicinity of the anhydride bond. For this reason, the accuracy of the activation reaction is crucial in governing the overall amino acid specifity of the tRNA aminoacylation.

In some cases, the amino acid specificity of aaRS was shown to be enhanced by a proofreading step (for recent reviews, see Freist, 1989; Jakubowski and Goldman, 1992). Correction of errors occurs either through the hydrolysis of the misactivated amino acid (e.g., Fersht, 1977; Fersht and Dingwall, 1979b; Jakubowski and Fersht, 1981) or through the deacylation of a misesterified tRNA (e.g., Igloi *et al.*, 1977; Fersht and Dingwall, 1979c). The necessity of such proofreadings was early predicted (Pauling, 1957), in particular in the case of aminoacids differing by a few methyl groups (for instance, isoleucine *vs*. valine). Accordingly, *E. coli* IleRS efficiently edits misactivated valine (Fersht, 1977; Freist *et al.*, 1988). Nevertheless, some aaRS do not require proofreading to achieve a high level of discrimination between amino acids. This is the case for aaRS corresponding to an amino acid whose side chain contains a specific group that can be bound by the enzyme with a high release of free energy (for instance cysteine) (Fersht and Dingwall, 1979a). Therefore, the molecular bases for the amino acid specificity of aaRS may be searched first at the level of the formation of the reactive complex enzyme:ATP-Mg^{2+}:amino acid.

Detailed kinetic studies were performed using either the [^{32}P]-PPi-ATP isotopic exchange reaction (*e.g.*, Cole and Schimmel, 1970; Santi and Pena, 1971; Blanquet *et al.*, 1974) or the direct monitoring by fluorescence of aminoacyl adenylate synthesis using the stopped flow technique (Blanquet *et al.*, 1972; Holler and Calvin, 1972; Fersht *et al.*, 1975; Hyafil *et al.*, 1976; Merle *et al.*, 1986). In the studied cases, the rapid formation of an enzyme:ATP-Mg^{2+}:amino acid ternary complex could be evidenced, followed by a rate-limiting isomerization into an enzyme:aminoacyl adenylate:PPi-Mg^{2+} species, and by the rapid dissociation of PPi-Mg^{2+} (Scheme 1, from Hyafil *et al.*, 1976). In all studied cases, including TyrRS, the bindings of ATP-Mg^{2+} and of the amino acid appeared kinetically random-ordered. However, the three-dimensional structure of *B. stearothermophilus* TyrRS further indicated that tyrosine should bind first (Brick and Blow, 1987).

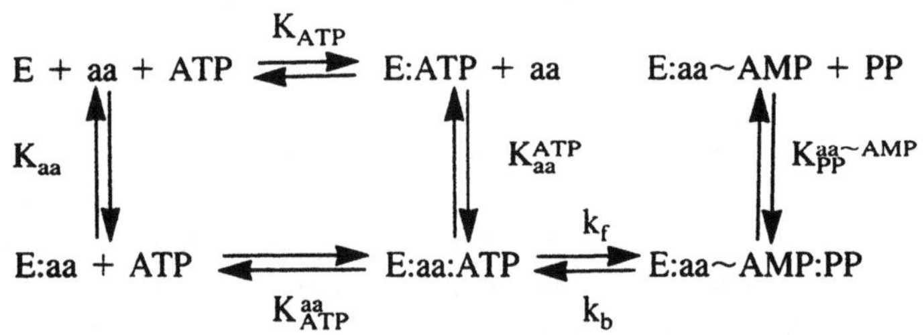

Scheme 1

The simultaneous binding of ATP-Mg^{2+} and of the amino acid in the active center raises an obvious problem. Indeed, since the α-phosphoryl group of ATP has to react with the carboxylate of the amino acid, one can expect that these negatively charged groups will be placed close to each other in the ternary complex in such a way that the binding of ATP-Mg^{2+} should be antagonistic with that of the amino acid, because of electrostatic repulsion between the carboxylate and the α-phosphoryl. However, in the case of the *E. coli* MetRS, measurement of the dissociation constants governing the formation of the enzyme:ATP-Mg^{2+}:Met complex showed that the binding of the amino acid was apparently indifferent to the occupation of the ATP-Mg^{2+} site (Hyafil *et al.*, 1976). This behavior was interpreted to reflect the compensation of electrostatic repulsion by positive coupling between the site of ATP-Mg^{2+} and that of the methionine. Indeed, synergistic effects could be made apparent by suppressing one negative charge carried by either the carboxylate of the amino acid or the α-phosphoryl of ATP. More precisely, the binding to MetRS of methioninol was found to be greatly enhanced upon occupation of the ATP-Mg^{2+} site (Blanquet *et al.*, 1975a; Fayat *et al.*, 1977). Symmetrically, the affinity of methionine for the enzyme is much increased in the presence of adenosine plus pyrophosphate, a combination mimicking an ATP molecule devoid of α-phosphoryl (Blanquet *et al.*, 1975b). Such couplings between the sites for ATP-Mg^{2+} and for the amino acid might be a general property of aaRS since they were also proved to occur in the cases of the *E. coli* IleRS, LeuRS, and PheRS (Holler *et al.*, 1975) and of the *B. stearothermophilus* MetRS (Kalogerakos *et al.*, 1980). Finally, synergism may be suspected to help the formation of the ternary complex in the cases of the *B. stearothermophilus* TyrRS (Fersht *et al.*, 1975) and of the *E. coli* TrpRS (Merle *et al.*, 1986), by considering the relatively high affinity of ATP-Mg^{2+} for the enzyme:amino acid complex.

2.3.2. The Active Center of Class I aaRS

Recent studies taking advantage of the site-directed mutagenesis technique (Hutchison *et al.*, 1978) as well as of the availability of three-dimensional structures (Brick and Blow, 1987; Brown *et al.*, 1987; Brick *et al.*, 1989) were aimed at understanding the synthesis of aminoacyl adenylate at the atomic level. In some cases, precise understanding of the role of residues in the catalysis could be obtained by comparing the formation of aminoacyl adenylate by enzyme variants (Fersht *et al.*, 1975; Hyafil *et al.*, 1976). Using presteady-state techniques, the whole set of kinetic and equilibrium parameters in scheme 1 can be determined and the contributions of the side chain of a studied residue along the reaction pathway can be evaluated, at least qualitatively (Fersht, 1988).

Thorough studies concerning the TyrRS from *B. stearothermophilus* have led to a nearly complete description of the enzyme:substrates interactions along the reaction pathway. Linear free-energy plots were used to detect artifacts that could lead to misinterpretation of the mutagenesis experiments (Fersht *et al.*, 1986; Fersht, 1988). These studies have pointed out the importance of an induced fit of the active site to achieve tight binding of the transition state (Fersht, 1987; Fersht *et al.*, 1988). Most of the interactions with the transition state are maintained in the enzyme:tyrosyl adenylate:PPi-Mg^{2+} complex. The results are summarized in Table I.

Interestingly, the two histidines of the HIGH sequence and the two lysines of the KMSKS resembling one (KFGKT in the case of TyrRS) were studied. The

Table I
Role in the Catalysis of Residues of the Active Site of *B. stearothermophilus* TyrRS[a]

Residue	Location	Target	Complexes	References
Asp78	CP1	α-amino of Tyr	All	Fersht, 1987
Tyr169	hD	α-amino of Tyr	All	Wells and Fersht, 1985
Gln173	hD	α-amino of Tyr	All	Fersht, 1987
Tyr34	βA	OH of Tyr	All	Wells and Fersht, 1985
Asp176[b]	hD	OH of Tyr	All	Brick and Blow, 1987
Lys82	Between βB and hC	β-phosphoryl of ATP	ETA, ETA‡, EAdP	Fersht *et al.*, 1988
Arg86	Between βB and hC	β-phosphoryl of ATP	ETA‡, EAdP	Fersht *et al.*, 1988
Lys230	KMSKS-like motif	β-phosphoryl of ATP	ETA‡, EAdP	Fersht *et al.*, 1988
Lys233	KMSKS-like motif	β-phosphoryl of ATP	ETA, ETA‡, EAdP	Fersht *et al.*, 1988
Thr40	Between βA and hB	γ-phosphoryl of ATP	ETA‡, EAdP	Leatherbarrow and Fersht, 1987
His45	HIGH motif	γ-phosphoryl of ATP	*TA‡, EAdP	Leatherbarrow and Fersht, 1987
His48	HIGH motif	0-1 of ribose	ETA‡, EAdP, EAd	Lowe *et al.*, 1985
Thr51	hB	OH on 3′ of ribose	ETA‡, EAdP, EAd	Ho and Fersht, 1986
Cys35	βA	OH on 3′ of ribose	ETA‡, EAdP, EAd	Wells and Fersht, 1986

[a]For each residue whose role was probed by site-directed mutagenesis, its location in the Rossman fold (see Fig. 1), the substrate group (target) to which it binds, and the complexes in which the contact occurs are indicated. The four considered complexes are TyrRS:Tyr:ATP-Mg^{2+} (ETA), TyrRS:Tyr:ATP-Mg^{2+} in the transition state (ETA‡), TyrRS:tyrosyl adenylate:PPi-Mg^{2+} (EAdP), and TyrRS:tyrosyl adenylate (EAd); "All" means that binding occurs in the four complexes.

[b]The role of Asp176 was deduced from the analysis of the three-dimensional structure, and indirect biochemical evidence has been obtained (de Prat Gay *et al.*, 1993).

two histidines, His45 and His48, mainly bind the transition state of the reaction, as well as the tyrosyl adenylate:PPi-Mg^{2+} product. More precisely, the ε-N of the imidazole ring of His45 binds the γ-phosphoryl of ATP (Leatherbarrow *et al.*, 1985; Leatherbarrow and Fersht, 1987), whereas the δ-N of His48 interacts with the O-1 position of the ribose of the nucleotide (Lowe *et al.*, 1985; Wells and Fersht, 1985, 1986). Mutation into an Asn of the first histidine in the HIGH sequence of the *B. stearothermophilus* ValRS revealed a behavior of the mutant enzyme identical to that of the His45 → Asn mutant of TyrRS, thereby suggesting a conserved role of the histidines of the HIGH motif in the class I aaRS (Borgford *et al.*, 1987). Such a conclusion is corroborated by the crystal structures of the GlnRS:ATP-Mg^{2+}:$tRNA^{Gln}$ (Rould and Steitz, 1992) and MetRS:ATP-Mg^{2+} (Brunie *et al.*, 1990) complexes, although, in these cases, observed interactions correspond to the ground–state complexes with ATP-Mg^{2+}. In addition, in the case of GlnRS, the invariant glycine of the HIGH motif (Gly42) was proposed to make a flat platform on which the adenine of ATP sits (Rould and Steitz, 1992).

Lys230 and Lys233 of the *B. stearothermophilus* TyrRS, i.e., the two lysines of the KMSKS resembling sequence, also play a crucial role in the tyrosine activation reaction. These two lysines form a salt bridge with the β-phosphoryl of ATP-Mg^{2+} in the transition state, with the interaction made by Lys233 occurring also in the ground–state complex formed by the enzyme with tyrosine and ATP-Mg^{2+} (Fersht *et al.*, 1988). Again, an involvement of residues belonging to the KMSKS sequence in the catalytic step of the activation reaction might be a feature common to the class I aaRS. Indeed, Lys335, the second lysine of KMSKS in the *E. coli* MetRS was demonstrated to bind the pyrophosphate group of ATP-Mg^{2+} in the transition state and in the enzyme:methionyl adenylate:PPi-Mg^{2+} complex (Mechulam *et al.*, 1991). In this case, however, the position in the crystalline enzyme of the lysine with regard to the ATP molecule designates the γ-phosphoryl group rather than that in position β (Brunie *et al.*, 1990).

In both cases of MetRS and TyrRS, the KMSKS motifs are located in mobile loops, movements of which have been proposed to produce a tight interaction with the substrates during the transition state (Fersht *et al.*, 1988; Mechulam *et al.*, 1991). In this context, it is noticeable that the KMSKS sequence is often flanked by glycine residues, which might be important to confer loop mobility. In the case of GlnRS, the second lysine of the KMSKS signature sequence, Lys270, appears well positioned to interact with the phosphates of ATP (Rould and Steitz, 1992). The same authors noticed that the two sequences HIGH and KMSKS interacted together, thereby giving a structural role for the simultaneous presence of these sequences in the active site of class I aaRS.

Recognition of the amino acid is probably more idiosyncratic than that of ATP, and little information is available about the interactions allowing discrimination. In TyrRS, the phenolic group of tyrosine fits in a slot at the active site (Brick and Blow, 1987). Discrimination toward phenylalanine appears to be the consequence of the capacity of the hydroxyl group of the tyrosine substrate to form hydrogen bonds with the side chains of Tyr34 and Asp176 (Fersht *et al.*, 1985; Brown *et al.*, 1987; de Prat Gay *et al.*, 1993). In MetRS, a 7-Å cavity in the vicinity of the ATP binding site was proposed to contain the recognition site of methionine (Brunie *et al.*, 1990). Indeed, mutation of Trp305, a residue located in the border of this cavity, causes a loss of the enzyme affinity for methionine (Fourmy *et al.*, 1991; Ghosh *et al.*, 1991b).

2.3.3. The Active Center of Class II aaRS

The structure–function relationships of the class II aaRS are presently understood with less details than those of the class I enzymes. Fortunately, the crystallographic structure of the *Thermus thermophilus* SerRS (Belrhali *et al.*, 1994) and that of the yeast AspRS (Cavarelli *et al.*, 1994), each complexed with both tRNA and ATP, will be soon available.

It is likely that residues belonging to motifs 2 and 3 play an important role in catalysis. This was suggested by the analysis of the three-dimensional structure of the *E. coli* SerRS, showing that the two universally conserved arginines in motifs 2 and 3, as well as the R/H one in motif 3, point inside the active site crevice (Cusack *et al.*, 1991). Moreover, a disordered loop comprising residues of motif 2 forms a flexible flap over the active center. An involvement in the catalysis of residues in motifs 2 and 3 was confirmed by site-directed mutagenesis of *E. coli* AsnRS (Anselme and Härtlein, 1991) and of yeast AspRS (Cavarelli *et al.*, 1994).

3. MOTIFS IN RNA STRUCTURES RECOGNIZED BY aaRS

According to the genetic code, the incorporation of a given amino acid in a nascent protein is specified by one or several synonymous codons. Because a tRNA can be used in translation whatever the nature of the esterified amino acid (Chapeville *et al.*, 1962), the molecular bases of the genetic code correspond to the set of mechanisms which insure the rate and fidelity of aminoacylation of tRNAs at the steady state *in vivo*. These mechanisms can be listed as follows:

1. The *in vivo* competition between cognate and noncognate tRNAs for the binding to a same aaRS.
2. The *in vivo* efficiencies of aminoacylation of free tRNA molecules by cognate and noncognate aaRSs.
3. The respective intracellular concentrations of aaRS and of tRNAs, and the mechanisms controlling their concentrations.
4. The relative turnover of consumption on the ribosome of each aminocyl–tRNA, on which the concentrations of free tRNAs depend.

All of these steps participate to the relative level of aminoacylation by a given amino acid of a given tRNA *in vivo*, and therefore to a functional definition of the identity of a tRNA. This definition includes a qualitative aspect, which is related to the nature of the esterified amino acids, as well as a quantitative one, which is connected to the rates at which aminoacyl–tRNAs are regenerated.

To achieve univocal genetic codes, strong discrimination mechanisms must obviously occur at the level of the aminoacylation reaction catalyzed by aaRS. Amino acid proofreading mechanisms may account for a part of such a discrimination (see Section 2.3.1.; for a review see Jakubowski and Goldman, 1992). However, this section of the review will only focus on the specificity of tRNA binding to aaRS. A tRNA molecule is expected to contain structural information, allowing an efficient aminoacylation by the cognate aaRS and preventing the aminoacylation by noncognate aaRS. The combination of such positive or negative determinants (or identity elements) is likely to greatly account for the tRNA identity problem.

3.1. Strategies for the Identification of Identity Elements in tRNAs

At first glance, the tRNA anticodon appears as a logical identity element since, in most cases, thanks to the wobble (Crick, 1966), one tRNA molecule, and therefore only one anticodon sequence, is enough to designate one amino acid. In turn, in cases where the degeneracy of the genetic code is strong (Arg, Leu, Ser, Thr systems, for instance), one aaRS has to aminoacylate several tRNA isoacceptors, with a maximum of one base in common within their respective anticodons. Clearly, in such a situation, the anticodon is not likely to give a tRNA its identity, because other noncognate tRNAs also will possess the same base in common. For instance, the anticodon sequences of the four *E. coli* $tRNAs^{Arg}$ only share the central C_{35} base. This base is also present in the cases of $tRNA^{Cys}$, $tRNA^{S3r}{}_3$, $tRNA^{Sel}$, $tRNA^{Trp}$, and the three $tRNAs^{Gly}$. Consequently, alternative or additional determinants must occur to account for the identity of these tRNAs. In the case of the five serine isoacceptors, whose anticodon sequences have no base in common, identity elements are likely to be found outside of the anticodon.

Strategies aimed at defining the identity elements have involved (1) *in vitro* studies of the aminoacylation by a given aaRS of tRNAs having different origins or different identities (Kern *et al.*, 1972; Roe *et al.*, 1973; Giegé *et al.*, 1974; Guillemaut and Weil, 1975; Kumazawa *et al.*, 1991; for compilations, see Jacobson, 1971; Giegé *et al.*, 1993) and (2) tRNA sequence alignments, sometimes computer-aided, to identify structural features shared only by tRNA molecules aminoacylated by a same aaRS (McClain and Nicholas, 1987; Steinberg and Kisselev, 1993). For this purpose, tRNA sequences compilations are of great help (Sprinzl *et al.*, 1987). On these bases, and taking into account the tertiary structure of the tRNA molecule (Fig. 1B), a set of potential identity elements can be proposed. To verify the role of the involved nucleotides, the methodology consisted in assaying the effects of their modifications through *in vitro* and/or *in vivo* measurements.

3.1.1. Mapping the Identity Elements of Transfer RNAs

The classical cloverleaf structure of tRNAs, the designation of the different parts, and the canonical numbering, which are used in the present review, are shown in Fig. 1A.

3.1.1a. *In Vivo* Approaches. Twenty years ago, mutations rendering tRNA molecules able to translate stop codons were discovered. These mutant tRNAs, or suppressor tRNAs, were shown to carry a mutation within their anticodon sequence (Goodman *et al.*, 1968). Many of such *E. coli* tRNA amber suppressor molecules could be obtained *in vivo* (reviewed in Eggertsson and Söll, 1988), while the genes for other ones could be synthesized *in vitro* (Normanly *et al.*, 1986a,b; Kleina *et al.*, 1990).

Suppression assays were of great help for the finding of tRNA identity elements. Such assays are based on the use of bacterial or yeast strains expressing reporter genes whose translation is interrupted by a stop codon. In the absence of aminoacylation of the suppressor tRNA, the corresponding mRNA is only partially translated and the reporter gene product is not active. In turn, if the tRNA variant can be aminoacylated, the translation of the reporter gene will occur in a proportional manner. Consequently, the effect on aminoacylation of a change at the level of the sequence of a suppressor tRNA can be measured as a function of the active amount of the reporter gene product. Noteworthy, the nature of the amino acid inserted in the protein in response to the suppressed codon must be verified. This could be achieved through genetic analysis (Murgola, 1985) or sequencing of the indicator protein (see for instance Normanly *et al.*, 1986b).

Using natural suppressors, it was early described that single substitutions outside of the anticodon could change the identity of a tRNA. For instance, *E. coli* amber $tRNA^{Tyr}$ charges Gln instead of Tyr upon a single mutation within the acceptor stem sequence (Hooper *et al.*, 1972; Shimura *et al.*, 1972; Smith and Celis, 1973). Later, Normanly *et al.* (1986b) reported that the identity of a tRNA could be switched from leucine to serine by changing only 12 bases. In addition, Hou and Schimmel (1988) and McClain and Foss (1988b) showed that the identity of amber derivatives of several tRNAs could be changed into an alanine by mutating a single base pair (3–70) near the acceptor stem end. This result was in good agreement with a previous work showing that missense suppressor $tRNA^{Lys}$ inserted alanine or glycine when the 3–70 pair was changed into G-U (Prather *et al.*, 1984).

Suppression assays rapidly indicated that base substitutions within the anticodon itself could also change the identity of a tRNA. For instance, the amber derivative of $tRNA^{Trp}$ incorporated glutamine (Yaniv *et al.*, 1974). A systematic survey of the amino acid inserted by each of 20 amber derivatives of *E. coli* tRNAs revealed that ten accepted lysine or glutamine (Normanly *et al.*, 1990).

In the systems where the anticodon is mainly involved, identity elements cannot be searched for by using the above suppressor assay strategy, based on the use of amber derivatives of tRNA. To get around this obstacle, another *in vivo* assay has been recently proposed. Chattapadhyay *et al.*, (1990) and Varshney and RajBhandary (1990) showed that an initiator $tRNA^{Met}{}_f$, whose CAU anticodon was modified, could initiate protein synthesis in *E. coli,* provided the initiation codon on a reporter gene was made complementary to the new anticodon sequence. By determining which amino acid has been introduced at the first position of the reporter protein, one can therefore evaluate to what extent a given anticodon is used as identity element. An assay with dihydrofolate reductase as reporter protein was developed (Pallanck and Schulman, 1991) and showed that, in various *E. coli* systems (Asn, Cys, Ile, Phe, Val), the anticodon played a major role (see references in Table II). This system can also be used to measure the importance of nucleotides located outside the anticodon. However, such an approach is weakened by the very recent demonstration that a number of nucleotides in the acceptor and anticodon stems of $tRNA^{Met}{}_f$ (Guillon *et al.*, 1993; Varshney *et al.*, 1993) play a role in initiation of translation beyond the action of aaRS. Changes at such positions may therefore affect the capacity of the $tRNA^{Met}{}_f$ variant to initiate the translation of the indicator protein, rather than the efficiency of its aminoacylation.

To summarize, *in vivo* assays have been powerful tools in *E. coli* to establish that a very limited number of the nucleotides of a tRNA molecule might dictate its identity toward its cognate synthetase.

3.1.1b. *In Vitro* Approaches. An alternative to the *in vivo* studies of the identity of tRNAs was to directly measure the catalytic ability of aaRS to aminoacylate mutant tRNAs. This implies the measurement of the catalytic parameters of the aminoacylation reaction of a purified tRNA in the presence of purified aaRS (k_{cat} and K_m), and the comparisons of the values obtained upon mutations of given sets of nucleotides to the parameters associated to the authentic tRNA. Obviously, one difficulty is to use *in vitro* assay conditions that are as close as possible to the cellular ones. Noteworthy, to evidence negative determinants, aminoacylation assays in the presence of the 19 other purified aaRS should be performed also with the same mutated tRNAs. The *in vitro* methodology was therefore of much use in defining only the positive identity elements, *i.e.*, the elements that are used by a given aaRS to efficiently aminoacylate a tRNA molecule.

Mutant tRNA molecules could be synthesized by *in vitro* transcription using T_7 RNA polymerase (Sampson and Uhlenbeck, 1988). However, there are several drawbacks to this technique. First, modified bases cannot be incorporated and, in several systems, the posttranscriptional modifications were shown to be major positive (Muramatsu *et al.*, 1988; Tamura *et al.*, 1992; Sylvers *et al.*, 1993) or negative (Perret *et al.*, 1990) determinants. Second, the first base of the transcript must be a G. Fortunately, most tRNAs start with G_1C_{72} or, otherwise, substituting the first nucleotide by a G appears not to influence the catalytic parameters of the aminoacylation reaction (see for instance Jahn *et al.*, 1991; Pütz *et al.*, 1991). To probe the role of the 5′ base, it is possible to force the incorporation of non-G bases at the first position of the transcript (Himeno *et al.*, 1989; Perret *et al.*, 1990). An alternative to this *in vitro* technique is to overproduce the tRNA variant *in vivo* and to purify it to homogeneity (Seong and RajBhandary, 1987; Hou and Schimmel, 1988; Meinnel *et al.*, 1988; Perona *et al.*, 1988). This can lead to the easy and quick obtention of milligram amounts of most mutant tRNAs (Guillon *et al.*, 1992). In some cases, however, the overproduction of a mutated tRNA could not be achieved (Meinnel *et al.*, 1992; Varshney *et al.*, 1993).

To conclude, positive tRNA identity determinants appear to be much more easily evidenced through *in vitro* measurements. Nevertheless, such determinants do not necessarily account for the complete set of identity elements. To find the negative determinants, *in vivo* assays might be more suitable.

3.1.2. The Efficiency of Aminoacylation and the Identity of RNA

Measurements of the aminoacylation capacities of aaRS indicate that their intracellular concentrations are just enough to insure the protein synthesis required for the optimal growth of *E. coli* (Jakubowski and Goldman, 1984). Indeed, inactivation of one of the two LysRS genes in *E. coli,* which causes a decrease of the cellular concentration of LysRS, leads to a significant increase of the doubling time of the culture (Lévêque *et al.*, 1991). This effect is likely to be due to a reduced level of lysyl-$tRNA^{Lys}$ in the bacterium. Therefore, the efficiency of aminoacylation of a tRNA by its cognate aaRS has to be considered as part of its identity. This problem is linked to the rapid turnover of aminoacyl-tRNAs in the cell.

Interestingly, *in vivo* as well as *in vitro* assays have shown that the three anticodon bases often played a major role in tRNA identity (Table II). According to *in vivo* assays, the anticodon sequence is sufficient to specify tRNA identity in the *E. coli* isoleucine, phenylalanine, or valine systems (Pallanck and Schulman, 1991). Nevertheless, *in vitro* studies have revealed that, in the same three systems, other bases were required to make a noncognate tRNA substrate as efficient as the cognate tRNA (Meinnel *et al.*, 1991b; Tamura *et al.*, 1991b; Peterson and Uhlenbeck, 1992). These non–anticodon bases should be regarded as part of the identity element set of a tRNA. A similar situation was encountered in the *E.*

coli alanine system, where the G_3-U_{70} pair plays a major role, as determined through *in vivo* suppression assays (Hou and Schimmel, 1988; McClain and Foss, 1988b). Nevertheless, additional determinants, located in the acceptor stem, appear to participate in the optimization of the aminoacylation reaction (see references in Table II). Care should therefore be taken in interpreting the *in vivo* experiments. They appear to be poorly sensitive to the aminoacyl–tRNA turnover, because of the use of a single codon as reporter. Consequently, the *in vivo* approach may mask the role of nucleotides mainly involved in the kinetical control of the aminoacylation reaction.

Similarly, suboptimal *in vitro* assay conditions can prevent the identification of several important bases. For instance, conditions mimicking the *in vivo* ionic environment were necessary to evidence the important role of the whole acceptor stem sequence of $tRNA^{Met}$ in optimizing the aminoacylation in the *E. coli* methionine system (Meinnel *et al.*, 1995).

3.1.3. Design of Minisubstrates Mimicking Transfer RNA

The tRNA acceptor stem appears to be important for the specificity of the aminoacylation reactions. The idea that minisubstrates, only comprising this part of the tRNA molecule, could be specifically charged has therefore emerged (Imura *et al.*, 1969; Chambers, 1971; Thiebe *et al.*, 1972; Renaud *et al.*, 1981). Francklyn and Schimmel (1989) synthesized *in vitro* a 24-bases stem-and-loop RNA comprising the acceptor stem and arm of $tRNA^{Ala}$, and called it a *microhelix*. This molecule was still a substrate of AlaRS, although less efficient. Later, an RNA substrate of only 16 bases could be shown to be a substrate of the same synthetase (Shi *et al.*, 1992). It was observed that changes at the level of the identity elements of a microhelix abolished the aminoacylation. This spectacular result, showing the specific aminoacylation of microhelixes, was reproduced in many other systems, including histidine, valine, methionine, and glycine (Francklyn and Schimmel, 1990; Frugier *et al.*, 1992; Martinis and Schimmel, 1992; Shi *et al.*, 1992). Finally, chemically synthesized RNA duplexes could be also specifically aminoacylated (Musier-Forsyth *et al.*, 1991a; Martinis and Schimmel, 1992; Musier-Forsyth and Schimmel, 1993).

It appears therefore that, at least in the above systems, the acceptor stem sequence of tRNAs carries enough information to direct specific aminoacylation by aaRS. Nevertheless, in most studied cases, the aminoacylation efficiency is much lower than in the presence of the native tRNA molecule. Whereas k_{cat}/K_m values are only reduced by two orders of magnitude in the case of alanine microhelixes, as compared to $tRNA^{Ala}$ (Francklyn and Schimmel, 1989), they are reduced by six to seven orders of magnitude in the case of methionine microhelixes compared to $tRNAs^{Met}$ (Martinis and Schimmel, 1992). Despite this, thanks to this approach, the understanding of tRNA identity elements has benefited from several major observations. First, overlapping determinants in the acceptor stem, including the second and third base pairs (2-71 and 3-70) as well as base 73, which act as identity elements in all assayed aaRS systems, participate in the specificity of the aminoacylation reaction by preventing cross-aminoacylations (Francklyn *et al.*, 1992). Second, it is possible to analyze at the atomic level the relative importance of chemical groups of a given nucleoside (Musier-Forsyth *et al.*, 1991b; Musier-Forsyth and Schimmel, 1992). Third, negative determinants within the tRNA acceptor stem seem to occur and to participate in the overall specificity of the aminoacylation reaction (Martinis and Schimmel, 1993).

3.2. Identity Elements among the Various RNAs

Since the pioneer work of Schulman and co-workers that showed that the first base of the anticodon acted as a major positive determinant of the aminoacylation of $tRNA^{Met_f}$ by MetRS (Schulman and Goddard, 1973; Schulman, 1980), much effort from many laboratories has focused toward the definition of identity elements among RNAs. This field was already reviewed in detail by Normanly and Abelson (1989) and Schulman (1991). The updated results are summarized in Table II; the changes that cause variations of the aminoacylation efficiency (k_{cat}/K_m) by more than three orders of magnitude are considered major and are boldfaced in the table. Note that in the case of *E. coli,* all systems have been studied. However, it should be underlined that at the present time a systematic mutagenesis analysis of a tRNA molecule has never been described, so that identity elements can miss.

3.2.1. The Anticodon Sequence as Major Identity Element

At the very beginning of the 1970s, the anticodon was already recognized as a major identity element, thanks to many *in vitro* (reviewed in Kisselev, 1985) and *in vivo* investigations (Yaniv *et al.*, 1974). At present, it appears that a minimum of one base in the anticodon sequence plays a major role in nearly all systems (see Table II). The exceptions are the *E. coli* alanine, the *E. coli* and bovine mitochondrial serine, and the *E. coli* leucine systems. Notably, in many cases changes at the level of the anticodon bases were shown to be enough to allow, *in vivo* as well as *in vitro,* a switch of the chargeability of a tRNA from an aaRS to another one (Muramatsu *et al.*, 1988; Schulman and Pelka, 1988, 1990; Meinnel *et al.*, 1991b; Pallanck and Schulman, 1991; Pak *et al.*, 1992). However, systematic mutagenesis in the sequence of such tRNAs revealed that other determinants, although less important, were involved in the improvement of the aminoacylation reaction (see for instance Meinnel *et al.*, 1993). Conse-

Table II
tRNA Identity Elements[a]

Organism	tRNA species	Identity elements[b]: Anticodon bases	Base 73	Acceptor stem	Additional nucleotides	References[c]
E. coli	Ala	—	A	2–71, **3–70,** 4–69	20	1–8
Yeast	Ala	*	*	**3–70**		9
E. coli	Arg	**35, 36**	A or G	*	20	10–12
E. coli	Asn	**34, 35, 36**	G	*		13, 14
E. coli	Asp	**34, 35,** 36	**G**	2–71	38	15, 16
Yeast	Asp	**34, 35, 36**	**G**	*	10	17
E. coli	Cys	**34, 35, 36**	U	*		18
E. coli	Gln	**34, 35, 36**	A or G	**2–71,** 3–70	10, 37, 38	19, 20
E. coli	Glu	**34**				21
E. coli	Gly	**35**	U	1–72, 2–71, 3–70		8, 22
E. coli	His	*	**C**	2–71, 3–70	**−1**	8, 23, 24
E. coli	Ile	**34, 35, 36**	A	*		13, 25–27
E. coli	Leu	—	**A**	*	V-loop, 8, **14**	29
E. coli	Lys	**34, 35, 36**	A	*		11, 12
E. coli	Met	**34, 35, 36**	A	2–71, 3–70	32, 33, 37	25, 26, 30–35
Yeast	Met	**34, 35, 36**	A	*	AC-loop	36, 37
Rabbit	Met	**34, 35, 36**	*	*	AC-loop	38
E. coli	Phe	**34, 35, 36**	A	*	20, 10–25, 26, 44, 45, 59	39

[a] tRNA identity elements which were known in June 1993 are reported. The origin and amino acid acceptances are indicated. The numbering corresponds to that shown in Fig. 1, panel A.

[b] nucleotides whose substitutions lead to a decrease of catalytic efficiency of aminoacylation (k_{cat}/K_m) of the cognate aaRS by a factor of more than three orders of magnitude are boldfaced. A dash means that changes of the base(s) did not modify the catalytic efficiency of aminoacylation. AC-loop is used for anticodon loop. * Means that changes at this location in the tRNA have not yet been studied.

[c] reference numbers correspond to: 1. Hou and Schimmel (1988); 2. McClain and Foss (1988b); 3. Shi *et al.* (1990); 4. McClain *et al.* (1991a); 5. Musier-Forsyth *et al.* (1991b); 6. Shi and Schimmel (1991); 7. Tamura *et al.* (1991a); 8. Francklyn *et al.* (1992); 9. Trézeguet *et al.* (1991); 10. Schulman and Pelka (1989); 11. McClain *et al.* (1990); 12. Tamura *et al.* (1992); 13. Shimizu *et al.* (1992); 14. Li *et al.* (1993); 15. Hasegawa *et al.* (1989); 16. Nameki *et al.* (1992); 17. Pütz *et al.* (1991); 18. Pallanck *et al.* (1992); 19. Jahn *et al.* (1991); 20. Hayase *et al.* (1992); 21. Sylvers *et al.* (1993); 22. McClain *et al.* (1991b); 23. Himeno *et al.* (1989); 24. Francklyn and Schimmel (1990); 25. Muramatsu *et al.* (1988); 26. Meinnel *et al.* (1991b); 27. Muramatsu *et al.* (1992); 29. Asahara *et al.* (1993); 30. Uemura *et al.* (1982); 31. Schulman and Pelka (1983); 32. Schulman and Pelka (1988); 33. Meinnel *et al.* (1991a); 34. Lee *et al.* (1992); 35. Meinnel *et al.* (1993); 36. Cigan *et al.* (1988); 37. Senger *et al.* (1992); 38. Meinnel *et al.* (1992); 39. Peterson and Uhlenbeck (1992); 40. Moor *et al.* (1992); 41. Sampson and Uhlenbeck (1988); 42. Sampson *et al.* (1989); 43. Sampson *et al.* (1990); 44. Sampson *et al.* (1992); 45. Nazarenko *et al.* (1992); 46. Hasegawa *et al.* (1993); 47. Himeno *et al.* (1990); 48. Normanly *et al.* (1992); 49. Ueda *et al.* (1992); 50. Schulman and Pelka (1990); 51. Hasegawa *et al.* (1992); 52. Himeno *et al.* (1991); 53. Pak *et al.* (1992); 54. Hou and Schimmel (1989); 55. Bordonne *et al.* (1987); 56. Bare and Uhlenbeck (1986); 57. Lee and RajBhandary (1991); 58. Tamura *et al.* (1991b).

(*continued*)

Table II (*Continued*)

Organism	tRNA species	Identity elements[b]: Anticodon bases	Base 73	Acceptor stem	Additional nucleotides	References[c]
Thermus	Phe	**34, 35, 36**	A	*		40
Yeast	Phe	**34, 35, 36**	A	*	20, 59	41–44
Human	Phe	**34, 35, 36**	A	*	20, 30–40, 31–39	45
E. coli	Pro	**35, 36**	A	**72**	49	13, 46
E. coli	Ser	—	(G)	72, 2–71, 3–70	11–24, **V-loop**	47, 48
Bovine (mito)	Ser	—	—	*	T-loop	49
E. coli	Thr	**35, 36**	—	1–72, 2–71		50, 51
E. coli	Trp	**34, 35, 36**	G	1–72		52, 53
E. coli	Tyr	34, **35**	A	*	**V-loop**	47, 54
Yeast (mito)	Tyr	*	*	*	V-loop	55
Yeast (cyto)	Tyr	34, 35	A	1–72		56, 57
E. coli	Val	**34, 35, 36**	A	3–70, 4–69	32, 38	32, 58

quently, the anticodon sequence alone cannot fully account for the identity of a tRNA, despite its predominant role.

3.2.2. The Role of the Acceptor Stem and Arm Sequences

As a general rule, bases at the top of the acceptor extremity play an important role in the aminoacylation of tRNA molecules. This conclusion meets that obtained through the study of RNA microhelixes (see Section 3.1.3).

Base 73, the base upstream from the conserved CCA trinucleotide at the end of the tRNA acceptor arm, was early on proposed as a discriminator position (Crothers *et al.*, 1972). Systematic studies (reviewed in Shimizu *et al.*, 1992) confirm that this base indeed plays an important role, with few exceptions (see Table II). In the case of $tRNAs^{His}$, C_{73} makes a base-pairing with an additional 5′-G (numbered −1). In this particular system, C_{73}, which is unique among *E. coli* tRNAs, acts as a major positive determinant (Himeno *et al.*, 1989). Such a role also occurs in eukaryotic cytoplasmic $tRNAs^{His}$ (Rudinger, 1992), the supplementary G_{-1} being posttranscriptionally added (Pande *et al.*, 1991).

The effects of changes at the level of the acceptor stem sequence have not yet been systematically investigated in all systems (see Table II). This is especially the case of the 1-72 base pair, because of the difficulty in starting transcription with a non-G base, when tRNA mutants are synthesized *in vitro* by means of T_7 RNA polymerase-driven transcription (see Section 3.1.1b). Despite this, two of the four first base pairs of the acceptor stem were often shown to participate to tRNA identity. As a general rule, base pairs 2-71 and 3-70 appear to be involved. An extreme situation is encountered in the case of the alanine system where the unusual G_3-U_{70} base pair is enough per se to dictate the esterification by alanine (see references in Table II).

3.2.3. Other Identity Elements

In many systems, nucleotides outside the anticodon or the acceptor stem play an important role (Table II). The involved regions are the long V-loop of class II tRNAs, base 20 of the D-loop, and several non–anticodon bases in the anticodon loop. Except for the case of the V-loop in the serine, leucine, and tyrosine systems, none of these determinants appear to be critical.

3.3. tRNA-like Structures Recognized by aaRS

Several tRNA-like structures are known to specifically interact or to be aminoacylated by aaRS (for a review see Mans *et al.*, 1991). The description of the structural elements on these RNA molecules capable to direct their recognition by aaRS is likely to be of great help for the understanding of the structural network required by this family of enzymes to bind and/or aminoacylate their RNA substrates.

3.3.1. Viral RNAs

From the early demonstration that the 3′ end of the 6.3-kilo base pair (kbp) RNA genome of the turnip yellow mosaic virus (TYMV) could be valylated in the presence of yeast ValRS, numerous other plant viral RNAs were shown to be chargeable *in vitro* with either valine, histidine, or tyrosine (a recent compilation is available in Mans *et al.*, 1991). Whether or not there exists a physiological relevance *in vivo* to such an aminoacylability of RNA viral genomes has still to be established (see, for instance, Dreher *et al.*, 1989; Tsai and Dreher, 1991). The 3′ end of plant viral RNAs is believed to fold into a tRNA-like structure. In the case of the TYMV RNA, a 112-base RNA derived from the genomic RNA adopts a tertiary structure close to that of a tRNA, including an anticodon loop centered around a CAC valine anticodon and a pseudoknotting structure underlying the existence of an acceptor end (Florentz *et al.*, 1982; Rietveld *et al.*, 1983; Dumas *et al.*, 1987; for a recent review on pseudoknots, see ten Dam *et al.*, 1992).

In good agreement with the three-dimensional model of the 3′ end of TYMV genomic RNA, deletion analysis shows that the last 82 nucleotides are sufficient to direct the aminoacylation by ValRS (Joshi *et al.*, 1982). However, additional bases, located upstream from nucleotide 82, are necessary to achieve the most efficient aminoacylation rate (Dreher *et al.*, 1988). This conclusion is consistent with protection experiments carried out in the presence of ValRS, using a 159-nucleotide 3′ terminal TYMV RNA fragment (Florentz and Giegé, 1986). Finally, mutational analysis of the three bases of the anticodonlike structure of TYMV RNA has established their important role in the aminoacylation by wheat germ or yeast ValRS (Florentz *et al.*, 1991; Dreher *et al.*, 1992). In addition, strong decreases of aminoacylation efficiency were observed upon changes of the A_4 or the C_{53} nucleotide in the viral RNA, which correspond respectively to bases A_{73} and U_{38} in the numbering of a canonical tRNA (Florentz *et al.*, 1991; Dreher *et al.*, 1992) (Fig. 1A). Because the corresponding bases participate in the identity of *E. coli* $tRNA^{Val}$ and, taking into account the fact that identity elements are well-conserved across the phyla (see Table II), these observations reinforce again the idea that the 3′ end of TYMV viral RNA adopts a tRNA-like structure.

Little is known about the structural features required for the aminoacylation of other viral RNAs. Nevertheless, in the case of brome mosaic virus, a 134-nucleotide-long RNA derived from the 3′ end of the genomic RNA, was shown to be the shortest tyrosylatable derivative of this RNA. This RNA fragment is believed to fold into a tRNA-like structure, albeit no anticodonlike structure could be recognized (Dreher and Hall, 1988a; Perret *et al.*, 1989). Moreover, tyrosylation of this fragment is prevented by mutations affecting the aminoacyl acceptor-end (Dreher and Hall, 1988b).

3.3.2. Operators on mRNAs

Putative tRNA-like structures have been noticed in the leader sequences of the *E. coli his* operon (Ames *et al.*, 1983), as well as in those of the *metG* (Dardel *et al.*, 1990; Nureki *et al.*, 1991) and *thrS* genes (Lestienne *et al.*, 1984; Springer *et al.*, 1986). The occurrence of such structures is believed to be linked to some control of the expression of the above genes through binding of the aaRS to the mRNA. Such an interaction was demonstrated in the case of the *thrS* gene, which encodes ThrRS. It causes the expression of ThrRS to be autogeneously controlled at the translational level (Lestienne *et al.*, 1984; Butler *et al.*, 1986). This control is mediated by the operator region on the mRNA, which folds into a tRNA-like structure and is the target of ThrRS (Springer *et al.*, 1986; Moine *et al.*, 1988).

Mutants affecting the expression of *thrS* have been isolated. A first category corresponds to cis-acting mutations that map within the tRNA-like structure (Springer *et al.*, 1989). In particular, mutations found at the level of the anticodonlike structure led to a reorganization of the operator structure (Moine *et al.*, 1988). A second category are trans-acting mutations that map within the ThrRS open reading frame itself. The resulting mutated ThrRS are affected also in their ability to aminoacylate wild-type $tRNA^{Thr}$ (Springer *et al.*, 1989). The occurrence of such a correlation between the capabilities of ThrRS to regulate its own expression and to aminoacylate its cognate tRNA substrate strongly favored the idea of a direct interaction between the synthetase and the tRNA-like structure. ThrRS was later shown to actually complex with the mRNA operator region or the *E. coli* $tRNAs^{Thr}$ with close affinity constants (Moine *et al.*, 1990). In addition, $tRNA^{Thr}$ is a competitive inhibitor of the binding of the protein to the mRNA operator region (Moine *et al.*, 1990). When bound to the operator, ThrRS protects both the mRNA anticodonlike and the acceptor endlike structures. The involvement of these two structures in the translational control was later confirmed by mutagenesis experiments directed toward the operator region (Brunel *et al.*, 1992).

The aminoacylability of $tRNA^{Thr}$ by *E. coli* ThrRS is governed by nucleotides belonging to the anticodon and acceptor stem sequences (Schulman and Pelka, 1990; Hasegawa *et al.*, 1992) (Table II). By following the rules of tRNA identity in *E. coli*, the translation of the *thrS* mRNA could be made sensitive to the concentration of MetRS instead of ThrRS *in vivo* (Graffe *et al.*, 1992). This was rendered possible by changing the CGU threonine anticodonlike sequence of the operator into a CAU one, the main determinant recognized by *E. coli* MetRS. This identity switch could be confirmed *in vitro* by comparing the binding constants of the two aaRS in the presence of various mutated operator regions (Romby *et al.*, 1992). Very recently, it was proposed that the operator region of *thrS* could form, in fact, two tRNA-like structures (Bedouelle, 1993). The occurrence of two such structures, close to each other within the operator region, would correspond to the capacity of the dimeric ThrRS to bind two equivalent tRNA binding sites at the same time.

3.3.3. Group I Introns

Group I introns are mobile elements comprising most of the introns of fungal mitochondrial genomes and are composed of series of RNA secondary structures required for splicing activity. While a number of group I introns are capable of self-splicing *in vitro*, genetic analyses have revealed that proteins are systematically required to achieve splicing *in vivo* (see Lambowitz and Perlman, 1990, for a review). In particular, LeuRS from *Saccaromyces* spp. and TyrRS from *Neurospora crassa* and *Podospora anserina* assist the splicing of such introns (Akins and Lambowitz, 1987; Herbert *et al.*, 1988; Kämper *et al.*, 1992). Reciprocally, the reintegration of one of these introns into a rRNA was recently shown to also be allowed by an aaRS (Mohr and Lambowitz, 1991).

Neurospora crassa TyrRS was proved to strongly bind to a number of group I introns with low sequence homology (Collins and Lambowitz, 1985; Guo and

Lambowitz, 1992; Mohr *et al.*, 1992). Conserved nucleotides of the P4-P5-P6-P6A-P7 secondary structures of the intron core (Cech, 1990) appear to be recognized by the synthetase. Accordingly, *N. crassa* TyrRS binds a 75-nucleotide-long RNA encompassing the above structures (Guo and Lambowitz, 1992). Interestingly, $tRNA^{Tyr}$ is a competitive inhibitor for the binding of TyrRS to group I introns.

The three-dimensional model of the structure of group I introns, as proposed by Michel and Westhof (1990), indeed share close similarities with that of a tRNA. This indicates that *N. crassa* TyrRS might assist the splicing activity by direct binding to an intron core, which could adopt a tRNA-like structure. The comparison between *Neurospora* mitochondrial $tRNA^{Tyr}$ and the intron core indeed shows sequence identities, in particular at the level of the long V-loop of $tRNA^{Tyr}$, which corresponds to P7 in the intron (Guo and Lambowitz, 1992). Interestingly, the V-loop is an important identity element of $tRNAs^{Tyr}$ (Table 2).

The study of the mechanisms by which tRNA-like structures are recognized by aaRS strongly suggests that these structures and tRNAs share common strategies to interact with the synthetases. One may therefore expect that, whatever the nature of the RNA, the synthetase recognizes an RNA ligand through the same binding sites.

4. tRNA BINDING SITES ON THE aaRS

The transfer of the aminoacid onto a tRNA may occur if the acceptor hydroxyl (at the 2′ or 3′ position of the terminal adenosine) becomes correctly positioned with respect to the active site of the enzyme. According to this view, the specifity of an aaRS toward a tRNA combines two steps (1) the formation of the initial enzyme:tRNA complex and (2) structural rearrangements beyond the binding step, which enable the adequate positioning of the terminal adenosine in the vicinity of the aminoacyl adenylate. Obviously, the identity elements of a tRNA can influence each of these steps.

Up to now, the dynamic aspects of the aminoacylation reaction have not been thoroughly studied. However, structural rearrangements upon complex formation between an aaRS and a tRNA were evidenced by using fluorescence-monitored rapid kinetics methods (Krauss *et al.*, 1976) and small-angle neutron scattering (Zaccaï *et al.*, 1979) or by following the reaction of truncated tRNA substrates (Renaud *et al.*, 1981). In many cases, the structural changes could not be observed upon the binding of noncognate tRNAs to the studied enzyme, thus indicating the importance of rearrangements in the expression of the enzyme specificity.

Studies of tRNA binding to aaRS have allowed the determination of the stoichiometries of the corresponding complexes (see a compilation in Meinnel *et al.*, 1995). All studied monomeric enzymes bind one tRNA molecule. Dimeric or pseudodimeric [i.e., of the $(\alpha\beta)_2$ type] aaRS display either one, two equivalent, or two nonequivalent tRNA binding sites. Moreover, in the case of some dimeric enzymes, the whole tRNA molecule is bound by a single subunit (e.g., *E. coli* MetRS) (Cassio and Waller, 1971; Blanquet *et al.*, 1973), whereas in the case of other dimeric aaRS, tRNA binds across the two subunits (Dessen *et al.*, 1982; Carter *et al.*, 1986). From this, it appears that the recognition of tRNAs by aaRS is much more idiosyncratic than the activation of the aminoacid. Moreover, the various observed behaviours cannot yet be related to the two class-partitioning of aaRS.

4.1. Structure of tRNA:aaRS Crystalline Complexes

A major breakthrough in the understanding of the recognition of a tRNA by an aaRS came from the elucidation of the three-dimensional structures of two crystallized complexes containing tRNA, that of the class I *E. coli* GlnRS complexed with ATP and $tRNA^{Gln}$ (Rould *et al.*, 1989) and that of the class II yeast AspRS complexed with $tRNA^{Asp}$ (Ruff *et al.*, 1991). These two structures display both common and distinctive features.

As could be expected from the general organization of aaRS deduced from biochemical studies (see Section 2.2), the two subunits of these enzymes are built of roughly two domains, the active center being involved in the binding of the tRNA acceptor stem and the additional domain recognizing the anticodon of the tRNA. Notably, in these systems, the acceptor stem as well as the anticodon loop contain major identity elements (see Table II). Interestingly, preliminary results coming from the study of crystals of *E. coli* SerRS with $tRNA^{Ser}$ (Price *et al.*, 1993) show that the long helical coiled-coil domain of the enzyme interacts with the long V-loop of the tRNA molecule. In this particular system, the V-loop is a major identity element. A common structural organization of all aaRS:tRNA complexes, with the involvement of two functional domains, appears therefore to emerge. One domain would bind the anticodon (or the long V-loop in class II tRNAs) and the other domain would interact with the acceptor stem. An additional binding area between bases of the D-stem and the enzyme region connecting the two domains was observed in the complexes of yeast AspRS or of *E. coli* GlnRS with their cognate tRNAs. In fact, base pair 10-25 of the D-stem has already been proposed to make part of the set of identity elements carried by yeast $tRNA^{Asp}$ or *E. coli* $tRNA^{Gln}$.

The gross three-dimensional structures of the tRNA molecules are unmodified upon complexation with their cognate synthetase (compare Fig. 1B and Fig. 4). However, the parts of the tRNA in close contact with the enzyme appear

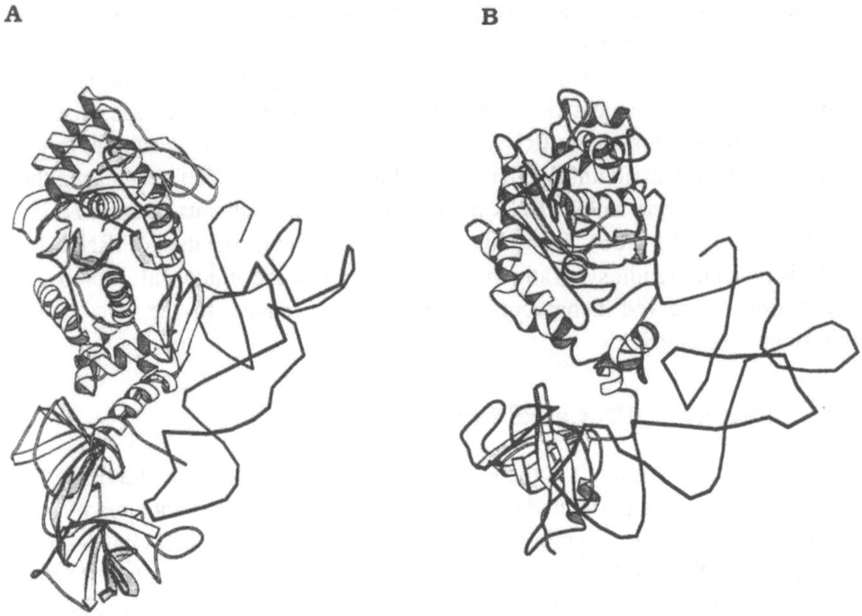

FIGURE 4. Folding of the two solved aminoacyl–tRNA synthetases:tRNA complexes. Shown are the three-dimensional structures of the complexes of (A) *E. coli* GlnRS (Rould *et al.*, 1989, 1991) and (B) yeast AspRS (Ruff *et al.*, 1991; Cavarelli *et al.*, 1993) with their cognate tRNA. The two pictures were drawn using the Molscript program (Kraulis, 1991). The peptidic backbones of the enzymes and phosphate backbones of the tRNAs only are shown.

to undergo significant conformational changes. In the two solved structures, the three anticodon bases are not stacked anymore and become buried within the anticodon binding domain of the enzyme (Rould *et al.*, 1991; Cavarelli *et al.*, 1993). Furthermore, the anticodon stem of enzyme-bound $tRNA^{Gln}$ is extended by two additional non-Watson–Crick pairs (Rould and Steitz, 1992) and the A_1U_{72} base pair is disrupted. Apart from this, there are two major differences between the structural organizations of these two aaRS:tRNA complexes. The first comes from the positioning of the tRNA acceptor stem with respect to the enzyme: in the aspartate system, the tRNA contacts AspRS from the major groove, while the recognition arises from the minor groove in the glutamine system. The second difference is a consequence of the first one: the acceptor arm of $tRNA^{Gln}$ bends to enter the active center of GlnRS, while that of $tRNA^{Asp}$ continues in a helical structure (Fig. 4). Interestingly, these two distinct positionings of the tRNA acceptor stems might be related to the belonging of GlnRS and AspRS to class I and class II, respectively. Structural similarities between the active site regions of *E. coli* GlnRS and MetRS have suggested a common orientation of tRNA binding (Perona *et al.*, 1991). From this, it could be suggested that the minor groove approach of the tRNA to the enzyme might be a general feature of the class I aaRS systems. In addition, the fact that the major groove of $tRNA^{Asp}$ interacts with a loop connecting the two β-strands of the conserved motif 2 (Cavarelli and Moras, 1993; Cavarelli *et al.*, 1993b), might indicate a common orientation of tRNA acceptor stems in their complexes with class II aaRS.

Examination of the GlnRS:$tRNA^{Gln}$ structural model yielded explanations concerning the precise role of the tRNA identity elements located at the level of the acceptor stem. The G_2C_{71} and G_3C_{70} pairs are recognized by the enzyme through base-specific contacts in the minor groove. An interesting case is that of the G_{73} identity element which does not contact the enzyme. In turn, the 2-amino group of G_{73} interacts with the phosphate oxygen of the phosphodiester linkage between C_{71} and A_{72}, this contact being crucial to stabilize the above-mentioned hairpin conformation of the acceptor arm. This conformation is also assisted by enzyme interactions with the C_{74} and A_{76} bases, universally conserved among tRNAs. In the case of the aspartic system, recognition of the crucial G_{73} identity element is achieved through hydrogen bonds with residues belonging to a variable part of the conserved motif 2. These residues make also hydrogen bonds with the U_1 and A_{72} bases.

Regarding the tRNA anticodon region, its recognition by GlnRS or AspRS involves β-barrel structures (Rould *et al.*, 1991; Cavarelli *et al.*, 1993) (Fig. 4). Despite this similarity, such a structural arrangement cannot be generalized to the recognition of tRNA anticodons by other aaRS, since in *E. coli* MetRS the anticodon region is likely to be bound by an α-helical domain (Ghosh *et al.*, 1990; Meinnel *et al.*, 1991a,b; Perona *et al.*, 1991; Schmitt *et al.*, 1993). Nevertheless, close examination of the three-dimensional structures of the two solved aaRS:tRNA complexes indicates common features for the binding of the anticodon loop. The deciphering of each of the anticodon bases clearly lies on interactions between base-specific chemical groups on one hand and enzyme side-chains or backbone atoms on the other (Fig. 5). In addition, both complexes are further stabilized by interactions with ribose or phosphate groups (e.g., interaction of the phosphate 36 of $tRNA^{Gln}$ with the guanidium ion of Arg520 of GlnRS, or interaction of O4′ of U_{35} of $tRNA^{Asp}$ with the Nϵ of Gln121 of AspRS).

To conclude, as a general rule, the major identity elements of tRNAs appear to directly interact with specific groups of the enzyme. However, conformational changes of the structure of the complexed tRNA molecule can occur as a consequence of the presence of specific nucleotides which are not necessarily in contact with the enzyme.

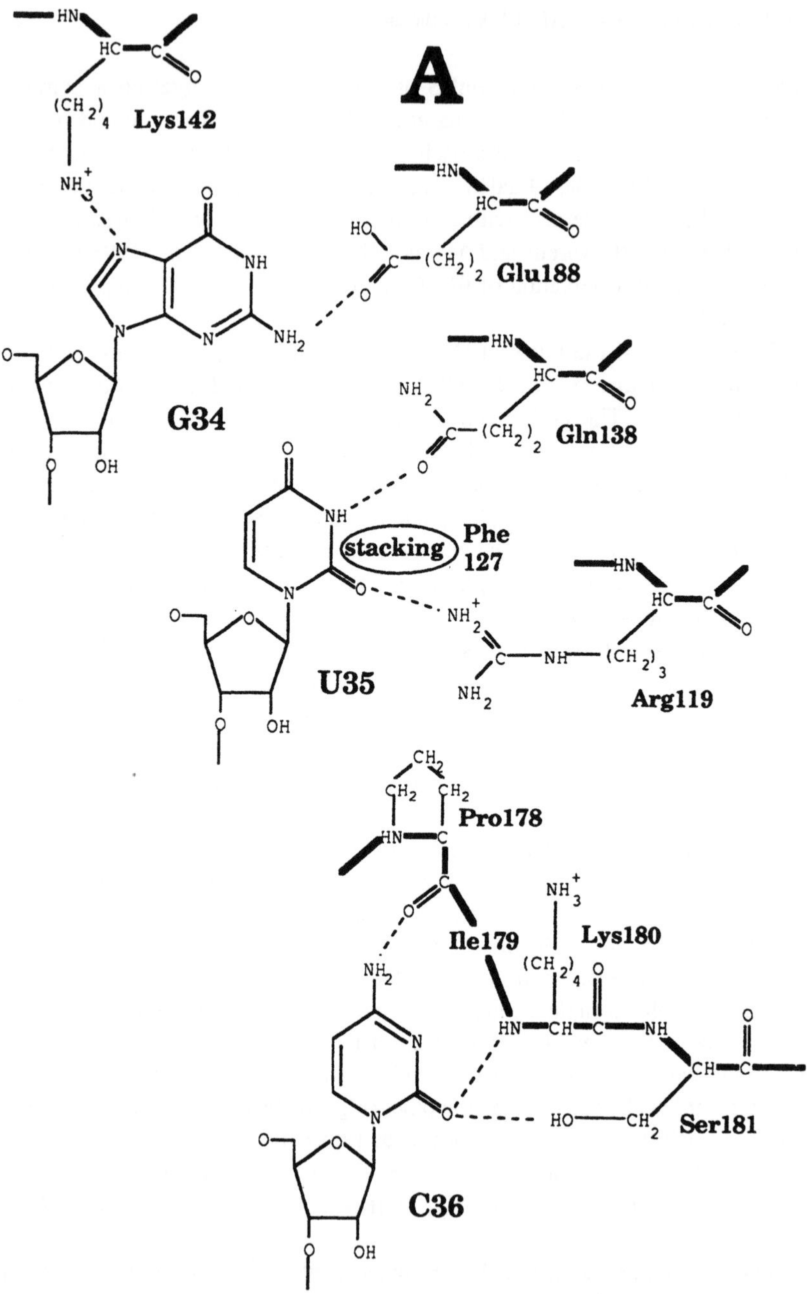

FIGURE 5. Amino acid residues involved in the deciphering of the anticodon bases of $tRNA^{Asp}$ or of $tRNA^{Gln}$. Shown are the amino acids of either (A) yeast AspRS or (B) *E. coli* GlnRS, which directly interact with the three bases of the anticodon of $tRNA^{Asp}$ or of $tRNA^{Gln}$. Hydrogen bonds are figured with a dashed line. (A) Bases G_{34}, U_{35}, and C_{36} of the anticodon of yeast $tRNA^{Asp}$ are schematized. Stacking of the aromatic side-chain of Phe127 with U_{35} is indicated. This scheme was deduced from the data in Cavarelli *et al.* (1993). (B) Bases C_{34}, U_{35}, and G_{36} of the anticodon of *E. coli* $tRNA^{Gl}{}_{1}$ are schematized according to the data in Rould *et al.* (1991).

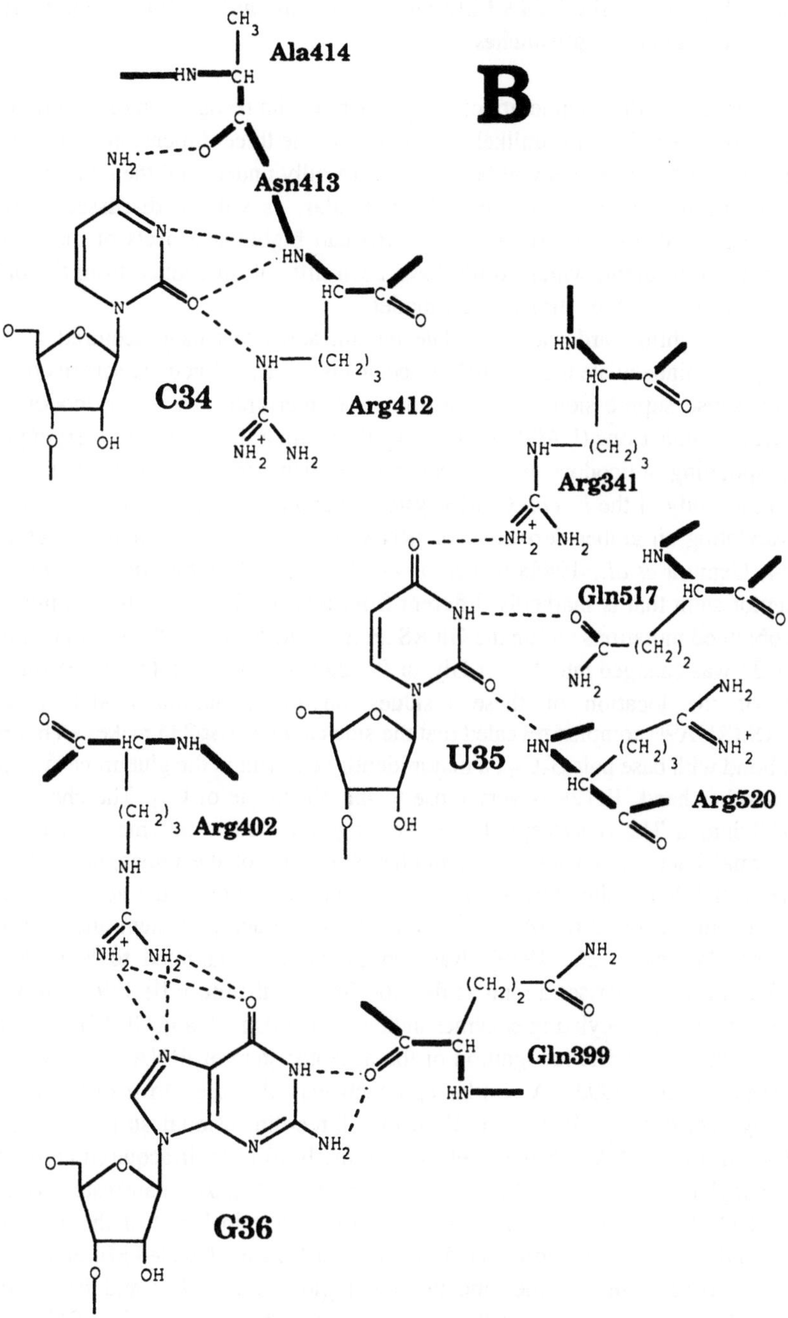

FIGURE 5. (*Continued*)

4.2. Mapping of the tRNA-Binding Area on the aaRS Using Genetic and Biochemical Approaches

Because of the complexity of the problem of the recognition or rejection of tRNAs by aaRS, it seems unlikely that knowing the three-dimensional structures of aaRS:tRNA complexes will be sufficient to fully understand the atomic bases of the specificity of this process. In particular, as will be discussed below, changing the tRNA specificity of an aaRS can highlight aspects of the tRNA recognition problem, which would have been difficult to deduce from the only examination of a three-dimensional model.

A straightforward way to isolate mutant aaRS that have acquired a new ability to aminoacylate a given tRNA species is the use of genetic screens based on nonsense suppression. An indicator gene interrupted by a stop codon, in general an amber one (UAG), is used, together with a tRNA species carrying the corresponding anticodon. A first example of such screening procedures came from the study of the *E. coli* GlnRS system. Variants of this enzyme, capable of misacylating an amber derivative of $tRNA^{Tyr}$, were obtained (Inokuchi *et al.*, 1984; Uemura *et al.*, 1988a,b; Perona *et al.*, 1989). Notably, $tRNA^{Tyr}$ has an acceptor stem that is markedly different from that of $tRNA^{Gln}$. The mapping of the obtained mutations within the GlnRS amino acid sequence showed that either Asp235 was changed into Asn or Gly, or Ile129 was changed into Thr. Examination of the location of these residues on the structural model of the GlnRS:$tRNA^{Gln}$ complex revealed that the side chain of Asp235 makes an hydrogen bond with base pair G_3C_{70}, a major identity element in the glutamine system. On another hand, Ile129 is very close to the phosphate of C_{74}. The change of Ile129 into a Thr is susceptible to allowing the mutant enzyme to make an additional interaction with this particular phosphate of the noncognate tRNA, thereby stabilizing the above-mentioned hairpin structure of the acceptor arm, although the required $tRNA^{Gln}$ determinants of the acceptor stem and arm are missing (Perona *et al.*, 1989). Random mutagenesis targeted to helix E of GlnRS, which comprises amino acids 126-138 (and therefore Ile 129), followed by selection of misacylating enzymes indicated that Arg130 and Glu131 could be critical for the specific recognition of the acceptor stem of $tRNA^{Gln}$ (Weygand-Durasevic *et al.*, 1993). A similar approach was also used to map the tRNA binding region of AlaRS, an aaRS whose three-dimensional structure is still unknown. Mutated AlaRS were selected on the basis of their acquired ability to aminoacylate a variant of $tRNA^{Ala}{}_{amber}$, carrying a G_3C_{70} base pair instead of the crucial G_3U_{70}. Most mutations mapped in the N-terminal part of the enzyme, close to the zinc binding domain (Miller *et al.*, 1991a,b). This led Miller and coworkers to propose that the zinc binding region of AlaRS could be closely involved in the deciphering of the major identity element of alanine tRNAs.

In the case of *E. coli* MetRS, a similar approach was used to identify the region of the enzyme responsible for anticodon recognition. An amber derivative of elongator $tRNA^{Met}$ was expressed in an indicator strain. To allow the aminoacylation of this mutated tRNA, several mutations, scattered within a limited region of the C-terminal domain of MetRS, were found to be necessary (Meinnel *et al.*, 1991b). This region (also called the 460 region) perfectly matched that already deduced from the chemical cross-linking of MetRS with a tRNA rendered reactive at the level of its anticodon (Leon and Schulman, 1987; Schulman *et al.*, 1987). Graphic modeling of the binding of the tRNA to MetRS further reinforced the involvement of the 460 region as anticodon binding site (Perona *et al.*, 1991). Additional analysis of this region by site-directed mutagenesis showed that two residues belonging to the 460 region, Trp461 (Ghosh *et al.*, 1990; Meinnel *et al.*, 1991b) and Asn452 (Schmitt *et al.*, 1993) are involved in the binding to the cognate methionine anticodon. It should be mentioned, however, that site-directed mutagenesis studies of MetRS have also designated Arg395, which does not belong to the 460 region, as part of the tRNA anticodon binding area (Ghosh *et al.*, 1991a). A more detailed analysis of the anticodon binding site on *E. coli* MetRS could be performed using the technique of saturation mutagenesis targeted to the 460 region (Schmitt *et al.*, 1993). Selection of several dozen variants of MetRS able to aminoacylate elongator $tRNA^{Met}{}_{amber}$ highlighted two Asp residues (Asp449 and Asp456). The substitution of any of these two Asp residues into a nonacidic one is sufficient to allow the enzyme to aminoacylate the mutated tRNA. The properties of such mutant enzymes were further studied *in vitro*. Variants of MetRS with a nonacidic side chain at either position 449 or 456 were shown to be able to aminoacylate not only $tRNA^{Met}{}_{amber}$ but also several other $tRNA^{Met}$ derivatives carrying various anticodons. Interestingly, the same variants were unaffected in their ability to aminoacylate or bind wild-type $tRNA^{Met}$. This analysis indicated therefore that the two Asp residues were likely to contribute to the overall specificity of the enzyme by rejecting tRNAs with noncognate anticodons (Schmitt *et al.*, 1993).

The role of an acidic residue of an aaRS acting as negative discriminant toward the aminoacylation of tRNAs was also proposed in the case of *B. stearothermophilus* TyrRS (Vidal-Cros and Bedouelle, 1992) or of *E. coli* GlnRS (Uemura *et al.*, 1988b). However, in these systems, the acidic side chain interferes with the recognition of the tRNA acceptor stem. It can therefore be proposed that the discrimination among tRNAs by an aaRS results from the combination of two effects: (1) the matching of the enzyme with the identity elements of the cognate tRNAs, and (2) the rejection of noncognate tRNAs, as a consequence of the location of some residues at the surfaces of the enzyme and/or of the tRNA.

5. CONCLUSIONS

Aminoacyl-tRNA synthetases form a class of isofunctional enzymes that are featured by a large diversity at the level of sizes, primary structures, and subunit organizations (for an extensive review see Meinnel *et al.*, 1995). Despite this, the currently increasing information concerning the primary and tertiary structures of these enzymes reveals that actually only two types of active site organization might occur. Accordingly, amino acid activation can be achieved through two alternative structural solutions, one of which involves a Rossman fold. Although the available data also suggest two binding modes of the tRNA acceptor stem, the problem of the recognition of tRNAs, a family of molecules with close three-dimensional structures (see Fig. 1, and for a review, Rich and Raj-Bhandary, 1976), remains puzzling. In particular, one can wonder if, similar to DNA binding proteins (for reviews see Schleif, 1988; Harrison and Aggarwal, 1990; Freemont *et al.*, 1991), the aaRS have adopted a limited number of solutions to decipher the major tRNA identity elements located outside the acceptor stem, i.e., the anticodon and variable loops. At the present time, two types of binding strategies already emerge: one involves β-barrels (AspRS and GlnRS) and the other involves α-helixes (SerRS and MetRS). The determination of the structures of other aaRS:tRNA complexes will be necessary to know whether other structural solutions may occur. For instance, a few aaRS contain a tightly bound zinc atom. In the case of *E. coli* AlaRS, the zinc binding site was mapped close to the tRNA acceptor arm binding region and proposed to be involved in RNA recognition (Miller *et al.*, 1991b). A question to ask is whether zinc-containing motifs may participate to tRNA binding, in a manner similar to the zinc fingers of DNA-binding proteins (Miller and Schimmel, 1992). Recent studies of the role of the zinc atom in *E. coli* MetRS rather suggested that the metal participates in the folding of the active site rather than in the formation of the tRNA binding site (Fourmy *et al.*, 1993b). It cannot be excluded, however, that the zinc binding region might help the correct positioning of the acceptor arm during the transfer of the amino acid onto the tRNA. In this context, it is remarkable that the three-dimensional structure of the isolated zinc domain of *E. coli* MetRS (Fourmy *et al.*, 1993a) displays significant similarity with that of *gag* retroviral proteins (Summers *et al.*, 1990), which are known to interact with the viral RNA genome.

To conclude, in spite of spectacular recent advances, the structural bases of RNA-specific recognition by aaRS are still far to be fully described. The problem may remain complex for a while because of the variety characterizing both this class of enzymes and their RNA substrates. In turn, such a difficulty is greatly counterbalanced by the possibility of making comparisons within the family of aaRS and of drawing general conclusions regarding the rules and the evolution of RNA:protein interaction.

ACKNOWLEDGMENTS. The authors wish to thank Drs. S. Brunie, S. Cusack, and D. Moras for the communication of the coordinates of MetRS, SerRS, and AspRS, respectively, and their colleagues at the Ecole Polytechnique for stimulating discussions. They are particularly indebted to Drs. F. Dardel, D. Fourmy, and E. Schmitt for help with the preparation of figures.

6. REFERENCES

Akins, R. A., and Lambowitz, A. M., 1987, A protein required for splicing group I introns in *Neurospora* mitochondria is mitochondrial tyrosyl-tRNA synthetase or a derivative thereof, *Cell* **50**:331–345.

Allen, E. H., and Schweet, R. S., 1960, Role of transfer ribonucleic acid in hemoglobin synthesis, *Biochim. Biophys. Acta* **39**:185–187.

Ames, B. N., Tsang, T. H., Buck, M., and Christman, M. F., 1983, The leader mRNA of the histidine attenuator region resembles $tRNA^{His}$: Possible regulatory implications, *Proc. Natl. Acad. Sci. USA* **80**:5240–5242.

Anselme, J., and Härtlein, M., 1991, Tyr-426 of the *Escherichia coli* asparaginyl-tRNA synthetase, an amino acid in a C-terminal conserved motif, is involved in ATP binding, *FEBS Lett* **280**:163–166.

Asahara, H., Himeno, H., Tamura, K., Hasagawa, T., Watanabe, K., and Shimizu, M., 1993, Recognition nucleotides of *Escherichia coli* $tRNA^{Leu}$ and its elements facilitating discrimination from $tRNA^{Ser}$ and $tRNA^{Tyr}$, *J. Mol. Biol.* **231**:219–229.

Bange, F. C., Flohr, T., Buwitt, U., and Böttger, E. C., 1992, An interferon-induced protein with release factor activity is a tryptophanyl-tRNA synthetase, *FEBS Lett* **300**:162–166.

Bare, L. A., and Uhlenbeck, O. C., 1986, Specific substitution into the anticodon loop of yeast tyrosine transfer RNA, *Biochemistry* **25**:5825–5830.

Barker, D. G., and Winter, G., 1982, Conserved cysteine and histidine residues in the structures of the tyrosyl and methionyl-tRNA synthetases, *FEBS Lett.* **145**:191–193.

Bedouelle, H., 1993, Symmetrical interactions between the translational operator of the *thrS* gene and dimeric threonyl transfer RNA synthetase, *J. Mol. Biol.* **230**:704–708.

Belrhali, H., Yaremchuk, A., Tukalo, M., Larsen, K., Berthet-Colominas, C., Leberman, R., Beijer, B., Sproat, B., Als-Nielsen, J., Grubel, G., Legrand, J.-F., Lehman, M., and Cusack, S., 1994, Crystal structures at 2.5 Ångstrom resolution of seryl–tRNA synthetase complexed with two analogues of seryl adenylate, *Science* **263**:1432–1436.

Berg, P., 1956, Acyl adenylates: The interaction of adenosine triphosphate and L-methionine, *J. Biol. Chem.* **222**:1025–1035.

Berg, P., 1961, Specificity in protein synthesis, *Annu. Rev. Biochem.* **30**:293–324.

Berg, P., and Ofengand, E. J., 1958, An enzymatic mechanism for linking amino acids to RNA, *Proc. Natl. Acad. Sci. USA* **44**:78–86.

Berg, P., Bergmann, F. H., Ofengand, E. J., and Diekmann, M., 1961, The enzymic synthesis of amino acyl derivatives of ribonucleic acid. I. The mechanism of leucyl-, valyl-, isoleucyl-, and methionyl-ribonucleic acid formation, *J. Biol. Chem.* **236**:1726–1734.

Bergmann, F. H., Berg, P., and Diekmann, M., 1961, The enzymic synthesis of amino acyl derivatives of ribonucleic acid. II. The preparation of leucyl-, valyl-, isoleucyl-, and methionyl-ribonucleic acid synthetases from *Escherichia coli*, *J. Biol. Chem.* **236**:1735–1740.

Bhat, T. N., Blow, D. M., and Brick, P., 1982, Tyrosyl-tRNA synthetase forms a mononucleotide-binding fold, *J. Mol. Biol.* **158**:699–709.

Blanquet, S., Fayat, G., Waller, J. P., and Iwatsubo, M., 1972, The mechanism of action of methionyl-tRNA synthetase from *Escherichia coli*—Interaction with ligands of the amino-acid-activation reaction, *Eur. J. Biochem.* **24:**461–469.

Blanquet, S., Iwatsubo, M., and Waller, J. P., 1973, The mechanism of action of methionyl-tRNA synthetase from *Escherichia coli*—1. Fluorescence studies on $tRNA^{Met}$ binding as a function of ligands, ions and pH, *Eur. J. Biochem.* **36:**213–226.

Blanquet, S., Fayat, G., and Waller, J. P., 1974, The mechanism of action of methionyl-tRNA synthetase from *Escherichia coli.* Mechanism of the amino acid activation reaction catalyzed by the native and the trypsin-modified enzymes, *Eur. J. Biochem.* **44:**343–351.

Blanquet, S., Fayat, G., Poiret, M., and Waller, J. P., 1975a, The mechanism of action of methionyl-tRNA synthetase from *Escherichia coli*—Inhibition by adenosine and 8-aminoadenosine of the amino-acid activation reaction, *Eur. J. Biochem.* **51:**567–571.

Blanquet, S., Fayat, G., and Waller, J. P., 1975b, The amino acid activation reaction catalyzed by methionyl-transfer RNA synthetase: Evidence for synergistic coupling between the sites for methionine, adenosine and pyrophosphate, *J. Mol. Biol.* **94:**1–15.

Blow, D. M., Bhat, T. N., Metcalfe, A., Risler, J. L., Brunie, S., and Zelwer, C., 1983, Structural homology in the amino-terminal domains of two aminoacyl-tRNA synthetases, *J. Mol. Biol.* **171:**571–576.

Bochner, B. R., Lee, P. C., Wilson, S. C., Cutler, C. W., and Ames, B. N., 1984, AppppA and related adenylylated nucleotides are synthesized as a consequence of oxidation stress, *Cell* **37:**225–232.

Bordonne, R., Bandlow, W., Dirheimer, G., and Martin, R. P., 1987, A single base change in the extra-arm of yeast mitochondrial tyrosine tRNA affects its conformational stability and impairs aminoacylation, *Mol. Gen. Genet.* **206:**498–504.

Borgford, T. J., Gray, T. E., Brand, N. J., and Fersht, A. R., 1987, Site-directed mutagenesis reveals transition-state stabilization as a general catalytic mechanism for aminoacyl-tRNA synthetases, *Biochemistry* **26:**7246–7250.

Brevet, A., Chen, J., Lévêque, F., Plateau, P., and Blanquet, S., 1989, *In vivo* synthesis of adenylated bis(5′-nucleosidyl) tetraphosphates (Ap_4N) by *Escherichia coli* aminoacyl-tRNA synthetases, *Proc. Natl. Acad. Sci USA* **86:**8275–8279.

Brick, P., and Blow, D. M., 1987, Crystal structure of a deletion mutant of a tyrosyl-tRNA synthetase complexed with tyrosine, *J. Mol. Biol.* **194:**287–297.

Brick, P., Bhat, T. N., and Blow, D. M., 1989, Structure of tyrosyl-tRNA synthetase refined at 2.3 Å resolution. Interaction of the enzyme with the tyrosyl adenylate intermediate, *J. Mol. Biol.* **208:**83–98.

Brown, K. A., Brick, P., and Blow, D. M., 1987, Structure of a mutant of tyrosyl-tRNA synthetase with enhanced catalytic properties, *Nature* **326:**416–418.

Brunel, C., Caillet, J., Lesage, P., Graffe, M., Dondon, J., Moine, H., Romby, P., Ehresmann, C., Ehresmann, B., Grunberg-Manago, M., and Springer, M., 1992, Domains of the *Escherichia coli* threonine-tRNA synthetase translational operator and their relation to threonine tRNA isoacceptors, *J. Mol. Biol.* **227:**621–634.

Brunie, S., Mellot, P., Zelwer, C., Risler, J.-L., Blanquet, S., and Fayat, G., 1987, Structure–activity relationships of methionyl-tRNA synthetase: Graphics modelling and genetic engineering, *J. Mol. Graph.* **5:**16–28.

Brunie, S., Zelwer, C., and Risler, J.-L., 1990, Crystallographic study at 2.5 Å resolution of the interaction of methionyl-tRNA synthetase from *Escherichia coli* with ATP, *J. Mol. Biol.* **216:**411–424.

Burbaum, J. J., and Schimmel, P., 1991, Structural relationships and the classification of aminoacyl-tRNA synthetases, *J. Biol. Chem.* **266:**16965–16968.

Burbaum, J. J., Starzyk, R. M., and Schimmel, P., 1990, Understanding structural relationships in proteins of unsolved three-dimensional structure, *Proteins* **7:**99–111.

Butler, J. S., Springer, M., Dondon, J., and Grunberg-Manago, M., 1986, Posttranslational autoregulation of *Escherichia coli* threonyl-tRNA synthetase expression *in vivo*, *J. Bacteriol.* **165:**198–203.

Carter, P., Bedouelle, H., and Winter, G., 1986, Construction of heterodimer tyrosyl-tRNA synthetase shows $tRNA^{Tyr}$ interacts with both subunits, *Proc. Natl. Acad. Sci. USA* **83:**1189–1192.

Cassio, D., and Waller, J.-P., 1971, Modification of methionyl-tRNA synthetase by proteolytic cleavage and properties of the trypsin-modified enzyme, *Eur. J. Biochem.* **20:**283–300.

Cavarelli, J., and Moras, D., 1993, Recognition of tRNAs by aminoacyl-tRNA synthetases, *FASEB J.* **7:**79–86.

Cavarelli, J., Rees, B., Ruff, M., Thierry, J. C., and Moras, D., 1993, Yeast transfer RNA(Asp) recognition by Its cognate class-II aminoacyl-transfer RNA synthetase, *Nature* **362:**181–184.

Cavarelli, J., Eriani, G., Rees, B., Ruff, M., Boeglin, M., Mitschler, A., Martin, F., Gangloff, J., Thierry, J. C., and Moras, D., 1994, The active site of yeast aspartyl–tRNA synthetase: Structural and functional aspects of the aminoacylation reaction, *EMBO J.* **13:**327–337.

Cech, T. R., 1990, Self splicing of group I introns, *Annu. Rev. Biochem.* **59:**543–568.

Cerini, C., Kerjan, P., Astier, M., Gratecos, D., Mirande, M., and Sémériva, M., 1991, A component of the multisynthetase complex is a multifunctional aminoacyl-tRNA synthetase, *EMBO J.* **10:**4267–4277.

Chambers, R. W., 1971, On the recognition of tRNA by its aminoacyl-tRNA ligase, *Prog. Nucleic Acid Res. Mol. Biol.* **11:**489–525.

Chapeville, F., Lipmann, F., von Ehrenstein, G., Weisblum, B., Ray, W. J. J., and Benzer, S., 1962, On the role of soluble ribonucleic acid in coding for aminoacids, *Proc. Natl. Acad. Sci. USA* **48:**1086–1092.

Chattapadhyay, R., Pelka, H., and Schulman, L. H., 1990, Initiation of protein synthesis with non-methionine amino acids, *Biochemistry* **29:**4263–4268.

Cigan, A. M., Feng, L., and Donahue, T. F., 1988, $tRNA_i^{Met}$ functions in directing the scanning ribosome to the start site of translation, *Science* **242:**93–97.

Cole, F. X., and Schimmel, P., 1970, On the rate law and mechanism of the adenosine triphosphate-pyrophosphate isotope exchange reaction of amino acyl transfer ribonucleic acid synthetases, *Biochemistry* **9:**480–489.

Cole, R. D., Coote, J., and Work, T. S., 1957, Activation of amino-acids by soluble enzymes from pancreas and other tissues, *Nature* **179:**199.

Collins, R. A., and Lambowitz, A. M., 1985, RNA splicing in *Neurospora crassa* mitochondria: Defective splicing of mitochondrial mRNA precursors in the nuclear mutant *cyt-18-1*, *J. Mol. Biol.* **184:**413–428.

Crick, F. H. C., 1958, On protein synthesis, *Symp. Soc. Exp. Biol.* **12:**138–153.

Crick, F. H. C., 1966, Codon-anticodon pairing: The wobble hypothesis, *J. Mol. Biol.* **19:**548–555.

Crothers, D. M., Seno, T., and Söll, D. G., 1972, Is there a discriminator site in transfer RNA, *Proc. Natl. Acad. Sci. USA* **69:**3063–3067.

Cusack, S., Berthet-Colominas, C., Härtlein, M., Nassar, N., and Leberman, R., 1990, A second class of synthetase structure revealed by X-ray analysis of *Escherichia coli* seryl-tRNA synthetase at 2.5 Å, *Nature* **347:**249–255.

Cusack, S., Härtlein, M., and Leberman, R., 1991, Sequence, structural and evolutionary relationships between class 2 aminoacyl-tRNA synthetases, *Nucleic Acids Res.* **19:**3489–3498.

Dardel, F., Panvert, M., and Fayat, G., 1990, Transcription and regulation of expression of the *Escherichia coli* methionyl-tRNA synthetase gene, *Mol. Gen. Genet.* **223:**121–133.

de Prat Gay, G., Duckworth, H. W., and Fersht, A. R., 1993, Modification of the aminoacid specificity of tyrosyl-tRNA synthetase by protein engineering, *FEBS Lett* **318:**167–171.

Dessen, P., Zaccai, G., and Blanquet, S., 1982, Neutron scattering studies of *Escherichia coli* tyrosyl-tRNA synthetase and of its interaction with $tRNA^{Tyr}$, *J. Mol. Biol.* **159:**651–664.

Dreher, T. W., and Hall, T. C., 1988a, Mutational analysis of the sequence and structural requirements in brome mosaic virus RNA for minus strand promoter activity, *J. Mol. Biol.* **201:**31–40.

Dreher, T. W., and Hall, T. C., 1988b, Mutational analysis of the tRNA mimicry of brome mosaic virus RNA. Sequence and structural requirements for aminoacylation and 3′-adenylation, *J. Mol. Biol.* **201:**41–55.

Dreher, T. W., Florentz, C., and Giegé, R., 1988, Valylation of tRNA-like transcripts from cloned cDNA of turnip yellow mosaic virus RNA that the L-shaped region at the 3′ end of the viral RNA is not sufficient for optimal aminoacylation, *Biochimie* **70:**1719–1727.

Dreher, T. W., Tsai, C. H., and Hall, T. C., 1989, Replication *in vivo* of mutant brome mosaic virus RNAs defective in aminoacylation, *J. Mol. Biol.* **206:**425–438.

Dreher, T. W., Tsai, C. H., Florentz, C., and Giegé, R., 1992, Specific valylation of turnip yellow mosaic virus RNA by wheat germ valyl-tRNA synthetase determined by three anticodon loop nucleotides, *Biochemistry* **31:**9183–9189.

Dube, S. K., Rudland, P. S., Clark, B. F. C., and Marcker, K. A., 1969, A structural requirement for codon anticodon interaction on the ribosome, *Cold Spring Harbor Symp. Quant. Biol.* **34:**161–166.

Dumas, P., Moras, D., Florentz, C., Giegé, R., Verlaan, P., van Belkum, A., and Pleij, C. W. A., 1987, 3-D graphics modelling of the tRNA-like 3′ end of turnip yellow mosaic virus RNA: Structural and functional implications, *J. Biomol. Struct. Dyn.* **4:**707–728.

Eggertsson, G., and Söll, D., 1988, Transfer ribonucleic acid-mediated suppression of termination codons in *Escherichia coli*, *Microbiol. Rev.* **52:**354–374.

Eriani, G., Delarue, M., Poch, O., Gangloff, J., and Moras, D., 1990, Partition of tRNA synthetases into two classes based on mutually exclusive sets of sequence motifs, *Nature* **347:**203–206.

Fayat, G., Fromant, M., and Blanquet, S., 1977, Couplings between the sites for methionine and adenosine 5′-triphosphate in the amino acid activation reaction catalyzed by trypsin-modified methionyl-transfer RNA synthetase from *Escherichia coli*, *Biochemistry* **16:**2570–2579.

Fayat, G., Hountondji, C., and Blanquet, S., 1979, Methionyl-tRNA synthetase from *Escherichia coli*. Inactivation and labeling by periodate-treated initiator tRNA, *Eur. J. Biochem.* **9:**87–92.

Fersht, A. R., 1977, Editing mechanisms in protein synthesis: Rejection of valine by the isoleucyl-tRNA synthetase, *Biochemistry* **16:**1025–1030.

Fersht, A. R., 1987, Dissection of the structure and activity of the tyrosyl-tRNA synthetase by site-directed mutagenesis, *Biochemistry* **26:**8031–8037.

Fersht, A. R., 1988, Relationships between apparent binding energies measured in site-directed mutagenesis experiments and energetics of binding and catalysis, *Biochemistry* **27:**1577–1580.

Fersht, A. R., and Dingwall, C., 1979a, Cysteinyl-tRNA synthetase from *Escherichia coli* does not need an editing mechanism to reject serine and alanine. High binding energy of small groups in specific molecular interactions, *Biochemistry* **18:**1245–1249.

Fersht, A. R., and Dingwall, C., 1979b, An editing mechanism for the methionyl-tRNA synthetase in the selection of aminoacids in protein synthesis, *Biochemistry* **18:**1250–1256.

Fersht, A. R., and Dingwall, C., 1979c, Establishing the misaminoacylation/deacylation of the tRNA pathway for the editing mechanism of prokaryotic and eukaryotic valyl-tRNA synthetases, *Biochemistry* **18:**1238–1245.

Fersht, A. R., Mulvey, R. S., and Koch, G. L. E., 1975, Ligand binding and enzymic catalysis coupled through subunits in tyrosyl-tRNA synthetase, *Biochemistry* **14:**13–18.

Fersht, A. R., Shi, J.-P., Knill-Jones, J., Lowe, D. M., A. J., W., Blow, D. M., Brick, P., Carter, P., Waye, M. M. Y., and Winter, G., 1985, Hydrogen bonding and biological specificity analyzed by protein engineering, *Nature* **314:**235–238.

Fersht, A. R., Leatherbarrow, R. J., and Wells, T. N. C., 1986, Quantitative analysis of structure–activity relationships in engineered proteins by linear free-energy relationships, *Nature* **322:**284–286.

Fersht, A. R., Knill-Jones, J. W., Bedouelle, H., and Winter, G., 1988, Reconstruction by site-directed mutagenesis of the transition state for the activation of tyrosine by the tyrosyl-tRNA synthetase: A mobile loop envelopes the transition state in an induced-fit mechanism, *Biochemistry* **27:**1581–1587.

Florentz, C., and Giegé, R., 1986, Contact areas of the turnip yellow mosaic virus tRNA-like structure interacting with yeast valyl-tRNA synthetase, *J. Mol. Biol.* **191:**117–130.

Florentz, C., Briand, J. P., Romby, P., Hirth, L., Ebel, J. P., and Giegé, R., 1982, The tRNA-like structure of turnip yellow mosaic virus RNA: Structural organization of the last 159 nucleotides from the 3′ OH terminus, *EMBO J.* **1:**269–276.

Florentz, C., Dreher, T., Rudinger, J., and Giegé, R., 1991, Specific valylation identity of turnip yellow mosaic virus RNA by yeast valyl-tRNA synthetase is directed by the anticodon in a kinetic rather than affinity based discrimination, *Eur. J. Biochem.* **195:**229–234.

Fourmy, D., Mechulam, Y., Brunie, S., Blanquet, S., and Fayat, G., 1991, Identification of residues involved in the binding of methionine by *Escherichia coli* methionyl-tRNA synthetase, *FEBS Lett* **292:**259–263.

Fourmy, D., Dardel, F., and Blanquet, S., 1993a, Methionyl-tRNA synthetase zinc binding site. 3-D structure and homology with rubredoxin and *gag* retroviral proteins, *J. Mol. Biol.* **232:**1078–1089.

Fourmy, D., Meinnel, T., Mechulam, Y., and Blanquet, S., 1993b, Mapping of the zinc binding domain of *Escherichia coli* methionyl-tRNA synthetase, *J. Mol. Biol.* **232:**1068–1077.

Francklyn, C., and Schimmel, P., 1989, Aminoacylation of RNA minihelices with alanine, *Nature* **337:**478–481.

Francklyn, C., and Schimmel, P., 1990, Enzymatic aminoacylation of an eight-base-pair microhelix with histidine, *Proc. Natl. Acad. Sci. USA* **87:**8655–8659.

Francklyn, C., Shi, J. P., and Schimmel, P., 1992, Overlapping nucleotide determinants for specific aminoacylation of RNA microhelices, *Science* **255:**1121–1125.

Freemont, P. S., Lane, A. N., and Sanderson, M. R., 1991, Structural aspects of protein–DNA recognition, *Biochem J.* **278:**1–23.

Freist, W., 1989, Mechanisms of aminoacyl–tRNA synthetases: Critical consideration of recent results, *Biochemistry* **28:**6787–6795.

Freist, W., Sternbach, H., and Cramer, F., 1988, Isoleucyl–tRNA synthetase from baker's yeast and from *Escherichia coli* MRE 600. Discrimination of 20 amino acids in aminoacylation of $tRNA^{Ile}$-C-C-A, *Eur. J. Biochem.* **173:**27–34.

Frolova, L., Fleckner, J., Justesen, J., Timms, K. M., Tate, W. P., Kisselev, L. L., and Haenni, A. L., 1993, Are the tryptophanyl–tRNA synthetase and the peptide-chain-release factor from higher eukaryotes one and the same protein? *Eur. J. Biochem.* **212:**457–466.

Frugier, M., Florentz, C., and Giegé, R., 1992, Anticodon-independent aminoacylation of an RNA minihelix with valine, *Proc. Natl. Acad. Sci. USA* **89:**3990–3994.

Ghosh, G., Pelka, H., and Schulman, L. H., 1990, Identification of the tRNA anticodon recognition site of *Escherichia coli* methionyl–tRNA synthetase, *Biochemistry* **29:**2220–2225.

Ghosh, G., Kim, H. Y., Demaret, J.-P., Brunie, S., and Schulman, L. H., 1991a, Arginine 395 is required for efficient *in vivo* and *in vitro* aminoacylation of tRNAs by *Escherichia coli* methionyl–tRNA synthetase, *Biochemistry* **30:**11767–11774.

Ghosh, G., Pelka, H., Schulman, L. H., and Brunie, S., 1991b, Activation of methionine by *Escherichia coli* methionyl–tRNA synthetase, *Biochemistry* **30:**9569–9575.

Giegé, R., Kern, D., Ebel, J. P., Grosjean, H., De Henau, S., and Chantrenne, H., 1974, Incorrect aminoacylations involving tRNAs or valyl–tRNA synthetase from *Bacillus stearothermophilus*, *Eur. J. Biochem.* **45:**351–362.

Giegé, R., Puglisi, J. D., and Florentz, C., 1993, tRNA structure and aminoacylation efficiency, *Prog. Nucleic Acid Res. Mol. Biol.* **45:**129–206.

Goodman, H. M., Abelson, J., Landy, A., Brenner, S., and Smith, J. D., 1968, Amber suppression: A nucleotide change in the anticodon of a tyrosine transfer RNA, *Nature* **217**:1019–1024.

Graffe, M., Dondon, J., Caillet, J., Romby, P., Ehresmann, C., Ehresmann, B., and Springer, M., 1992, The specificity of translational control switched with transfer RNA identity rules, *Science* **255**:994–996.

Guillemaut, P., and Weil, J. H., 1975, Aminoacylation of *Phaseolus vulgaris* cytoplasmic, chloroplactic and mitochondrial tRNAs^{Met} and of *Escherichia coli* tRNAs^{Met} by homologous and heterologous enzymes, *Biochim. Biophys. Acta* **407**:240–248.

Guillon, J. M., Meinnel, T., Mechulam, Y., Lazennec, C., Blanquet, S., and Fayat, G., 1992, Nucleotides of tRNA governing the specificity of *Escherichia coli* methionyl-tRNA$^{Met}{}_f$ formyltransferase, *J. Mol. Biol.* **224**:359–367.

Guillon, J. M., Mechulam, Y., Blanquet, S., and Fayat, G., 1993, Importance of formilability and anticodon stem sequence to give tRNA^{Met} an initiator identity in *Escherichia coli*, *J. Bacteriol.* **175**:4507–4514.

Guo, Q. B., and Lambowitz, A. M., 1992, A tyrosyl–transfer RNA synthetase binds specifically to the group-I intron catalytic core, *Genes Dev.* **6**:1357–1372.

Harrison, S. C., and Aggarwal, A. K., 1990, DNA recognition by proteins with the helix-turn-helix motif, *Annu. Rev. Biochem.* **59**:933–969.

Hasegawa, T., Himeno, H., Ishikura, H., and Shimizu, M., 1989, Discriminator base of tRNA^{Asp} is involved in amino acid acceptor activity, *Biochem. Biophys. Res. Commun.* **163**:1534–1538.

Hasegawa, T., Miyano, M., Himeno, H., Sano, Y., Kimura, K., and Shimizu, M., 1992, Identity elements of *E. coli* threonine tRNA, *Biochem. Biophys. Res. Commun.* **184**:478–484.

Hasegawa, T., Sano, Y., Asahara, H., Himeno, H., Tamura, K., Nameki, N., Kimura, K., and Shimizu, M., 1993, Recognition sites of *E. coli* proline tRNA for prolyl–tRNA synthetase, 15th International tRNA Workshop, Cap d'Agde, France, May 30–June 4.

Hayase, Y., Jahn, M., Rogers, M. J., Sylvers, L. A., Koizumi, M., Inoue, H., Ohtsuka, E., and Söll, D., 1992, Recognition of bases in *Escherichia coli* tRNA^{Gln} by glutaminyl–tRNA synthetase: a complete identity set, *EMBO J.* **11**:4159–4165.

Hecht, L. I., Stephenson, M. L., and Zamecnik, P. C., 1959, Binding of amino acids to the end group of a soluble ribonucleic acid, *Proc. Natl. Acad. Sci. USA* **45**:505–518.

Herbert, C. J., Labouesse, M., Dujardin, G., and Slonimski, P. P., 1988, The NAM2 proteins from *S. cerevisiae* and *S. douglasii* are mitochondrial leucyl–tRNA synthetases, and are involved in mRNA splicing, *EMBO J.* **7**:473–483.

Himeno, H., Hasegawa, T., Ueda, T., Watanabe, K., Miura, K.-i., and Shimizu, M., 1989, Role of the extra G-C pair at the end of the acceptor stem of tRNA^{His} in aminoacylation, *Nucleic Acids Res.* **17**:7855–7863.

Himeno, H., Hasegawa, T., Ueda, T., Watanabe, K., and Shimizu, M., 1990, Conversion of aminoacylation specificity from tRNA^{Tyr} to tRNA^{Ser} *in vitro*, *Nucleic Acid Res.* **18**:6815–6819.

Himeno, H., Hasegawa, T., Asahara, H., Tamura, K., and Shimizu, M., 1991, Identity determinants of *E. coli* tryptophan tRNA, *Nucleic Acids Res.* **19**:6379–6382.

Ho, C. K., and Fersht, A. R., 1986, Internal thermodynamics of position 51 mutants and natural variants of tyrosyl-tRNA synthetase, *Biochemistry* **25**:1891–1897.

Hoagland, M. B., 1955, An enzymic mechanism for amino acid activation in animal tissues, *Biochim. Byophys. Acta* **16**:288–289.

Hoagland, M. B., Stephenson, M. L., Scott, J. F., Hecht, L. I., and Zamecnik, P. C., 1958, A soluble ribonucleic acid intermediate in protein synthesis, *J. Biol. Chem.* **231**:241–257.

Holler, E., and Calvin, M., 1972, Isoleucyl transfer ribonucleic acid synthetase of *Escherichia coli* B. A rapid kinetic investigation of the L-isoleucine-activating reaction, *Biochemistry* **11**:3741–3752.

Holler, E., Hammer-Raber, B., Hanke, T., and Bartmann, P., 1975, The catalytic mechanism of aminoacid:tRNA ligases. Synergism and formation of the ternary enzyme–amino acid–ATP complex, *Biochemistry* **14**:2496–2503.

Holley, R. W., Apgar, J., Everett, G. A., Madison, J. T., Marquisee, M., Merrill, S. H., Penswick, J. R., and Zamir, A., 1965, Structure of a ribonucleic acid, *Science* **147**:1462–1465.

Hooper, M. L., Russell, R. L., and Smith, J. D., 1972, Mischarging in mutant tyrosine transfer RNAs, *FEBS Lett* **22**:149–155.

Hou, Y.-M., and Schimmel, P., 1988, A simple structural feature is a major determinant of the identity of a transfer RNA, *Nature* **333**:140–145.

Hou, Y.-M., and Schimmel, P., 1989, Modeling with *in vitro* kinetic parameters for the elaboration of transfer RNA identity *in vivo, Biochemistry* **28**:4942–4947.

Hou, Y. M., Shiba, K., Mottes, C., and Schimmel, P., 1991, Sequence determination and modeling of structural motifs for the smallest monomeric aminoacyl–tRNA synthetase, *Proc. Natl. Acad. Sci. USA* **88**:976–980.

Hountondji, C., Fayat, G., and Blanquet, S., 1979, Complete inactivation and labeling of methionyl–tRNA synthetase by periodate-treated initiator tRNA in the presence of sodium cyanohydridoborate, *Eur. J. Biochem.* **102**:247–250.

Hountondji, C., Blanquet, S., and Lederer, F., 1985, Methionyl–tRNA synthetase from *Escherichia coli.* Primary structure at the binding site for the 3′-end of $tRNA_f^{Met}$, *Biochemistry* **24**:1175–1180.

Hountondji, C., Dessen, P., and Blanquet, S., 1986a, Sequence similarities among the family of aminoacyl-tRNA synthetases, *Biochimie* **68**:1071–1078.

Hountondji, C., Lederer, F., Dessen, P., and Blanquet, S., 1986b, *Escherichia coli* tyrosyl– and methionyl–tRNA synthetases display sequence similarity at the binding site for the 3′-end of tRNA, *Biochemistry* **25**:16–21.

Hountondji, C., Schmitter, J. M., Fukui, T., Tagaya, M., and Blanquet, S., 1990, Affinity labeling of aminoacyl–tRNA synthetases with adenosine triphosphopyridoxal: Probing Lys-Met-Ser-Lys-Ser signature sequence as the ATP-binding site in *Escherichia coli* methionyl– and valyl–tRNA synthetases, *Biochemistry* **29**:11266–11273.

Hutchison, C. A., Phillips, S., Edgell, M. H., Gillam, S., Jahnke, P., and Smith, M., 1978, Mutagenesis at a specific position in a DNA sequence, *J. Biol. Chem.* **253**:6551.

Hyafil, F., Jacques, Y., Fayat, G., Fromant, M., Dessen, P., and Blanquet, S., 1976, Methionyl-tRNA synthetase from *Escherichia coli:* Active stoichiometry and stopped-flow analysis of methionyl adenylate formation, *Biochemistry* **15**:3678–3685.

Igloi, G. L., von der Haar, F., and Cramer, F., 1977, Hydrolytic action of aminoacyl–tRNA synthetases from baker's yeast. Chemical proofreading of Thr–$tRNA^{Val}$ by valyl–tRNA synthetase studied with modified $tRNA^{Val}$ and amino acid analogues, *Biochemistry* **16**:1696–1702.

Imura, N., Schwam, H., and Chambers, R. W., 1969, Transfer RNA III. Reconstitution of alanine acceptor activity from fragments produced by specific cleavage of $tRNA^{Ala}{}_{II}$ at its anticodon, *Proc. Natl. Acad. Sci. USA* **62**:1203–1209.

Inokuchi, H., Hoben, P., Yamao, F., Ozeki, H., and Söll, D., 1984, Transfer RNA mischarging mediated by a mutant *Escherichia coli* glutaminyl–tRNA synthetase, *Proc. Natl. Acad. Sci. USA* **81**:5076–5080.

Jacobson, K. B., 1971, Reaction of aminoacyl–tRNA synthetases with heterologous tRNAs, *Prog. Nucleic Acid Res. Mol. Biol.* **11**:461–488.

Jahn, M., Rogers, M. J., and Söll, D., 1991, Anticodon and acceptor stem nucleotides in $tRNA^{Gln}$ are major recognition elements for *E. coli* glutaminyl–tRNA synthetase, *Nature* **352**:258–260.

Jakubowski, H., and Fersht, A. R., 1981, Alternative pathways for editing noncognate amino acids by aminoacyl–tRNA synthetases, *Nucleic Acids Res.* **9**:3105–3117.

Jakubowski, H., and Goldman, E., 1984, Quantities of individual aminoacyl–tRNA families and their turnover in *Escherichia coli, J. Bacteriol.* **158**:769–776.

Jakubowski, H., and Goldman, E., 1992, Editing of errors in selection of amino acids for protein synthesis, *Microbiol. Rev.* **56:**412–429.

Jasin, M., Regan, L., and Schimmel, P., 1983, Modular arrangement of functional domains along the sequence of an aminoacyl tRNA synthetase, *Nature* **306:**441–447.

Jones, D. H., McMillan, A. J., and Fersht, A. R., 1985, Reversible dissociation of dimeric tyrosyl–tRNA synthetase by mutagenesis at the subunit interface, *Biochemistry* **24:**5852–5857.

Jordana, X., Chatton, B., Paz-Weisshaar, M., Buhler, J. M., Cramer, F., Ebel, J. P., and Fasiolo, F., 1987, Structure of the yeast valyl–tRNA synthetase gene (VAS1) and the homology of its translated amino acid sequence with *Escherichia coli* isoleucyl–tRNA synthetase, *J. Biol. Chem.* **262:**7189–7194.

Joshi, S., Chapeville, F., and Haenni, A. L., 1982, Length requirements for tRNA-specific enzymes and cleavage specificity at the 3′ end of turnip yellow mosaic virus RNA, *Nucleic Acids Res.* **10:**1947–1962.

Kalogerakos, T., Dessen, P., Fayat, G., and Blanquet, S., 1980, Proteolytic cleavage of methionyl transfer ribonucleic acid synthetase from *Bacillus stearothermophilus*: Effects on activity and structure, *Biochemistry* **19:**3712–3723.

Kämper, U., Kück, U., Cherniack, A. D., and Lambowitz, A. M., 1992, The mitochondrial tyrosyl–tRNA synthetase of *Podospora anserina* is a bifunctional enzyme active in protein synthesis and RNA splicing, *Mol. Cell. Biol.* **12:**499–511.

Kern, D., Giegé, R., and Ebel, J. P., 1972, Incorrect aminoacylations catalysed by the phenylalanyl- and valyl-tRNA synthetase from yeast, *Eur. J. Biochem.* **31:**148–155.

Kisselev, L. L., 1985, The role of the anticodon in recognition of tRNA by aminoacyl–tRNA synthetases, *Prog. Nucleic Acid Res. Mol. Biol.* **32:**237–266.

Kittle, J. D. J., Mohr, G., Gianelos, J. A., Wang, H., and Lambowitz, A. M., 1991, The *Neurospora* mitochondrial tyrosyl–tRNA synthetase is sufficient for group I intron splicing *in vitro* and uses the carboxy-terminal tRNA-binding domain along with other regions, *Genes Dev.* **5:**1009–1021.

Kleina, G. L., Masson, J.-M., Normanly, J., Abelson, J., and Miller, J. H., 1990, Construction of *Escherichia coli* amber suppressor tRNA genes. II. Synthesis of additional tRNA genes and improvement of suppressor efficiency, *J. Mol. Biol.* **213:**705–717.

Kohda, D., Yokoyama, S., and Miyazawa, T., 1987, Functions of isolated domains of methionyl–tRNA synthetase from an extreme thermophile, *Thermus thermophilus* HB8, *J. Biol. Chem.* **262:**558–563.

Kraulis, P., 1991, A program to produce both detailed and schematic plots of protein structures, *J. Appl. Cryst.* **24:**946–950.

Krauss, G., Riesner, D., and Maass, G., 1976, Mechanism of discrimination between cognate and non-cognate tRNAs by phenylalanyl–tRNA synthetase from yeast, *Eur. J. Biochem.* **68:**81–93.

Kumazawa, Y., Himeno, H., Miura, K., and Watanabe, K., 1991, Unilateral aminoacylation specificity between bovine mitochondria and eubacteria, *J. Biochem.* **109:**421–427.

Lambowitz, A. M., and Perlman, P. S., 1990, Involvement of aminoacyl–tRNA synthetases and other proteins in group I and group II intron splicing, *Trends Biochem. Sci.* **15:**440–444.

Lapointe, J., Duplain, L., and Proulx, M., 1986, A single glutamyl–tRNA synthetase aminoacylates $tRNA^{Glu}$ and $tRNA^{Gln}$ in *Bacillus subtilis* and efficiently misacylates *Escherichia coli* $tRNA_1^{Gln}$ *in vitro*, *J. Bacteriol.* **165:**88–93.

Leatherbarrow, R. J., and Fersht, A. R., 1987, Investigation of transition-state stabilization by residues histidine-45 and threonine-40 in the tyrosyl–tRNA synthetase, *Biochemistry* **26:**8524–8528.

Leatherbarrow, R. J., Fersht, A. R., and Winter, G., 1985, Transition-state stabilization in the mechanism of tyrosyl–tRNA synthetase revealed by protein engineering, *Proc. Natl. Acad. Sci. USA* **82:**7840–7844.

Lee, C. C., Craigen, W. J., Muzny, D. M., Harlow, E., and Caskey, C. T., 1990, Cloning and expression of a mammalian peptide chain release factor with sequence similarity to trytophanyl–tRNA synthetases, *Proc. Natl. Acad. Sci. USA* **87**:3508–3512.

Lee, C. P., and RajBhandary, U. L., 1991, Mutants of *Escherichia coli* initiator tRNA that suppress amber codons in *Saccharomyces cerevisiae* and are aminoacylated with tyrosine by yeast extracts, *Proc. Natl. Acad. Sci. USA* **88**:11378–11382.

Lee, C. P., Dyson, M. R., Mandal, N., Varshney, U., Bahramian, B., and Rajbhandary, U. L., 1992, Striking effects of coupling mutations in the acceptor stem on recognition of transfer RNAs by *Escherichia coli* Met–transfer RNA synthetase and met–transfer RNA transformylase, *Proc. Natl. Acad. Sci. USA* **89**:9262–9266.

Leinfelder, W., Zehelein, E., Mandrand-Berthelot, M. A., and Böck, A., 1988, Gene for a novel tRNA species that accepts L-serine and cotranslationally inserts selenocysteine, *Nature* **331**:723–725.

Leon, O., and Schulman, L. H., 1987, tRNA recognition site of *Escherichia coli* methionyl–tRNA synthetase, *Biochemistry* **26**:5416–5422.

Lestienne, P., Plumbridge, J. A., Grunberg-Manago, M., and Blanquet, S., 1984, Autogenous repression of *Escherichia coli* threonyl–tRNA synthetase expression *in vitro, J. Biol. Chem.* **259**:5232–5237.

Lévêque, F., Gazeau, M., Fromant, M., Blanquet, S., and Plateau, P., 1991, Control of *Escherichia coli* lysyl–tRNA synthetase expression by anaerobiosis, *J. Bacteriol.* **173**:7903–7910.

Li, G., Herbert, C. J., Labouesse, M., and Slonimski, P. P., 1992, *In vitro* mutagenesis of the leucyl–tRNA synthetase of *S. cerevisiae* reveals residues critical for its *in vivo* activities, *Curr. Genet.* **22**:69–74.

Li, S., Pelka, H., and Schulman, L. H., 1993, The anticodon and discriminator base are important for aminoacylation of *Escherichia coli:* $tRNA^{ASN}$, *J. Biol. Chem.* **268**:18335–18339.

Lotfield, R. B., 1972, The mechanism of aminoacylation of transfer RNA, *Prog. Nucleic Acids Res. Mol. Biol.* **12**:87–128.

Lowe, D. M., Fersht, A. R., and Wilkinson, A. J., 1985, Probing histidine-substrate interactions in tyrosyl-tRNA synthetase using asparagine and glutamine replacements, *Biochemistry* **24**:5106–5109.

Mans, R. M. W., Pleij, C. W. A., and Bosh, L., 1991, tRNA-like structures: Structure, function and evolutionary significance, *Eur. J. Biochem.* **201**:303–324.

Martinis, S. A., and Schimmel, P., 1992, Enzymatic aminoacylation of sequence-specific RNA minihelices and hybrid duplexes with methionine, *Proc. Natl. Acad. Sci. USA* **89**:65–69.

Martinis, S. A., and Schimmel, P., 1993, Microhelix aminoacylation by a class-I transfer RNA synthetase—Non-conserved base pairs required for specificity, *J. Biol. Chem.* **268**:6069–6072.

McClain, W. H., and Foss, K., 1988a, Changing the acceptor identity of a transfer RNA by altering nucleotides in a "variable pocket," *Science* **241**:1804–1807.

McClain, W. H., and Foss, K., 1988b, Changing the identity of a tRNA by introducing a G-U wobble pair near the 3′ acceptor end, *Science* **240**:793–796.

McClain, W. H., and Nicholas Jr, H. B., 1987, Differences between transfer RNA molecules, *J. Mol. Biol.* **194**:635–642.

McClain, W. H., Foss, K., Jenkins, R. A., and Schneider, J., 1990, Nucleotides that determine *Escherichia coli* $tRNA^{Arg}$ and $tRNA^{Lys}$ acceptor identities revealed by analyses of mutant opal and amber suppressor tRNAs, *Proc. Natl. Acad. Sci. USA* **87**:9260–9264.

McClain, W. H., Foss, K., Jenkins, R. A., and Schneider, J., 1991a, Four sites in the acceptor helix and one site in the variable pocket of $tRNA^{Ala}$ determine the molecule's acceptor identity, *Proc. Natl. Acad. Sci. USA* **88**:9272–9276.

McClain, W. H., Foss, K., Jenkins, R. A., and Schneider, J., 1991b, Rapid determination of nucleotides that define $tRNA^{Gly}$ acceptor identity, *Proc. Natl. Acad. Sci. USA* **88**:6147–6151.

Mechulam, Y., Dardel, F., LeCorre, D., Blanquet, S., and Fayat, G., 1991, Lysine 335, part of the

KMSKS signature sequence, plays a crucial role in the amino acid activation catalysed by the methionyl-tRNA synthetase from *Escherichia coli*, *J. Mol. Biol.* **217**:465–475.

Meinnel, T., Mechulam, Y., and Fayat, G., 1988, Fast purification of a functional elongator $tRNA^{Met}$ expressed from a synthetic gene *in vivo*, *Nucleic Acids Res.* **16**:8095–8096.

Meinnel, T., Mechulam, Y., Blanquet, S., and Fayat, G., 1991a, Binding of the anticodon domain of $tRNA_f^{Met}$ to *Escherichia coli* methionyl-tRNA synthetase, *J. Mol. Biol.* **220**:205–208.

Meinnel, T., Mechulam, Y., LeCorre, D., Panvert, M., Blanquet, S., and Fayat, G., 1991b, Selection of suppressor methionyl-tRNA synthetases: Mapping the tRNA anticodon binding site, *Proc. Natl. Acad. Sci. USA* **88**:291–295.

Meinnel, T., Mechulam, Y., Fayat, G., and Blanquet, S., 1992, Involvement of the size and sequence of the anticodon loop in tRNA recognition by mammalian and *E. coli* methionyl–tRNA synthetases, *Nucleic Acids Res.* **20**:4741–4746.

Meinnel, T., Mechulam, Y., and Blanquet, S., 1995, Aminoacyl–tRNA synthetases: General features, in *Transfer RNAs* (D. Söll and U. RajBhandary, eds.), pp. 251–292, American Society for Microbiology, Washington, D.C.

Meinnel, T., Mechulam, Y., Lazennec, C., Blanquet, S., and Fayat, G., 1993, Critical role of the acceptor stem of $tRNAs^{Met}$ in their aminoacylation by *Escherichia coli* methionyl–tRNAsynthetase, *J. Mol. Biol.* **229**:26–36.

Mellot, P., Mechulam, Y., LeCorre, D., Blanquet, S., and Fayat, G., 1989, Identification of an amino acid region supporting specific methionyl–tRNA synthetase:tRNA recognition, *J. Mol. Biol.* **208**:429–443.

Merle, M., Trézéguet, V., Graves, P. V., Andrews, D., Muench, K. H., and Labouesse, B., 1986, Tryptophanyl adenylate formation by tryptophanyl–tRNA synthetase from *Escherichia coli*, *Biochemistry* **25**:1115–1123.

Michel, F., and Westhof, E., 1990, Modelling of the three-dimensional architecture of Group I catalytic introns based on comparative sequence analysis, *J. Mol. Biol.* **216**:585–610.

Miller, W. T., and Schimmel, P., 1992, A metal-binding motif implicated in RNA recognition by an aminoacyl–tRNA synthetase and by a retroviral gene product, *Mol. Microbiol.* **6**:1259–1262.

Miller, W. T., Hill, K. A. W., and Schimmel, P., 1991a, Evidence for a "cysteine–histidine box" metal-binding site in an *Escherichia coli* aminoacyl–tRNA synthetase, *Biochemistry* **30**:6970–6976.

Miller, W. T., Hou, Y.-M., and Schimmel, P., 1991b, Mutant aminoacyl–tRNA synthetase that compensates for a mutation in the major identity determinant of its tRNA, *Biochemistry* **30**:2635–2641.

Mohr, G., and Lambowitz, A. M., 1991, Integration of a group I intron into a ribosomal RNA sequence promoted by a tyrosyl-tRNA synthetase, *Nature* **354**:164–167.

Mohr, G., Zhang, A., Gianelos, J. A., Belfort, M., and Lambowitz, A. M., 1992, The *Neurospora* CYT-18 protein suppresses defects in the phage T4 *td* intron by stabilizing the catalytically active structure of the intron core, *Cell* **69**:483–494.

Moine, H., Romby, P., Springer, M., Grunberg-Manago, M., Ebel, J. P., Ehresmann, C., and Ehresmann, B., 1988, Messenger RNA structure and gene regulation at the translational level in *Escherichia coli:* The case of threonine:$tRNA^{Thr}$ ligase, *Proc. Natl. Acad. Sci. USA* **85**:7892–7896.

Moine, H., Romby, P., Springer, M., Grunberg-Manago, M., Ebel, J. P., Ehresmann, C., and Ehresmann, B., 1990, *Escherichia coli* threonine–tRNA synthetase and $tRNA^{Thr}$ modulate the binding of the ribosome to the translational initiation site of the *ThrS* mRNA, *J. Mol. Biol.* **216**:299–310.

Moor, N., Nazarenko, I., Ankilova, V., Khodyreva, S., and Lavrik, O., 1992, Determination of $tRNA^{Phe}$ recognition nucleotides for phenylalanyl–tRNA synthetase from *Thermus thermophilus*, *Biochimie* **74**:353–356.

Muramatsu, T., Nishikawa, K., Nemoto, F., Kuchino, Y., Nishimura, S., Miyazawa, T., and

Yokoyama, S., 1988, Codon and amino-acid specificities of a transfer RNA are both converted by a single post-transcriptional modification, *Nature* **336:**179–181.

Muramatsu, T., Miyazawa, K., and Yokoyama, S., 1992, Recognition of the nucleoside in the first position of the anticodon of isoleucine tRNA by isoleucyl–tRNA synthetase from *Escherichia coli*, *Nucleoside Nucleotide* **11:**719–730.

Murgola, E. J., 1985, tRNA, suppression, and the code, *Annu. Rev. Genet.* **19:**57–80.

Musier-Forsyth, K., and Schimmel, P., 1992, Functional contacts of a transfer RNA synthetase with 2′-hydroxyl groups in the RNA minor groove, *Nature* **357:**513–515.

Musier-Forsyth, K., and Schimmel, P., 1993, Aminoacylation of RNA oligonucleotides–Minimalist structures and origin of specificity, *FASEB J.* **7:**282–289.

Musier-Forsyth, K., Scaringe, S., Usman, N., and Schimmel, P., 1991a, Enzymatic aminoacylation of single-stranded RNA with an RNA cofactor, *Proc. Natl. Acad. Sci. USA* **88:**209–213.

Musier-Forsyth, K., Usman, N., Scaringe, S., Doudna, J., Green, R., and Schimmel, P., 1991b, Specificity for aminoacylation of an RNA helix: An unpaired, exocyclic amino group in the minor groove, *Science* **253:**784–786.

Nagel, G. M., and Doolittle, R. F., 1991, Evolution and relatedness in two aminoacyl–tRNA synthetase families, *Proc. Natl. Acad. Sci. USA* **88:**8121–8125.

Nakamura, Y., Kawakami, K., and Mikuni, O., 1990, Alternative translation and functional diversity of release factor 2 and lysyl–tRNA synthetase, in *Posttranscriptional Control of Gene Expression* (J. E. G. McCarthy and M. F. Tuite, eds.), pp. 455–464, Springer-Verlag, Berlin, Heidelberg.

Nameki, N., Tamura, K., Himeno, H., Asahara, H., Hasegawa, T., and Shimizu, M., 1992, *Escherichia coli* $tRNA^{Asp}$ recognition mechanism differing from that of the yeast system, *Biochem. Biophys. Res. Commun.* **189:**856–862.

Natsoulis, G., Hilger, F., and Fink, G. R., 1986, The *HTS1* gene encodes both the cytoplasmic and mitochondrial histidine tRNA synthetases of *S. cerevisiae*, *Cell* **46:**235–243.

Nazarenko, I. A., Peterson, E. T., Zakharova, O. D., Lavrik, O. I., and Uhlenbeck, O. C., 1992, Recognition nucleotides for human phenylalanyl–tRNA synthetase, *Nucleic Acids Res.* **20:**475–478.

Normanly, J., and Abelson, J., 1989, tRNA identity, *Annu. Rev. Biochem.* **58:**1029–1049.

Normanly, J., Masson, J.-M., Kleina, L. G., Abelson, J., and Miller, J. H., 1986a, Construction of two *Escherichia coli* amber suppressor genes: $tRNA^{Phe}CUA$ and $tRNA^{Cys}CUA$, *Proc. Natl. Acad. Sci. USA* **83:**6548–6552.

Normanly, J., Ogden, R. C., Horvath, S. J., and Abelson, J., 1986b, Changing the identity of a transfer RNA, *Nature* **321:**213–219.

Normanly, J., Kleina, G. L., Masson, J.-M., Abelson, J., and Miller, J. H., 1990, Construction of *Escherichia coli* amber suppressor tRNA genes. III. Determination of tRNA specificity, *J. Mol. Biol.* **213:**719–726.

Normanly, J., Ollick, T., and Abelson, J., 1992, Eight base changes are sufficient to convert a leucine-inserting tRNA into a serine-inserting tRNA, *Proc. Natl. Acad. Sci. USA* **89:**5680–5684.

Novelli, G. D., 1967, Aminoacid activation for protein synthesis, *Annu. Rev. Biochem.* **36:**449–484.

Nureki, O., Muramatsu, T., Suzuki, K., Kohda, D., Matsuzawa, H., Ohta, T., Miyazawa, T., and Yokoyama, S., 1991, Methionyl–tRNA synthetase gene from an extreme thermophile, *Thermus thermophilus* HB8. Molecular cloning, primary-structure analysis, expression in *Escherichia coli*, and site-directed mutagenesis, *J. Biol. Chem.* **266:**3268–3277.

Okimoto, R., and Wolstenholme, D. R., 1990, A set of tRNAs that lack either the TΨC arm or the dihydrouridine arm: Towards a minimal tRNA adaptor, *EMBO J.* **9:**3405–34011.

Pak, M., Pallanck, L., and Schulman, L. H., 1992, Conversion of a methionine initiator tRNA into a tryptophan-inserting elongator tRNA *in vivo*, *Biochemistry* **31:**3303–3309.

Pallanck, L., and Schulman, L. H., 1991, Anticodon-dependent aminoacylation of a noncognate tRNA with isoleucine, valine, and phenylalanine *in vivo, Proc. Natl. Acad. Sci. USA* **88**:3872–3876.

Pallanck, L., Li, S., and Schulman, L. H., 1992, The anticodon and discriminator base are major determinants of cysteine tRNA identity *in vivo, J. Biol. Chem.* **267**:7221–7223.

Pande, S., Jahn, D., and Söll, D., 1991, Histidine tRNA guanylyltransferase from *Saccharomyces cerevisiae*. I. Purification and physical properties, *J. Biol. Chem.* **266**:22826–22831.

Pauling, L., 1957, The probability of errors in the process of synthesis of protein molecules, in *Festschrift für Pr. Dr. Arthur Stoll*, pp. 597–602, Birkhauser Verlag, Basel.

Perona, J. J., Swanson, R., Steitz, T. A., and Söll, D., 1988, Overproduction and purification of *Escherichia coli* $tRNA^{Gln_2}$ and its use in crystallization of the glutaminyl-tRNA synthetase–$tRNA^{Gln}$ complex, *J. Mol. Biol.* **202**:121–126.

Perona, J. J., Swanson, R. N., Rould, M. A., Steitz, T. A., and Söll, D., 1989, Structural basis for misaminoacylation by mutant *E. coli* glutaminyl–tRNA synthetase enzymes, *Science* **246**:1152–1154.

Perona, J. J., Rould, M. A., Steitz, T. A., Risler, J.-L., Zelwer, C., and Brunie, S., 1991, Structural similarities in glutaminyl- and methionyl–tRNA synthetases suggest a common overall orientation of tRNA binding, *Proc. Natl. Acad. Sci. USA* **88**:2903–2907.

Perret, V., Florentz, C., Dreher, T., and Giegé, R., 1989, Structural analogies between the 3′ tRNA-like structure of brome mosaic virus and yeast $tRNA^{Tyr}$ revealed by protection studies with tyrosyl–tRNA synthetase, *Eur. J. Biochem.* **185**:331–339.

Perret, V., Garcia, A., Grosjean, H., Ebel, J.-P., Florentz, C., and Giégé, R., 1990, Relaxation of a transfer RNA specificity by removal of modified nucleotides, *Nature* **344**:787–789.

Peterson, E. T., and Uhlenbeck, O. C., 1992, Determination of recognition nucleotides for *Escherichia coli* phenylalanyl–transfer RNA synthetase, *Biochemistry* **31**:10380–10389.

Plateau, P., and Blanquet, S., 1992, Synthesis of Np_n' (n = 3 or 4) *in vitro* and *in vivo*, in *Ap_4A and other Dinucleoside Polyphosphates* (A. G. McLennan, ed.), pp. 63–79, CRC Press, London.

Plateau, P., Fromant, M., and Blanquet, S., 1987, Heat shock and hydrogen peroxide responses of *Escherichia coli* are not changed by dinucleoside tetraphosphate hydrolase overproduction, *J. Bacteriol.* **169**:3817–3820.

Prather, N. E., Murgola, E. J., and Mims, B. H., 1984, Nucleotide substitution in the amino acid acceptor stem of lysine transfer RNA causes missense suppression, *J. Mol. Biol.* **172**:177–184.

Preiss, J., Berg, P., Ofengand, E. J., Bergmann, F. H., and Dieckmann, M., 1959, The chemical nature of the RNA–amino acid compound formed by amino acid-activating enzymes, *Proc. Natl. Acad. Sci. USA* **45**:319–328.

Price, S., Cusack, S., Borel, F., Berthet-Colominas, C., and Leberman, R., 1993, Crystallization of the seryl–tRNA synthetase:$tRNA^{Ser}$ complex of *Escherichia coli, FEBS Lett.* **324**:167–170.

Putney, S. D., Royal, N. J., de Vegvar, H. N., Herlihy, W. C., Biemann, K., and Schimmel, P., 1981a, Primary structure of a large aminoacyl–tRNA synthetase, *Science* **213**:1497–1501.

Putney, S. D., Sauer, R. T., and Schimmel, P. R., 1981b, Purification and properties of alanine tRNA synthetase from *Escherichia coli, J. Biol. Chem.* **256**:198–204.

Pütz, J., Puglisi, J. D., Florentz, C., and Giegé, R., 1991, Identity elements for specific aminoacylation of yeast $tRNA^{Asp}$ by cognate aspartyl–tRNA synthetase, *Science* **252**:1696–1699.

Pützer, H., Gendron, N., and Grunberg-Manago, M., 1992, Co-ordinate expression of the two threonyl–tRNA synthetase genes in *Bacillus subtilis*: Control by transcriptional antitermination involving a conserved regulatory sequence, *EMBO J.* **11**:3117–3127.

Regan, L., Bowie, J., and Schimmel, P., 1987, Polypeptide sequences essential for RNA recognition by an enzyme, *Science* **235**:1651–1653.

Reid, B. R., Koch, G. L. E., Boulanger, Y., Hartley, B. S., and Blow, D. M., 1973, Crystallization and preliminary X-ray diffraction studies on tyrosyl–transfer RNA synthetase from *Bacillus sterothermophilus, J. Mol. Biol.* **80**:199–201.

Renaud, M., Bacha, H., Rémy, P., and Ebel, J.-P., 1981, Conformational activation of the yeast phenylalanyl–tRNA synthetase catalytic site induced by $tRNA^{Phe}$ interaction: Triggering of adenosine or CpCpA trinucleoside diphosphate aminoacylation upon binding of $tRNA^{Phe}$ lacking these residues, *Proc. Natl. Acad. Sci. USA* **78:**1606–1608.

Rich, A., and RajBhandary, U. L., 1976, Transfer RNA: molecular structure, sequence and properties, *Annu. Rev. Biochem.* **45:**805–860.

Rietveld, K., Pleij, C. W. A., and Bosch, L., 1983, Three-dimensional models of the tRNA-like 3′ termini of some plant viral RNAs, *EMBO J.* **2:**1079–1085.

Risler, J. L., Zelwer, C., and Brunie, S., 1981, Methionyl–tRNA synthetase shows the nucleotide binding fold observed in dehydrogenases, *Nature* **292:**384–386.

Roe, B., Sirover, M., and Dodock, B., 1973, Kinetics of homologous and heterologous aminoacylation with yeast phenylalanyl transfer ribonucleic acid synthetase, *Biochemistry* **12:**4146–4154.

Romby, P., Brunel, C., Caillet, J., Springer, M., Grunberg-Manago, M., Westhof, E., Ehresman, C., and Ehresman, B., 1992, Molecular mimicry in translational control of *E. coli* threonyl–tRNA synthetase gene. Competitive inhibition in tRNA aminoacylation and operator–repressor recognition switch using identity rules, *Nucleic Acids Res.* **20:**5633–5640.

Rossman, M. G., Moras, D., and Olsen, K. W., 1974, Chemical and biological evolution of a nucleotide-binding protein, *Nature* **250:**194–199.

Rould, M. A., and Steitz, T. A., 1992, Structure of the glutaminyl–tRNA synthetase–$tRNA^{Gln}$–ATP complex, in *Nucleic Acids and Molecular Biology* (F. Eckstein and D. M. J. Lilley, eds.), pp. 225–245, Springer-Verlag, Berlin.

Rould, M. A., Perona, J. J., Söll, D., and Steitz, T. A., 1989, Structure of *E. coli* glutaminyl–tRNA synthetase complexed with $tRNA^{Gln}$ and ATP at 2.8 Å resolution, *Science* **246:**1135–1142.

Rould, M. A., Perona, J. J., and Steitz, T. A., 1991, Structural basis of anticodon loop recognition by glutaminyl–tRNA synthetase, *Nature* **352:**213–218.

Rudinger, J., 1992, Thesis, Université Louis Pasteur, Strasbourg, France.

Ruff, M., Krishnaswamy, S., Boeglin, M., Poterszman, A., Mitschler, A., Podjarny, A., Rees, B., Thierry, J. C., and Moras, D., 1991, Class II aminoacyl transfer RNA synthetases: Crystal structure of yeast aspartyl–tRNA synthetase complexed with $tRNA^{Asp}$, *Science* **252:**1682–1689.

Sampson, J. R., and Uhlenbeck, O. C., 1988, Biochemical and physical characterization of an unmodified phenylalanine transfer RNA transcribed *in vitro*, *Proc. Natl. Acad. Sci. USA* **85:**1033–1037.

Sampson, J. R., DiRenzo, A. B., Behlen, L. S., and Uhlenbeck, O. C., 1989, Nucleotides in yeast $tRNA^{Phe}$ required for the specific recognition by its cognate synthetase, *Science* **243:**1363–1366.

Sampson, J. R., DiRenzo, A. B., Behlen, L. S., and Uhlenbeck, O. C., 1990, Role of tertiary nucleotides in the interaction of yeast phenylalanine tRNA with its cognate synthetase, *Biochemistry* **29:**2523–2532.

Sampson, J. R., Behlen, L. S., DiRenzo, A. B., and Uhlenbeck, O. C., 1992, Recognition of yeast $tRNA^{Phe}$ by its cognate yeast phenylalanyl–tRNA synthetase: An analysis of specificity, *Biochemistry* **31:**4161–4167.

Santi, D. V., and Pena, V. A., 1971, Order of substrate binding to tyrosyl-tRNA synthetase of *Escherichia coli B*, *FEBS Lett.* **13:**157–160.

Schimmel, P., 1987, Aminoacyl tRNA synthetases: General scheme of structure–function relationships in the polypeptides and recognition of transfer RNAs, *Annu. Rev. Biochem.* **56:**125–158.

Schimmel, P. R., and Söll, D., 1979, Aminoacyl–tRNA synthetases: General features and recognition of transfer RNAs, *Annu. Rev. Biochem.* **48:**601–648.

Schleif, 1988, DNA binding by proteins, *Science* **241:**1182–1187.

Schmitt, E., Meinnel, T., Panvert, M., Mechulam, Y., and Blanquet, S., 1993, Two acidic residues of *Escherichia coli* methionyl–tRNA synthetase are negative discriminants towards the binding of non-cognate tRNA anticodons, *J. Mol. Biol.* **233**:615–628.

Schön, A., Kannangara, C. G., Gough, S., and Söll, D., 1988, Protein biosynthesis in organelles requires misaminoacylation of tRNA, *Nature* **331**:187–190.

Schulman, L. H., 1980, Chemical approaches to the study of protein–tRNA recognition, in *Transfer RNA: Biological Aspects* (D. Söll, J. N. Abelson, and P. R. Schimmel, eds.), pp. 311–324, Cold Spring Harbor Laboratory, Cold Spring Harbor.

Schulman, L. H., 1991, Recognition of tRNAs by aminoacyl–tRNA synthetases, *Prog. Nucleic Acid Res. Mol. Biol.* **41**:23–87.

Schulman, L. H., and Goddard, J. P., 1973, Loss of methionine acceptor activity from a base change in the anticodon of *Escherichia coli* formylmethionine transfer ribonucleic acid, *J. Biol. Chem.* **248**:1341–1345.

Schulman, L. H., and Pelka, H., 1983, Anticodon loop size and sequence requirements for recognition of formylmethionine tRNA by methionyl–tRNA synthetase, *Proc. Natl. Acad. Sci. USA* **80**:6755–6759.

Schulman, L. H., and Pelka, H., 1988, Anticodon switching changes the identity of methionine and valine transfer RNAs, *Science* **242**:765–768.

Schulman, L. H., and Pelka, H., 1989, The anticodon contains a major element of the identity of arginine transfer RNAs, *Science* **246**:1595–1597.

Schulman, L. H., and Pelka, H., 1990, An anticodon change switches the identity of *E. coli* $tRNA_m^{Met}$ from methionine to threonine, *Nucleic Acids Res.* **18**:285–289.

Schulman, L. H., Pelka, H., and Leon, O., 1987, Peptides at the tRNA binding site of the crystallizable monomeric form of *E. coli* methionyl–tRNA synthetase, *Nucleic Acids Res* **15**:10523–10529.

Schweet, R. S., Bovard, F. C., Allen, E., and Glassman, E., 1958, The incorporation of amino acids into ribonucleic acid, *Proc. Natl. Acad. Sci. USA* **44**:173–177.

Senger, B., Despons, L., Walter, P., and Fasiolo, F., 1992, The anticodon triplet is not sufficient to confer methionine acceptance to a transfer RNA, *Proc. Natl. Acad. Sci. USA* **89**:10768–10771.

Seong, B. L., and RajBhandary, U. L., 1987, *Escherichia coli* formylmethionine tRNA: Mutations in GGG/CCC sequence conserved in anticodon stem of initiator tRNAs affect initiation of protein synthesis and conformation of anticodon loop, *Proc. Natl. Acad. Sci. USA* **84**:334–338.

Shi, J.-P., and Schimmel, P., 1991, Aminoacylation of alanine minihelices. "Discriminator" base modulates transition state of single turnover reaction, *J. Biol. Chem.* **266**:2705–2708.

Shi, J.-P., Francklyn, C., Hill, K., and Schimmel, P., 1990, A nucleotide that enhances the charging of RNA minihelix sequence variants with alanine, *Biochemistry* **29**:3621–3626.

Shi, J. P., Martinis, S. A., and Schimmel, P., 1992, RNA tetraloops as minimalist substrates for aminoacylation, *Biochemistry* **31**:4931–4936.

Shimizu, M., Asahara, H., Tamura, K., Hasegawa, T., and Himeno, H., 1992, The role of anticodon bases and the discriminator nucleotide in the recognition of some *E. coli* transfer RNAs by their aminoacyl–transfer RNA synthetases, *J. Mol. Evol.* **35**:436–443.

Shimura, Y., Aono, H., Ozeki, H., Sarabhai, A., Lamfrom, H., and Abelson, J., 1972, Mutant tyrosine tRNA of altered aminoacid specificity, *FEBS Lett.* **22**:144–148.

Smith, J. D., and Celis, J. E., 1973, Mutant tyrosine transfer RNA that can be charged with glutamine, *Nature New Biol.* **243**:66–71.

Söll, D., and Schimmel, P. R., Aminoacyl–tRNA synthetases, 1974, *Enzymes* **10**:489–537.

Springer, M., Graffe, M., Butler, J. S., and Grunberg-Manago, M., 1986, Genetic definition of the translational operator of threonyl-tRNA ligase in *Escherichia coli*, *Proc. Natl. Acad. Sci. USA* **83**:4384–4388.

Springer, M., Graffe, M., Dondon, J., and Grunberg-Manago, M., 1989, tRNA-like structures and gene regulation at the translational level: A case of molecular mimicry in *Escherichia coli*, *EMBO J.* **8**:2417–2424.

Sprinzl, M., Hartmann, T., Weber, J., Blank, J., and Zeidler, R., 1987, Compilation of tRNA sequences and sequences of tRNA genes, *Nucleic Acids Res.* **17**:r1–r172.

Starzyk, R. M., Webster, T. A., and Schimmel, P., 1987, Evidence for dispensable sequences inserted into a nucleotide fold, *Science* **237**:1614–1618.

Steinberg, S. V., and Kisselev, L. L., 1993, Mosaic tile model for tRNA–enzyme recognition, *Nucleic Acids Res.* **21**:1941–1947.

Summers, M. F., South, T. L., Kim, B., and Hare, D. R., 1990, High-resolution structure of an HIV zinc fingerlike domain via a new NMR-based distance geometry approach, *Biochemistry* **29**:329–340.

Sylvers, L. A., Rogers, K. C., Shimizu, M., Ohtsuka, E., and Söll, D., 1993, A 2-thiouridine derivative in transfer RNA(Glu) is a positive determinant for aminoacylation by *Escherichia coli* glutamyl–transfer RNA synthetase, *Biochemistry* **32**:3836–3841.

Tagaya, M., and Fukui, T., 1986, Modification of lactate dehydrogenase by pyridoxal phosphate and adenosine polyphosphopyridoxal, *Biochemistry* **25**:2958–2964.

Tamura, K., Asahara, H., Himeno, H., Hasegawa, T., and Shimizu, M., 1991a, Identity elements of *Escherichia coli* $tRNA^{Ala}$, *J. Mol. Recog.* **4**:129–132.

Tamura, K., Himeno, H., Asahara, H., Hasegawa, T., and Shimizu, M., 1991b, Identity elements of *E. coli* $tRNA^{Val}$, *Biochem. Biophys. Res. Commun.* **177**:619–623.

Tamura, K., Himeno, H., Asahara, H., Hasegawa, T., and Shimizu, M., 1992, *In vitro* study of *E. coli* $tRNA^{Arg}$ and $tRNA^{Lys}$ identity elements, *Nucleic Acids Res.* **20**:2335–2339.

ten Dam, E., Pleij, K., and Draper, D., 1992, Structural and functional aspects of RNA pseudoknots, *Biochemistry* **31**:11665–11676.

Thiebe, R., Harbers, K., and Zachau, H. G., 1972, Aminoacylation of fragment combinations from yeast $tRNA^{Phe}$, *Eur. J. Biochem.* **26**:144–152.

Tocchini-Valentini, G. D., and Abelson, J. N., 1993, The identity set of *E. coli* leucine tRNAs, 15th International tRNA Workshop, Cap d'Agde, France, May 30–June 4.

Trézeguet, V., Edwards, H., and Schimmel, P., 1991, A single base pair dominates over the novel identity of an *Escherichia coli* tyrosine tRNA in *Saccharomyces cerevisiae*, *Mol. Cell. Biol.* **11**:2744–2751.

Tsai, C. H., and Dreher, T. W., 1991, Turnip yellow mosaic virus RNAs with anticodon loop substitutions that result in decreased valylation fail to replicate efficiently, *J. Virol.* **65**:3060–3067.

Tzagoloff, A., Gatti, D., and Gampel, A., 1990, Mitochondrial aminoacyl–tRNA synthetases, *Prog. Nucleic Acid Res. Mol. Biol.* **39**:129–158.

Ueda, T., Yotsumoto, Y., Ikeda, K., and Watanabe, K., 1992, The T-loop region of animal mitochondrial $tRNA^{Ser}$(AGY) is a main recognition site for homologous seryl–tRNA synthetase, *Nucleic Acids Res.* **20**:2217–2222.

Uemura, H., Imai, M., Ohtsuka, E., Ikehara, M., and Söll, D., 1982, *E. coli* initiator tRNA analogs with different nucleotides in the discriminator base position, *Nucleic Acids Res.* **10**:6531–6539.

Uemura, H., Conley, J., Yamao, F., Rogers, J., and Söll, D., 1988a, *Escherichia coli* glutaminyl–tRNA synthetase: A single amino acid replacement relaxes tRNA specificity, *Protein Seq. Data Anal.* **1**:479–485.

Uemura, H., Rogers, M. J., Swanson, R., Watson, L., and Söll, D., 1988b, Site-directed mutagenesis to fine-tune enzyme specificity, *Protein Eng.* **2**:293–296.

Varshney, U., and RajBhandary, U. L., 1990, Initiation of protein synthesis from a termination codon, *Proc. Natl. Acad. Sci. USA* **87**:1586–1590.

Varshney, U., Lee, C. P., and Rajbhandary, U. L., 1993, From elongator transfer RNA to initiator transfer RNA, *Proc. Natl. Acad. Sci. USA* **90**:2305–2309.

Vidal-Cros, A., and Bedouelle, H., 1992, Role of residue Glu152 in the discrimination between transfer RNAs by tyrosyl–tRNA synthetase from *Bacillus stearothermophilus*, *J. Mol. Biol.* **223**:801–810.

Waller, J. P., Risler, J. L., Monteilhet, C., and Zelwer, C., 1971, Crystallisation of trypsin-modified methionyl–tRNA synthetase from *Escherichia coli*, *FEBS Lett.* **16:**186–188.

Waye, M. M. Y., Winter, G., Wilkinson, A. J., and Fersht, A. R., 1983, Deletion mutagenesis using an "M13 splint": The N-terminal structural domain of tyrosyl–tRNA synthetase (*B. stearothermophilus*) catalyses the formation of tyrosyl adenylate, *EMBO J.* **2:**1827–1829.

Webster, T., Tsai, H., Kula, H. T. M., Mackie, G. A., and Schimmel, P., 1984, Specific sequence homology and three-dimensional structure of an aminoacyl transfer RNA synthetase, *Science* **226:**1315–1317.

Wells, T. N. C., and Fersht, A. R., 1985, Hydrogen bonding in enzymatic catalysis analysed by protein engineering, *Nature* **316:**656–657.

Wells, T. N. C., and Fersht, A. R., 1986, Use of binding energy in catalysis analyzed by mutagenesis of the tyrosyl–tRNA synthetase, *Biochemistry* **25:**1881–1886.

Weygand-Durasevic, I., Schwob, E., and Söll, D., 1993, Acceptor end binding domain interactions ensure correct aminoacylation of transfer RNA, *Proc. Natl. Acad. Sci. USA* **90:**2010–2014.

Yaniv, M., Folk, W. R., Berg, P., and Soll, L., 1974, A single mutational modification of a tryptophan-specific transfer RNA permits aminoacylation by glutamine and translation of the codon UAG, *J. Mol. Biol.* **86:**245–260.

Zaccaï, G., Morin, P., Jacrot, B., Moras, D., Thierry, J. C., and Giegé, R., 1979, Interactions of yeast valyl–tRNA synthetase with RNAs and conformational changes of the enzyme, *J. Mol. Biol.* **129:**483–500.

Zachau, H. G., Acs, G., and Lipmann, F., 1958, Isolation of adenosine amino acid esters from a ribonuclease digest of soluble, liver ribonucleic acid, *Proc. Natl. Acad. Sci. USA* **44:**885–889.

Zamecnik, P. C., Stephenson, M. L., Janaway, C. M., and Randerath, K., 1966, Enzymatic synthesis of diadenosine tetraphosphate and diadenosine triphosphate with a purified lysyl–sRNA synthetase, *Biochem. Biophys. Res. Commun.* **24:**91–105.

Chapter 12

Functions and Structures of Ribonuclease H Enzymes

Shigenori Kanaya and Morio Ikehara

1. INTRODUCTION

Ribonuclease H* (RNase H, EC 3.1.26.4) is an endonuclease that specifically hydrolyzes an RNA hybridized to a complementary DNA to produce an oligoribonucleotide with 5′-phosphate and 3′-hydroxyl groups (Fig. 1). It requires divalent cations, such as Mg^{2+} and Mn^{2+}, for activity. The enzyme was first isolated from calf thymus (Stein and Hausen, 1969; Hausen and Stein, 1970). Since then, the enzyme has been shown to be present in viruses, phages, and various organisms from *Escherichia coli* to human (Crouch and Dirksen, 1982; Wintersberger, 1990).

Enormous progress in the structural and functional studies of RNase H has been made in the last few years. The determinations of the crystal structures of RNase HI from *E. coli* (Katayanagi *et al.*, 1990, 1992; Yang *et al.*, 1990b) and reverse transcriptase (RT) from human immunodeficiency virus type 1 (HIV-1)

*Abbreviations: RNase H, ribonuclease H; RT, reverse transcriptase; HIV, human immunodeficiency virus; Mo-MuLV, Moloney murine leukemia virus; AMV, avian myeloblastosis virus; AKR-MuLV, AKR murine leukemia virus; GdnHCl, guanidine hydrochloride.

Shigenori Kanaya and Morio Ikehara Protein Engineering Research Institute, Osaka 565, Japan.

Subcellular Biochemistry, Volume 24. Proteins: Structure, Function, and Engineering, edited by B. B. Biswas and Siddhartha Roy. Plenum Press, New York, 1995.

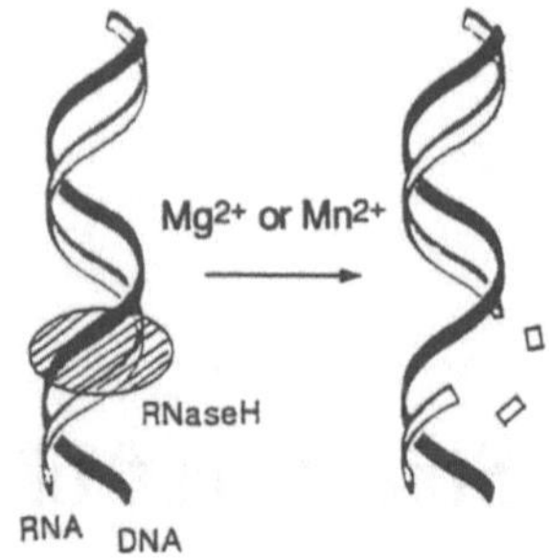

FIGURE 1. Enzymatic action of RNase H.

(Arnold *et al.*, 1992; Kohlstaedt, *et al.*, 1992; Jacobo-Molina *et al.*, 1993), in which RNase H is present at the carboxy-terminus as a separate domain, are responsible for this progress in understanding the functions of RNase H. Detailed structural information is essential to study the precise *in vivo* function of the enzyme. In addition, it may help in designing specific inhibitors of the RNase H activity, which may be effective as drugs against AIDS. In this chapter, we summarize the structure–function relationships of *E. coli* RNase HI, which include the proposed mechanisms for the substrate recognition and the catalytic reaction of the enzyme. We also briefly summarize the recent progress in studies of RNase H enzymes from other organisms, and the protein engineering of *E. coli* RNase HI.

2. *E. Coli* RNase HI

2.1. Physiological Functions

E. coli RNase HI, which was simply designated as RNase H until *E. coli* RNase HII was found (Itaya, 1990), was first described by Miller *et al.* (1973), Berkower *et al.* (1973), and Henry *et al.* (1973). The enzyme prefers Mg^{2+} over Mn^{2+} as a cofactor and is sensitive to *N*-ethylmaleimide. It cleaves RNA endonucleolytically to produce a 5′-phosphate, but cannot cleave the phosphodiester bond covalently linking ribonucleotides to DNA. The *rnhA* gene, encoding *E. coli* RNase HI, is located close to the *dnaQ* gene (Horiuchi *et al.*, 1981) at 5.1 min on the chromosome (Carl *et al.*, 1980). The *dnaQ* gene encodes the epsilon subunit of the DNA polymerase III holoenzyme. The promoter regions for the *rnhA* and *dnaQ* genes overlap in opposite directions (Maki *et al.*, 1983; Nomura *et al.*, 1985; Cox and Horner, 1986). The meaning of this promoter overlap in the promoter regions, however, remains to be determined.

The physiological functions of *E. coli* RNase HI have not been conclusively determined. However, the enzyme has been suggested to be involved in DNA replication and repair. Itoh and Tomizawa (1980) showed that the enzyme is required for the initiation of ColE1 DNA replication *in vitro*. They also suggested that after the RNA has served as a primer, it is eliminated from the product by RNase H. However, it was shown that the enzyme is dispensable for *in vivo* initiation of ColE1 DNA replication, because ColE1 DNA replicates in *E. coli rnhA* mutants defective in RNase H activity (Horiuchi *et al.*, 1984; Ogawa and Okazaki, 1984; Naito and Uchida, 1986). Dasgupta *et al.* (1987) suggested that the *rnhA*$^-$ and *rnhA*$^+$ strains use different mechanisms for the initiation of ColE1 DNA replication. A ColE1-type plasmid can replicate in *rnhA*$^-$ mutants even in the absence of DNA polymerase I, which is required for the replication of this plasmid (Kogoma, 1984). This result, as well as those of Ohmori *et al.* (1987), suggest that *E. coli* RNase HI suppresses the initiation of DNA replication of the ColE1-type plasmid at sites other than the Ori site. The finding that ColE1 mutants, which are unable to replicate in *rnhA*$^+$ strains, can replicate in *rnhA*$^-$ strains (Naito *et al.*, 1984) supports this possibility.

The involvement of *E. coli* RNase HI in chromosomal DNA replication has also been described (Ogawa *et al.*, 1984; Kogoma *et al.*, 1985; Horiuchi *et al.*, 1984; Meyenburg *et al.*, 1987; Hong and Kogama, 1993). These investigators have suggested that the enzyme suppresses the initiation of chromosomal DNA replication at regions other than OriC. *In vivo*, at least four such initiation regions, collectively termed OriK, have been localized in *rnhA*$^-$ strains (de Massy and Fayet, 1984). The involvement of RNase H in DNA repair systems has been described. The absence of RNase HI contributes to the recovery of DNA synthesis in UV-irradiated *E. coli* cells (Casaregola *et al.*, 1987). Constitutive expression of SOS function lead to a strong reduction in *rnhA* transcription (Quinones *et al.*, 1987). In accordance with these results, an increase in the level of the RNase H activity in cells results in an increase in the sensitivity to DNA damage exerted by UV-irradiation (Bockrath *et al.*, 1987).

In conclusion, *E. coli* RNase HI seems to be involved in DNA replication and repair, but in an auxiliary manner. It is known that at least two enzymes in *E. coli*, DNA polymerase I and exonuclease III, in addition to *E. coli* RNase HII, can cleave the RNA of DNA/RNA hybrids (Crouch and Dirksen, 1982). These enzymes may function in place of RNase HI in *rnhA*$^-$ strains. The *rnhA*$^-$ strains that have been isolated to analyze the role of RNase HI in DNA replication or repair grow normally. However, such *rnhA*$^-$ strains require a functional RecBCD protein and are lethal when a *recBCD* mutation is additionally introduced (Kanaya and Crouch, 1984; Itaya and Crouch, 1991a,b). The reason why these *rnhA*$^-$ strains require a protein involved in DNA repair and recombination remains to be determined.

2.2. Enzymatic Activity

The enzyme exhibits no strong base specificities. It has been reported that the enzyme cleaves the phosphodiester bond adjacent to pyrimidines with a slight preference (Robertson and Dunn, 1975). However, studies of the recognition mode of the enzyme using synthetic oligonucleotides have shown that the enzyme cleaves the phosphodiester bond adjacent to purines as well (Inoue *et al.*, 1987; Shibahara *et al.*, 1987; Atabekov *et al.*, 1988; Metelev *et al.*, 1988). These studies, as well as that of Donis-Keller (1979), suggest that the minimum size of the DNA/RNA hybrid recognized by the enzyme is a tetramer. The complete digestion of poly(rA)poly(dT) yields oligoribonucleotides with various chain lengths ranging from monomer to hexamer (Berkower *et al.*, 1973) or dimer to nonamer (Crouch, 1981). The major products are $(pA)_3$ and $(pA)_4$ in both cases. AMP was detected, with poor yield (4%), only by Berkower *et al.* (1973), suggesting that the enzyme does not exhibit an exonuclease activity.

The optimal conditions for the enzymatic activity have been basically described by Berkower *et al.* (1973). The enzyme possesses a broad pH optimum between 7.5 and 9.1. It requires 2–4 mM $MgCl_2$, 50–100 mM NaCl, and sulfhydryl reagents for maximal activity. Later, it was shown that the sulfhydryl reagent could be omitted without affecting the enzymatic activity (Kanaya *et al.*, 1990b). The requirement for $Mg2^+$ is only partially replaced by Mn^{2+} and the enzyme exhibits 2–3% of the maximal activity in the presence of 0.2 mM $MnCl_2$ (Berkower *et al.*, 1973). The enzymatic activity is inhibited by dextran (Dirksen and Crouch, 1981).

Various methods have been described to determine RNase H activities (Crouch and Dirksen, 1982). Hydrolysis of a radioactively labeled substrate with the enzyme followed by measurement of the radioactivity in acid (ethanol) soluble or insoluble materials is generally used for RNase H assay. The former represents the amount of product released from the substrate and the latter represents the amount of the remaining substrate. Many laboratories have reported different specific activities for the enzyme, ranging from 43,200 to 870,000 units/mg, using a poly(rA)poly(dT) duplex as a substrate (Berkower *et al.*, 1973; Arendes *et al.*, 1982; Itoh and Tomizawa, 1982; Kanaya and Crouch, 1983a; Kanaya *et al.*, 1989; Hogrefe *et al.*, 1990). However, it seems unclear whether such differences in the specific activity of the enzyme reflect differences in the purity of the enzyme or alternative definitions of the enzymatic activity. If one unit of activity is defined as the amount of enzyme producing one μmole of acid soluble material per min (Kanaya *et al.*, 1991a), the specific activities reported by Ito and Tomizawa (1982) and Hogrefe *et al.* (1990) become 23.5 units/mg (30 °C) and 29 units/mg (37 °C), respectively, which are comparable to those reported with an M13 DNA/RNA hybrid as a substrate (20 units/mg at 30 °C and 36 units/mg at 37 °C) (Kanaya *et al.*, 1991a,c; Kimura *et al.*, 1992a,b).

Table I
Kinetic Parameters of *E. coli* RNase HI for Different Substrates

Substrates[a]	Temperature (°C)	K_m (μM)	k_{cat} (min^{-1})	V_{max} (μ/mg)
M13 DNA/RNA[b]	30	0.11		20
Chimeric d9-mer/r9-mer[c]	30	0.53	90	
DNA/RNA–DNA/DNA[d]	37	4.2	425	

[a]The concentration of the M13 DNA/RNA hybrid represents the concentration of the RNA mononucleotide phosphate, and those of the other two substrates represent the concentrations of the r9-mer and the DNA/RNA–DNA.
[b]Data from Oda *et al.* (1993a).
[c]Data from Kanaya *et al.* (1990a).
[d]Data from Hogrefe *et al.* (1990).

The kinetic parameters of the enzyme have been determined using an M13 DNA/RNA hybrid (Kanaya *et al.*, 1991a,b), a chimeric nonanucleotide (chimeric d9-mer)/nonaribonucleotide (r9-mer) duplex (Kanaya *et al.*, 1990a, 1991b), or a DNA/RNA–DNA/DNA duplex (Hogrefe *et al.*, 1990) as substrates. The M13 DNA/RNA hybrid is cleaved by the enzyme at multiple sites, whereas the chimeric d9-mer/r9-mer and the DNA-RNA–DNA/DNA duplex are cleaved at a single site. The chimeric d9-mer consists of tetradeoxyribonucleotides flanked on either side by 2′-*O*-methyl-ribonucleotides. The enzyme cleaves an RNA hybridized to such a chimeric nonanucleotide specifically at the 3′-end of the DNA/RNA hybrid region (Inoue *et al.*, 1987; Shibahara *et al.*, 1987; Hayase *et al.*, 1990). The DNA–RNA–DNA strand consists of tetraribonucleotides flanked on either side by DNA. In the presence of the complementary DNA strand, this substrate is cleaved in the middle of the tetraribonucleotide (Hogrefe *et al.*, 1990). The K_m and the V_{max} or k_{cat} values of the enzyme with these substrates are summarized in Table I.

2.3. Overproduction

Due to the low content of the enzyme in normal cells, the purification of the enzyme in amounts sufficient for physicochemical studies has been difficult. The amount of the enzyme purified from 1 g of wet cells (normal cells) varied from 0.06 μg to 10 μg (Berkower *et al.*, 1973; Darlix, 1975; Arendes *et al.*, 1982; Itoh and Tomizawa, 1982). The production level of the enzyme in normal cells was estimated to be at most 4–5 μg/g wet cells from the total RNase H activities in the crude lysate of cells (Itoh and Tomizawa, 1982). Because RNase HI is not the only one of several enzymes that exhibit RNase H activity in cells, the production level of the enzyme in normal cells is probably 2–3 μg/g wet cells

(5–10 μg/liter culture). In this case, each normal cell contains approximately 100 RNase HI molecules.

To improve the production level of the enzyme in cells, a series of plasmids have been constructed. Plasmids pSK750 and pSK760 bear the *rnhA* gene under the control of its own weak promoter (Kanaya and Crouch, 1983a). Plasmid pDR600 (Kanaya *et al.*, 1990a) and plasmids pPL801 (Kanaya *et al.*, 1989) and pSK58 (Yang *et al.*, 1990a) bear the *rnhA* gene under the control of the tac promoter and the bacteriophage λ promoter P_L, respectively. Plasmid pJAL 600 bears the *rnhA* gene under the control of the bacteriophage λ promoters P_R and P_L in tandem, as well as the cI^{ts857} gene and the bacteriophage fd transcription terminator (Kanaya *et al.*, 1993). The enzyme is constitutively overproduced in *E. coli* LE392, harboring either plasmid pSK750 or pSK760. In contrast, it is overproduced in *E. coli* JM109, harboring pDR600 on induction with IPTG, and *E. coli* N4830-1, harboring pPL801 or pSK58, or *E. coli* HB101, harboring pJAL600 on thermal induction. The production levels of the enzyme in LE392/pSK750 or pSK760, JM109/pDR600, N4830-1/pPL801 or pSK58, and HB101/pJAL600 are at least 0.1, 5, 15, and 50 mg/liter culture, respectively. It has previously been shown that the production level in LE392/pSK750 or pSK760, which is 15 times the normal level of RNase H, is not detrimental to cell growth (Kanaya and Crouch, 1983a). However, those in other overproducing strains, which are more than 1000 times the normal level of RNase H, are detrimental to cell growth.

2.4. Physicochemical Properties

The enzyme accumulates in cells of overproducing strains in a soluble form and is easily purified with a yield of more than 80% (Kanaya *et al.*, 1989, 1993). The solubility of the enzyme is highest at around pH 5 and dramatically decreases as the pH is raised beyond 7. The $A_{280}{}^{0.1\%}$ value of the enzyme is 2.01 (Kanaya *et al.*, 1990b). The molar absorption coefficient, ϵ, of the enzyme is therefore calculated as $3.5 \times 10^4\ M^{-1}\cdot cm^{-1}$. This value compares reasonably well with the theoretical value of 3.9×10^4 calculated, using $\epsilon = 1576\ M^{-1}\cdot cm^{-1}$ for tyrosine (×5) and $\epsilon = 5225\ M^{-1}\cdot cm^{-1}$ for tryptophan (×6) at 280 nm (Goodwin and Morton, 1946). The enzyme consists of a single polypeptide chain with 155 amino acid residues, with a molecular weight of 17,559 (Kanaya and Crouch, 1983a). The good correlation in the molecular weights estimated under denaturing and nondenaturing conditions indicates that the enzyme acts as a monomer (Darlix, 1975; Itoh and Tomizawa, 1982; Arendes *et al.*, 1982; Kanaya and Crouch, 1983a). The enzyme purified from overproducing strains yielded one peak on isoelectric focusing, with a pI value of 9.0 (Kanaya *et al.*, 1989). In contrast, the enzyme purified from normal strains gave at least two peaks on isoelectric focusing with pI values of 9.0 and 9.6 (Arendes *et al.*,

1982). The peak with the pI value of 9.6 may represent a modified or degraded form of the enzyme.

The enzyme is stable between pH 1.66 and pH 10.12 at 10 °C, as judged by the circular dichroic (CD) spectra (Kanaya *et al.*, 1993). The unfolding of the enzyme with thermal denaturation is fully reversible between pH 1.66 and 4.02. The temperature of the midpoint of the transition, T_m, is 65–66 °C at pH 4.02, but decreases as the pH is lowered. The T_m of the enzyme is 49.4–49.8 °C at pH 3.0 (Kimura *et al.*, 1992a,b; Kanaya *et al.*, 1993). At pH 5.5, the enzyme is more stable, but it reversibly unfolds in a single cooperative fashion in the presence of 1 M guanidine hydrochloride (GdnHCl). The T_m of the enzyme is 50.5–52.7 °C under these conditions (Kanaya *et al.*, 1991a, 1993; Kimura *et al.*, 1992b). At an alkaline pH, the unfolding of the enzyme is not reversible, even in the presence of denaturants, and therefore the T_m cannot be determined. However, the apparent T_m values decrease as the pH is raised.

The unfolding of the enzyme by denaturation with GdnHCl or urea is also fully reversible (Kanaya *et al.*, 1991a,b; Kanaya and Itaya, 1992; Kimura *et al.*, 1992a,b). The denaturant concentrations for the midpoints of the transitions, $D_{1/2}$ are 1.8–1.9 M for GdnHCl and 5.0 M for urea at pH 5.5. The good coincidence between the parameters characterizing the GdnHCl denaturation, as determined by fluorescence measurements and those determined by CD measurements, suggests that the enzyme unfolds in a single cooperative manner with GdnHCl.

2.5. Three-Dimensional Structure

The enzyme was crystallized at a concentration of 7.5 mg/ml at pH 9.0 with a low salt concentration and in the absence of Mg^{2+} (Kanaya *et al.*, 1989). The crystals were further grown as tetragonal bipyramids with dimensions of 0.7 × 0.4 × 0.4 mm by successive seeding. These crystals belong to the space group $P2_12_12_1$, with unit cell parameters of a = 44.1 Å, b = 86.9 Å, and c = 35.5 Å, and contain one molecule per asymmetric unit. The crystal structure of the enzyme was first determined at 1.8 Å resolution (Katayanagi *et al.*, 1990), and its refinement has been further improved to a crystallographic R-factor of 0.196 at 1.48 Å resolution (Katayanagi *et al.*, 1992). The enzyme with complete selenomethionines has also been overproduced and isomorphously crystallized by vapor diffusion in hanging droplets at pH 8.8 in the presence of EDTA, dithiothreitol, ammonium sulfate, and PEG (Yang *et al.*, 1990a). This crystal (type I), with dimensions of 0.62 × 0.26 × 0.12 mm, belongs to the space group $P2_12_12_1$, with unit cell parameters of a = 41.8 Å, b = 86.4 Å, and c = 36.4 Å. Type II crystals with similar cell units as those reported by Kanaya *et al.* (1989) were also obtained upon replacement of sulfate with NaCl. These crystals could be converted to type I by soaking in phosphate buffer. The three-dimensional structure of the type I crystals has been determined by the multiwavelength anomalous diffraction technique (MAD analysis) at 2 Å resolution (Yang *et al.*, 1990b). These two crystal structures agree well with each other, with few exceptions. The major difference was observed in the conformation of the loop containing His^{124}. The imidazole group of His^{124} is located relatively close to the Asp^{10} and Asp^{70} carboxylates according to the crystal structure of Yang *et al.* (1990b), but is not according to that of Katayanagi *et al.* (1990, 1992). This discrepancy is probably due to crystal packing effects. In this chapter, we use the crystal structure and terminology reported by Katayanagi *et al.* (1990, 1992).

The enzyme is an $\alpha + \beta$ protein and consists of a five-stranded β-sheet and five α-helixes. The schematic representation of the three-dimensional structure of the enzyme is shown in Fig. 2. The single binding site for Mg^{2+}, which is required for the enzymatic activity, has been identified by soaking RNase H crystals in a solution containing Mg^{2+} ion followed by calculation of the difference Fourier maps (Katayanagi *et al.*, 1990). More recently, the position of the Mg^{2+} binding site was confirmed by determining the three-dimensional structure of the enzyme in complex with Mg^{2+} at 2.8 Å resolution (Katayanagi *et al.*, 1993b). The enzyme–Mg^{2+} complex was grown in the presence of 100 mM $MgSO_4$. The backbone structure of the enzyme was basically unchanged on binding of Mg^{2+}. According to the crystal structure of the enzyme–Mg^{2+} complex, the $O/_{\delta d2}$ atom of Asp^{10}, the main chain carbonyl oxygen of Gly^{11}, the O/ϵ_{e2} atom of Glu$_{48}$, and the O/δ_{d1} atom of Asn^{44} are located close to the Mg^{2+} ion with the distances of 2.1, 2.4, 2.4, and 3.0 Å, respectively, indicating that these residues, except for Asn^{44}, directly coordinate with the Mg^{2+}. The localizations of the Mg^{2+} and the amino acid residues that provide the coordinates for the binding of Mg^{2+} are also shown in Fig. 2. The dissociation constant of Mg^{2+} has been determined as 60 μM from [^{25}Mg]-NMR and 200 μM from calorimetry at pH 7.6 (Huang and Cowan, 1994) and 4.6 mM from NMR at pH 5.5 (Oda *et al.*, 1991). The dissociation constants of metal ions have also been determined as 740 μM for Mg^{2+} and 36 μM for Mn^{2+} at pH 9.0 in the presence of 2.8 M urea

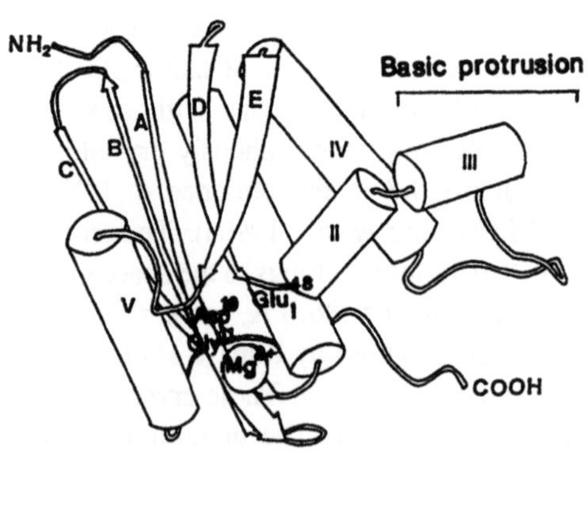

FIGURE 2. Schematic representation of the three-dimensional structure of *E. coli* RNase HI. The α helixes I–V and the β-strands A–E are represented by cylinders and arrows, respectively. The localizations of the Mg^{2+} ion, which is required for activity, and the coordinating amino acid residues are also shown.

by examining the effect of metal ions on the stability of the enzyme (Y-Y Liu, M. Oobatake, and S. Kanaya, personal communication). In either case, only one Mg^{2+} ion was shown to bind to the enzyme.

The characteristic features of the crystal structure of the enzyme are as follows. A large cleftlike depression, in which a DNA/RNA hybrid is expected to bind, is located in the middle of the molecule. The amino acid residues included in αII, αIII, the C-terminal portion of βA, the N-terminal portion of αI, the loops between βC and αI, between αIII and αIV, and between βE and αV, constitute this cleftlike depression. The extensive hydrophobic core is formed by amino acid residues located in βA, βD, βE, αI, αII, and αIV. Two parallel α-helixes, αI and αIV, are tightly associated by a ladder of leucine and isoleucine interactions. The αIII helix and the following loop form the relatively independent structure termed the *handle region* (Yang *et al.*, 1990b) or the *basic protrusion* (Katayanagi *et al.*, 1992). This densely charged basic protrusion, which is located at the edge of the cleftlike depression, is in sharp contrast to the region rich in negative charges around the Mg^{2+} binding site. Three tryptophan residues clustered in this region, as well as Trp^{104} in the αIV helix, seem to fix the conformation of this basic protrusion through aromatic interactions. The high values in the mean temperature factors in the αIII helix suggest that this helix is rather flexible.

2.6. Alignment of Secondary Structure Elements

Comparison of the amino acid sequence of *E. coli* RNase HI with those of other RNase H enzymes allows us to identify geometrical correspondences of amino acid residues in the two molecules. Such information is helpful to understand the structure–function relationships of RNase H enzymes. Geometrical alignments of secondary structural elements in selected RNase H enzymes are shown in Fig. 3. These enzymes are expected to share a common ancestor because of similarity in the three-dimensional structure and/or the primary structure. The secondary structural elements in *E. coli* RNase HI (Katayanagi *et al.*, 1990), *T. thermophilus* RNase H (Ishikawa *et al.*, 1993b), and the RNase H domain of HIV-1 RT (Davies *et al.*, 1991) have been assigned from the crystal structures of these enzymes determined by X-ray analyses. The secondary structural elements in the carboxyl terminal half of the 39-kDa yeast RNase H and the RNase H domain of Mo-MuLV RT are predicted from the comparison of their amino acid sequences with that of *E. coli* RNase HI. All the enzymes listed in Fig. 3 exhibit RNase H activities *in vitro*. In addition to those RNase H sequences, many RNase H sequences in various retroviral RTs and RT-like entities have been shown to have similarity with the *E. coli* RNase HI sequence (Doolittle *et al.*, 1989). However, they may include pseudo-sequences of RNase H enzymes that are not functional either *in vivo* or *in vitro*.

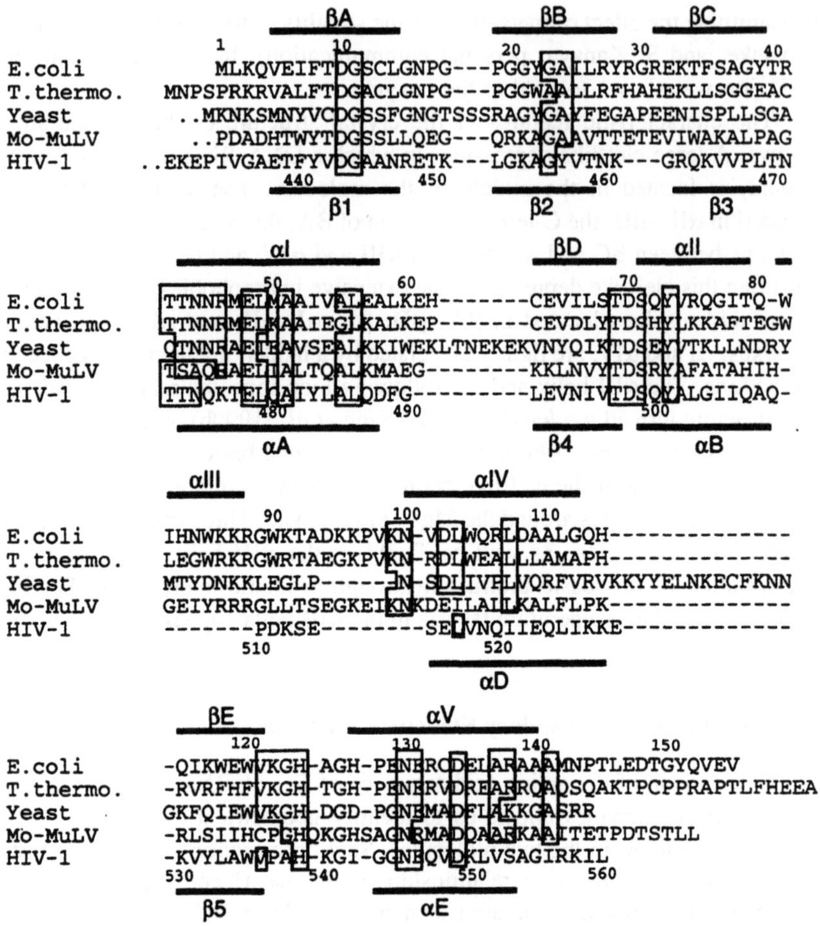

FIGURE 3. Alignment of amino acid sequences of selected RNase H enzymes. These sequences are from *E. coli* RNase HI (Kanaya and Crouch, 1983a), *T. thermophilus* RNase H (Itaya and Kondo, 1991; Kanaya and Itaya, 1992), yeast *Saccharomyces cerevisiae* RNase H (Itaya *et al.*, 1991), and reverse transcriptases from Mo-MuLV (Shinnick *et al.*, 1981) and HIV-1 (Ratner *et al.*, 1985). The ranges of the secondary structured elements for *E. coli* RNase HI and the HIV-1 RNase H domain are shown along the corresponding sequences. Numbers along these sequences represent the positions of amino acids in these sequences. Highly or relatively well conserved residues are boxed.

T. thermophilus RNase H is composed of 166 amino acid residues and has a molecular weight of 18,279 (Kanaya and Itaya, 1992). When compared to the *E. coli* RNase HI sequence, the *T. thermophilus* RNase H sequence contains additional residues: four at the N-terminus, six at the C-terminus, and the insertion of Gly^{80b} (Fig. 3). The positions of the amino acid residues in the *T. thermophilus*

RNase H sequence are numbered based on the *E. coli* RNase HI sequence instead of from its N-terminus. The crystal structure of this enzyme, which has been determined at 2.8 Å resolution (Ishikawa *et al.*, 1993b), includes the amino acid residues from Arg^{2} to Thr^{147}. Five residues at the N-terminus and 14 residues at the C-terminus are poorly defined in the electron density map, probably due to structural disorder. The crystal structure of *T. thermophilus* RNase H is almost identical to that of *E. coli* RNase HI, with few exceptions. A large structural difference was observed in the loop between βB and βC, which might arise from the difference in the crystal packing. In addition, the βE-strand and the αII-helix of the thermophilic enzyme are internally and externally shifted by 1.0 Å, respectively, when compared to those of the *E. coli* enzyme. These shifts have been suggested to arise from the replacements of amino acid residues in the hydrophobic core of the thermophilic enzyme, such as Ile^{116} by Val, Trp^{118} by Phe, Trp^{120} by Phe, Val^{74} by Leu, and Ile^{78} by Phe (Ishikawa *et al.*, 1993b). The conformational strains caused by such shifts seem to be alleviated by the insertion of Gly^{80b} between the αII and αIII helixes. From the comparison with the crystal structure of *E. coli* RNase HI, Ishikawa *et al.* (1993b) have suggested that the expansion of an aromatic cluster and an increased number of salt-bridges additively contribute to the unusually high stability of the thermophilic RNase H.

The crystal structure of the RNase H domain of HIV-1 RT (Tyr-Ala-Ser-Arg-Tyr^{427} to Leu^{560}), which has little RNase H activity, has been determined at 2.4 Å resolution (Davies *et al.*, 1991). A similarity was observed in the crystal structures of this molecule and *E. coli* RNase HI, as shown in Fig. 4, but to a smaller extent than that between the *E. coli* and *T. thermophilus* enzymes. The αIII helix and the following loop in *E. coli* RNase HI (basic protrusion) are missing in the

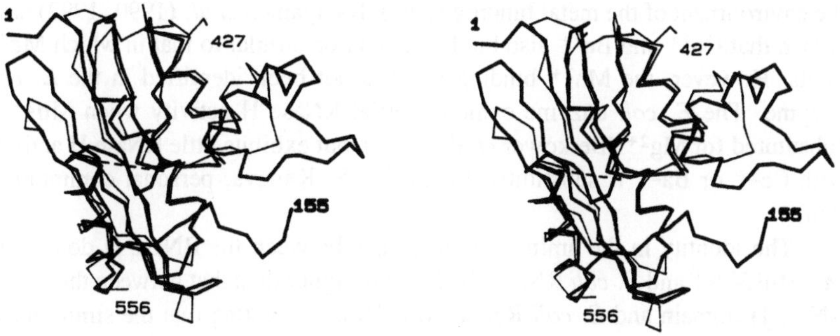

FIGURE 4. Stereo view of α-carbon chains of *E. coli* RNase HI (thick lines) and the HIV-1 RNase H domain (thin lines) superimposed for 88 α-carbon atoms in the central core. The thin dashed line indicates the disordered loop containing His^{539} in the structure of the HIV-1 RNase H domain. The crystal structures of *E. coli* RNase HI (Katayanagi *et al.*, 1992) and the HIV-1 RNase H domain (Davies *et al.*, 1991) are deposited in the Protein Data Bank as 2RN2 and 1HRH, respectively. This figure was kindly provided by Dr. K. Katayanagi.

crystal structure of the HIV-1 RNase H domain, in which a five-residue loop with an extended conformation connects αB and αD (corresponding to αII and αIV of *E. coli* RNase HI) (Fig. 3). A three-residue deletion in the connection between β2 and β3 in the HIV-1. RNase H domain results in a two-residue reduction in the size of both strands. The conformations of the connections between β1 and β2, β2 and β3, αA and β4, and αD and β5 in the HIV-1 RNase H domain also differ from those in *E. coli* RNase HI. Furthermore, the loop containing His^{539} (between β5 and αE) is disordered, although it is ordered and interacts with the thumb subdomain of p51 in the crystal structure of the HIV-1 RT heterodimer (Kohlstaedt *et al.*, 1992). Nevertheless, the secondary structural elements in *E. coli* RNase HI are well conserved in the HIV-1 RNase H domain. The geometrical arrangements of the four acidic amino acid residues, which are fully conserved in 25 RT and RT-related sequences, as well as in the *E. coli* RNase HI sequence (Doolittle *et al.*, 1989), are quite similar in these two enzymes (Fig. 5). They are Asp^{10}, Glu^{48}, Asp^{70}, and Asp^{134} in *E. coli* RNase HI, which correspond to Asp^{443}, Glu^{478}, Asp^{498}, and Asp^{549} in the HIV-1 RNase H domain, respectively. These residues contribute to the formation of metal binding site(s) (Katayanagi *et al.*, 1990, 1992, 1993b; Davies *et al.*, 1991). It must be noted, however, that there is a discrepancy in the number of the metal ions bound to the enzyme. *E. coli* RNase HI has a single Mg^{2+} binding site, in which Asp^{10} and Glu^{48} directly coordinate with Mg^{2+} (Katayanagi *et al.*, 1993b). In contrast, the HIV-1 RNase H domain has two Mn^{2+} binding sites (sites A and B), in which Glu^{478} and Asp^{498} interact exclusively with Mn^{2+} at site A, Asp^{549} interacts with Mn^{2+} at site B, and Asp^{443} interacts with both sites (Davies *et al.*, 1991). It remains to be determined whether this discrepancy in the metal binding sites of these two enzymes reflects a difference in the metals examined or a difference in the environment of the metal binding site(s). Katayanagi *et al.* (1990, 1992) have shown that Ca^{2+} and Ba^{2+} also bind to a position similar to that in which Mg^{2+} binds. However, the Mn^{2+} binding site has not been identified in the *E. coli* enzyme. The *E. coli* enzyme exhibits partial RNase H activity when Mn^{2+} is substituted for Mg^{2+} (Berkower *et al.*, 1973), but exhibits little RNase H activity with Ca^{2+} or Ba^{2+} is substituted for Mg^{2+} (S. Kanaya, personal communication).

The identity in the amino acid sequence between the RNase H domain of Mo-MuLV RT and *E. coli* RNase HI (25%) is higher than that between the HIV-1 RNase H domain and *E. coli* RNase HI (21%), suggesting that the similarity in the three-dimensional structures of the Mo-MuLV and *E. coli* enzymes is higher than that between the HIV-1 and *E. coli* enzymes. Comparison of the primary structures suggests that neither a three-residue deletion in the connection of β2 and β3 nor a deletion of the basic protrusion is observed in the Mo-MuLV RNase H domain. In contrast to the prediction for the secondary structure elements in the Mo-MuLV RNase H sequence, which seems to be essentially reliable, that for the yeast RNase H sequence includes some ambiguities. The amino acid

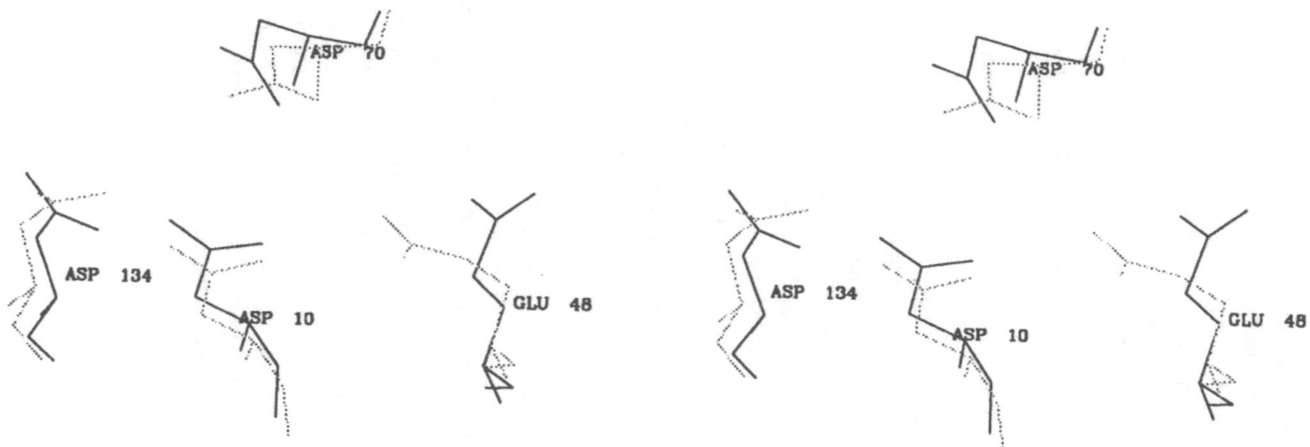

FIGURE 5. Stereo view of the conformations of the four acidic amino acid residues that are fully conserved in various RNase H and RNase H-related sequences. Asp^{10}, Glu^{48}, Asp^{70}, and Asp^{134} in the crystal structure of *E. coli* RNase HI (————————) are superimposed with Asp^{443}, Glu^{478}, Asp^{498}, and Asp^{549} in the crystal structure of the HIV-1 RNase H domain (···········). The crystal structures of these enzymes are described in the legend of Fig. 4. This figure was kindly provided by Mr. S. Kawakita.

sequence connecting αIII and βE of yeast RNase H is aligned so that the amino acid residues involved in the αIV helix are well conserved, and insertions or deletions within α helixes and β-strands are avoided. According to the secondary structural elements in the yeast RNase H sequence, which are predicted from the alignment of the amino acid sequence, yeast RNase H has a three-residue insertion between βA and βB, seven-residue insertion between αI and βD, a 15-residue insertion between αIV and βE, and a six-residue deletion between αIII and αIV, as compared to *E. coli* RNase HI. The identity of the amino acid sequence between the *E. coli* and yeast enzymes is 28% and that between the *E. coli* and *T. thermophilus* enzymes is 52%.

The four acidic amino acid residues that are fully conserved in 26 RNase H-related sequences (Doolittle *et al.*, 1989) are also conserved in the yeast and *T. thermophilus* RNases H. In addition, Gly^{11} in *E. coli* RNase HI, which directly coordinates with Mg^{2+}, is also conserved in these enzymes, suggesting that the Mg^{2+} binding sites of these cellular enzymes are quite similar to that of *E. coli* RNase HI. The role of other conserved residues in *E. coli* RNase HI on the enzymatic function will be described in the Section 2.8.

The hydrophobic core of *E. coli* RNase HI is formed by Ile^{7}, Ala^{55}, Ala^{58}, Val^{65}, Leu^{67}, Val^{74}, Ile^{116}, and Trp^{118}. The corresponding residues in other RNase H enzymes are well conserved in their hydrophobicities, except for the residue at position 58. Likewise, a ladder of hydrophobic interactions between αI and αIV (49-103, 53-107, and 56-111 in *E. coli* RNase HI) is also well conserved in other RNase H enzymes, with few exceptions. These results strongly suggest that the hydrophobic core formed by these residues contributes to maintain the molecular architecture of the enzyme. The ionic interaction between Arg^{46} and Asp^{102} and the hydrogen bond between Arg^{46} and Asn^{100} also participate (Katayanagi *et al.*, 1992). These three residues are well conserved in other RNase H enzymes, although Asp^{102} is replaced by Glu in retroviral RNases H and Arg^{46} is replaced by Lys in HIV-1 RNase H. The N-terminal residue of the αI-helix is conserved as Thr or Ser, and this residue has been suggested to contribute to the stabilization of the αI-helix by forming the N-cap (Davies *et al.*, 1991). The aromatic interactions among Trp^{81}, Trp^{85}, Trp^{90}, and Trp^{104}, which are expected to fix the conformation of the basic protrusion in *E. coli* RNase HI, are conserved in *T. thermophilus* RNase H, but not in the yeast and Mo-MuLV RNases H. The amino acid sequences connecting αII and αIV may have unique conformations in the yeast and Mo-MuLV RNases H.

2.7. NMR Studies

2.7.1. Assignments

Nuclear magnetic resonance (NMR) is one of the most efficient techniques for clarifying the structure–function relationships of proteins in solution. To investigate the structure–function relationships of a given protein by NMR, each resonance peak in the NMR spectrum should first be completely assigned to a specific atom of this protein molecule. The backbone nuclei of *E. coli* RNase HI have been completely assigned by heteronuclear three-dimensional NMR spectroscopy, with the help of amino acid-specific ^{15}N labeling (Nagayama *et al.*, 1990; Yamazaki *et al.*, 1991). Amino acid-selective ^{1}H labeling of a fully deuterated protein sample also allowed the assignments of the backbone nuclei in the labeled residues (Oda *et al.*, 1992). More recently, all types of ^{1}H, ^{15}N, and ^{13}C magnetic resonances for the enzyme have been completely assigned by double- and triple-resonance two- and three-dimensional NMR spectroscopy (Yamazaki *et al.*, 1993). The protein samples, which are uniformly labeled with ^{15}N or ^{13}C, were prepared using the overproducing strain (N4830-1/pPL801) in minimal medium with [^{15}N]-H_4Cl as the sole nitrogen source, or with [^{13}C]glucose as the sole carbon source, respectively (Yamazaki *et al.*, 1991). The fully deuterated protein sample was obtained from the same overproducing strain grown in minimal medium prepared in D_2O, including either deuterated glucose or sodium succinate as a carbon source (Oda *et al.*, 1992). The protein samples specifically labeled with ^{15}N-labeled amino acids were prepared in minimal medium with ^{15}N-labeled amino acids, which were added after the overproduction of the enzyme was induced. These protein samples were dissolved in 0.1 M deuterated acetate buffer, pH 5.5, with a protein concentration of 1–2 mM, and used for NMR experiments.

2.7.2. pKa Values

The pKa values of ionizable groups in proteins are generally affected by their environment. Determination of the pKa values of ionizable groups therefore provides information on their interactions with other residues in the solution structure of the enzyme. The pKa values for the imidazole groups of all the His residues of *E. coli* RNase HI were determined by analyzing the pH titration shifts of the C2H proton resonances of the imidazole groups (Oda *et al.*, 1993a). In these experiments, the C2H proton resonance of each His residue was assigned by comparing the [^{1}H]-NMR spectrum of the wild-type protein to the spectra of five mutant proteins, each of which lacked a different His residue (Kanaya *et al.*, 1991b). The pKa values of all the His residues accurately reflected their environments in the crystal structure. The pKa value of His^{83} (approximately 5.5) is significantly lower than the normal value, because it interacts with basic amino acid residues. The pKa value of His^{114} (<5) is unusually low, because this imidazole group is buried in the protein molecule and forms hydrogen bonds with the main chain carbonyl oxygen atoms of Ala^{110} and Glu^{61}. The pKa value of His^{127} (7.9) is significantly higher than the normal value, because it forms a salt bridge with Glu^{119}. The pKa value of His^{124}, which has been suggested to be located at the flexible loop (Oda *et al.*, 1993), is 7.1, which is only slightly higher than the normal value. These results suggest that His[124] is located in an acidic environment, but does not interact directly with a carboxyl group in the solution structure. In conclusion, the determination of the pKa values of all the His residues strongly suggests that the solution structure of the enzyme is similar to the crystal structure.

The pKa values for the side-chain carboxyl groups of all the Asp and Glu residues have also recently been determined (Oda *et al.*, 1994). Among four conserved acidic residues, Asp[10] and Asp[70] show unusual ionization behaviors, as reported for the catalytic carboxylate pairs in pepsinlike aspartic protease (Hsu *et al.*, 1977; Blundell *et al.*, 1985) and hen egg lysosyme (Parsons and Raftery, 1972). The pKa values of 6.1 for Asp[10] and 2.6 Asp[70] are considerably higher and lower, respectively, than the normal value of 4.0 (Fersht, 1985). These pKa values, however, shift to 4.2 for Asp[10] and 3.4 for Asp[70], but that of Glu[48] does not, when Mg^{2+} ion binds to the enzyme. These results suggest that Glu[48] is not involved in the Mg^{2+} binding site, which seems surprising because determination of the crystal structure of the enzyme–Mg^{2+} complex clearly shows that Glu[48] directly coordinates with Mg^{2+} ion.

2.7.3. Binding of Metal Ions and Substrates

The metal ion binding site of *E. coli* RNase HI and the binding constant were determined by monitoring chemical shift changes during the titration of the enzyme with divalent cations, using $^{1}H-^{15}N$ heteronuclear two-dimensional NMR (Oda *et al.*, 1991). The uniformly ^{15}N-labeled protein sample was used for these experiments. On addition of $MgCl_2$, $CaCl_2$, and $BaCl_2$, the backbone amide chemical shifts of Ser[12], Ser[71], and Gln[72] moved significantly. In addition, those of Gly[11], Glu[48], Asp[70], Tyr[73], Cys[133], Asp[134], Gln[152], and Val[153] were affected by the addition of these metal ions. Because all these residues, except for Gln[152] and Val[153], are located close to the active site of the enzyme, the position of the metal binding site of the enzyme in solution is consistent with that determined by X-ray crystallographic studies (Katayanagi *et al.*, 1993b). From the titration analysis, a single Mg^{2+}, Ca^{2+}, or Ba^{2+} ion has been shown to bind to the enzyme at pH 5.5, in the absence of substrate, with dissociation constants of 4.6 mM, 2.9 mM, and 2.2 mM, respectively. These values are considerably higher than those determined by NMR or calorimetry at pH 7.6 (Huang and Cowan, 1994) or that determined by monitoring the change in protein stability at pH 9.0 (Y-Y Liu, M. Oobatake, and S. Kanaya, personal communication), probably because Asp[10], which directly coordinates with Mg^{2+} ion, is protonated at pH 5.5 (Oda *et al.*, 1994).

The amino acid residues that interact with the substrate have also been identified by titration of the chimeric nonanucleotide duplex, while changes in the chemical shifts of the nitrogen and hydrogen nuclei of the backbone amides were monitored (Nakamura *et al.*, 1991b). The chimeric nonanucleotide duplex used in this experiment has previously been shown to be cleaved at a single site (Kanaya *et al.*, 1990a). The amino acid residues with backbone amide chemical shifts that were affected by the addition of this substrate are clustered around the Mg^{2+} binding site of the enzyme. Such information, as well as that obtained from site-directed mutagenesis studies, is quite useful to predict the substrate-binding site of the enzyme. The enzyme–substrate binding constant has not been determined from these experiments, because a large amount of precipitate appeared when the molar ratio of the substrate to the enzyme exceeded 0.3 at pH 5.5. In addition to the DNA/RNA hybrid, the DNA/DNA and RNA/RNA duplexes have recently been shown to bind the enzyme by heteronuclear 2D NMR (Oda *et al.*, 1993b). The binding affinities of these duplexes, however, seem to be much weaker than that of the DNA/RNA hybrid to the enzyme.

2.7.4. Folding Pathways

Characterization of transient conformers in refolding studies using stopped-flow methods, coupled with NMR measurements of hydrogen exchange, have allowed identification of the folding domains of proteins such as barnase (Matouschek *et al.*, 1990) and lysozyme (Miranker *et al.*, 1991). The folding domain of *E. coli* RNase HI may be identified by these methods. *E. coli* RNase HI has been shown to be partially denatured at an extremely low pH (M. Oobatake, personal communication). The determination of the remaining secondary or tertiary structure in this partially denatured protein using hydrogen exchange measurements with NMR may help to reveal the folding pathway of the enzyme. Such methods have been shown to effectively provide structural information about incompletely folded proteins, such as phage 434 repressor (Neri *et al.*, 1992a,b) and ubiquitin (Stockman *et al.*, 1993).

2.8. Role of Amino Acid Residues

2.8.1. Site-Directed Mutagenesis

The role of each amino acid residue in the function and stability of *E. coli* RNase HI has been investigated by site-directed mutagenesis experiments, which have been carried out by the procedure developed by Taylor *et al.* (1985) and later with PCR as described by Higuchi (1990). The mutant proteins were purified to homogeneity and analyzed for their enzymatic activities and stabilities. *E. coli* MIC3001, with the *rnh::cat* and *recB270*(ts) mutations (Itaya and Crouch, 1991a,b), allowed the efficient cloning of the structural genes of yeast RNase H (Itaya *et al.*, 1991) and *T. thermophilus* RNase H (Itaya and Kondo, 1991). This strain has also been used to examine whether the mutant proteins constructed by site-directed mutagenesis are functional *in vivo* (Oda *et al.*, 1993), because the production of the functional mutant *E. coli* RNase HI proteins suppresses its RNase H-dependent temperature-sensitive growth phenotype. The relative k_{cat} and K_m values of the various mutant proteins with single amino acid substitutions are summarized in Table II.

2.8.2. Conserved Residues

Among the various highly conserved amino acid residues with functional side chains, Asp^{10}, Glu^{48}, and Asp^{70} were first identified as active-site residues, because the replacement of any of these residues with Asn or Gln and with Ala almost abolishes the enzymatic activity (Kanaya *et al.*, 1990a). However, the crystal structures of the inactive mutant proteins, D10N, E48Q, and D70N, in which Asp^{10}, Glu^{48}, and Asp^{70} are replaced by Asn, Gln, and Asn, respectively, suggest that freezing or alteration in the geometrical arrangement of these residues, due to the formation of hydrogen bond networks, prevents Mg^{2+} binding and thereby inactivates the enzyme (Katayanagi *et al.*, 1993a). These results suggest that Asp^{10}, Glu^{48}, and Asp^{70} might not be required for the catalytic function, but are required for the formation of the Mg^{2+} binding site. Unlike Asp^{10} and Glu^{48}, Asp^{70}, which is located 4.4 Å away from the Mg^{2+} ion bound to the enzyme (Katayanagi *et al.*, 1993b), does not directly interact with the metal cation. Asp^{70} may be responsible to maintain the repulsion of the negatively charged carboxyl groups clustered around the Mg^{2+} binding site and thereby keep the conformation of the Mg^{2+} binding site functional. Alternatively, it may function to activate an attacking H_2O molecule as a general base. It is also possible that Asp^{70} may have such dual functions in the catalytic reaction.

It has recently been shown that a negative-charge repulsion in the active-site of the enzyme, which results from ionizations of Asp^{10} and Asp^{70}, destabilizes the enzyme by 3 kcal/mol in ΔG (Y-Y Liu, M. Oobatake, and S. Kanaya, personal communication). Such a negative-charge repulsion is fully canceled when Mg^{2+} ion binds to the enzyme. By monitoring the change in thermal stability of proteins, Mg^{2+} ion was found to bind to all the active-site mutant proteins as well, but with different affinities (dissociation constants of 0.74 mM for WT, 10 mM for D10N, 1.1 mM for E48Q, and 19 mM for D70N). Thus, the dissociation constants increased dramatically for mutant proteins, D10N and D70N, but not for E48Q. These results as well as those reported by Oda *et al.* (1994) that the pKa values of Asp^{10} and Asp^{70} change considerably when Mg^{2+} ion binds to the enzyme but that of Glu^{48} does not strongly suggest that Glu^{48} is not required for the binding of Mg^{2+} ion. Because it seems likely that Asp^{10} and Asp^{70} are required to form the Mg^{2+} binding site, Glu^{48}, instead of Asp^{70}, may be directly involved in the catalytic function.

Table II
Relative Kinetic Parameters of *E. coli* RNase HI Mutants

Mutants[a]	Relative k_{cat}[b]	Relative K_m	Mutants[a]	Relative k_{cat}[b]	Relative K_m
WT	1.0	1.0			
K3E[1]	0.6	0.8	17C	0.9*	
T9A	0.4*		D10N[2]	<0.001	
D10E[2]	0.08	1.0	D10A[2]	<0.001	
C13A[3]	4.9	14	N16G	1.3	33
N16A[3]	3.5	34	G23A	1.0*	
R29P	1.2*		G30P	0.9*	
T43C[5]	0.14	20	N44C[5]	0.6	8.0
N44A[3]	1.0	7.5	N45A[3]	2.0	20
E48Q[2]	<0.001		E48D[2]	0.04	1.0
E48A[2]	<0.001		H62A[6]	1.0	1.0
H62P[7]	1.1*		C63A[5]	1.0	1.0
S68A[8]	1.0*		S68L[8]	0.4*	
D70N[2]	<0.001		D70E[2]	0.01	1.0
D70A[2]	<0.001		S71A[2]	0.6	0.6
Q72A[3]	1.7	4.7	Y73L[2]	0.8	2.0
V74L[9]	1.1	1.0	Q76A[3]	0.9	2.0
G77A[10]	0.3	1.5	178F	0.6*	
G80b[10]	0.1	50	A80b[10]	0.1	50
Q80A[3]	1.0	1.1	H83A[6]	2.0	4.8
N84A[1]	0.8	1.4	W85A[1]	0.06	7.8
K86A[1]	0.6	1.3	K87A[1]	0.7	2.9
R88A[1]	0.6	4.7	G89A[1]	0.9	1.3
W90A[1]	0.2	60	K91A[1]	0.9	4.7
T92A[1]	1.3	1.0	D94A[1]	1.0	0.9
K95A[1]	1.3	6.3	K95G[11]	1.2*	
K96A[1]	1.1	4.3	P97A[1]	1.5	1.2
V98A[1]	0.8	0.9	K99A[1]	1.3	5.0
Q113P[7]	1.0*		H114A[6]	0.7	2.2
Q115R	1.3.		W118F	0.4*	
K122N	0.3	5.3	H124A[6]	0.03	2.2
H124Q[12]	0.01	0.9	H124E[12]	0.02	2.5
H124K[12]	0.04	1.5	A125T	1.6*	
H127A[6]	0.2	1.4	N130D[2]	0.09	7.0
E131C	0.6*		C133S[4]	0.5*	
C133A[5]	1.0	1.0	D134N[2]	1.2	1.0
D134H[14]	0.6*		D134E[14]	0.03	4.1
D134Q[14]	0.02	3.7	D134S[14]	0.007	1.6
D134T[14]	0.0009	3.6	D134L[14]	<0.0001	
D134A[14]	<0.0001		E135C[13]	1.0	3.1
R138C	0.2*		M142C	1.0*	
P144L[15]	0.2*				

[a] Superscript numbers indicate data from 1. Kanaya *et al.* (1991c); 2. Kanaya *et al.* (1990a); 3. Nakamura *et al.* (1991b); 4. Kanaya *et al.* (1990b); 5. Kanaya *et al.* (1991a); 6. Kanaya *et al.* (1991b); 7. Kimura *et al.* (1992b); 8. Kimura *et al.* (1992a); 9. Ishikawa *et al.* (1993c); 10. Ishikawa *et al.* (1993d); 11. Kimura *et al.* (1992c); 12. Oda *et al.* (1993); 13. Kanaya *et al.* (1993); 14. Haruki *et al.* (1994); 15. Kanaya and Crouch (1983b). All other data are personal communications from S. Kanaya, C. Nakai, and S. Kimura.

[b] Asterisk indicates relative specific activity.

His^{124} and Asp^{134} have not been considered to be catalytic residues, because the $His^{124} \rightarrow$ Ala mutation reduced both the substrate-binding affinity and the hydrolysis rate and the $Asp^{134} \rightarrow$ Asn mutation has no effect on the enzymatic activity (Kanaya *et al.*, 1990a). However, further mutagenesis experiments revealed that the imidazole group of His^{124} is not essential for the enzymatic activity, but is required to enhance the catalytic efficiency of the enzyme (Oda *et al.*, 1993a). Likewise, Asp^{134} has also been shown to contribute to the acceleration of the hydrolysis rate of the enzyme (Haruki *et al.*, 1994). Because the replacement of this residue with Asn or His has no serious effect on the enzymatic activity, the polar atoms at the $\delta 1$ position of Asp^{134} may be required to position an attacking H_2O molecule or to keep the conformation of the active site functional.

The $Asn^{44} \rightarrow$ Ala mutation decreased the substrate-binding affinity without affecting the hydrolysis rate (Table II). This result is inconsistent with the finding that Asn^{44} is important for the binding of the Mg^{2+} ion (Katayanagi *et al.*, 1993b). The distance between the $O/_{\delta 1}$ atom of Asn^{44} and the Mg^{2+} ion is 3.0 Å according to the crystal structure of the enzyme complexed with Mg^{2+}, which does not seem to be close enough for direct coordination. The mutagenesis experiment suggests that Asn^{44} is involved in the substrate binding instead of the Mg^{2+} binding.

Other conserved residues besides those that are involved in the catalytic reaction or the metal binding, or those that apparently contribute to maintain the molecular architecture of the enzyme, are Thr^{42}, Thr^{43}, Asn^{45}, Thr^{69}, Ser^{71}, Tyr^{73}, Asn^{130}, and Glu^{131}. Thr^{43}, Asn^{45}, and Asn^{130} are probably involved in the substrate-binding site, because mutations of these residues increase the K_m values (Table II). The hydroxyl group of Thr^{43} may form a hydrogen bond with the backbone phosphate group of the RNA strand. This residue is exposed to the solvent and faces the sulfate-binding position, which is presumed to be a strong binding site of the phosphate group in the substrate (Yang *et al.*, 1990b). Thr^{42} may be structurally important, because the mutant protein T42C did not accumulate in cells (S. Kanaya, personal communication). The reason why Ser^{71}, Tyr^{73}, and Glu^{131} are highly conserved remains to be determined, because mutations of these residues have no serious effect on the enzymatic activity. The role of Thr^{69} has not been analyzed.

The effects of mutations of the conserved residues in HIV-1 RT on its RNase H and polymerase activities have also been analyzed. These two enzymatic activities have been usually determined for the RT, because the RT consists of the polymerase and RNase H domains. Asp^{443} and Glu^{478} are required for the RNase H activity but not for the polymerase activity, because the $Asp^{443} \rightarrow$ Asn (Mizrahi *et al.*, 1990) or $Glu^{478} \rightarrow$ Gln (Schatz *et al.*, 1989) mutation impairs the RNase H activity without affecting the polymerase activity. The replacement of Asp^{498} by Asn yielded an unstable protein that could not be enzymatically characterized (Mizrahi *et al.*, 1990). These three acidic residues, as well as

Asp^{549}, provide Mn^{2+} binding sites in the HIV-1 RNase H domain (Davies *et al.*, 1991). Like the corresponding mutations in the *E. coli* enzyme, these mutations might induce a conformational change around the Mn^{2+} binding site probably due to the formation of new hydrogen bond networks, thereby preventing Mn^{2+} binding or destabilizing the protein. The instability caused by the Asp^{498} mutation was relieved by the introduction of a second distal Asn^{443} substitution (Mizrahi *et al.*, 1990), probably because unfavorable hydrogen bond networks formed by the $Asp^{498} \rightarrow Asn$ mutation would be rearranged to favorable ones by the additional introduction of the $Asp^{443} \rightarrow Asn$ mutation. No mutational analysis has been done for Asp^{549}. The importance of the two aspartates in Mo-MuLV RT, which correspond to Asp^{443} and Asp^{498} of HIV-1 RT, for the RNase H activity has also been reported (Repaske *et al.*, 1989).

His^{539} (corresponding to His^{124} in *E. coli* RNase HI) has been suggested to be essential for HIV-1 replication, because the $His^{539} \rightarrow Asn$ mutation impairs virus infectivity (Tisdale *et al.*, 1991). Like the mutation of His^{124} in the *E. coli* enzyme, the $His^{539} \rightarrow Phe$, Asn, or Asp mutation did not completely abolish the RNase H activity but dramatically reduced it (Schatz *et al.*, 1989; Tisdale *et al.*, 1991). However, unlike His^{124} in *E. coli* RNase HI, which has been suggested to be involved in the catalytic reaction (Oda *et al.*, 1993a), His^{539} in HIV-1 RT has been suggested to be required to enhance the substrate binding affinity (Wohrl *et al.*, 1991). An assay of binding of a synthetic DNA/RNA hybrid to the mutant protein by measuring the alteration in Trp^{535} fluorescence intensity supports such a role for His^{539} in HIV-1 RT (Cirino *et al.*, 1993). The mutant protein H539F, as well as E478Q, failed to bind the DNA/RNA hybrid, although they are structurally similar to the wild-type protein. In addition, unlike the $His^{124} \rightarrow Ala$ mutation in *E. coli* RNase HI, which does not affect the thermal stability of the protein (Kanaya *et al.*, 1991b), the $His^{539} \rightarrow Asn$ mutation resulted in a dramatic decrease in the thermal stability of the RNase H activity (Tisdale *et al.*, 1991). This discrepancy in the role of the conserved His residue may reflect the difference in the structural independence of the loop containing this residue. In *E. coli* RNase HI, the His^{124}-containing loop is exposed to the solvent (Katayanagi *et al.*, 1992), whereas in HIV-1 RT, the His^{539}-containing loop interacts with a portion of the thumb subdomain of p51 (Kohlstaedt *et al.*, 1992).

2.8.3. Basic Protrusion

In the basic protrusion of the enzyme, seven basic amino acid residues are clustered. The involvement of this region in the substrate-binding site was originally proposed by Kanaya *et al.* (1990a), because side chains with positive charges are generally required to interact with the phosphate backbone of the nucleotide. Alanine-scanning mutagenesis, followed by the determination of the kinetic parameters of each mutant protein, strongly suggests that the cluster of positive charges in this region is important for the effective binding of the substrate (Kanaya *et al.*, 1991c). Interestingly, the RNase H domain of HIV-1 RT lacks this basic protrusion (Davies *et al.*, 1991). There is a controversy as to whether the isolated RNase H domain of HIV-1 RT is enzymatically active or not (see Section 3.1). Some groups have reported that it is inactive (Hostomska *et al.*, 1991; Hostomsky *et al.*, 1991; Becerra *et al.*, 1990). Superposition of the crystal structure of *E. coli* RNase HI on that of the heterodimer of HIV-1 RT indicates that the basic protrusion of *E. coli* RNase HI is located at the position in which a part of the connection and the thumb subdomains of p66 are located (Kohlstaedt *et al.*, 1992). Because these polymerase subdomains contribute to provide a DNA/RNA binding site in HIV-1 RT, the reduction in the binding affinity to the substrate may account for the poor RNase H activity of the isolated HIV-1 RNase H domain (Kanaya *et al.*, 1991c; Nakamura *et al.*, 1991a; Kohlstaedt *et al.*, 1992). The disorder in the conformation of the loop containing His^{539} may also account for its poor RNase H activity (Davies *et al.*, 1991; Kohlstaedt *et al.*, 1992).

The basic protrusion seems to be quite variable in size and overall charge in various RNase H enzymes. In contrast to the HIV-1 RNase H domain, the Mo-MuLV RNase H domain has a portion of the cluster of positive charges in this region. The result that the isolated RNase H domain of Mo-MuLV RT exhibits enzymatic activity *in vitro* (Tanese and Goff, 1988) supports the importance of the positive charge cluster for the binding of the substrate. However, Telesnitsky *et al.* (1992) have shown that this positive charge cluster is not involved in the substrate binding site of Mo-MuLV RT. Deletion of 11 amino acid residues, which are predicted to constitute the αC-helix in the Mo-MuLV RNase H domain, affects neither the RNase H nor the polymerase activities of Mo-MuLV RT. This region may have a different function in Mo-MuLV RT, because its deletion abolishes viral infectivity (Telesnitsky *et al.*, 1992).

2.8.4. Cysteine Residues

The enzyme is sensitive to the thiol-blocking reagent, *N*-ethylmaleimide (NEM), and the enzymatic activity is completely inhibited in the presence of 10 mM NEM (Berkower *et al.*, 1973). The enzyme contains three free Cys residues at positions 13, 63, and 133. Site-directed mutagenesis and chemical modification experiments have clearly shown that the inactivation of the enzyme arises not from the blocking of the thiol groups but from the steric hindrance by the modifying group incorporated at Cys^{13} or Cys^{133} (Kanaya *et al.*, 1990b). Loya and Hizi (1993) have recently reported, however, that *E. coli* RNase HI is resistant to NEM, whereas HIV-1 RT RNase H is sensitive to it. NEM failed to inhibit *E. coli* RNase HI, probably because either Cys^{13} or Cys^{133} of *E. coli* RNase HI was not modified by NEM under their conditions. In fact, the reaction of these residues, which are somewhat buried within the protein molecule, with thiol-blocking reagents is relatively slow as compared to that of Cys^{63}, which is fully exposed to the solvent (Kanaya *et al.*, 1990b). There is no doubt that HIV-1 RT RNase H is more sensitive to NEM than *E. coli* RNase HI. Like the *E. coli* enzyme, the inactivation of HIV-1 RT RNase H by a thiol blocking reagent has been shown to arise not from the blocking of the thiol group but from the significant alteration in the structure of the RT induced by the incorporation of the bulky groups at Cys^{280} (Hizi *et al.*, 1992; Loya *et al.*, 1992).

2.8.5. Other Residues

Cys^{13}, Asn^{16}, Gln^{72}, His^{83}, Trp^{85}, Trp^{90}, and Lys^{122} seem to be functionally or structurally important for the substrate binding, because mutations of these residues resulted in a considerable increase in the K_m values (Table II). Asn^{16}, Gln^{72}, and Lys^{122} may directly interact with the substrate, because these residues are exposed to the solvent and localized along the groovelike depression, in which the Mg^{2+} binding site is involved. The amide groups in the side chains of Asn^{16} and Gln^{72}, as well as Thr^{43}, Asn^{44}, and Asn^{45}, are likely to form hydrogen bonds with the substrate, which may facilitate specific enzyme–substrate binding. Lys^{122} may electrostatically interact with the phosphate backbone of the substrate. Cys^{13}, His^{83}, Trp^{85}, and Trp^{90} may contribute indirectly to the substrate binding. Because Cys^{13} is somewhat buried within the protein molecule and His^{83} is located near the positive charge cluster, mutations of these residues seem to induce an unfavorable conformational change around the substrate-binding site. Likewise, the mutations of Trp^{85} and Trp^{90} might substantially alter the conformation of the basic protrusion and thereby dramatically reduce both the substrate binding affinity and the hydrolysis rate, because these residues, as well as Trp^{81} and Trp^{104}, are required to maintain the conformation of the basic protrusion through aromatic interactions. In fact, the mutant proteins W85A and W90A gave distinctly different far- and near-ultraviolet CD spectra from those of the wild-type protein (Kanaya *et al.*, 1991c).

The insertion of Gly^{80b} or Ala^{80b} between Gln^{80} and Trp^{81} of *E. coli* RNase HI seriously affects the substrate binding of the enzyme (Ishikawa *et al.*, 1993d). The crystal structure of the mutant protein, in which Gly^{80b} is inserted, indicated that Gly^{80b} allows the formation of a "paperclip" motif between the αII and αIII helixes (Ishikawa *et al.*, 1993d). This slight shift in the position of the αII and αIII helixes or the rigidity at this junction may impede the substrate binding affinity of the enzyme.

2.9. Mechanisms for Substrate-Binding and Hydrolysis

2.9.1. Model for the Enzyme–Substrate Complex

A model for the complex between the enzyme and a DNA/RNA hybrid was originally constructed by Yang *et al.* (1990b). They docked a hybrid model (decamer) constructed from an A-type mixed hybrid (Wang *et al.*, 1982) onto the free structure of the enzyme by simply superimposing a phosphate from the DNA strand onto the sulfate position and allowing the RNA strand to interact with the active site of the enzyme. According to this model, a DNA/RNA hybrid fits well into the large groovelike depression of the protein molecule. A similar model for the complex between the enzyme and a chimeric nonanucleotide duplex has been constructed by Nakamura *et al.* (1991b) by docking the free structure of the enzyme onto the free structure of the chimeric nonanucleotide duplex. These structures were independently determined by X-ray crystallography and NMR distance geometry, respectively, with experimental supports, such as site-directed mutagenesis and substrate titration with NMR. This model, however, does not show the interaction between the basic protrusion of the enzyme and the substrate, because the chimeric nonanucleotide duplex does not seem to be large enough to interact with both the Mg^{2+} binding site and the basic protrusion. Models for the complex between the enzyme and a substrate large enough to cover the entire interface of the enzyme have been constructed (Kanaya *et al.*, 1991c; Nakamura *et al.*, 1991a; Katayanagi *et al.*, 1992). According to these models, the interface of the enzyme covers roughly 1.5 turns of the hybrid, and the basic protrusion electrostatically interacts with the RNA strand approximately one turn away toward the 5′-direction from the scissile bond that interacts with the Mg^{2+} binding site. This model is schematically shown in Fig. 6. The recent observation that a double-stranded DNA, which is not a substrate for RNase H, contacts αA, αB, the β1-β2 hairpin, and the His^{539}-containing loop of the RNase H domain in HIV-1 RT complexed with the double-stranded DNA (Jacobo-Molina *et al.*, 1993) is consistent with this model.

A model proposed by Nakamura *et al.* (1991b) allows detailed analysis of the mechanism of substrate recognition by the enzyme. If the positions of the ribonucleotides in the RNA strand of the chimeric nonanucleotide duplex are numbered as 1, 2, and so on from the scissile bond to the 3′-direction and vice versa, the main-chain carbonyl of Cys^{13}, the amide group of Asn^{44}, and the amide group of Gln^{72} are candidates to form hydrogen bonds with the 2′-OH groups of the ribonucleotides at positions 1, −1, and −2, respectively. The formation of these hydrogen bonds may facilitate the unique substrate recognition by the enzyme.

Asn^{16} can interact with the DNA strand, instead of the RNA strand. This residue has been suggested to be required to sterically lock the position of the chimeric nonanucleotide duplex to the enzyme (Nakamura *et al.*, 1991b). In addition, it may form a hydrogen bond, probably with the backbone phosphate group. It is noted that Asn^{16} is located in a long flexible loop, rich in Gly residues, between βA and βB (Leu^{14}-Gly^{15}-Asn^{16}-Pro^{17}-Gly^{18}-Pro^{19}-Gly^{20}-Gly^{21}). This Gly-rich loop, which is located near the Mg^{2+} binding site, may be involved in the binding of the nucleotides, as suggested for various ATP-

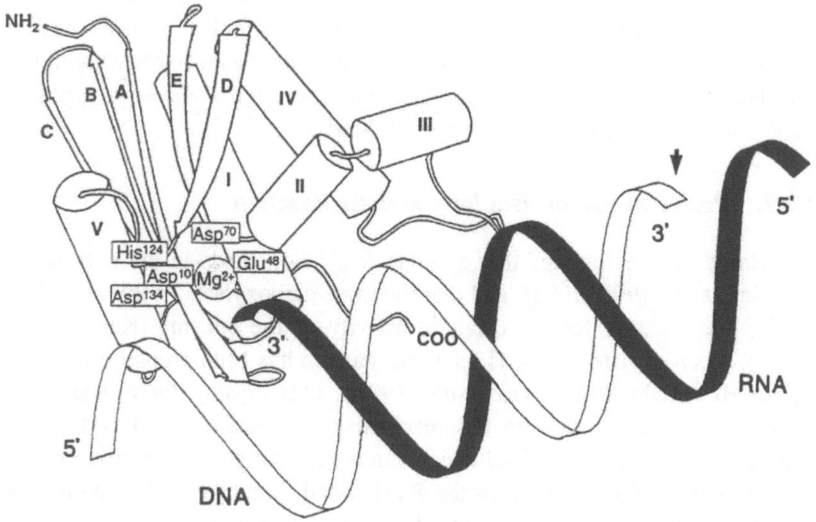

FIGURE 6. Schematic representation of the model for the interaction between a DNA/RNA hybrid and *E. coli* RNase HI. The three-dimensional structure of *E. coli* RNase HI is described in the legend of Fig. 2. The localizations of the amino acid residues (Asp^{10}, Glu^{48}, Asp^{70}, His^{124}, and Asp^{134}) with side chains that are involved in the catalytic function or the Mg^{2+} binding site are indicated. If the model for the complex between HIV-1 RT and a DNA/RNA hybrid were constructed such that the RNase H domain of HIV-1 RT and *E. coli* RNase HI are superimposed, the active site of the DNA polymerase domain of HIV-1 RT would contact the DNA strand at the position indicated by an arrow.

requiring enzymes (Walker *et al.*, 1982). The importance of such Gly-rich loops for the binding of nucleotides has been analyzed by site-directed mutagenesis. A typical example was seen in porcine adenylate kinase, which has a Gly-rich loop with the sequence Gly^{15}-Gly^{16}-Pro^{17}-Gly^{18}-Ser^{19}-Gly^{20}-Lys^{21}-Gly^{22}. Either the mutation of Pro^{17} (Reinstein *et al.*, 1988; Tagaya *et al.*, 1989), or Gly^{15} or Gly^{20} (Yoneya *et al.*, 1989) in this sequence reduced the binding affinity of the enzyme.

The model by Nakamura *et al.* (1991b) cannot account for the involvement of Asn^{45} in the substrate binding. The amide group of Asn^{45} does not interact with a DNA/RNA hybrid and can only form a hydrogen bond with the purine ring of the DNA strand when the hybrid bends on binding to the enzyme. This result suggests that the hybrid is kinked upon binding to the enzyme. Recently, the solution structure of the DNA/RNA hybrid was determined by NMR spectroscopy (Fedoroff *et al.*, 1993). Interestingly, the DNA/RNA hybrid was shown to assume neither the A-type nor the B-type geometry. Such a characteristic feature in the structure of the DNA/RNA hybrid was suggested to be responsible for the discrimination between DNA/RNA hybrids and pure RNA duplexes by the enzyme. According to the model of Fedoroff *et al.* (1993) for the enzyme–substrate complex, the DNA/RNA hybrid can contact with Asn^{45} without bending through the formation of hydrogen bond between DNA sugar phosphate backbone and side-chain amide group of Asn^{45}.

2.9.2. Proposed Mechanism for Catalytic Reaction

Based on the findings that a single Mg^{2+} ion binds to the enzyme (Katayanagi *et al.*, 1990, 1992), and that the three carboxylates, Asp^{10}, Glu^{48}, and Asp^{70}, are essential for the catalytic activity of the enzyme (Kanaya *et al.*, 1990a), a carboxylate-hydroxyl relay mechanism has been proposed for *E. coli* RNase HI activity (Nakamura *et al.*, 1991b; Oda *et al.*, 1993a; Kanaya and Ikehara, 1993). According to this mechanism, at least one carboxyl group is required to bind the Mg^{2+} ion, which can attract the phosphate group of RNA electrostatically. For cleavage of the P-O3′ bond, one carboxyl group acts as a proton acceptor for H_2O to release a hydroxyl ion, which subsequently attacks the phosphate group nucleophilically, and one carboxylate acts as a proton donor. This mechanism is similar to that proposed for DNase I activity (Suck and Oefner, 1986; Lahm *et al.*, 1991), in which an imidazole group, instead of a carboxyl group, activates an attacking H_2O molecule, and Ca^{2+}, instead of Mg^{2+}, interacts with the phosphate group of the substrate. This carboxylate-hydroxyl relay mechanism has been modified such that two carboxylates (Asp^{10} and Glu^{48}), as well as Gly^{11} and Asn^{44}, directly coordinate with the Mg^{2+} ion (Katayanagi *et al.*, 1993b). Because Asp70 is the only carboxylate that does not coordinate with the Mg^{2+} ion and is essential for the catalytic activity of the enzyme, this residue is currently proposed to activate a H_2O molecule as a proton acceptor for the cleavage of the P-O3′ bond (Katayanagi *et al.*, 1993a,b). However, the pH titration experiments with NMR in the presence or absence of Mg^{2+} ion (Oda *et al.*, 1994) and the determination of the dissociation constants for the binding of Mg^{2+} ion to active-site mutant proteins (Y-Y. Liu, M. Oobatake, and S. Kanaya, personal communication) suggest that Glu^{48}, rather than Asp^{70}, is required to activate an attacking H_2O molecule as a general base for the cleavage of the P-O3′ bond. Comparison of the crystal structure of the enzyme complexed with Mg^{2+} ion with that of the enzyme free from Mg^{2+} ion indicated that the position of Glu^{48} shifts on the binding of Mg^{2+} so that it can coordinate with Mg^{2+} ion. However, the possibility cannot be ruled out that Glu^{48} shifts the position on the binding of Mg^{2+} ion due to the electrostatic attraction.

A schematic drawing for the revised model for the catalytic mechanism of the enzyme is shown in Fig. 7. According to this mechanism, Glu^{48} accepts a proton from an attacking H_2O molecule. Asp^{70} governs the conformation of

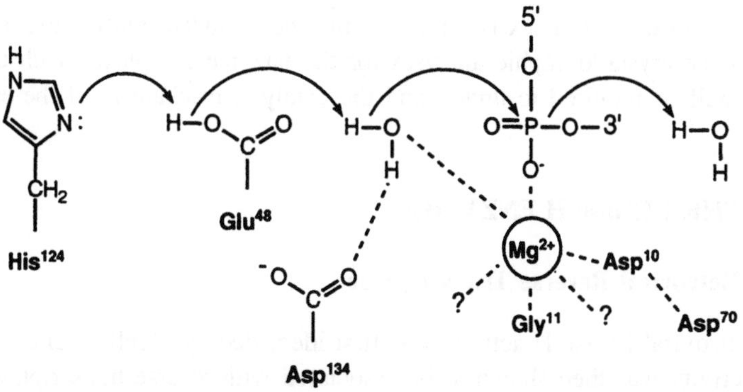

FIGURE 7. A proposed mechanism for the catalytic reaction of *E. coli* RNase HI. One possible role of Asp[134] in a carboxylate-hydroxyl relay mechanism is indicated. Alternatively, it may be required to hold an H_2O that supplies a proton to the 3′-oxyanion released after the cleavage of the RNA.

Asp[10] through electrostatic repulsion. His[124] accelerates the proton transfer by accepting a proton from the resultant-COOH group of Glu[48] and acting as a proton pump, and thereby enhances the catalytic efficiency of the enzyme (Oda *et al.*, 1993a). A possible role for Asp[134], which may be required to hold the attacking H_2O molecule or an H_2O molecule that supplies a proton to the 3′-oxyanion released after the cleavage of the P-O3′ bond of the RNA, is also indicated. There is an alternative possibility, however, that the attacking hydroxyl ion is activated by Mg^{2+}, instead of Glu[48], as suggested for the catalytic mechanism of the hammerhead ribozyme (Koizumi and Ohtsuka, 1991), the tetrahymena ribozyme (Grosshans and Cech, 1989), and an RNA component of RNase P (Guerrie-Takada *et al.*, 1986), although the P-O5′ bond, instead of the P-O3′ bond, is cleaved by these ribozymes. In this case, Glu[48] may only be required to keep the conformation of the active site functional and His[124] may assist the activation of the attacking hydroxyl ion by accepting a proton from a water molecule.

The two-metal ion mechanism, which is basically proposed for the 3′ → 5′ exonuclease activity of *E. coli* DNA polymerase I (Beese and Steitz, 1989,1991), has been proposed for *E. coli* RNase HI activity (Yang *et al.*, 1990b), because of the similarity in the geometrical conformations of the catalytically essential carboxylates between these two enzymes. However, no evidence to support this two-metal ion mechanism has so far been provided. Crystallographic (Katayanagi *et al.*, 1993b) and NMR (Oda *et al.*, 1991) studies clearly indicate that only one Mg^{2+} ion binds to the free-form of the enzyme. However, the possibility that the second Mg^{2+} ion binds to the enzyme only when the enzyme–substrate complex is formed cannot be excluded. Further experiments, such as the crystallographic analyses for the enzyme complexed with the substrate, will be required to understand the catalytic mechanism of the enzyme.

3. OTHER RNase H ENZYMES

3.1. Retroviral Reverse Transcriptase

Retroviral RNase H activity was first identified by Moelling *et al.* (1971). This activity was then shown to be associated with reverse transcriptase (RT), which is required to convert a single-stranded genomic RNA to a double-stranded DNA (for a review, see Crouch and Dirksen, 1982; Varmus, 1988; Goff, 1990; Crouch, 1990; Wintersberger, 1990). According to these reviews, the main physiological functions of RNase H are the degradation of the RNA genome during the synthesis of the minus-DNA strand and the generation of RNA primers for the synthesis of the plus-DNA strand. In addition, the RNase H activities associated with RT are required for tRNA-primer excision. A requirement for the RNase H activity of RT for the transfer of the minus-strand strong-stop DNA to the 3′-end of the genome is proposed for HIV-1 RT (Garces and Wittek, 1991; Peliska and Benkovic, 1992) and Mo-MuLV RT (Tanese *et al.*, 1991). The involvement of the RNase H activities of HIV-1 RT in the reinitiation of DNA synthesis at specific pause sites along the RNA template has also been proposed (Dudding *et al.*, 1991). The functions of RNase H do not basically vary for different RT enzymes, except for a few exceptions. While the AMV-RNase H hydrolyzes the phosphodiester bond at the DNA–RNA junction, the HIV-1- and Mo-MuLV-RNases H do not (Furfine and Reardon, 1991b). Because the RNase H activities in reverse transcriptases are required for the proliferation of retroviruses (Repaske *et al.*, 1989; Schatz *et al.*, 1989; Tisdale *et al.*, 1991), retroviral RNase H is regarded as an important target for antiretroviral therapy, especially in connection with HIV infections and AIDS (Hostomsky *et al.*, 1993; Nanni *et al.*, 1993).

Many RT and RT-related sequences have been identified in assorted retroviruses and retroviruslike entities. Computer analyses revealed that the COOH-termini of these sequences have similarity to the *E. coli* RNase HI sequence (Johnson *et al.*, 1986; Doolittle *et al.*, 1989), indicating that retroviral RNase H resides in the COOH-terminal domain of the RT. The DNA polymerase, which consists of approximately 250 amino acid residues and is located in the NH_2-terminal domain, is separated from the RNase H domain, which consists of approximately 150 amino acid residues, by a "tether" of approximately 150 amino acid residues.

The crystal structure of HIV-1 RT, complexed either with the nonnucleotide inhibitor neviapine (Kohlstaedt *et al.*, 1992) or with double-stranded DNA and a noninhibitory monoclonal antibody Fab fragment (Arnold *et al.*, 1992; Jacobo-Molina *et al.*, 1993), has recently been determined at 3.5 and 3.0 Å resolution, respectively (the structure of Arnold *et al.*, is at 7 Å resolution). These structures resemble each other, except for the interfaces between the enzyme and the nucleic acid substrate or the Fab fragment. HIV-1 RT is composed of two subunits of 66 kDa (p66) and 51 kDa (p51). The p51 subunit is identical to the p66 subunit, except that it lacks the C-terminal RNase H domain of the p66 subunit. The RNase H domain, as well as other subdomains of the DNA polymerase domain (termed *fingers, palm, thumb,* and *connection*), form relatively independent structures. The RNase H domain contacts the p66 connection and p51 thumb subdomains. Enzymatic analyses of the polymerase and RNase H activities of HIV-1 RT have shown that these activities are functionally distinct as well (Starnes and Cheng, 1989; Tan *et al.*, 1991). The RNase H activity exhibits a broader Mg^{2+} optimum, a lower KCl optimum, and prefers a more alkaline pH than the DNA polymerase activity.

However, data have suggested that the polymerase and RNase H domains are structurally and functionally interdependent. RNase H activity is associated only with the dimeric forms of HIV-1 (Restle *et al.*, 1992). The isolated RNase H domain itself was enzymatically inactive, but its activity was restored when it was combined with the polymerase domain *in vitro* (Hostomsky *et al.*, 1991). Deletions of the connection subdomain or replacement of the RNase H domain with *E. coli* RNase HI resulted in a qualitative and quantitative alteration in the polymerase and RNase H activities of AKR-MuLV (Post *et al.*, 1993). Furthermore, mutations in one domain of Mo-MuLV RT (Telesnitsky *et al.*, 1992; Telesnitsky and Goff, 1993) or HIV-1 RT (Prasad and Goff, 1989; Hizi *et al.*, 1990, 1991; Boyer *et al.*, 1992) affect the function of the other domain. Interference of dimerization due to the alteration in the structure of RNase H has been suggested to be a reason why mutations in the RNase H domain affect the polymerase activity (Telesnitsky and Goff, 1993).

A controversy exists as to whether the HIV-1 RNase H domain is enzymatically active when it is separated from the polymerase domain. The RNase H domain of HIV-1 RT (Tyr-Ala-Ser-Arg-Tyr^{427} to Leu^{560}) expresses little activity when it is released from a recombinant DHFR-RNase H fusion protein by the action of HIV-1 protease (Hostomsky *et al.*, 1991; Hostomska *et al.*, 1991) or when an RNase H domain (Met-Asn-Glu-Leu-Tyr^{427} to Leu^{560}) is directly expressed in *E. coli* (Becerra *et al.*, 1990). In contrast, an RNase H domain consisting of amino acids 441 to 560 (p15), which is derived from p66 HIV-1 RT by the cleavage with HIV-1 protease, exhibits RNase H activity as a complex with a monoclonal antibody (Hansen *et al.*, 1988) or by itself (Schulze *et al.*, 1991). Hafkemeyer *et al.* (1991) have reported that this RNase H domain exhibits DNA polymerase activity as well. In addition, the recombinant HIV-1 RNase H domain consisting of amino acids 441 to 560 of RT, with a N-terminal tag for purification by immobilized metal affinity chromatography (Evans *et al.*, 1991), or that consisting of amino acids 427 to 560 of RT, with an N-terminal poly-histidine tag (Smith and Roth, 1993), are enzymatically active when expressed in *E. coli*. Because RNase H domains without these tags are inactive, these tags have been suggested to be required for the binding and alignment of the DNA/RNA hybrid or for the stabilization of the C-terminus, which includes the loop containing His^{539}. The disorder of the His^{539} loop has been suggested to be a reason for the inactivity of the isolated RNase H domain without tags (Davies *et al.*, 1991; Kohlstaedt *et al.*, 1992; Powers *et al.*, 1992). The mode of action of RNase H has been reported to be changed (Hansen *et al.*, 1988) or unchanged (Smith and Roth, 1993) when the RNase H domain is separated from the polymerase domain. The RNase H domain of Mo-MuLV RT, expressed in *E. coli* as a fusion protein with *trpE*, has also been reported to be enzymatically active (Tanese and Goff, 1988).

The polymerase and RNase H activities of AMV RT (Oyama *et al.*, 1989) and HIV-1 RT (Krug and Berger, 1989; Schatz *et al.*, 1990; Wohrl and Moelling, 1990; Furfine and Reardon, 1991a; Lederer *et al.*, 1992; Volkmann *et al.*, 1993) are functionally coupled in the process of reverse transcription, but not in a strict way (DeStefano *et al.*, 1991a,b). RNase H cleavage proceeds toward the 5′-end of RNA concomitantly with the synthesis of the minus-DNA strand, as shown in Fig. 8. These laboratories have reported different numbers of bases, varying from 7 to 20, for the distance between the polymerase and RNase H active sites. The simultaneous action of both activities is supported by the determination of

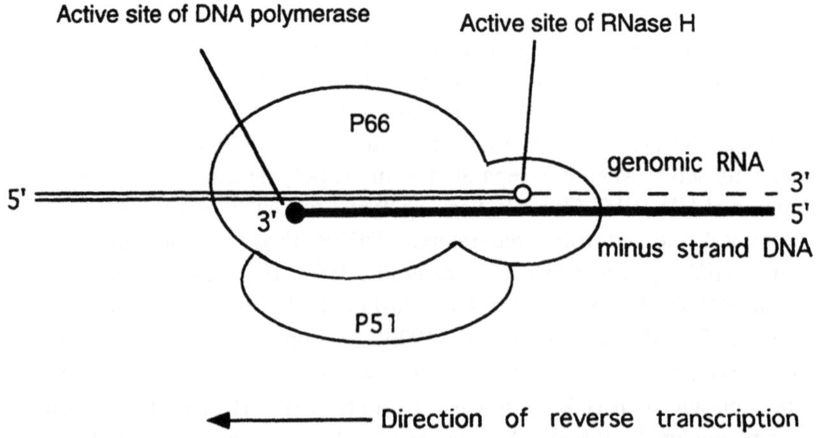

FIGURE 8. A model for the coupling of DNA polymerase and RNase H activities in the reverse transcription by HIV-1 RT.

the crystal structure of HIV-1 RT complexed with the DNA template-primer (Jacobo-Molina *et al.*, 1993).

The 19-mer/18-mer template-primer, which binds to a large cleft formed by the fingers, palm, and thumb subdomains of the p66 polymerase domain, also contacts αA, αB, the β1-β2 hairpin, and the loop containing His^{539} in the RNase H domain (Jacobo-Molina *et al.*, 1993). The distance between the polymerase active site, which resides in the palm subdomain, and the RNase H active site was shown to be 17–18 bases. The striking feature in the conformation of the DNA template-primer in the complex is that it assumes an A-form near the polymerase active site and a B-form near the RNase H active site. These two regions of different forms bend toward each other with an angle of 40–45°. The DNA/RNA hybrid, which is a substrate of RNase H, assumes an A-form. It remains to be determined whether such a bend is also observed when a DNA/RNA hybrid, instead of a DNA/DNA duplex, binds to the enzyme, because double-stranded DNA is not a substrate of RNase H. Nevertheless, the determination of the crystal structure of HIV-1 RT complexed with an inhibitor or DNA template-primer at high resolution would certainly accelerate the development of drugs for AIDS therapy.

3.2. Bacterial RNases H

Three genes encoding bacterial RNase H enzymes, in addition to that encoding *E. coli* RNase HI (Horiuchi *et al.*, 1981), have been cloned from *E. coli* (RNase HII, Itaya, 1990), *Salmonella typhimurium* (Itaya *et al.*, 1991), and *Thermus thermophilus* HB8 (Itaya and Kondo, 1991). In addition, the gene encoding bacteriophage T4 RNase H has been cloned (Hollingsworth and Nossal, 1991). The sizes of these enzymes are approximately 18 kDa for *Salmonella* RNase HI, 19 kDa for *T. thermophilus* RNase H, 24 kDa for *E. coli* RNase HII, and 35 kDa for T4 RNase H. There is little sequence similarity between T4 RNase H or *E. coli* RNase HII and *E. coli* RNase HI. However, considerable sequence similarities have been observed between *Salmonella* RNase H or *T. thermophilus* RNase H and *E. coli* RNase HI. T4 RNase H was shown to remove the RNA primers from the DNA chain initiated by the T4 primase-helicase (Hollingsworth and Nossal, 1991). The physiological functions of the other enzymes remain to be determined.

T. thermophilus RNase H has been overexpressed in *E. coli*, purified, and well characterized (Kanaya and Itaya, 1992). Like *E. coli* RNase HI, the enzyme reversibly unfolds with thermal denaturation. The melting temperature of *T. thermophilus* RNase H is, however, 34 °C higher than that of *E. coli* RNase HI at pH 5.5 in the presence of 1.2 M GdnHCl. The availability of such a thermophilic RNase H enzyme might serve as a tool for recombinant DNA technologies. The recognition properties of the RNase H enzymes have allowed their applications *in vitro*, such as for the removal of mRNA during the second strand of cDNA synthesis (Gubler and Hoffman, 1983; Okayama and Berg, 1982), the removal of the poly(A) tail from mRNA in the presence of oligo(dT) (Crouch, 1981), and the site-specific fragmentation of RNA (*in vitro* RNA editing) (Donis-Keller, 1979; Shibahara *et al.*, 1987; Hayase *et al.*, 1990). Currently, *E. coli* RNase HI is used for these purposes. However, the instability of the enzyme to heat inactivation limits the extent of its application. In addition, *T. thermophilus* RNase H may be effective to evaluate the dependence of antisense-arrested translation on the RNase H activity by introducing it together with the antisense oligonucleotide into cells. Only a few reports have shown that RNase H is involved in antisense-arrested translation (Minshull and Hunt, 1986; Dash *et al.*, 1987; Walder and Walder, 1988; Saison-Behmoaras *et al.*, 1991). It should be noted that RNase H, which shares sequence similarity with *E. coli* RNase HI or *T. thermophilus* RNase H, has recently been cloned from a lower eukaryote parasite (Campbell and Ray, 1993).

3.3. Yeast RNases H

From the yeast, *Saccharomyces cerevisiae*, many RNase H enzymes with different molecular weights, in the range of 21 to 68 kDa, and apparently different physiological functions have been isolated (Wyers *et al.*, 1973, 1976a,b; Huet *et al.*, 1976, 1977; Karwan *et al.*, 1983, 1986; Karwan and Wintersberger, 1986, 1988). The enzyme, which consists of a single polypeptide chain with a molecular weight of 68 kDa, copurifies with the replicative DNA polymerase (Karwan *et al.*, 1983). RNase H activity associated with RNA polymerase has also been described (Huet *et al.*, 1976, 1977). Wyers *et al.* (1976a,b) reported that RNase H_1 (48 kDa) and RNase H_2 (21 kDa) are physically and functionally distinct proteins. RNase H_1 is active at an acidic pH and is not sensitive to *N*-ethylmaleimide, whereas RNase H_2 is active at alkaline pH and is inhibited by *N*-ethylmaleimide. Both enzymes require divalent cations, preferably Mg^{2+}. Karwan and Kindas-Mugge (1989) reported that the 55 kDa RNase H enzyme, which is identical to the human Sm antigen, is associated with small RNAs, including tRNAs and spliceosomal RNAs. These results suggest that multiple structural genes encoding RNase H enzymes exist in yeast.

One of the structural genes encoding RNase H enzymes from yeast, *RNH1*, was originally cloned as that encoding a 21 kDa protein that has sequence similarity with *E. coli* RNase HI and expresses the RNase H activity *in vitro* (Itaya *et al.*, 1991). Later, the *RNH1* gene encoding this 21 kDa protein was shown to represent the 3′-terminal portion of the entire *RNH1* gene, encoding a 39 kDa protein, RNase HI (Cerritelli *et al.*, 1993). When yeast RNase HI is expressed in *E. coli*, the 30 kDa protein, which results from proteolytic cleavage near the amino-terminus of the 39-kDa protein, accumulates in the cells (Cerritelli *et al.*, 1993). Like the 21-kDa protein, this 30-kDa protein has RNase H activity *in vitro*. These results suggest that yeast RNase HI is susceptible to proteolysis. Some of the RNase H enzymes from yeast, which have been purified so far, may therefore represent degraded forms of native RNase H enzymes. In fact, it has been suggested that the 55-kDa RNase H is proteolytically processed to the 42-kDa RNase H in yeast (Karwan and Wintersberger, 1988).

3.4. RNases H from Higher Eukaryotes

Higher eukaryotic cells contain at least two distinct RNase H enzymes, which are classified as RNase HI, with a higher native molecular weight (>80 kDa), and RNase HII, with a lower native molecular weight (35–52 kDa) (Vonwirth *et al.*, 1989). The RNase HI enzymes have an acidic isoelectric point and require either Mn^{2+} or Mg^{2+} for activity. In contrast, the RNase HII enzymes have a higher isoelectric point and require only Mg^{2+} for activity. The RNase HI enzymes seem to play a role in DNA replication, whereas the RNase HII enzymes appear to be involved in transcription.

Several RNase HI enzymes have been highly purified from calf thymus (Busen, 1980, 1982; Rong and Carl, 1990), mouse FM3A cells (Masutani *et al.*, 1990), and K562 human erythroleukemic cells (Eder and Walder, 1991). RNase HI from the mouse cell line has been reported to be associated with DNA primase stimulatory factor and was suggested to be responsible for the removal of RNA primers from DNA chains during DNA replication (Masutani *et al.*, 1990). This enzyme consists of three subunits with molecular weights of 42, 41, and 27 kDa. Calf thymus RNase HI was also suggested to be involved in DNA replication (Busen *et al.*, 1977). There are conflicting reports on the molecular weight and subunit composition of this enzyme. The enzyme was originally reported to have a native molecular weight of 80 kDa, consisting of four subunits with molecular weights of 31.6, 26.6, 24.8, and 24.3 kDa (Busen, 1980). Later, the proposed subunit structure of this enzyme was modified, so that it had a trimeric structure consisting of two 31.6-kDa polypeptides and one 24.8-kDa polypeptide (Busen, 1982). More recently, however, Rong and Carl (1990) reported that the enzyme consists of a homodimer of 68-kDa subunits. They suggested that all of the polypeptides previously identified may reflect degraded forms of this native polypeptide. RNase HI from the human cell line has been suggested to play a role in ribonucleotide excision from genomic DNA during DNA replication (Eder and Walder, 1991). This enzyme has a native molecular weight of 89 kDa, consisting of a homodimer of 44-kDa subunits.

A mammalian RNase HII enzyme (RNase HIIb) has been highly purified only from calf thymus to date (Vonwirth *et al.*, 1989). The enzyme has a monomeric structure with molecular weight of 52 kDa, an isoelectric point of 7.0, and an optimum pH at 9.0–9.5. It requires Mg^{2+} for activity and cleaves RNA exonucleolytically instead of endonucleolytically. Enzymes similar to calf thymus RNase HIIb have been reported in chick embryo (Kitahara *et al.*, 1982), rat liver (Tashiro and Ueno, 1978a,b), Krebs II ascites cells (Cathala *et al.*, 1979), and HeLa cells (Kane, 1988). The enzyme from Hela cells was shown to have a renaturase activity, which facilitates transcript displacement by mammalian RNA polymerase II *in vitro* (Kane, 1988).

In addition to these two classes of RNase H enzymes, a distinct form of RNase H has been highly purified from calf thymus, which consists of a single polypeptide chain with a molecular weight of 78 kDa (Hagemeier and Grosse, 1989). An RNase H enzyme has also been purified from *Drosophila melanogaster* embryos, with an approximate molecular weight of 180 kDa consisting of two 49- and two 39-kDa subunits (DiFrancesco and Lehman, 1985). These enzymes are different from other RNase HI enzymes in their sensitivity to *N*-ethylmaleimide and in their preference for the divalent cation. While the other eukaryotic RNase HI enzymes prefer Mn^{2+} over Mg^{2+}, both of these enzymes prefer Mg^{2+}. Unlike the RNase HI enzymes, the 78-kDa enzyme from calf thymus is sensitive to *N*-ethylmaleimide. Like the 68-kDa yeast RNase H, both of these enzymes appear to stimulate the DNA polymerase–primase complex, and thereby increase the rate of DNA synthesis. They are expected to be involved in the removal of RNA primers during DNA replication (Hagemeier and Grosse, 1989), as suggested for RNase HI from the mouse cell line (Masutani *et al.*, 1990). Although considerable work has gone into identifying and characterizing RNase H enzymes from higher eukaryotes, this field is still in its early stage. It may be a long while until the functional roles of the various RNases from higher eukaryotes are clearly understood.

4. PROTEIN ENGINEERING OF RNase H

4.1. DNA:Protein Hybrid RNase H

Extensive knowledge on the structure and functions of *E. coli* RNase HI allowed the construction of an RNase H–DNA hybrid enzyme that cleaves single-stranded RNA with sequence-specific recognition (Kanaya *et al.*, 1992; Nakai *et al.*, 1994). In this hybrid enzyme, the DNA portion governs sequence-specific recognition and directs the protein portion to the RNA to be cleaved (Fig. 9). A dramatic decrease in the stability of the hybrid between the oligo-DNA adduct and the RNA after the cleavage of the RNA within the hybrid region allows the hybrid enzyme to perform multiple turnovers. Hybrid enzymes with various sizes and sequences of oligodeoxyribonucleotides may therefore function as RNA restriction enzymes and may be useful for structural and functional studies of RNA.

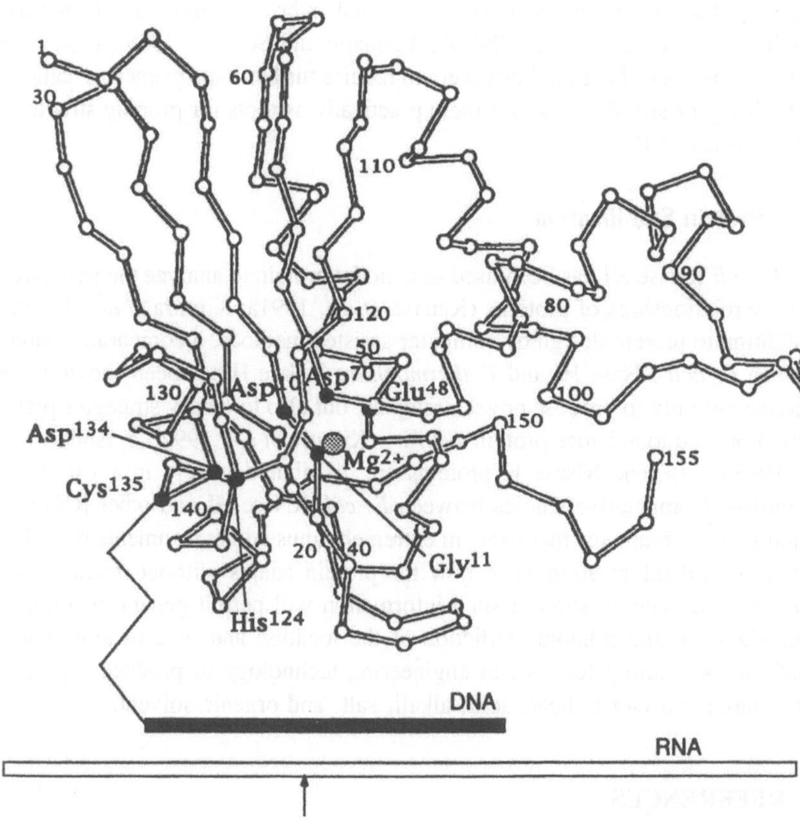

FIGURE 9. Schematic representation of a model for the sequence-specific cleavage of RNA with a DNA:protein hybrid RNase H. The backbone structure of *E. coli* RNase HI and the position of Mg^{2+} are shown. Oligo-DNA, which is covalently attached to Cys^{135} through a linker, directs the RNase H portion to the RNA strand to be cleaved. An arrow indicates the position at which the RNA is cleaved. The α-carbons of the amino acid residues with side chains that are involved in the catalytic function or the Mg^{2+} binding site, as well as that of Cys^{135}, are represented by solid circles. Numbers indicate the positions of amino acids in the *E. coli* RNase HI sequence.

RNase H and either oligo-DNA or modified oligo-DNA have been shown to be useful for the sequence-specific cleavage of RNA (Donis-Keller, 1979; Inoue *et al.*, 1987; Shibahara *et al.*, 1987; Metelev *et al.*, 1988; Atabekov *et al.*, 1988; Hayase *et al.*, 1990). All of these methods consist of two successive procedures: the annealing of an oligo-DNA or chimeric oligonucleotide to the RNA to be cleaved and the digestion with RNase H. In addition, several other approaches to sequence-specific RNA cleavage, such as cleavages with ribozymes (Haseloff and Gerlach, 1988), Fe(II)-bleomycin (Carter *et al.*, 1990), hybrid RNase A protein (Zuckermann and Schultz, 1988), and hybrid staphylococcal nuclease protein (Zuckermann *et al.*, 1988; Zuckermann and Schultz, 1989), have been studied. However, these methods seem to require further improvement in catalytic efficiency or specificity to use them practically as tools for probing structures and functions of RNA.

4.2. Protein Stabilization

E. coli RNase HI has been used as a model protein to analyze the structure–stability relationships of proteins (Kanaya *et al.*, 1991a; Kimura *et al.*, 1992a). In addition to protein design by computer-assisted methods, a comparative study between *E. coli* RNase HI and *T. thermophilus* RNase H has been shown to be effective not only to propose novel strategies, but also to justify strategies previously proposed to enhance protein stability (Kimura *et al.*, 1992b,c; Ishikawa *et al.*, 1993a,c,d). The RNase H proteins are distributed widely in a variety of organisms. Comparative studies between *E. coli* RNase HI and other RNase H proteins from organisms that grow in extremely unusual environments may thus provide useful information as to how the protein adapts without impairing its function. The accumulation of such information will permit general theoretical comprehension and reliable prediction of the location and type of amino acid substitutions required for protein engineering technology to produce a protein that is more resistant to heat, acid, alkali, salt, and organic solvent.

5. REFERENCES

Arendes, J., Carl, P. L., and Sugino, A., 1982, A mutation in the *rnh*-locus of *Escherichia coli* affects the structural gene for RNase H, *J. Biol. Chem.* **257**:4719–4722.

Arnold, E., Jacobo-Molina, A., Nanni, R. G., Williams, R. L., Lu, X., Ding, J., Clark, A. D., Jr., Zhang, A., Ferris, A. L., Clark, P., Hizi, A., and Hughes, S. H., 1992, Structure of HIV-1 reverse transcriptase/DNA complex at 7 Å resolution showing active site locations, *Nature* **357**:85–89.

Atabekov, K. J., Tyulkina, L. G., Karpova, O. V., Metelev, V. G., Rodionova, N. P., Shabarova, Z. A., and Atabekov, J. G., 1988, Site-specific enzymatic cleavage of TMV RNA directed by deoxyribo- and chimeric(deoxyribo-ribo)oligonucleotides, *FEBS Lett.* **232**:96–98.

Becerra, S. P., Clore, G. M., Gronenborn, A. M., Karlstrom, A. R., Stahl, S. J., Wilson, S. H., and Wingfield, P. T., 1990, Purification and characterization of the RNase H domain of HIV-1 reverse transcriptase expressed in recombinant *E. coli*, *FEBS Lett.* **270**:76–80.

Beese, L. S., and Steitz, T. A., 1989, Structure of *E. coli* DNA polymerase I, large fragment, and its functional implications, in *Nucleic Acids and Molecular Biology* (F. Eckstein and D. M. J. Lilley, eds.), Vol. 3, pp. 28–43, Springer-Verlag Berlin Heidelberg.

Beese, L. S., and Steitz, T. A., 1991, Structural basis for the 3′-5′ exonuclease activity of *Escherichia coli* DNA polymerase I: A two metal ion mechanism, *EMBO J.* **10**:25–33.

Berkower, I., Leis, J., and Hurwitz, J., 1973, Isolation and characterization of an endonuclease from *Escherichia coli* specific for ribonucleic acid in ribonucleic acid–deoxyribonucleic acid hybrid structure, *J. Biol. Chem.* **248**:5914–5921.

Blundell, T., Jenkins, J., Pearl, L., Sewell, T., and Pedersen, V., 1985, The high resolution structure of endothiapepsin, in *Aspartic Proteinases and Their Inhibitors* (V. Kostka, ed.), pp. 151–161, Walter de Gruyter, Berlin.

Bockrath, R., Wolff, L., Farr, A., and Crouch, R. J., 1987, Amplified RNase H activity in *Escherichia coli* B/r increases sensitivity to ultraviolet radiation, *Genetics* **115**:33–40.

Boyer, P. L., Ferris, A. L., and Hughes, S. H., 1992, Cassette mutagenesis of the reverse transcriptase of human immunodeficiency virus type 1, *J. Virol.* **66**:1031–1039.

Busen, W., 1980, Purification, subunit structure and serological analysis of calf thymus ribonuclease HI, *J. Biol. Chem.* **255**:9434–9443.

Busen, W., 1982, The subunit structure of calf thymus ribonuclease HI as revealed by immunological analysis, *J. Biol. Chem.* **257**:7106–7108.

Busen, W., Peters, J. H., and Hausen, P., 1977, Ribonuclease H levels during the response of bovine lymphocytes to concanavalin A, *Eur. J. Biochem.* **74**:203–208.

Campbell, A. G., and Ray, D. S., 1993, Functional complementation of an *Escherichia coli* ribonuclease H mutation by a cloned genomic fragment from trypanosomatid *Crithidia fasciculata*, *Proc. Natl. Acad. Sci. USA* **90**:9350–9354.

Carl, P. L., Bloom, L., and Crouch, R. J., 1980, Isolation and mapping of a mutation in *Escherichia coli* with altered levels of ribonuclease H, *J. Bacteriol.* **144**:28–35.

Carter, B. J., DeVroom, E., Long, E. C., Van Der Marel, G. A., Van Boom, J. H., and Hecht, S. M., 1990, Site-specific cleavage of RNA by Fe(II)-bleomycin, *Proc. Natl. Acad. Sci. USA* **87**:9373–9377.

Casaregola, S., Khidhir, M., and Holland, I. B., 1987, Effects of modulation on RNase H production on the recovery of DNA synthesis following UV-irradiation in *Escherichia coli*, *Mol. Gen. Genet.* **209**:494–498.

Cathala, G., Rech, J., Huet, J., and Jeanteur, R., 1979, Isolation and characterization of two types of ribonucleases H in Krebs II ascites cells, *J. Biol. Chem.* **254**:7353–7361.

Cerritelli, S. M., Shin, D. Y., Chen, H. C., Gonzales, M., and Crouch, R. J., 1993, Proteolysis of *Saccharomyces cerevisiae* RNase HI in *E. coli*, *Biochimie* **75**:107–111.

Cirino, N. M., Kalayjian, R. C., Jentoft, J. E., and LeGrice, S. F. J., 1993, Fluorimetric analysis of recombinant p15 HIV-1 RNase H, *J. Biol. Chem.* **268**:14743–14749.

Cox, E. C., and Horner, D. L., 1986, DNA sequence and coding properties of mutD (dnaQ) a dominant *Escherichia coli* mutator gene, *J. Mol. Biol.* **190**:113–117.

Crouch, R. J., 1981, Analysis of nucleic acid structure by RNase H, in *Gene Amplification and Analyses* (J. G. Chirikjian and T. S. Papas, eds.), Vol. 2, pp. 218–228, Elsevier/North-Holland, New York.

Crouch, R. J., 1990, Ribonuclease H: From discovery to 3D structure, *New Biol.* **2**:771–777.

Crouch, R. J., and Dirksen, M.-L., 1982, Ribonuclease H, in *Nuclease* (S. M. Linn and R. J. Roberts, eds.), pp. 211–241, Cold Spring Harbor Laboratory, Cold Spring Harbor, New York.

Darlix, J., 1975, Simultaneous purification of Escherichia coli termination factor rho, RNase III and RNAse H, *Eur. J. Biochem.* **51**:369–376.

Dasgupta, S., Masukata, H., and Tomizawa, J., 1987, Multiple mechanisms for initiation of ColE1 DNA replication: DNA synthesis in the presence and absence of ribonuclease H, *Cell* **51**:1113–1122.

Dash, P., Lotan, I., Knapp, M., Kandel, E. R., and Goelet, P., 1987, Selective elimination of mRNAs *in vivo:* Complementary oligodeoxynucleotides promote RNA degradation by an RNase H-like activity, *Proc. Natl. Acad. Sci. USA* **84**:7896–7900.

Davies, J. F., Hostomska, Z., Hostomsky, Z., Jordan, S. R., and Matthews, D. A., 1991, Crystal structure of the ribonuclease H domain of HIV-1 reverse transcriptase, *Science* **252**:88–95.

DeMassy, B., and Fayet, O., 1984, Multiple origin usage for DNA replication in *sdrA(rnh)* mutations of *Escherichia coli* K-12, *J. Mol. Biol.* **178**:227–236.

DeStefano, J. J., Buiser, R. G., Mallaber, L. M., Myers, T. W., Bambara, R. A., and Fay, P. J.,

1991a, Polymerization and RNase H activities of the reverse transcriptases from avian myeloblastosis, human immunodeficiency, and Moloney murine leukemia viruses are functionally uncoupled, *J. Biol. Chem.* **266:**7423–7431.

DeStefano, J. J., Buiser, R. G., Mallaber, L. M., Bambara, R. A., and Fay, P. J., 1991b, Human immunodeficiency virus reverse transcriptase displays a partially processive 3′ to 5′ endonuclease activity, *J. Biol. Chem.* **266:**24295–24301.

DiFrancesco, R. A., and Lehman, I. R., 1985, Interaction of ribonuclease H from *Drosophila melanogaster* embryos with DNA polymerase-primase, *J. Biol. Chem.* **260:**14764–14770.

Dirksen, M.-L., and Crouch, R. J., 1981, Selective inhibition of RNase H by dextran, *J. Biol. Chem.* **256:**11569–11573.

Donis-Keller, H., 1979, Site specific enzymatic cleavage of RNA, *Nucleic Acids Res.* **7:**179–192.

Doolittle, R., Feng, D. F., Johnson, M. S., and McClure, M. A., 1989, Origins and evolutionary relationships of retroviruses, *Q. Rev. Biol.* **64:**1–30.

Dudding, L. R., Nkabinde, N. C., and Mizrahi, V., 1991, Analysis of the RNA- and DNA-dependent DNA polymerase activities of point mutants of HIV-1 reverse transcriptase lacking ribonuclease H activity, *Biochemistry* **30:**10498–10506.

Eder, P. S., and Walder, J. A., 1991, Ribonuclease H from K562 human erythroleukemia cells: Purification, characterization, and substrate specificity, *J. Biol. Chem.* **266:**6472–6479.

Evans, D. B., Brawn, K., Deibel, M. R., Tarpley, Jr. W. G., and Sharma, S. K., 1991, A recombinant ribonuclease H domain of HIV-1 reverse transcriptase that is enzymatically active, *J. Biol. Chem.* **266:**20583–20585.

Fedoroff, O.-Y., Salazar, M., and Reid, B. R., 1993, Structure of a DNA:RNA hybrid duplex: Why RNase H does not cleave pure RNA, *J. Mol. Biol.* **233:**509–523.

Fersht, A., 1985, *Enzyme Structure and Mechanism*, 2nd ed., W. H. Freeman, New York.

Furfine, E. S., and Reardon, J. E., 1991a, Reverse transcriptase RNase H from the human immunodeficiency virus: Relationship of the DNA polymerase and RNA hydrolysis activities, *J. Biol. Chem.* **266:**406–412.

Furfine, E. S., and Reardon, J. E., 1991b, Human immunodeficiency virus reverse transcriptase RNase H: Specificity of $tRNA^{Lys3}$-primer excision, *Biochemistry* **30:**7041–7046.

Garces, J., and Wittek, R., 1991, Reverse-transcriptase-associated RNase H activity mediates template switching during reverse transcription *in vitro*, *Proc. R. Soc. Lond. B* **243:**235–239.

Goff, S. P., 1990, Retroviral reverse transcriptase: Synthesis, structure, and function, *J. Acquired Immune Defic. Syndr.* **3:**817–831.

Goodwin, T. W., and Morton, R. A., 1946, The spectrophotometric determination of tyrosine and tryptophan in proteins, *Biochem. J.* **40:**628–632.

Grosshans, C. A., and Cech, T. R., 1989, Metal ion requirements for sequence-specific endoribonuclease activity of the *Tetrahymena* ribozyme, *Biochemistry* **28:**6888–6894.

Gubler, U., and Hoffman, B. J., 1983, Simple and very efficient method for generating cDNA libraries, *Gene* **25:**263–269.

Guerrier-Takada, C., Haydock, K., Allen, L., and Altman, S., 1986, Metal ion requirements and other aspects of the reaction catalyzed by M1 RNA, the RNA subunit of ribonuclease P from *Escherichia coli*, *Biochemistry* **25:**1509–1515.

Hafkemeyer, P., Ferrari, E., Brecher, J., and Hubscher, U., 1991, The p15 carboxyl-terminal proteolysis product of the human immunodeficiency virus type 1 reverse transcriptase P66 has DNA polymerase activity, *Proc. Natl. Acad. Sci. USA* **88:**5262–5266.

Hagemeier, A., and Grosse, F., 1989, A distinct form of ribonuclease H from calf thymus simulates its homologous DNA-polymerase-α-primase complex, *Eur. J. Biochem.* **185:**621–628.

Hansen, J., Schulze, T., Mellert, W., and Moelling, K., 1988, Identification and characterization of HIV-specific RNase H by monoclonal antibody, *EMBO J.* **7:**239–243.

Haruki, M., Noguchi, E., Nakai, C., Liu, Y. Y., Oobatake, M., Itaya, M., and Kanaya, S., 1994,

Investigating the role of conserved residue Asp134 in *Escherichia coli* ribonuclease HI by site-directed random mutagenesis, *Eur. J. Biochem.* **220:**623–631.

Haseloff, J., and Gerlach, W. L., 1988, Simple RNA enzymes with new and highly specific endoribonuclease activity, *Nature* **334:**585–591.

Hausen, P., and Stein, H., 1970, Ribonuclease H. an enzyme degrading the RNA moiety of DNA-RNA hybrids, *Eur. J. Biochem.* **14:**278–283.

Hayase, Y., Inoue, H., and Ohtsuka, E., 1990, Secondary structure in formylmethionine tRNA influences the site-directed cleavage of ribonuclease H using chimeric 2′-*O*-methyl oligodeoxyribonucleotides, *Biochemistry* **29:**8793–8797.

Henry, C. M., Ferdinand, F.-J., and Knippers, R., 1973, A hybridase from *Escherichia coli*, *Biochem. Biophys. Res. Commun.* **50:**603–611.

Higuchi, R., 1990, Recombinant PCR, in *PCR Protocols; A Guide to Methods and Applications* (M. Innis, D. H. Gelfand, J. J. Sninsky, and T. J. White, eds.), pp. 177–183, Academic Press, New York.

Hizi, A., Hughes, S. H., and Shaharabany, M., 1990, Mutational analysis of the ribonuclease H activity of human immunodeficiency virus 1 reverse transcriptase, *Virology* **175:**575–580.

Hizi, A., Tal, R., and Hughes, S. H., 1991, Mutational analysis of the DNA polymerase and ribonuclease H activities of human immunodeficiency virus type 2 reverse transcriptase expressed in *Escherichia coli*, *Virology* **180:**339–346.

Hizi, A., Shaharabany, M., Tal, R., and Hughes, S. H., 1992, The effects of cysteine mutations on the reverse transcriptases of human immunodeficiency virus types 1 and 2, *J. Biol. Chem.* **267:**1293–1297.

Hogrefe, H. H., Hogrefe, R. I., Walder, R. Y., and Walder, J. A., 1990, Kinetic analysis of *Escherichia coli* RNase H using DNA-RNA–DNA/DNA substrates, *J. Biol. Chem.* **265:**5561–5566.

Hollingsworth, H. C., and Nossal, N. G., 1991, Bacteriophage T4 encodes an RNase H which removes RNA made by the T4 DNA replication system *in vitro*, *J. Biol. Chem.* **266:**1888–1897.

Hong, X., and Kogoma, T., 1993, Absence of a direct role for RNase HI in initiation of DNA replication at the *oriC* site on the *Escherichia coli* chromosome, *J. Bacteriol.* **175:**6731–6734.

Horiuchi, T., Maki, H., Maruyama, M., and Sekiguchi, M., 1981, Identification of the *dnaQ* gene product and location of the structural gene for RNase H of *Escherichia coli* by cloning of the genes, *Proc. Natl. Acad. Sci. USA* **78:**3770–3774.

Horiuchi, T., Maki, H., and Sekiguchi, M., 1985, RNase H-defective mutants of *Escherichia coli*: A possible discriminatory role of RNase H in initiation of DNA replication, *Mol. Gen. Genet.* **195:**17–22.

Hostomska, Z., Matthews, D. A., Davies, J. F., Nodes, B. R., and Hostomsky, Z., 1991, Proteolytic release and crystallization of the RNase H domain of human immunodeficiency virus type 1 reverse transcriptase, *J. Biol. Chem.* **266:**14697–14702.

Hostomsky, Z., Hostomska, Z., Hudson, G. O., Moomaw, E. W., and Nodes, B. R., 1991, Reconstitution *in vitro* of RNase H activity by using purified N-terminal and C-terminal domains of human immunodeficiency virus type 1 reverse transcriptase, *Proc. Natl. Acad. Sci. USA* **88:**1148–1152.

Hostomsky, Z., Hostomska, Z., and Matthews, D. A., 1993, Ribonuclease H, in *Nuclease*, 2nd ed. (S. M. Linn and R. J. Roberts, eds.), pp. 341–376, Cold Spring Harbor Laboratory, Cold Spring, New York.

Hsu, I-N., Delbaere, L. T. J., James, M. N. G., and Hofmann, T., 1977, Penicillopepsin from *Penicillium janthinellum* crystal structure at 2.8 Å and sequence homology with porcine pepsin, *Nature* **266:**140–145.

Huang, H.-W., and Cowan, J. A., 1994, Metallobiochemistry of the magnesium ion: Characterization of the essential metal-binding site in *Escherichia coli* ribonuclease H, *Eur. J. Biochem.* **219:**253–260.

Huet, J., Wyers, F., Buhler, J.-M., Sentenac, A., and Fromageot, P., 1976, Association of RNase H activity with yeast RNA polymerase A, *Nature* **261:**431–433.

Huet, J., Buhler, J.-M., Sentenac, A., and Fromageot, P., 1977, Characterization of ribonuclease H activity associated with yeast RNA polymerase A, *J. Biol. Chem.* **252:**8848–8855.

Inoue, H., Hayase, Y., Iwai, S., and Ohtsuka, E., 1987, Sequence-dependent hydrolysis of RNA using modified oligonucleotide splints and RNase H, *FEBS Lett.* **215:**327–330.

Ishikawa, K., Kimura, S., Kanaya, S., Morikawa, K., & Nakamura, H., 1993a, Structural study of mutants of *Escherichia coli* ribonuclease HI with enhanced thermostability, *Protein Eng.* **6:**85–91.

Ishikawa, K., Okumura, M., Katayanagi, K., Kimura, S., Kanaya, S., Nakamura, H., and Morikawa, K., 1993b, Crystal structure of ribonuclease H from *Thermus thermophilus* HB8 refined at 2.8 Å resolution, *J. Mol. Biol.* **230:**529–542.

Ishikawa, K., Nakamura, H., Morikawa, K., and Kanaya, S., 1993c, Stabilization of *Escherichia coli* ribonuclease HI by cavity-filling mutations within a hydrophobic core, *Biochemistry* **32:**6171–6178.

Ishikawa, K., Nakamura, H., Morikawa, K., Kimura, S., and Kanaya, S., 1993d, Cooperative stabilization of *Escherichia coli* ribonuclease HI by insertion of Gly-80b and Gly-77 → Ala substitution, *Biochemistry* **32:**7136–7142.

Itaya, M., 1990, Isolation and characterization of a second RNase H (RNase HII) of *Escherichia coli* K-12 encoded by the *rnhB* gene, *Proc. Natl. Acad. Sci. USA* **87:**8587–8591.

Itaya, M., and Crouch, R. J., 1991a, A combination of RNase H (*rnh*) and *recBCD* or *sbcB* mutations in *Escherichia coli* K12 adversely affects growth, *Mol. Gen. Genet.* **227:**424–432.

Itaya, M., and Crouch, R. J., 1991b, Correlation of activity with phenotypes of *Escherichia coli* partial function mutants of *rnh*, the gene encoding RNase H, *Mol. Gen. Genet.* **227:**433–437.

Itaya, M., and Kondo, K., 1991, Molecular cloning of a ribonuclease H (RNase HI) gene from an extreme thermophile *Thermus thermophilus* HB8: A thermostable RNase H can functionally replace the *Escherichia coli* enzyme *in vivo*, *Nucleic Acids Res.* **19:**4443–4449.

Itaya, M., McKelvin, D., Chatterjie, S. K., and Crouch, R. J., 1991, Selective cloning of genes encoding RNase H from *Salmonella typhimurium*, *Sacchromyces cerevisiae* and *Escherichia coli rnh* mutants, *Mol. Gen. Genet.* **227:**438–445.

Itoh, T., and Tomizawa, J., 1980, Formation of an RNA primer for initiation of replication of ColE! DNA by ribonuclease H, *Proc. Natl. Acad. Sci. USA* **77:**2450–2454.

Itoh, T., and Tomizawa, J., 1982, Purification of ribonuclease H as a factor for initiation of *in vitro* ColEI DNA replication, *Nucleic Acids Res.* **10:**5949–5965.

Jacobo-Molina, A., Ding, J., Nanni, R. G., Clark, A. D. Jr., Lu, X., Tantillo, C., Williams, R. L., Kamer, G., Ferris, A. L., Clark, P., Hizi, A., Hughes, S. H., and Arnold, E., 1993, Crystal structure of human immunodeficiency virus type 1 reverse transcriptase complexed with double-stranded DNA at 3.0 Å resolution shows bent DNA, *Proc. Natl. Acad. Sci. USA* **90:**6320–6324.

Johnson, M. S., McClure, M. A., Feng, D. F., Gray, J., and Doolittle, R. F., 1986, Computer analysis of retroviral *pol* genes: Assignment of enzymatic functions to specific sequences and homologies with nonviral enzymes, *Proc. Natl. Acad. Sci. USA* **83:**7648–7652.

Kanaya, S., and Crouch, R. J., 1983a, DNA sequence of the gene coding for *Escherichia coli* ribonuclease H, *J. Biol. Chem.* **258:**1276–1281.

Kanaya, S., and Crouch, R. J., 1983b, Low levels of RNase H activity in *Escherichia coli* FB2 *rnh* result from a single-base change in the structural gene of RNase H, *J. Bacteriol.* **154:**1021–1026.

Kanaya, S., and Crouch, R. J., 1984, The *rnh* gene is essential for growth of *Escherichia coli*, *Proc. Natl. Acad. Sci. USA* **81:**3447–3451.

Kanaya, S., and Ikehara, M., 1993, Structure and function of ribonuclease H from *Escherichia coli*, in *Nucleic Acids and Molecular Biology* (F. Eckstein and D. M. J. Lilley, eds.). Vol. 7, pp. 285–302, Springer-Verlag, Berlin Heidelberg.

Kanaya, S., and Itaya, M., 1992, Expression, purification, and characterization of a recombinant ribonuclease H from *Thermus thermophilus* HB8, *J. Biol. Chem.* **267**:10184–10192.

Kanaya, S., Kohara, A., Miyagawa, M., Matsuzaki, T., Morikawa, K., and Ikehara, M., 1989, Overproduction and preliminary crystallographic study of ribonuclease H from *Escherichia coli*, *J. Biol. Chem.* **264**:11546–11549.

Kanaya, S., Kohara, A., Miura, Y., Sekiguchi, A., Iwai, S., Inoue, H., Ohtsuka, E., and Ikehara, M., 1990a, Identifications of amino acid residues involved in an active site of *Escherichia coli* ribonuclease H by site-directed mutagenesis, *J. Biol. Chem.* **265**:4615–4621.

Kanaya, S., Kimura, S., Katsuda, C., and Ikehara, M., 1990b, Role of cysteine residues in ribonuclease H from *Escherichia coli*: Site-directed mutagenesis and chemical modification, *Biochem. J.* **271**:59–66.

Kanaya, S., Katsuda, C., Kimura, S., Nakai, T., Kitakuni, E., Nakamura, H., Katayanagi, K., Morikawa, K., and Ikehara, M., 1991a, Stabilization of *Escherichia coli* ribonuclease H by introduction of an artificial disulfide bond, *J. Biol. Chem.* **266**:6038–6044.

Kanaya, S., Katayanagi, K., Morikawa, K., Inoue, H., Ohtsuka, E., and Ikehara, M., 1991b, Effect of mutagenesis at each of five histidine residues on enzymatic activity and stability of ribonuclease H from *Escherichia coli*, *Eur. J. Biochem.* **198**:437–440.

Kanaya, S., Katsuda-Nakai, C., and Ikehara, M., 1991c, Importance of the positive charge cluster in *Escherichia coli* ribonuclease H for the effective binding of the substrate, *J. Biol. Chem.* **266**:11621–11627.

Kanaya, S., Nakai, C., Konishi, A., Inoue, H., Ohtsuka, E., and Ikehara, M., 1992, A hybrid ribonuclease H: A novel RNA cleaving enzyme with sequence-specific recognition, *J. Biol. Chem.* **267**:8492–8498.

Kanaya, S., Oobataka, M., Nakamura, H., and Ikehara, M., 1993, pH-dependent thermostabilization of *Escherichia coli* ribonuclease HI by histidine to alanine substitutions, *J. Biotech.* **28**:117–136.

Kane, C. M., 1988, Renaturase and ribonuclease H: A novel mechanism that influences transcript displacement by RNA polymerase II *in vitro*, *Biochemistry* **27**:3187–3196.

Karwan, R., and Kindas-Mugge, I., 1989, Identification of a yeast ribonuclease H as an Am antigen, *Eur. J. Biochem.* **179**:549–555.

Karwan, R., and Wintersberger, U., 1986, Yeast ribonuclease H(70) cleaves RNA–DNA junctions, *FEBS Lett.* **206**:189–192.

Karwan, R., and Wintersberger, U., 1988, In addition to RNase H(70) two other proteins of *Saccharomyces cerevisiae* exhibit ribonuclease H activity, *J. Biol. Chem.* **263**:14970–14977.

Karwan, R., Blutsch, H., and Wintersberger, U., 1983, Physical association of a DNA polymerase stimulating activity with ribonuclease H purified from yeast, *Biochemistry* **22**:5500–5507.

Karwan, R., Kuhne, C., and Wintersberger, U., 1986, Ribonuclease H(70) from *Saccharomyces cerevisiae* possesses cryptic reverse transcriptase activity, *Proc. Natl. Acad. Sci. USA* **83**:5919–5923.

Katayanagi, K., Miyagawa, M., Matsushima, M., Kanaya, S., Ikehara, M., Matsuzaki, T., and Morikawa, K., 1990, Three-dimensional structure of ribonuclease H from *Escherichia coli*, *Nature* **347**:306–309.

Katayanagi, K., Miyagawa, M., Matsushima, M., Ishikawa, M., Kanaya, S., Nakamura, H., Ikehara, M., Matsuzaki, T., and Morikawa, K., 1992, Structural details of ribonuclease H from *Escherichia coli* as refined to an atomic resolution, *J. Mol. Biol.* **223**:1029–1052.

Katayanagi, K., Ishikawa, M., Okumura, M., Ariyoshi, M., Kanaya, S., Kawano, Y., Suzuki, M., Tanaka, I., and Morikawa, K., 1993a, Crystal structures of ribonuclease HI active site mutants from *Escherichia coli*, *J. Biol. Chem.* **268**:22092–22099.

Katayanagi, K., Okumura, M., and Morikawa, K., 1993b, Crystal structure of *Escherichia coli*

RNase HI in complex with Mg^{2+} at 2.8 Å resolution: Proof for a single Mg^{2+}-binding site, *Proteins* **17**:337–346.

Kimura, S., Oda, Y., Nakai, T., Katayanagi, K., Kiyakuni, E., Katsuda-Nakai, C., Nakamura, H., Ikehara, M., and Kanaya, S., 1992a, Effect of cavity-modulating mutations on the stability of *Escherichia coli* ribonuclease H, *Eur. J. Biochem.* **206**:337–343.

Kimura, S., Nakamura, H., Hashimoto, T., Oobatake, M., and Kanaya, S., 1992b, Stabilization of *Escherichia coli* ribonuclease HI by strategic replacement of amino acid residues with those from the thermophilic counterpart, *J. Biol. Chem.* **267**:21535–21542.

Kimura, S., Kanaya, S., and Nakamura, H., 1992c, Thermostabilization of *Escherichia coli* ribonuclease HI by replacing left-handed helical Lys^{95} with Gly or Asn, *J. Biol. Chem.* **267**:22014–22017.

Kitahara, N., Sawai, Y., and Tsukada, K., 1982, Purification and properties of magnesium- and manganese-dependent ribonuclease H from chick embryo, *J. Biochem.* **92**:855–864.

Kogoma, T., 1984, Absence of RNase H allows replication of pBR322 in *Escherichia coli* mutants lacking DNA polymerase I, *Proc. Natl. Acad. Sci. USA* **81**:7845–7849.

Kogoma, T., Subia, N. L., and Meyenberg, K., 1985, Function of ribonuclease H in initiation of DNA replication in *Escherichia coli* K-12, *Mol. Gen. Genet.* **200**:103–109.

Kohlstaedt, L. A., Wang, J., Friedman, J. M., Rice, P. A., and Steitz, T. A., 1992, Crystal structure at 3.5 Å resolution of HIV-1 reverse transcriptase complexed with an inhibitor, *Science* **256**:1783–1790.

Koizumi, M., and Ohtsuka, E., 1991, Effects of phosphorothioate and 2-amino groups in hammerhead ribozymes on cleavage rates and Mg^{2+} binding, *Biochemistry* **30**:5145–5150.

Krug, M. S., and Berger, S. L., 1989, Ribonuclease H activities associated with viral reverse transcriptases are endonucleases, *Proc. Natl. Acad. Sci. USA* **86**:3539–3543.

Lahm, A., Weston, S. A., and Suck, D., 1991, Structure of DNase I, in *Nucleic Acids and Molecular Biology* (F. Eckstein and D. M. J. Lilley, eds.), Vol. 5, pp. 171–186, Springer-Verlag, Berlin Heidelberg.

Lederer, H., Schatz, O., May, R., Crespi, H., Darlix, J-L., LeGrice, S. F. J., and Heumann, H., 1992, Domain structure of the human immunodeficiency virus reverse transcriptase, *EMBO J.* **11**:1131–1139.

Loya, S., and Hizi, A., 1993, The interaction of illimaquinone, a selective inhibitor of the RNase H activity, with the reverse transcriptases of human immunodeficiency and murine leukemia retroviruses, *J. Biol. Chem.* **268**:9323–9238.

Loya, S., Tal, R., Hughes, S. H., and Hizi, A., 1992, The effects of cysteine mutations on the catalytic activities of the reverse transcriptase of human immunodeficiency virus type-1, *J. Biol. Chem.* **267**:13879–13883.

Maki, H., Horiuchi, T., and Sekiguchi, M., 1983, Structure and expression of the *dnaQ* mutator and the RNase H genes of *Escherichia coli*: Overlap of the promoter regions, *Proc. Natl. Acad. Sci. USA* **80**:7137–7141.

Masutani, C., Enomoto, T., Suzuki, M., Hanaoka, F., and Ui, M., 1990, DNA primase stimulatory factor from mouse FM3A cells has an RNase H activity: Purification of the factor and analysis of the stimulation, *J. Biol. Chem.* **265**:10210–10216.

Matouschek, A., Kellis, J. T., Serrano, L., Bycroft, M., and Fersht, A. R., 1990, Transient folding intermediates characterized by protein engineering, *Nature* **346**:440–445.

Metelev, V. G., Zayakina, G. V., Ryabushenko, I. L., Krynetskaya, N. F., Romanova, E. A., Oretskaya, T. S., and Shabarova, Z. A., 1988, Influence of probe structure on unique (regiospecific) cleavage of RNA by RNase H, *FEBS Lett.* **226**:232–234.

Meyenburg, K. von, Boye, E., Skarstad, K., Koppes, L., & Kogoma, T., 1987, Mode of initiation of constitutive stable DNA replication in RNase H-defective mutants of *Escherichia coli* K12, *J. Bacteriol.* **169**:2650–2658.

Miller, H. I., Riggs, A. D., and Gill, G. N., 1973, Ribonuclease H (hybrid) in *Escherichia coli*: Identification and characterization, *J. Biol. Chem.* **248**:2621–2624.

Minshull, J., and Hunt, T., 1986, The use of single-stranded DNa and RNase H to promote quantitative "hybrid arrect of translation" of mRNA/DNA hybrids in reticulocyte lysate cell-free translations, *Nucleic Acids Res.* **14**:6433–6451.

Miranker, A., Radford, S. E., Karplus, M., and Dobson, C. M., 1991, Demonstration by NMR of folding domains in lysozyme, *Nature* **349**:633–636.

Mizrahi, V., 1989, Analysis of the ribonuclease H activity of HIV-1 reverse transcriptase using RNA·DNA hybrid substrates derived from the *gag* region of HIV-1, *Biochemistry* **28**:9088–9094.

Mizrahi, V., Usdin, M. Y., Harington, A., and Dudding, L. R., 1990, Site-directed mutagenesis of the conserved Asp-443 and Asp-498 carboxy-terminal residues of HIV-1 reverse transcriptase, *Nucleic Acids Res.* **18**:5359–5363.

Moelling, K., Bolognesi, D. P., Bauer, H., Busen, W., Plassmann, H. W., and Hausen, P., 1971, Association of viral reverse transcriptase with an enzyme degrading the RNA moiety of RNA-DNA hybrids, *Nature New Biol.* **234**:240–243.

Nagayama, K., Yamazaki, T., Yoshida, M., Kanaya, S., and Nakamura, H., 1990, Combination of heteronuclear ^{1}H-^{15}N and ^{1}H-^{13}C three-dimensional nuclear magnetic resonance experiments for amide-directed sequential assignment in larger proteins, *J. Biochem.* **108**:149–152.

Naito, S., and Uchida, H., 1986, RNase H and replication of ColE1 DNa in *Escherichia coli, J. Bacteriol.* **166**:143–147.

Naito, S., Kitani, T., Ogawa, T., Okazaki, T., and Uchida, H., 1984, *Escherichia coli* mutants suppressing replication-defective mutations of the ColE1 plasmid, *Proc. Natl. Acad. Sci. USA* **81**:550–554.

Nakai, C., Konishi, A., Komatsu, Y., Inoue, H., Ohtsuka, E., and Kanaya, S., 1994, Sequence-specific cleavage of RNA by a hybrid ribonuclease H, *FEBS Lett.* **339**:67–72.

Nakamura, H., Katayanagi, K., Morikawa, K., and Ikehara, M., 1991a, Structural models of ribonuclease H domains in reverse transcriptases from retroviruses, *Nucleic Acids Res.* **19**:1817–1823.

Nakamura, H., Oda, Y., Iwai, S., Inoue, H., Ohtsuka, E., Kanaya, S., Kimura, S., Katsuda, C., Katayanagi, K., Morikawa, K., Miyashiro, H., and Ikehara, M., 1991b, How does RNase H recognize a DNA·RNA hybrid? *Proc. Natl. Acad. Sci. USA* **88**:11535–11539.

Nanni, R. G., Ding, J., Jacobo-Molina, A., Hughes, S. H., and Arnold, E., 1993, Review of HIV-1 reverse transcriptase three-dimensional structure: Implications for drug design, *Perspect. Drug Discov. Design* **1**:129–150.

Neri, D., Wider, G., and Wuthrich, K., 1992a, Complete ^{15}N and ^{1}H NMR assignments for the amino-terminal domain of the phage 434 repressor in the urea-unfolded form, *Proc. Natl. Acsd. Sci. USA* **89**:4397–4401.

Neri, D., Billeter, M., Wider, G., and Wuthrich, K., 1992b, NMR determination of residual structure in a urea-denatured protein, the 434-repressor, *Science* **257**:1559–1563.

Nomura, T., Aiba, H., and Ishihama, A., 1985, Transcriptional organization of the convergent overlapping *dnaQ-rnh* genes of *Escherichia coli, J. Biol. Chem.* **260**:7122–7125.

Oda, Y., Nakamura, H., Kanaya, S., and Ikehara, M., 1991, Binding of metal ions to *E. coli* RNase HI observed by ^{1}H-^{15}N heteronuclear 2D NMR, *J. Biomol. NMR* **1**:247–255.

Oda, Y., Nakamura, H., Yamazaki, T., Nagayama, K., Yoshida, M., Kanaya, S., and Ikehara, M., 1992, ^{1}H NMR studies of deuterated ribonuclease HI selectively labeled with protonated amino acids, *J. Biomol. NMR* **2**:137–147.

Oda, Y., Yoshida, M., and Kanaya, S., 1993a, Role of His^{124} in the catalytic function of ribonuclease HI from *Escherichia coli, J. Biol. Chem.* **268**:88–92.

Oda, Y., Iwai, S., Ohtsuka, E., Ishikawa, M., Ikehara, M., and Nakamura, H., 1993b, Binding of nucleic acids to *E. coli* RNase HI observed by NMR and CD spectroscopy, *Nucleic Acids Res.*, **21**:4690–4695.

Oda, Y., Yamazaki, T., Nagayama, K., Kanaya, S., Kuroda, Y., and Nakamura, H., 1994, Individual ionization constants of all the carboxyl groups in ribonuclease HI from *Escherichia coli* determined by NMR, *Biochemistry* **33**:5275–5284.

Ogawa, T., and Okazaki, T., 1984, Function of RNase H in DNA replication revealed by RNase H defective mutants of *Escherichia coli*, *Mol. Gen. Genet.* **193**:231–237.

Ogawa, T., Pickett, G. G., Kogoma, T., and Kornberg, A., 1984, RNase H confers specificity in the *dnaA*-dependent initiation of replication at the unique origin of the *Escherichia coli* chromosome *in vivo* and *in vitro*, *Proc. Natl. Acad. Sci. USA* **81**:1040–1044.

Ohmori, H., Murakami, Y., and Nagata, T., 1987, Nucleotides sequences required for a ColE1-type plasmid to replicate in *Escherichia coli* cells with or without RNase H, *J. Mol. Biol.* **198**:223–234.

Okayama, H., and Berg, P., 1982, High-efficiency cloning of full-length cDNA, *Mol. Cell. Biol.* **2**:161–170.

Oyama, F., Kikuchi, R., Crouch, R. J., and Uchida, T., 1989, Intrinsic properties of reverse transcriptase in reverse transcription: Associated RNase H is essentially regarded as an endonuclease, *J. Biol. Chem.* **264**:18808–18817.

Parsons, S. M., and Raftery, M. A., 1972, Ionization behavior of the catalytic carboxyls of lysozyme. Effects of ionic strength, *Biochemistry* **11**:1623–1629.

Peliska, J. A., and Benkovic, S. J., 1992, Mechanism of DNA strand transfer reactions catalyzed by HIV-1 reverse transcriptase, *Science* **258**:1112–1118.

Post, K., Guo, J., Kalman, E., Uchida, T., Crouch, R. J., and Levin, J. G., 1993, A large deletion in the connection subdomain of murine leukemia virus reverse transcriptase or replacement of the RNase H domain with *Escherichia coli* RNase H results in altered polymerase and RNase H activities, *Biochemistry* **32**:5508–5517.

Powers, R., Clore, G. M., Stahl, S. J., Wingfield, P. T., and Gronenborn, A., 1992, Analysis of the backbone dynamics of the ribonuclease H domain of the human immunodeficiency virus reverse transcriptase using ^{15}N relaxation measurements, *Biochemistry* **31**:9150–9157.

Prasad, V. R., and Goff, S. P., 1989, Linker insertion mutagenesis of the human immunodeficiency virus reverse transcriptase expressed in bacteria: Definition of the minimal polymerase domain, *Proc. Natl. Acad. Sci. USA* **86**:3104–3108.

Quinones, A., Kucherer, C., Piechocki, R., and Messer, W., 1987, Reduced transcription of the *rnh* gene in *Escherichia coli* mutants expressing the SOS regulon constitutively, *Mol. Gen. Genet.* **206**:95–100.

Ratner, L., Haseltine, W., Patarca, R., Livak, K. J., Starcich, B., Josephs, S. F., Doran, E. R., Rafalski, J. A., Whitehorn, E. A., Baumeister, K., Ivanoff, L., Petteway, S. R., Jr., Pearson, M. L., Lautenberger, J. A., Papas, T. S., Ghrayeb, J., Chang, N. T., Gallo, R. C., and Wong-Staal, F., 1985, Complete nucleotide sequence of the AIDS virus, HTLV-III, *Nature* **313**:277–283.

Reinstein, J., Brune, M., and Wittinghofer, A., 1988, Mutations in the nucleotide binding loop of adenylate kinase of *Escherichia coli*, *Biochemistry* **27**:4712–4720.

Repaske, R., Hartley, J. W., Kavlick, M. F., O'Neill, R. R., and Austin, J. B., 1989, Inhibition of RNase H activity and viral replication by single mutations in the 3′ region of Moloney murine leukemia virus reverse transcriptase, *J. Virol.* **63**:1460–1464.

Restle, T., Muller, B., and Goody, R. S., 1992, RNase H activity of HIV reverse transcriptases is confined exclusively to the dimeric forms, *FEBS Lett* **300**:97–100.

Robertson, H. D., and Dunn, J. J., 1975, Ribonucleic acid processing activity of *Escherichia coli* by ribonuclease III, *J. Biol. Chem.* **250**:3050–3056.

Rong, Y. W., and Carl, P. L., 1990, On the molecular weight and subunit composition of calf thymus ribonuclease HI, *Biochemistry* **29**:383–389.

Saison-Behmoaras, T., Tocque, B., Rey, I., Chassignol, M., Thuong, N. T., and Helene, C., 1991, Short modified antisense oligonucleotides directed against Ha-*ras* point mutation induce selective cleavage of the mRNA and inhibit T24 cells proliferation, *EMBO J.* **10:**1111–1118.

Schatz, O., Cromme, F. V., Gruninger-Leitch, F., and Le Grice, S. F. J., 1989, Point mutations in conserved amino acid residues within the C-terminal domain of HIV-1 reverse transcriptase specifically repress RNase H function, *FEBS Lett.* **257:**311–314.

Schatz, O., Mous, J., and Le Grice, S. F. J., 1990, HIV-1 RT-associated ribonuclease H displays both endonuclease and 3′→5′ exonuclease activity, *EMBO J.* **9:**1171–1176.

Schulze, T., Nawrath, M., and Moelling, K., 1991, Cleavage of the HIV-1 p66 reverse transcriptase/RNase H by the p9 protease *in vitro* generates active p15 RNase H, *Arch. Virol.* **118:**179–188.

Shibahara, S., Mukai, S., Nishihara, T., Inoue, H., Ohtsuka, E., and Morisawa, H., 1987, Site-directed cleavage of RNA, *Nucleic Acids Res.* **15:**4403–4415.

Shinnick, T. M., Lerner, R. A., and Sutcliffe, J. G., 1981, Nucleotide sequence of Moloney murine leukemia virus, *Nature* **293:**543–548.

Smith, J. S., and Roth, M. J., 1993, Purification and characterization of an active human immunodeficiency virus type 1 RNase H domain, *J. Virol.* **67:**4037–4049.

Starnes, M. C., and Cheng, Y-C., 1989, Human immunodeficiency virus reverse transcriptase-associated RNase H activity, *J. Biol. Chem.* **264:**7073–7077.

Stein, H., and Hausen, P., 1969, Enzyme from calf thymus degrading the RNA moiety of DNA-RNA hybrids: Effect on DNA-dependent RNA polymerase, *Science* **166:**393–395.

Stockman, B. J., Euvrard, A., and Scahill, T. A., 1993, Heteronuclear three-dimensional NMR spectroscopy of a partially denatured protein: The A-state of human ubiquitin, *J. Biomol. NMR* **3:**285–296.

Suck, D., and Oefner, C., 1986, Structure of DNase I at 2.0 Å resolution suggests a mechanism for binding to and cutting DNA, *Nature* **321:**620–625.

Tagaya, M., Yagami, T., Noumi, T., Futai, M., Kishi, F., Nakazawa, A., and Fukui, T., 1989, Site-directed mutagenesis of Pro-17 located in the glycine-rich region of adenylate kinase, *J. Biol. Chem.* **264:**990–994.

Tan, C-K., Zhang, J., Li, Z-Y., Tarpley, W. G., Downey, K. M., and So, A. G., 1991, Functional characterization of RNA-dependent DNA polymerase and RNase H activities of a recombinant HIV reverse transcriptase, *Biochemistry* **30:**2651–2655.

Tanese, N., and Goff, S. P., 1988, Domain structure of the Moloney murine leukemia virus reverse transcriptase: Mutational analysis and separate expression of the DNA polymerase and RNase H activities, *Proc. Natl. Acad. Sci. USA* **85:**1777–1781.

Tanese, N., Telesnitsky, A., and Goff, S. P., 1991, Abortive reverse transcription by mutants of Moloney murine leukemia virus deficient in the reverse transcriptase-associated RNase H function, *J. Virol.* **65:**4387–4397.

Tashiro, F., and Ueno, Y., 1978a, Ribonuclease H from rat liver: I. Partial purification and characterization of nuclear ribonuclease H, *J. Biochem.* **84:**385–393.

Tashiro, F., and Ueno, Y., 1978b, Ribonuclease H from rat liver: II. Partial purification and characterization of cytosol ribonuclease H, *J. Biochem.* **84:**395–402.

Taylor, J. W., Ott, J., and Eckstein, F., 1985, The rapid generation of oligonucleotide-directed mutations at high frequence using phosphorothioate-modified DNA, *Nucleic Acids Res.* **13:**8764–8785.

Telesnitsky, A., and Goff, S. P., 1993, RNase H domain mutations affect the interaction between Moloney murine leukemia virus reverse transcriptase and its primer-template, *Proc. Natl. Acad. Sci. USA* **90:**1276–1280.

Telesnitsky, A., Blain, S. W., and Goff, S. P., 1992, Defects in Moloney murine leukemia virus replication caused by a reverse transcriptase mutation modeled on the structure of *Escherichia coli* RNase H, *J. Virol.* **66:**615–622.

Tisdale, M., Schulze, T., Larder, B. A., and Moelling, K., 1991, Mutations within the RNase H domain of human immunodeficiency virus type 1 reverse transcriptase abolish virus infectivity, *J. Gen. Virol.* **72**:59–66.

Varmus, H., 1988, Retroviruses, *Science* **240**:1427–1435.

Volkmann, S., Wohrl, B. M., Tisdale, M., and Moelling, K., 1993, Enzymatic analysis of two HIV-1 reverse transcriptase mutants with mutations in carboxyl-terminal amino acid residues conserved among retroviral ribonucleases H, *J. Biol. Chem.* **268**:2674–2683.

Vonwirth, H., Frank, P., and Busen, W., 1989, Serological analysis and characterization of calf thymus ribonuclease HIIb, *Eur. J. Biochem.* **184**:321–329.

Walder, R. Y., and Walder, J. A., 1988, Role of RNase H in hybrid-arrested translation by antisense oligonucleotides, *Proc. Natl. Acad. Sci. USA* **85**:5011–5015.

Walker, J. E., Saraste, M., Runswick, M. J., and Gay, N. J., 1982, Distantly related sequences in the α- and β-subunits of ATP synthase, myosin, kinases and other ATP-requiring enzymes and a common nucleotide binding fold, *EMBO J.* **1**:945–951.

Wang, A. H.-J., Fujii, S., van Boom, J. H., van der Marel, G. A., van Boeckel, S. A., and Rich, A., 1982, Molecular structure of r(GCG)d(TATACGC): A DNA-RNA hybrid helix joined to double helical DNA, *Nature* **299**:601–604.

Wintersberger, U., 1990, Ribonucleases H of retroviral and cellular origin, *Pharmacol. Ther.* **48**:259–280.

Wohrl, B. M., and Moelling, K., 1990, Interaction of HIV-1 ribonuclease H with polypurine tract containing RNA-DNA hybrids, *Biochemistry* **29**:10141–10147.

Wohrl, B. M., Volkmann, S., and Moelling, K., 1991, Mutations of a conserved residue within HIV-1 ribonuclease H affect its exo- and endonuclease activities, *J. Mol. Biol.* **220**:801–818.

Wyers, F., Sentenac, A., and Fromageot, P., 1973, Role of DNA-RNA hybrids in eukaryotes: Ribonuclease H in yeast, *Eur. J. Biochem.* **35**:270–281.

Wyers, F., Sentenac, A., and Fromageot, P., 1976a, Role of DNA-RNA hybrids in eukaryotes: Purification of two ribonucleases H from yeast cells, *Eur. J. Biochem.* **69**:377–383.

Wyers, F., Fuet, J., Sentenac, A., and Fromageot, P., 1976b, Role of DNA-RNA hybrids in eukaryotes: Characterization of yeast ribonucleases H_1 and H_2, *Eur. J. Biochem.* **69**:385–395.

Yamazaki, T., Yoshida, M., Kanaya, S., Nakamura, H., and Nagayama, K., 1991, Assignments of backbone ^{1}H, ^{13}C, and ^{15}N resonance and secondary structure of ribonuclease H from *Escherichia coli* by heteronuclear three-dimensional NMR spectroscopy, *Biochemistry* **30**:6036–6047.

Yamazaki, T., Yoshida, M., and Nagayama, K., 1993, Complete assignments of magnetic resonances of ribonuclease H from Escherichia coli by double- and triple-resonance 2D and 3D NMR spectroscopies, *Biochemistry* **32**:5656–5669.

Yang, W., Hendrickson, W. A., Kalman, E. T., and Crouch, R. J., 1990a, Expression, purification, and crystallization of natural and selenomethionyl recombinant ribonuclease H from *Escherichia coli*, *J. Biol. Chem.* **265**:13553–13559.

Yang, W., Hendrickson, W. A., Crouch, R. J., and Satow, Y., 1990b, Structure of ribonuclease H phased at 2 Å resolution by MAD analysis of the selenomethionyl protein, *Science* **249**:1398–1405.

Yoneya, T., Tagaya, M., Kishi, F., Nakazawa, A., and Fukui, T., 1989, Site-directed mutagenesis of Gly-15 and Gly-20 in the glycine-rich region of adenylate kinase, *J. Biochem.* **105**:158–160.

Zuckermann, R. N., and Schultz, P. G., 1988, H hybrid sequence-selective ribonuclease S, *J. Am. Chem. Soc.* **110**:6592–6594.

Zuckermann, R. N., and Schultz, P. G., 1989, Site-selective cleavage of structural RNA by a staphylococcal nuclease–DNA hybrid, *Proc. Natl. Acad. Sci. USA* **86**:1766–1770.

Zuckermann, R. N., Corey, D. R., and Schultz, P. G., 1988, Site-selective cleavage of RNA by a hybrid enzyme, *J. Am. Chem. Soc.* **110**:1614–1615.

Index

A-ring, 104
Absorbance, 32, 221, 235, 282
Absorption spectrum, 120, 122, 124, 129
AC, 104, 105
Acceptor arm, 335, 348, 353, 355, 358
Acceptor stem, 345, 346, 348, 349, 351, 354, 358–360
Acetylation, 263, 269
Acetyl tryptophan ethyl ester, 76
Acid-denatured state, 29, 30
Acid-unfolded state, 38, 42
Acrylamide, 106, 112
Acrylodan, 109, 110
Actin, 127, 128, 203
Actin-based filaments, 255
Activator of transcription, 303
Active center, 340
Active site, 235, 394
Adenine, 181, 201
Adenine nucleotides, 197
Adenosine, 337, 352
Adenosyl ribose, 324
Adenylated-bis-(nucleotidyl)polyphosphate, 327
Adenylate synthesis, 336
ADP, 181, 183, 185, 195, 196, 197, 198, 203
ADP-ribosylation, 263
Affinity cleavage, 183, 191, 192, 193, 195, 203
Affinity label, 178, 179, 183, 187, 191, 195, 198
Affinity labeling, 177, 179, 183, 195, 200, 202, 203, 329
Aflatoxin B1, 189
Aggregation, 30, 32, 137, 161, 165
AIDS, 378, 404
Alanine, 84, 148, 226, 230, 240, 326, 335, 344, 346, 349
Alanine scanning mutagenesis, 397
Alanyl-tRNA synthetase, 333–335, 358, 360
Alarmones, 327
Aldimine E, 221
All or none transition, 10
Allocolchicine, 104
Allocolchicinoids, 106
Allosteric changes, 303, 304, 313
Allosteric effectors, 102
Allosteric effects, 231
Allosteric enzyme, 195
Allosteric information, 235
Allosteric interactions, 228, 244
Allosteric ligands, 221, 238, 239
Allosteric properties, 238
Allosteric sites, 177
Amidation, 284
Amide exchange, 41
Amide modes, 66
Amide protection, 33
Amino acids, 2, 8, 12, 15, 44, 56, 57, 60, 83, 149, 177, 178, 182, 186, 192, 195, 196, 230, 233, 237, 261, 314, 324, 336, 339, 340, 341, 352, 356, 385, 393

Amino Acids (*cont.*)
15-N labeled, 391
Amino acid activation, 335, 353, 360
Amino acid replacement, 213
Amino acid sequence, 11, 18, 19, 42, 358
Amino acid specificity, 335, 336
Aminoacrylate, 209, 221, 222, 223, 227, 228, 236, 241, 242
Aminoacyl adenylate, 324, 329, 335–337, 352
Aminoacylation, 323, 324, 326, 328, 340–346, 348–350, 359
Aminoacyl tRNA, 340, 341, 345
Aminoacyl-tRNA synthetase, 323, 324, 326–328, 330, 332, 333–336, 340, 341, 346, 349, 350, 352–355, 360
Amino colchicinoids, 106
Ammonia, 223
AMP, 109, 380
AMPS-BOP, 181, 198
Amphipathy, 70, 75
Amphipathic helix, 278
Angle
chi, 3, 4
phi, 3
psi, 3
Anhydride bond, 335
Aniline, 221
Anion binding, 31
ANS, 29, 30, 31, 32, 38, 39, 40, 102, 109, 110, 311, 312
Antibody, 217, 266, 269, 273, 275, 280
monoclonal, 217, 218, 229, 275
Anticodon, 110, 324, 334, 341, 342, 343, 346, 349, 351, 353, 355, 356, 358, 359, 360
Anticodonlike structure, 350
Antigenic index, 271, 272
Antiparallel coiled-coil, 334
Antisense arrested translation, 408
Ap_4A hydrolase, 327
Ap_3N, 327
Ap_4N, 327
Apocytochrome C, 45
Apomyoglobin, 33, 35, 37, 42, 44, 45
Arc repressor, 39
Arginine, 43, 201, 229, 239, 329, 358, 359
Arginyl-tRNA synthetase, 110, 335
Aromatic amino acids, 64, 84, 235, 278
Aromatic clusters, 387
Aromatic side chains, 76
Arrhenius equation, 163
Ascorbate, 192
Asparagine, 338, 358
Asparaginyl-tRNA synthetase, 333, 340
Aspartame, 88
Aspartic acid, 57, 80, 95, 226, 231, 354, 358, 359
Aspartyl-tRNA synthetase, 330, 331, 333, 334, 340, 353, 354, 356, 360
Assembly, 216, 217
ATP, 48, 102, 109, 110, 181, 182, 183, 202, 203, 324, 327, 329, 332, 336, 337, 338, 339, 340, 353
analogs, 181
ATPase, 202, 203, 266, 291
ATP-pyrophospate exchange, 110, 335, 336
ε-ATP, 102
Average heat capacity, 138, 141
Axonal transport, 256

Ba^{+2}, 388, 392
Bacteriorhodopsin, 57, 65, 90, 92, 93, 105
Barnase, 33, 393
Basic protrusion, 385, 397, 398
α/β Barrel, 209, 210, 212
β Barrel, 333, 334, 355, 360
2-BDB-TA-5′DP, 180
6-BDB-TA-5′DP, 180
8-BDB-TA-5′DP, 180
8-BDB TATP, 186–188
2-BDB-T ε A 5′DP, 179, 184–188, 200
2-BDB-T ε A-2′5′DP, 199–201
Bending modes, 62
Bends, 67, 309
Benomyl, 285
Benzene, 73, 116
Benzoimidazole carbamates, 264
Bifunctional affinity label, 182, 198
Bilirubin, 88
Bis-ANS, 34, 102, 109
Boltzmann constant, 146
Boltzmann distribution, 61
Boltzmann formula, 15
Boltzmann statistics, 6
Boltzmann-like relationship, 6
Boltzmann-like statistics, 1, 3, 4, 6, 7, 15, 20, 24
2-BOP-T ε A-5′DP, 185, 186
BPMV subunits, 69
BPTI, 31, 45
Brillouin scattering, 58

Brome mosaic virus, 350
Bromodioxobutyl derivatives, 183, 189, 198, 203
Bromodioxobutyl nucleotides, 179, 180, 185
Bromoketo nucleotides, 183
Bromooxopropyl derivatives, 186, 203
Bromooxopropyl nuclosides, 180
Butanediol, 119

C-ring, 104
Ca^{+2}, 32, 127, 128, 268, 281–284, 388, 392, 402
Ca^{+2}-ATPase, 129, 165
cAMP, 303, 305, 307, 311–316, 318
cAMP–CRP complex, 303, 305, 306, 310–313
cAMP–CRP–DNA complex, 305, 307, 308, 310, 311, 313
cAMP independent mutant of CRP, 318
cAMP-induced allostery, 311
Cd^{+2}, 192
Calmodulin, 277
Calorimetry, 218, 229, 384, 392
Calorimetric enthalpy, 138, 139, 140, 234
Calorimetric studies, 238
Capsid protein, 89
Carbanion, 224, 244
Carbonic anhydrase, 39, 147
Carboxylate–hydroxyl relay, 402
Carboxypeptidase, 262, 266
Carboxypeptidase A, 165
Carboxypeptidase B, 165
Carboxypeptidase Y, 283, 284
Catalytic activity, 335
Catalytic base, 214
Catalytic center, 328
Catalytic sites, 179
Cavities, 3, 19
CCA trinucleotide, 348
CCD, 65
Cellular anisotropy, 256
Cellular motility, 256
Cellular polarity, 256
Centrioles, 256, 257
Chain configuration, 148
Channel, 235
Chaperones, 30, 48
Charge repulsion, 43, 124, 281, 283, 284, 394
Chemical baseline, 136, 138, 142, 164
Chemical modification, 177, 284, 398
Chemical quench flow, 236
Chemical rescue, 226
Chemical shift, 33
Chimeric-beta tubulin, 285
Chimeric duplex, 392, 393, 400
Chimeric nonanucleotides, 381
Chimeric oligonucleotide, 411
Chinese hamster ovary cell, 285
Chlorambucil, 189
β-Chloro-L-alanine, 219, 224, 226, 228
Chloroplastic aminoacyl-tRNA synthetase, 326
Cholera toxin, 166
Cholesterol, 88
Chromatin, 119
Chromosomal DNA, 379
Chromosome, 88
Chymotrypsin, 147, 270–272, 276, 311
Chymotrypsinogen, 41
Cilia, 256
Circular dichroism, 28, 30, 32, 110, 126, 212, 217, 280–282, 311, 383
  far UV, 38, 40, 41
  near UV, 37, 38
  stopped flow, 40, 41
Citrate, 193
Clausius inequality, 164
Cloverleaf structure, 325, 342
Coat proteins, 80, 83
Codons, 324, 326, 345
Colchicine, 103, 106, 112, 256, 264, 265, 268, 278, 282, 286, 289
Colchicinol, 104
Colchicinoids, 105
Cold denaturation, 145, 146
Cold-denatured state, 149
ColE1, 379
Colicin A, 47
Collisional quencher, 111
Combrestatin, 264
Compact intermediate states, 27
Compact states, 43
Concentration quenching, 117
Configurational entropy, 147, 153, 154, 172
Configurational freedom, 149
Configurational Gibbs energy, 148, 156
Confocal microscopy, 88
Conformational change, 87, 115, 127, 129, 133, 179, 217, 228, 229, 231, 234, 236, 238–242, 269, 303, 305, 310, 314, 354, 355

Conformational drift, 219
Conformational stability, 212
Conformational state, 37
Conformational switch, 230
Conformational temperature, 3, 4, 14, 17
Conformational transitions, 55
Constant heat capacity approximation, 151
Convergence temperature, 151–153, 156
Cooperativity, 150, 233
Cooperative binding, 303
Coumarin maleiimide, 108
p-Cresol, 73
CRP, 303, 305–307, 310–315, 317, 318
Crp* mutants, 313–317
Critical concentration, 284
Critical temperature, 9, 10, 21
Crystal structure, 117, 187, 201, 208, 212, 213, 224, 310, 377, 383–385, 387, 388, 392, 394, 399, 407
Crystallin, 129
Crystal packing effect, 384
Crystallographic studies, 211, 220, 232
C-terminal deletion, 277
C-terminal domain, 109, 212, 305, 311, 317, 333, 334, 359
C-terminal tail, 108, 280, 287, 289
C-terminus, 255, 261–263, 265, 266, 269, 273–275, 335
Cyclic dipeptide, 148, 152, 156, 157
Cysteine, 38, 39, 57, 62, 71, 72, 77, 78, 80, 81, 88, 89, 119, 125, 126, 127, 128, 129, 186, 187, 189, 190, 200, 336, 398
Cystine, 62, 80
Cytochrome C, 29, 31, 33, 39, 40, 147, 172
Cytochrome C oxidase, 165
Cytoplasmic aminoacyl-tRNA synthetase, 326
Cytoskeleton, 256, 265

Deacetamido colchicine, 104, 105
Deacetyl colchicine, 104, 105
N-deacetyl N-retinoyl colchicine, 104
Deconvolution, 142
Defect, 2, 3
  energetic, 18, 19
  entropic, 18, 19
  variety, 19
Dehydroalanine, 222
Dehydrogenase, 102, 328
Demecolchicine, 104, 105
Denaturation, 10, 21, 111, 127, 134, 142, 143, 147, 150, 155, 156, 232
  irreversible, 134, 137, 162
Denaturation equilibrium, 140
Denaturation heat capacity, 143, 145, 146, 147, 148, 151, 157
Denaturation mechanism, 137
Denaturation process, 137, 139, 142
Denaturation temperature, 138, 140, 145, 157
Denaturational change, 150, 151
Denaturational enthalpy, 151, 153, 164, 232
Denaturational entropy, 151, 153, 163, 164
Denaturational Gibbs energy, 157
Denatured state, 27, 28, 29, 30, 41, 44, 137, 143, 147–150, 166, 277, 180
Denaturing conditions, 27
Desolvation, 157
Detergent, 271
Deuteromethyl alanine, 83
Dextran, 380
Dialysis, 125
Dicyanovinyl julolidine, 109
Difference spectrum, 63, 78, 93
Differential equation, 160
Differential labeling, 232
Differential scanning calorimeter, 134, 144
Differential scanning calorimetry, 133, 134, 143, 145, 148, 155, 160, 212, 232–234
Diffusion, 236, 244
Diffusion-controlled collision, 116
Dihydrofolate reductase, 40, 343
Dimethylacetamide, 72
Diptheria toxin, 47
Dipyrenyl, 120
Direct titration, 102
Discriminator position, 348
Dissolution heat capacity, 148
Disulfide bond, 32, 38, 41, 80, 125, 128, 129, 311
Dithiobis nitrobenzoate, 311
D-loop, 349
DNA, 87, 88, 213, 304–307, 311, 312, 316, 377
DNA binding motif, 305
DNA binding protein, 360
DNA/DNA duplex, 393, 407
DNA-induced allostery, 310
Dna K, 30, 31, 33, 34, 35
DNA polymerase, 305

DNA polymerase I, 403
DNA polymerase III, 378
DNA repair, 379
DNA replication, 379, 409
DNA/RNA hybrid, 379–381, 385, 393, 397, 399–402, 406, 407
Domain, 46, 110, 111, 116, 143, 154, 212, 215, 233–235, 269, 276, 277, 287, 288, 305, 334, 353, 378, 387, 388, 393
Domain–domain alignment, 317
Domain–domain interaction, 143, 154
Donor quenching, 111, 112
DSC profile, 142, 143, 162, 163
DSC transition, 135, 139, 140, 143, 144, 158, 161–163, 165, 166, 169, 171
DSC thermogram, 136, 158
D-stem, 353
DTT, 122, 125
Dynamic instability, 264, 284
Dyenin, 265, 266, 279

Electric field strength, 61
Electric polarizability tensor, 61
Electron diffraction, 269
Electron donor, 193
Electron microscopy, 269, 286
π-Electrons, 75
Electron spin resonance, 38
Electrophoresis, 32, 193, 270, 271, 273, 274, 282, 309
Electrostatic interaction, 148
Electrostatic repulsion, 43, 336, 337, 403
β Elimination, 220, 223, 224, 227, 228
Ellipticity, 28, 33, 37, 41, 45
Emission spectra, 239
Endonuclease, 377
Endoprotease lys, 270
Endotherm, 234
Energy
  activation, 103, 164
  interaction, 12
  internal, 15
  mean, 19
  potential, 12
  threshold, 8, 10
  torsional, 4
Energy barrier, 29, 30
Energy dispersion, 19
Energy minimisation, 23
Energy of folds, 7, 12
Energy spectrum, 15, 19
5-Enolpyruvyl shikimate-3-phosphate synthase, 165
Enthalpy, 135, 137, 138, 139, 143, 145, 146, 147, 152, 155, 156, 157, 159, 170, 230
  activation, 104
Enthalpy/entropy compensation, 146
Enthalpy plot, 152, 153
Entropy, 3, 137, 145, 146, 147, 152, 157, 159
Entropy plot, 152, 153
Enzyme-reagent complex, 178, 185
Epitope, 273, 280, 289
Erythrocyte, 256, 261
Ester linkage, 324
Ethylation, 307
N-Ethylmaleimide, 378, 398, 399, 408, 410
Excess enthalpy, 138, 141, 170
Excess heat capacity, 138, 141, 146, 159, 162, 163
Exchange
  H/D, 32, 40, 75, 77, 88, 89
Excimer, 116, 117, 118, 119, 124, 126, 127, 128, 129
Exciplex, 106
Excited state, 6, 115
Excitation spectra, 122, 126, 129
Exonuclease III, 379

Fd bacteriophage, 83–87
Fd subunit, 69
Fe^{+2}-Bleomycin, 411
Fe^{+2}-EDTA, 192
Fe^{+2}-isocitrate, 192–195
Fermi doublet, 72, 75, 92
Fermi resonance, 72
Fiber diffraction, 86
Fiber system, 255
Filamentous bacteriophage, 83
Filamentous virus, 84
Flagella, 256
Flexible loop, 238
Flexibility, 147, 211, 235, 269, 286
Fluorescein maleimide, 108
Fluorescence, 32, 33, 34, 37, 57, 64, 101, 102, 103, 105, 106, 107, 108, 117, 118, 120, 122, 123, 124, 126, 127, 128, 185, 200, 217, 219, 242, 268, 282, 310–312, 336, 352, 383
  anisotropy, 107, 108, 268, 283

Fluorescence (*cont.*)
energy transfer, 29, 107, 110, 111, 127, 142
enhancement, 102, 105
lifetime, 126
polarization, 32, 103, 110, 111, 116
probe, 115
quenching, 102, 106
steady-state, 101
stopped flow, 40
time-resolved, 101, 111
5′-Fluorosulfonyl benzoyl adenosine, 181, 182, 196, 197, 203
Folding, 130, 148, 212, 216, 217, 269
Fold energy, 8, 12
Folding intermediates, 147
Folding pathway, 41
Folding patterns, 1, 2, 17, 19, 20
Folding topologies, 215
Folds
low energy, 10, 19, 20
lowest energy, 14, 18, 23, 24
non-compact, 20
stable, 10, 11, 12
unique, 10, 11, 12
Formaldehyde, 279
Fragment complementation, 212
Frequency shifts, 55
5′-FSBAzA, 182, 196–198
5′-FSB ϵA, 182, 183
5′-FSBG, 182, 183
FTIR, 30, 37, 56, 92
FT Raman, 65

β-Galactosidase, 312
Gal repressor, 109
Gas phase sequencing, 194
Gaussian distribution, 8
GDP, 263, 268, 283, 284
Gel chromatography, 312
Gel filtration, 103, 125
Gelsolin, 121, 128
Geminal diamine intermediate, 224, 225
Gene duplication, 216
Gene fusion, 216
Genetic code, 326, 340, 341
Genetic switch, 107
Gibbs energy, 137, 138, 141, 143, 145–147, 167, 168
activation, 168
Gibbs statistics, 20
Globular folds, 8
Globular proteins, 1, 3, 20, 24, 76, 133, 146–149, 152, 157, 269
Globular shape, 34
Globules, 6, 19, 21
β-Globulin, 39
Glutamate dehydrogenase, 183, 195–201
NADP dependent, 198, 202
Glutamic acid, 57, 80, 201, 263, 326, 358
Glutamine, 200, 201, 342, 343, 354
γ-Glutamylation, 263
Glutaminyl-tRNA synthetase, 109, 110, 326, 329, 332, 335, 339, 353–356, 358, 360
Glutamyl-tRNA synthetase, 110, 335
Glutaraldehyde fixation, 256
Glutathione conjugate, 189
Glutathione-S-transferase, 183, 187, 189–191
Glyceraldehyde-3-phosphate, 208, 238, 239, 241, 244
Glycerol-3-phosphate, 230–234, 236–240, 242
Glycerophosphate moiety, 215
Glycine, 229–231, 345, 386, 390
Glycolytic enzyme, 265
Glycosidic bond, 197, 198
Glycyl-tRNA synthetase, 333
GMPS-BOP, 180, 181, 198
G-proteins, 257, 265, 266
Greek key, 2
Ground state, 6
Group frequency, 60, 62
Group vibrations, 60
Growth hormone, 39
GTP, 181, 183, 195–198, 256, 263, 264, 268, 273, 284
azido analogs, 263
non-hydrolyzable analog, 264
Guanidine HCl, 30, 32, 33, 34, 39, 120, 121, 125, 143–145, 212, 383
Guanidinium ion, 355
Guanine, 180
Guaninidinium unfolded state, 38
GU pair, 345, 349, 355

Hairpin, 355, 358, 400
Handle region, 385
Heat absorption, 134
Heat capacity, 43–45, 134, 135, 137, 142–144, 148–150, 155, 164, 217, 218, 232, 234
Heavy isotope incorporation, 83

Helical bundles, 2
Helical domain, 67, 355
Helical folds, 280
Helical polypeptides, 60
Helical region, 278
Helical structure, 280
α-Helix, 2, 33, 37, 45, 70, 126, 210, 215, 280, 313–318, 328, 330, 331
Helix-coil transition, 4, 23, 148
Helix-turn-helix motif, 307, 313
Heme protein, 93
Hemocyanin, 165
Hemoglobin, 75, 93–95
  mutant, 95
  CO, 94, 95
Heteropolymer, 6, 7, 8, 10, 11, 14, 24
HIGH sequence, 329, 333, 337–339
HXGH sequence, 328, 329
Hinge, 229, 230, 305, 313–315, 317, 318
Histidine, 43, 71, 337, 345, 349, 384, 391, 392, 396
Histidyl-tRNA synthetase, 326, 333
Histone H3, 119, 128
HPLC, 30, 33, 34, 186, 187, 190, 237
Hsp 60, 48
Hsp 70, 31, 48
Hydration, 149
Hydration potential, 43
Hydrocarbon, 149
Hydrodynamic radius, 29, 31, 37
Hydrodynamic size, 34
Hydrogen bond, 22, 70–73, 75, 76, 80, 82, 85, 86, 89, 148, 150, 152, 154, 156, 172, 215, 317, 339, 358, 397, 401
Hydrogen exchange, 212, 393
Hydrophilicity, 271, 272
Hydrophobic, 15, 22, 29–34, 37, 39, 43, 45, 76, 93, 109, 110, 124, 129, 148, 157, 187, 191, 210, 235, 315, 316, 390
Hydrophobic collapse, 45
Hydrophobic effect, 148, 157
Hydrophobicity, 8, 75, 76, 148, 390
Hydroxylamine, 324
Hysteric behaviour, 219

Identity determinants, 344
Identity elements, 341, 342, 344–346, 349, 350, 352, 353, 355, 358–360
Identity rules, 327
Identity switch, 351
Imidazole, 391, 396, 402
Immobilization, 105
in vivo folding, 47
Inactivation temperature, 234
Indole, 75, 85, 208, 209, 214, 223, 227, 228, 231, 235–238, 244
Indole-3-glyceroyl phosphate, 208–209, 214, 215, 229–231, 236–242, 244
Indolenine, 215, 222, 223
Indole-3-propanol phosphate, 210, 211, 214, 229–231, 237, 239, 240
Indoline, 227
Induced fit, 337
Inelastic light scattering, 58
Infrared, 32, 60
Infrared spectra, 63
Infrared spectroscopy, 55
Initiation codon, 343
Insulin, 47
Interaction, 7, 8, 12, 13, 14, 15, 17, 18, 22, 32, 46
  cooperative, 6
  hydrophobic, 22, 37, 46
  low energy, 20
Interaction energy, 212
Interleukin-1-β, 40
Interleukin-2, 40
Intermediate, 22, 28, 29, 32, 33, 35, 36, 38, 39, 44, 46, 47, 221, 236, 237, 242
  compact, 41, 42, 43, 44, 45, 48
  early formed, 28
  molten globule like, 40
  transient, 28, 40, 221
Intermediate filaments, 255
Intermediate state, 37, 45, 110, 142, 154
Internal conversion, 105
Intersubunit communication, 231, 240, 244
Intragenic suppressor, 314
Intramolecular interaction, 240
Introns, 328, 351, 352
Iodoacetamide, 126, 128
Ion pair, 3, 231
Ionizable groups, 391
Ionic strength, 271, 284
Irreversible inactivation, 218
  substrate induced, 228
Irreversible modification, 185
Irreversibility, 161
Isoacceptors, 326, 341
Isocitrate, 191–193
Isocitrate dehydrogenase, 183, 192–194
  NADP dependent, 191, 192, 194
Isocolchicine, 104

Isoelectric focusing, 275
Isoleucine, 336, 344, 385
Isoleucyl-tRNA synthetase, 329, 336, 337
Isomerization, 229, 244
Isothermal calorimetric titrations, 230
Isotope effect, 223
Isotope exchange, 55, 56
Isotope substitution, 85
Isotopomer, 83–85, 87
Isotype, 255, 257, 261, 262, 278, 281, 284, 289
Isotype-specific residues, 286
Isozyme, 183, 190, 191

Jelly Roll, 2

kcat, 201, 202, 223, 343, 345, 346, 394
Kd, 240, 268, 307
Keratin, 88
α-Ketoglutarate, 191, 199–201
K_i, 240
Kinase, 265
  cMos, 265
  vMos, 265
Kinesin, 265, 266
Kinetics
  steady-state, 214, 221
  rapid, 214, 221, 235, 352
Kirchoff equation, 138
K_m, 200–202, 240, 343, 345, 346, 399
KMSKS motif, 329, 333, 337–399

Labeling, 391
  15-N, 391
Lac operon, 312
Lac repressor, 109
α-Lactalbumin, 29, 31, 32, 33, 40, 41, 42, 47
β-Lactamase, 31, 37, 42
β-Lactoglobulin, 40, 41, 149
Lambda repressor, 69, 102, 107, 108, 111, 277, 287
Laser, 64, 91
Laser raman spectroscopy, 55, 57
Lehrer plot, 106
Leucine, 239, 342, 346, 349, 385
Leucyl-tRNA synthetase, 327, 337, 351
Light scattering, 30, 34, 35, 37, 65, 268
Loop rigidity, 3
Loops, 2, 90, 210–213, 237–241, 273, 276, 277, 287, 309, 340, 384, 391, 401
Loops (*cont.*)
  bending, 2
  crossing, 2
Luciferase, 203
Lumry–Eyring model, 165, 167–169
Lysine, 43, 119, 126, 196, 197, 201, 220, 329, 343
Lysosomes, 47
Lysozyme, 40, 41, 42, 45, 89, 90, 147, 161, 172, 317, 393
Lysyl-tRNA synthetase, 327, 333, 344

M13, 83
MAD analysis, 384
Magnetic resonance spectroscopy, 57
Magnetization transfer, 33
Major groove, 309, 313
Malate, 193
Maleimides, 126
MAPs, 256, 265, 266, 278–280, 284, 289, 290
MAP1, 266, 279
MAP2, 266, 279
Maximum heat capacity, 138
Maytansine, 265
Mean fold energy, 8
Meiotic division, 285
Melphalan, 189
Melting, 21
Membrane, 257
Membrane tubulin, 256, 257
Mercaptoethanol, 221, 271
Metabolite channeling, 212, 228, 235
Metal catalyzed oxidation, 193
Metal ion affinity chromatography, 406
Metal-isocitrate complex, 193, 194
Methionine, 57, 90, 326, 337, 339, 345, 359
Methioninol, 337
Methionyl-tRNA synthetase, 328, 329, 331, 332, 334–337, 339, 346, 351, 355, 359, 360
N-Methylacetamide, 66, 68
Methylbenzimidazole carbamate, 285
N-Methyl demecolchicine, 104, 105
3-Methyl indole, 74
N-1 Methyl pyrene, 118
O-Methyl-DL-Serine, 241
Mg^{+2}, 106, 192, 256, 264, 281, 283, 336–339, 377, 378, 380, 384, 385, 388, 390, 392–394, 396, 400, 402, 409

Micelle, 105
Microcalorimetry, 217
Microhelix, 345, 348
Microtubule, 255–257, 262, 263, 265, 268, 269, 274, 276, 278, 284–287, 289–291
  hook, 283
  open, 283
  ribbon, 283
  sheets, 283
Microtubule organizing center, 257
Minisubstrates, 345
Minor groove, 355
Misacylation, 358
Missense mutants, 213
Missense substitution, 213
Missense suppressor, 342
Mitochondrial aaRS, 326
Mitotic division, 285
Mitotic spindle, 256, 257
Mn^{+2}, 185, 186, 192, 377, 378, 380, 388, 397
Molecular dynamics, 45
Molecular force field, 55, 56, 58
Molecular geometry, 56
Molecular orbital, 63
Molecular volume, 34
Molten globule, 21, 27, 29, 30, 31, 32, 33, 35, 37, 39, 41, 42, 44, 45, 46, 47, 110, 212, 217
Motif 2, 340, 355
Motif 3, 340
Motor proteins, 265, 266, 291
Multienzyme complex, 244
Multitude principle, 24
Mutagenesis, 183, 212, 334, 337, 351, 396
  random, 213, 232, 358
  random saturation, 214
  saturation, 359
  site directed, 29, 37, 38, 78, 127, 129, 224, 226, 227, 264, 314, 337, 340, 359, 393, 394, 398, 400
  systematic, 346
Mutant, 35, 183
Mutant aaRS, 358
Mutant repressor, 317
Mutational analysis, 350
Mutations, 7, 8, 37, 80, 212, 221, 227, 228, 232, 239–242, 316, 339, 342, 399
  site-directed, 125, 284, 285
  missense, 229
Myoglobin, 23, 43, 147, 151, 155, 156, 172
Myosin, 203

NAD, 199, 203
NAD/NADP dependent dehydrogenase, 181
NADH, 102, 181, 182, 195–199, 202
NADP, 199
NADPH, 102, 199–202
Native fold, 7, 12, 15, 16, 22, 28
Native state, 29, 30, 33, 34, 37, 38, 41–47, 142, 143, 146, 149, 277, 280
Native structure, 148, 149
Negative cooperativity, 229, 310
Negative determinants, 344, 346
Negative discriminant, 359
Neutron scattering, 352
Neviapine, 405
NH.Me.DAAC, 105
Nicotinamide ring, 182, 198
Nile red, 238, 239, 242, 268
6-Nitroindole, 240, 241
6-Nitro indole-3-glycerol phosphate, 241
NMe2.DAAC, 105
NMN analog, 181
NMNS-BOP, 181, 198
NMR, 28, 30, 32, 33, 37, 40, 57, 91, 101, 185, 212, 280, 281, 306, 312, 384, 390, 391, 393, 400, 402
  amide exchange, 37, 40
  heteronuclear two-dimensional, 392
  pulsed H/D exchange, 41
  stopped flow, 40
  structure, 6
  three dimensional, 391
  two-dimensional, 33, 281, 393
Non-adenylylated dinucleotides, 327
Noncovalent interaction, 289, 325
Nonsense suppression, 358
Nonspecific DNA, 310
Non-Watson-Crick pair, 354
Normal mode, 73–76
N-terminal domain, 109
Nuclear Overhauser effect, 33
Nucleophile, 221
Nucleophilic addition, 189
Nucleophilic attack, 335
Nucleosome, 128
Nucleotide site, 182
Nucleus, 256

Operator, 109, 350, 351
Operator region, 351

Operator structure, 350
Ori C, 379
Ori K, 379
Oxidative decarboxylation, 191
Oxoprolinase, 203
Oxygen, 183, 192
Oxyindolyl alanine, 222

p51, 397
P22 subunit, 69
P22 tailspike, 69, 80, 89
$(pA)_3$, 380
$(pA)_4$, 380
Papain, 147, 272, 287
Paperclip motif, 399
Parallel α-helices, 385
Partial heat capacity, 135
Parvalbumin, 41
Partition coefficient, 4
Partition function, 142, 169
Passive diffusion, 235
PCR, 393
Penicilin selection, 213
Pepsin, 270, 277, 287
Pepsinogen, 147
Peptide, 56, 60, 87–89, 194, 196, 200, 266
Pfl subunit, 69
Phase boundary, 31, 42
Phase diagram, 31, 37, 41, 42, 44
Phase transition, 35
Phenol, 116
  para substituted, 116
Phenylalanine, 57, 62, 94, 116, 339, 344
Phenylalanyl-tRNA synthetase, 333, 337
Phenylboronate column, 186, 187
2'Phosphoadenosine diphsopho ribose, 199
Phosphodiester bond, 380
Phosphoenol pyruvate, 183, 185, 186
Phosphoglycerate kinase, 39
Phospholipid vesicles, 47
Phosphorescence life time, 218
Phosphoribosyl anthranilate isomerase; indole-3-glycerol-phosphate synthetase, 210, 212
γ-Phosphoryl group, 329
Phosphorylation, 262, 263
Photoaffinity labeling, 263–265
Photoaffinity-labeled peptides, 265
Photolysis, 94
pKa, 391
Plasmid, 382
Plasminogen, 147
Podophyllotoxin, 104, 264, 284
Polarity-sensitive fluorescent probe, 238
Polarization, 61, 62
Polarizability, 57, 61, 62
Polarized raman methods, 57
Polarized raman spectra, 63
Poly A tail, 408
Polyalanine, 69
Polyethylene glycol, 230
Polyglutamate, 69
Poly histidine tag, 406
Polylysine, 69, 70
Polymer
  stochastic, 44
Polymer elasticity, 3
Polymerization, 261, 264, 273, 277–279, 284, 285, 289, 310
Polypeptide, 45, 48, 57, 66, 67, 69, 192, 209, 213, 216, 217, 235, 313, 326, 409
Poly(rA).Poly(dT), 380
Poly-L-Serine, 70
Positive control mutant, 308
Positive cooperativity, 310
Posttranscriptional modification, 344
Posttranslational modification, 257, 262
Primary structure, 15, 21, 360, 385
PRODAN, 102, 109
Pre-steady state techniques, 337
Prolate ellipsoid, 268
Proline 3, 4, 230, 231, 326
Prolyl-tRNA synthetase, 333
Promoter, 308, 378, 382
  P_L, 382
  type I, 309
  type II, 309
Promoter overlap, 378
Proofreading, 335, 336, 341
Protease, 269, 283, 289, 311
Protease K, 270, 271
Protease sensitivity, 277
Protein architecture, 1, 2, 6
Protein denaturation, 143–145, 150, 156–158
Protein engineering, 224, 238, 378, 410, 412
Protein folding, 21, 22, 27
Protein folds, 3, 6
Protein kinase, 107
Protein–protein interaction, 207, 208

Protein stability, 1, 3, 12, 22, 28, 44, 133, 144, 147, 150, 156, 157, 162, 213, 412
Protein stability curve, 143–145
Protein statistics,
computer modeling, 14
Protein structure, 1, 3, 77, 87
Protein structure prediction, 21, 22
Protein synthesis, 323, 326
Proteins,
all β, 40
folded, 29
fully deuterated, 391
unfolded, 45
Proteolysis, 216, 230, 238, 239, 265, 268–270, 272, 280, 282, 311, 312
Proteolytic cleavage, 273
Protofilaments, 256, 272, 290
Protoctists, 257, 261
Pseudoknot, 349
Purine, 187, 197, 198
Pyrene, 115, 116, 118, 119, 121, 122, 127, 129
Pyrene dithiothreitol, 122–124
Pyrene iodoacetamide, 119, 125, 126, 128
Pyrene maleimide, 119, 120, 122, 123, 125, 126, 128
Pyrimidines, 380
Pyruvate, 183, 223
Pyruvate kinase, 183–187
Pyridoxal phosphate, 208, 215–222, 224–227, 229, 233, 241
Pyrophosphate, 110, 181, 201, 336–338
Pyrrole, 74

Quarternary interaction, 129
Quarternary structure, 94
Quenching, 38, 115, 116, 121
static, 124, 129
Quinonoid, 237, 240, 244
Quinonoid intermediate, 221–223, 227, 235, 236

Radius of gyration, 31, 311
Raman amide bands, 67, 80
Raman analysis, 58, 71
Raman band, 55, 58, 62, 71, 73, 75, 76, 78, 79, 80, 82, 312
Raman band width, 73
Raman effect, 55, 56, 62
Raman frequency, 73, 77, 83
Raman fundamental mode, 72
Raman intensity, 71, 77
Raman microspectroscopy, 88
Raman scattering, 57, 58, 61, 62, 64, 65, 78
stokes, 61
anti-stokes, 61
Raman spectra, 56, 57, 59, 63–66, 71, 73–80, 83, 85, 86, 89, 90, 94, 96, 312
frequencies, 56, 58, 59, 60
infrared excitation, 64
intensities, 56, 58, 59, 60, 62, 63
polarizations, 56, 58, 59, 60
Raman spectroscopy, 55, 56, 57, 60, 63–66, 75, 80, 83, 87, 88, 90, 95
Raman spectrophotometer, 57, 65
Raman vibrational band, 60
Random-ordered kinetics, 336
Rapid-scanning stopped-flow spectra, 222
Rapid-scanning stopped-flow spectrophotometry, 221, 236
Rate limiting step, 212, 244
Rate saturation effect, 178, 185
Rayleigh scattering, 58, 65
α-Reaction, 208
β-Reaction, 208
α, β-Reaction, 209
Rec BCD protein, 379
Recognition helix, 313
Redox active metal ion, 183
Reducing agent, 183
Reductive methylation, 278, 289
Refolding, 27, 28, 31, 39, 41, 129, 217, 234, 393
Regulatory nucleotide, 196
Regulatory site, 179, 195
Reheating run, 136, 137, 161
Relaxation time, 159, 160
REM (random energy model), 7, 8, 13, 14
Renaturation, 127
β-Replacement, 209, 220, 223, 228
Reporter gene, 342, 343
434 Repressor, 393
Resonance raman, 56, 57, 62, 87, 90, 95
UV, 64, 65, 69, 70–76, 87, 90–95
Retinal chromophore, 65
Retinol binding protein, 39
Reverse titration, 102
Reverse transcriptase, 377, 385, 387, 388, 396–399, 404–406
Reverse turns, 67
Reversible allosteric transition, 303

Reversion, 213
Rhizoxin, 265
Rhodanase, 39
Rhodopsin, 57
Ribonuclease A, 31, 134, 147, 172, 411
Ribonuclease T1, 144
Ribonucleotide excision, 409
Ribose, 180–182, 187, 201, 335
Ribosomes, 47, 48, 217, 324, 340
Ribozyme, 403, 411
Ring formation, 283
RNA, 87, 323, 324, 340, 345, 346, 348, 349, 352, 377, 379, 396, 400, 403, 404, 409
mRNA, 324, 350, 351, 408
tRNA, 323, 324, 326, 327, 329, 333, 335, 340, 341, 342, 345, 346, 348, 349, 352, 353, 355, 358–360
  amber suppressor, 342
  cognate, 335, 344, 351, 353, 359
  mutant, 343
  noncognate, 341, 358, 359
  suppressor, 342
tRNA^{ala}, 335, 345
  amber, 358
tRNA^{arg}, 341
tRNA^{asp}, 331, 353–356
tRNA^{cys}, 341
tRNA^{gln}, 109, 110, 329, 353–356, 358
tRNA^{gly}, 341
tRNA^{his}, 348
tRNA^{lys}, 342, 344
tRNA^{met}, 345, 359
  amber, 359
tRNA^{metf}, 343, 346
tRNA^{sel}, 341
tRNA^{ser}, 353
tRNA^{s3r}, 341
tRNA^{thr}, 351
tRNA^{trp}, 341
tRNA^{tyr}, 328, 334, 352, 358
  amber, 342
tRNA^{val}, 350
tRNA like structure, 328, 349–359
RNA genome, 349, 360
RNA polymerase, 305, 309, 310
  T7, 343, 348
RNA polymerase II, 410
RNA protein interaction, 360
RNA recognition, 360
RNA restriction enzyme, 410
RNA/RNA duplex, 393
RNA substrate, 360
RNAase H, 377, 378, 380–382, 384, 387, 390, 396, 398, 399, 406, 407, 409–412
RNAase H I, 377–379, 381, 385, 387, 388, 390–394, 399, 401, 402, 404, 408–410, 412
  mutant, 395
RNAase H II, 379, 407, 409
RNAase H domain, 388, 397, 398, 400, 405
RNAase P, 403
Rossman fold, 328, 329, 333, 360
Rotamer, 6
Rotational correlational time, 107
Rotational diffusion, 124
Rotational relaxation time, 38

Salt bridge, 41, 381
Sanguinarine, 265
S-BDB-G, 189–191
S-butylglutathione, 190
Scattered frequencies, 64
Schiff base, 90, 209, 219, 221–224, 226, 233, 241
SDS, 193, 269
SDS gel, 271, 272, 275, 277
Secondary structure, 2, 3, 6, 21, 23, 27–29, 32–34, 37–40, 43, 45–47, 55, 66, 67, 80, 130, 217, 270, 272, 281, 325, 351, 385, 393
Sedimentation coefficient, 268
Selenocysteine, 326
Selenomethionine, 383
Self-organization, 21
Self-splicing, 351
Sequence alignment, 341
Sequence specific cleavage of RNA, 411
L-serine, 208, 209, 220–229, 231, 234, 240–242, 326, 342
Seryl-tRNA synthetase, 326, 329, 330, 331, 333, 334, 340, 353, 360
S-ethylglutathione, 190
4-β Sheets, 2
β Sheet domain, 67
β Sheets, 39, 45, 46, 70, 215, 317, 328, 330, 331, 333, 334
S-hexylglutathione, 189, 190
Single turnover condition, 235, 236, 241
Site-specific fragmentation, 408
Size exclusion chromatography, 30, 32, 33, 34, 268, 277
S-methylglutathione, 190

Snake venom toxin, 65
Sodium borohydride, 186, 187, 218, 219
Solution structure, 392
SOS function, 379
Specific enthalpy change, 152, 153
Specific entropy change, 152, 153
Specific heat capacity change, 147, 152, 153
Specific recognition, 358
Spectrum line, 8
Spectrum density, 9, 17
Spin label, 38
Splicing, 328, 351
Splicing activity, 351
S-propylglutathione, 190
Staphylococcal nuclease, 28, 39, 40, 41, 145–147, 412
Statistical mechanics, 23
Stefin B, 39
Steganacin, 264
Stochastic polymer, 44
Stokes radius, 30, 34, 35, 38, 268, 312
Stop codon, 342, 358
Stopped flow technique, 336, 393
β-Strands, 2, 80, 210, 330, 355, 390
  parallel, 2, 209
Streptomyces subtilisin inhibitor, 149
Stretching mode, 62
Stretching motion, 66
Stretching vibrations, 62, 71, 79
Subdomains, 38, 46, 388, 398
Subtilisin, 106, 265, 270–273, 276–280, 283, 311
Subunit interaction, 231
Sucrose gradient centrifugation, 229
Sulfhydryl group, 108
Surface probability plot, 271, 272
Syn conformation, 187

Tau, 266
Tautomer, 223, 227
Tautomerization, 215, 222
Taxol, 264, 272
Temperature integration, 135
Ternary complex, 336
Tertiary interaction, 33, 38, 46, 129
Tertiary structure, 21, 29, 38, 41, 47, 129, 130, 212, 257, 325, 341, 349, 360, 393
Thermal denaturation, 39, 133, 141, 143, 146, 168, 383, 407
Thermal inactivation, 233, 234
Thermal induction, 382
Thermal stability, 162, 213, 232, 328
Thermal unfolding, 219, 232, 234
Thermal vibrations, 6
Thermodynamics, 20, 21, 23, 35, 150
  equilibrium, 133, 166
Thermogram, 136, 137, 158
Thermolysine, 165, 168, 269, 270, 272
Thiobendazole, 285
Thiocolchicine, 104
Thioredoxin, 77–82
Three-dimensional structure, 186, 189, 215, 305, 328, 333, 337, 340, 353, 354, 358, 360, 384, 385
Threonine, 231, 240, 358
Threonyl-tRNA synthetase, 326, 333, 350, 351
TIM barrel, 2
T_m, 383
Toluene, 116
Torsion angle, 71, 74
Total heat-capacity change, 151
Transaldimination, 224
Transamination, 221, 224–226
Transcription, 309
Transcription activation, 308, 310, 311, 313, 314
Transcription terminator, 382
Transfer-free energy, 4
Transition state, 104, 167, 169, 337–339
Transition state theory, 168
Translation, 342, 351
Translational control, 351
Transmembrane helix, 129
N,N′,N′-triacetyl chiotriose, 90
Tricarbyllate, 193
Trifluoroacetic acid, 73
Trifluoroethanol, 280, 281
Trimethyl colchicinic acid, 104
Triose phosphate isomerase, 210, 211, 213, 238, 240
Tropolone, 103, 104
Tropomyosin, 119–122, 126, 127
Troponin C, 127
Troponin I, 127
Trp repressor, 317
Truncated tRNA substrate, 352
Trypsin, 147, 186, 187, 239, 270–272, 276, 311, 334
Trypsinolysis, 328
Tryptic cleavage, 238
Tryptic fragment, 216
Tryptophan, 32, 33, 34, 37, 57, 62, 69, 71,

Tryptophan (*cont.*)
74–76, 80, 83, 84, 87–90, 92–95, 102, 105, 106, 108–110, 112, 208, 209, 219–221, 224, 226, 227, 231, 234–238, 240, 244, 282, 310, 312, 317, 399
D-Tryptophan, 226, 234
Tryptophan synthase, 40, 41, 207, 211, 212, 214, 222–224, 228, 229, 232–235, 238, 315
  mutant β subunit, 208
  α subunit, 229, 277, 315
  β subunit, 217, 229, 231, 277
Tryptophanyl-tRNA synthetase, 327, 337
Tubulin, 103, 105, 106, 112, 129, 255, 257, 261–263, 265, 266, 269, 271, 273, 276, 278, 280, 281, 283, 286, 290
  α, 257, 266, 269, 271–273, 278
  β, 257, 264, 266, 269, 270, 272, 273, 278, 285, 286, 289
  α,β S, 274, 283, 284, 286
  γ, 257, 266
  membrane, 257
  polymerized, 257
  soluble, 257
Tubulin genes
  α, 261
  β, 261
Tubulin tyrosine ligase, 262
Tubulin-S, 106, 274, 275, 277, 281–284, 289
Tunnel, 208, 209, 227, 235, 236
Turnover number, 240
Turns, 80
Two-state transition, 234
TYMV tRNA, 349, 350
Type II β turn, 70
Tyrosine, 33, 34, 57, 62, 71–73, 80, 81, 87, 88, 92, 116, 186, 187, 190, 191, 196, 197, 334, 339, 349
Tyrosinlation, 263, 350
Tyrosyl adenylate, 329, 332, 337, 338
Tyrosyl-tRNA synthetase, 326, 328, 329, 331, 332, 334, 336, 337, 339, 351, 352, 359

Ubiquitin, 33, 393
Ultracentrifugation, 229, 268
Unfolded state, 27, 28, 29, 30, 33, 34, 35, 36, 38, 42, 43, 44, 137
Unfolding, 32, 33, 129, 233, 234, 383
Unfolding pathway, 234
Urea, 37, 111, 125, 145, 212, 383, 384
UV irradiation, 379
UV spectroscopy, 234

Valine, 336, 345, 349
Valyl-tRNA synthetase, 326, 329, 338, 344, 349, 350
Van der Waals interaction, 148, 156, 172
Van der Waals radius, 314
Vant' Hoff enthalpy, 139, 140
Vant' Hoff equation, 138
Vibrational energy level, 72
Vibrational frequency shift, 63
Vibrational modes, 60, 77
Vibrational spectroscopy, 63
Vibrational spectrum, 55
Vibrational state, 72
Vibrations
  in-plane, 62
  out of plane, 62
Vinblastine, 264, 265
Vincristine, 265
Viral RNA, 349, 350
Viruses, 83
Viscometry, 268
Viscosity, 30
Viscosity enhancing agents, 212
V-loop, 349, 352, 353
V8 protease, 270, 271, 311

Water accessible surface area, 149, 154
Wave number, 59

Xenobiotic compound, 189
X-ray crystallography, 57, 95, 101, 203, 214, 219, 312, 392, 400
X-ray crystal structure, 74
X-ray diffraction, 179, 191, 269, 276, 286, 287, 305
X-ray fiber diffraction, 268, 276
X-ray method, 91
X-ray scattering, 229, 276
X-ray structure, 127, 150, 309, 313

Zinc, 62, 192, 268, 358
Zinc containing motif, 360
Zinc domain, 360
Zinc sheets, 272

MIX
Papier aus verantwortungsvollen Quellen
Paper from responsible sources
FSC® C105338

If you have any concerns about our products,
you can contact us on
ProductSafety@springernature.com

In case Publisher is established outside the EU,
the EU authorized representative is:
Springer Nature Customer Service Center GmbH
Europaplatz 3, 69115 Heidelberg, Germany

Printed by Libri Plureos GmbH
in Hamburg, Germany